Green Chemistry: Advanced Principles and Techniques

Green Chemistry: Advanced Principles and Techniques

Edited by
Simon Doyle

CWILLFORD PRESS

www.willfordpress.com

Preface

Green chemistry is a field of chemistry and chemical engineering that deals with the designing of products. It focuses on the processes that eliminate the use and production of harmful substances. Green chemistry is concerned with the environmental impacts of chemistry. It uses technology to prevent pollution and reduce the consumption of non-renewable resources. The goals of green chemistry include making the designing of molecules, products, materials and processes more resource efficient and safe. There are various principles of green chemistry that address different ways to eliminate the environmental and health impacts of chemical production. It aims to produce less waste, avoid using substances that are toxic to humans and the environment and minimize the energy requirements. This book contains some path-breaking studies in the field of green chemistry. It presents researches and studies performed by experts across the globe. This book will help the readers in keeping pace with the rapid changes in this field.

The information contained in this book is the result of intensive hard work done by researchers in this field. All due efforts have been made to make this book serve as a complete guiding source for students and researchers. The topics in this book have been comprehensively explained to help readers understand the growing trends in the field.

I would like to thank the entire group of writers who made sincere efforts in this book and my family who supported me in my efforts of working on this book. I take this opportunity to thank all those who have been a guiding force throughout my life.

Editor

The Organic Pollutant Characteristics of Lurgi Coal Gasification Wastewater before and after Ozonation

Chunrong Wang ⓘ, Qi Zhang, Longxin Jiang, and Zhifei Hou

School of Chemical and Environmental Engineering, China University of Mining and Technology (Beijing), Beijing 100083, China

Correspondence should be addressed to Chunrong Wang; wcrzgz@126.com

Academic Editor: Sabrina Copelli

The effluent of distilled and extracted Lurgi coal gasification wastewater has been found to have low biodegradability and high toxicity, which inhibits further biodegradation. However, ozonation enhances the biodegradability and reduces the toxicity of this effluent, enabling further biological treatment and increased removal of organic materials. In this study, the dissolved organic matters in Lurgi coal gasification wastewater were isolated into six classes by resin adsorbents, after which TOC, UV_{254}, UV-Vis, and 3D EEM were employed to quantitatively and qualitatively analyze organic materials in each part of the fractionated samples. The HoA and HiN fraction accounted for large amounts of the Lurgi coal gasification wastewater, and their TOC values were about 380.21 mg·L^{-1} and 646.84 mg·L^{-1}, respectively. After ozonation, the TOC removal rates of HoA and HiN reached 42.85% and 67.13%, respectively. The UV_{254} of HoA was basically stable before and after ozonation, while that of HiN increased continuously because a portion of the humic macromolecular organic materials in HoA was oxidized to HiN. Additionally, UV-Vis analysis revealed that the larger molecular organics of HoA were oxidized during ozonation, resulting in high biodegradability. Finally, the 3D EEM spectra indicated that the macromolecular organics were oxidized to smaller molecules with the degradation of soluble microbial by-products.

1. Introduction

Coal exploitation and consumption have been growing rapidly and causing serious environmental pollution; accordingly, there is a demand for efficient, safe, and clean utilization of coal [1, 2]. Therefore, a new type of coal chemical engineering technology is needed to optimize the use of coal. As one of the most widely used coal gasification technologies, Lurgi gasification generates various contaminants during the cleaning and washing of gas [3]. After ammonia distillation and dephenolization, Lurgi coal gasification wastewater (LCGW) is still a complex industrial wastewater containing high concentrations of organic matter, ammonia, and prussiate [4], resulting in unsatisfactory effluent quality after anaerobic-anoxic-oxic (A-A-O) process and sequencing batch reactor (SBR) biological treatment process [5].

Most studies of gasification wastewater quality conducted in recent years have focused on its quality. The LCGW treatment process has been widely investigated, but these studies have investigated the COD, TOC, BOD$_5$, and other simple indicators, which showed that gasification wastewater was characterized by high levels of organic pollutants, low biodegradability, and high toxicity [6–8]. Moreover, previous studies have demonstrated that after ammonia distillation and dephenolization, the LCGW still contained phenolics, polycyclic aromatic hydrocarbons, and nitrogen heterocyclic compounds [9–11]. Because of the high amounts of common organic pollutants, the biological treatment of LCGW led to poor biodegradation. However, few studies have included systematic analyses of wastewater qualities of LCGW. Moreover, identification of methods for the transformation of organic pollutants during the treatment of LCGW is urgently needed.

Before biological treatment of LCGW, ammonia distillation and dephenolization were applied to remove the organic pollutants. Because of the limited removal efficiency of ammonia distillation and dephenolization, most organics were removed by biological treatments, such as hydrolyzation acidification, activated sludge processes, sequencing batch reactor (SBR), anoxic-oxic (A/O) processes, and anaerobic-anoxic-oxic (A^2/O) processes [12].

However, the main characteristics of the effluents after biological treatment were as follows: 100–200 mg·L^{-1} of COD, 0.01–0.08 of BOD$_5$/COD ratio (B/C), 40–70 mg·L^{-1} of total nitrogen (TN), 10–20 mg·L^{-1} of NH$_4^+$-N, and 90–120 mg·L^{-1} of TOC [13–16]. Moreover, most studies only used simple indexes to express the characteristics of LCGW, and none have analyzed the effects of qualities on treatment process or the transformation of organic compounds during treatment. According to the national LCGW discharge policy of China, especially the requirement of zero liquid discharge, more measures for the advanced biological treatment of LCGW effluent are needed [17]. Because the effluent after ammonia distillation and dephenolization has low biodegradability and high toxicity, ozonation is commonly applied to enhance the removal rate of organic materials because it improves biodegradability, enabling further biological treatment. Nevertheless, when compared with the removal rate, the transformation of organics during ozonation has received less attention. Moreover, the analysis of raw LCGW quality was difficult because of the complexity characteristic [18].

Therefore, this study was conducted to investigate the characteristics of LCGW after ammonia distillation, dephenolization, and organics transformation and to investigate the effects of organics on ozonation. Fractionation of the LCGW after ammonia distillation and dephenolization was conducted using resin adsorption, after which the dissolved organic carbon (DOC), UV$_{254}$, UV-Vis, and EEM were used to quantify organic matter in each part of the fractionated samples. Following ozonation, DOM was analyzed in the solution, and the effects of organics on ozonation were further examined. The results presented herein will serve as a reference for further engineering application, guiding the commission and normal operation of future treatment facilities.

2. Materials and Methods

2.1. Wastewater Samples. Lurgi coal gasification wastewater samples were obtained from the effluent of ammonia stripping and phenolic solvent extraction processes in a Lurgi coal gasification wastewater treatment plant (Henan Coal Chemical Industry Group Co., Ltd.). The raw water was filtered through a 0.45 μm cellulose membrane (PES) to obtain dissolved organic matters (DOMs). All experimental data were analyzed using SPSS 17.0. The data were analyzed as the means of three replicates, and differences were considered significant at a p value < 0.05. The characteristics of the coal gasification wastewater discharge are listed in Table 1.

TABLE 1: Main physical and chemical properties of LCGW discharge in this plant.

Parameter	Concentration	Unit
pH	8.0 ± 0.16	
COD	4066.67 ± 200	mg·L^{-1}
TOC	1214 ± 50	mg·L^{-1}
BOD$_5$	419 ± 10	mg·L^{-1}
NH$_4^+$-H	300 ± 20	mg·L^{-1}
CN$^-$	0.009 ± 0.001	mg·L^{-1}
Oil	58.34 ± 3	mg·L^{-1}
Benzopyrene	0.00082 ± 0.00003	mg·L^{-1}

2.2. Ozonation Equipment. The ozonation equipment is shown in Figure 1. Ozonation was conducted at room temperature in a semicontinuous model reactor with an effective volume of 1.5 L. The optimum reaction time was selected based on Figure 1S, while the optimum pH was based on Figure 2S. Ozone gas was generated using a VMUS-1SE laboratory ozone generator (AZCO Industries LTD, Canada) rated at 5.7 mg O$_3$ L^{-1} (based on Figures 3S and 4S). In a typical experiment, 1 L of LCGW was added into the borosilicate glass reactor (1000 mm tall and 50 mm in diameter); after which, ozone was continuously fed to the wastewater through a microporous titanium plate at the bottom of the reactor to obtain gas bubbles at the same time. The input gas pressure was 0.04 MPa, and the gas flow rate was 3 L min^{-1}. The excess ozone in the outlet gas was absorbed by 5% Na$_2$S$_2$O$_3$ solution. Total organic carbon (TOC), UV$_{254}$, UV-Vis, and EEM were determined at different times.

2.3. Analytical Methods. The concentration of ozone in ingas and off-gas was measured using an ozone analyzer (ZX-01, China). Gaseous ozone in off-gas was continuously introduced to the ozone analyzer, while the concentration of ozone dissolved in the aqueous phase was determined using the indigo method. Before the test of resin fractionation, the resins (Amberlite XAD-8 nonionic resin, Dowex 50WX2 H$^+$ cation exchange resin, and Amberlite IRA-958 (Cl$^-$) anion exchange resin) were Soxhlet-extracted with methanol for 24 h and then rinsed with acid-base solutions and Milli-Q water (Supplementary Material for details). After pretreatment of resins, DOM fractionation via the adsorbent resin method was employed to separate the dissolved organic matter into six groups: hydrophobic acid (HoA), hydrophobic neutral (HoN), hydrophobic base (HoB), hydrophilic acid (HiA), hydrophilic neutral (HiN), and hydrophilic base (HiB) [19]. In addition, the TOC in the water was determined using a TOC analyzer (Shimadzu, TOC-VCPN, Japan) after filtration through 0.45 μm cellulose membrane. The UV$_{254}$ in each degraded solution was monitored by UV-visible spectroscopy (Hach, DR5000, USA). Fluorescence was determined using a HITACHIF-7000 spectrofluorometer. The EEM was conducted using a 1500 W xenon lamp, PMT voltage of 500 V, excitation and emission of 200 to 450 nm and 220 to 600 nm, respectively, a scanning speed of 1200 nm·min^{-1}, and the slit widths of 5 nm. Before analysis, the Raman scattering and Rayleigh scatter effects were removed.

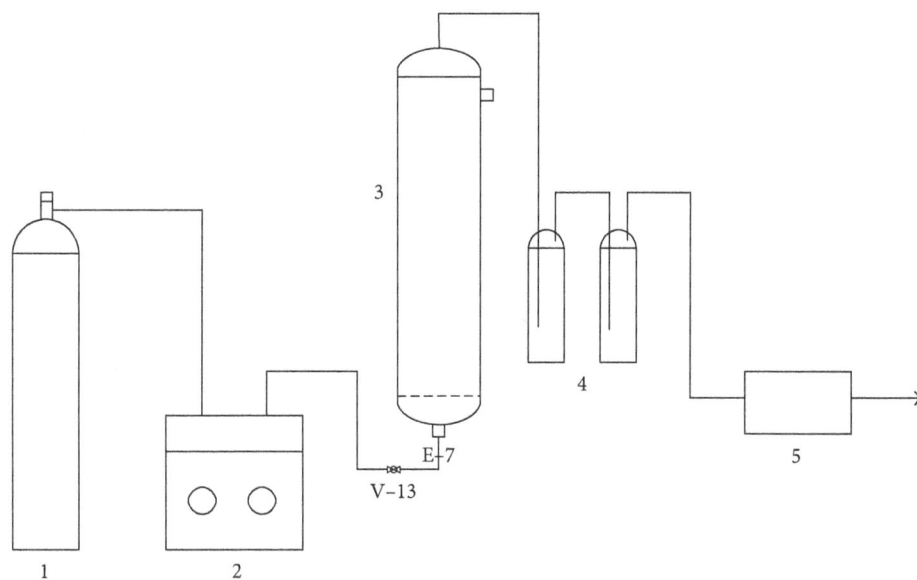

FIGURE 1: Schematic diagram of ozonation experiment. 1: oxygen cylinder, 2: ozone generator, 3: sample tap, 4: KI trap, and 5: ozone absorber.

3. Results and Discussion

3.1. Distribution Characteristics for TOC before and after Ozonation. The results revealed that TOC was present in each of the fractions; therefore, further analysis was conducted to determine the distribution of organic carbon in LCGW. The phenols and their derivatives in the LCGW effluent accounted for 53.30% of the total COD, which reached 2712.36 mg·L^{-1}. Furthermore, recalcitrant organic compounds such as pyridine, indole, quinolone, biphenyl, and other aromatic derivatives accounted for a high proportion of TOC in traditional LCGW. Ozonation effectively oxidized organic compounds with unsaturated bonds, especially in LCGW, because of its dipole structure. The TOC concentrations of the six separated components in the LCGW before and after ozonation are shown in Figure 2.

As shown in Figure 2, the main components of DOM in the current study were HoA and HiN with a concentration of 380.21 mg·L^{-1} and 646.84 mg·L^{-1}, respectively. The concentration of TOC decreased from 1214.51 mg·L^{-1} to 537.31 mg·L^{-1} after ozonation with a removal rate of 55.76%. The removal rates of HoA in 60 min and 120 min were 21.23% and 42.85%, respectively, while the TOC of HoB remained stable during the process. Although the removal rate of HoN reached as high as 95.38% in 120 min, it had little influence on total TOC because its initial concentration was only 5.40 mg·L^{-1}. Moreover, HiA and HiB reached a TOC removal rate of 56.00% and 50.28%, respectively, while the rate of HiN reached 56.01% in 60 min and 67.13% in 120 min. Taking together, these findings indicated that ozonation mainly showed a significant effect on HoA and HiN of TOC removal, while its effects on other components were lower.

3.2. Distribution Characteristics for UV$_{254}$ before and after Ozonation. The UV$_{254}$ value was measured using a colorimetric dish as the carrier for solutions. To accomplish this,

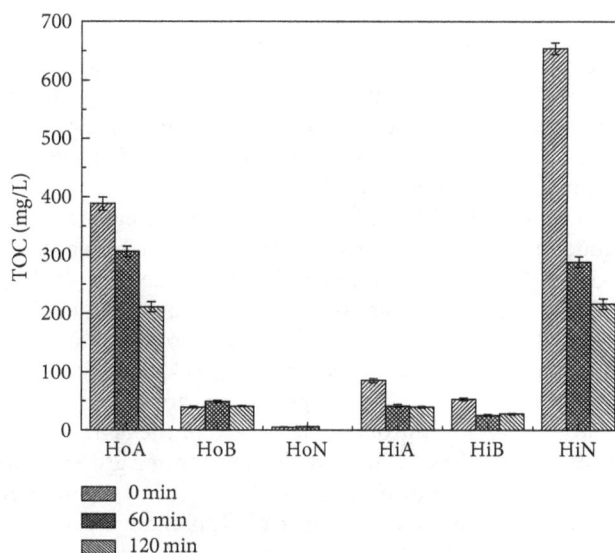

FIGURE 2: TOC of the components when the reaction time was 0 min, 60 min, and 120 min.

the absorbance of the sample at 254 nm was measured in an ultraviolet spectrophotometer. The UV$_{254}$ values can represent the concentration of TOC qualitatively to some extent. The polycyclic aromatic hydrocarbons and nitrogen heterocyclic compounds present in effluent are still difficult to degrade by biochemical treatment, despite the low level of phenols in the LCGW and good performance of biochemical technology at removing the phenols remaining in the effluent. The UV$_{254}$ value can also be used to measure the degree of humification as most DOMs in natural waters are humic substances. The ratio of UV$_{254}$ of the components in total DOM values is shown in Figure 3.

The main components of DOM were HoA and HiN, which accounted for 31.30% and 53.26% of the total, while

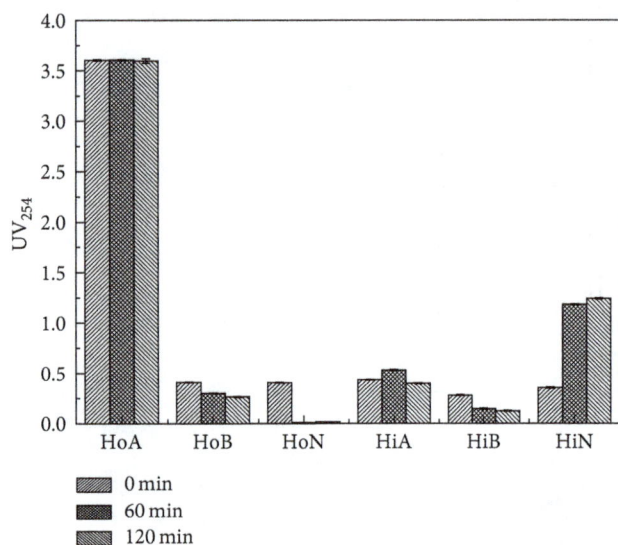

FIGURE 3: UV_{254} of the components when the reaction time is 0 min, 60 min, and 120 min.

HiA, HiB, HoB, and HoN accounted for only 7.20%, 4.46%, 3.33%, and 0.44% of the total, respectively. Among these, HoA had the highest ratio of UV_{254}, accounting for 65.66%, while the UV_{254} values in HiA, HiB, HiN, HoB, and HoN were 8.02%, 5.15%, 6.44%, 7.40%, and 7.33%, respectively, demonstrating that there were large amounts of aromatic compounds in HoA. These findings also indirectly explained the high TOC values in HoA.

Ozonation leads to different degrees of decrease for most components, with the highest removal rate being observed for HoN and the lowest for HiN. However, the UV_{254} values of HoA remained basically unchanged and that of HiN increased continuously. Based on previous studies, the reason for the increase in the UV_{254} value can be explained as follows: First, C=O was generated during ozonation because the hydroxy group was oxidized or the benzene ring was destroyed [20–22]. Second, the unsaturated group in the hydrophobic substances will be destroyed by ozonation, resulting in the material structure of HoA changing and the hydrophobic substances being transferred to other kinds of materials, leading to oxidation of the humic macromolecule organic materials in HoA generating HiN.

3.3. UV-Vis Spectral Analysis of DOM in LCGW before and after Ozonation.

UV-Vis spectra analysis can effectively be used for determination of specific functional groups in water samples and further indication of the characteristics of DOM. However, the number of chromophoric groups that generated absorption peaks in the spectrum was so large that it was difficult to identify groups clearly. Figure 4 shows the spectrogram of separated components before and after ozonation.

As shown in Figure 4, similar waveforms appeared in the spectrum with few distinct peaks and troughs. In addition, many peaks overlapped, which may be explained by the presence of humic substances. As shown in Figure 4(a), the maximum absorbance of each component except HoN was about 220 nm, while every component except HoA

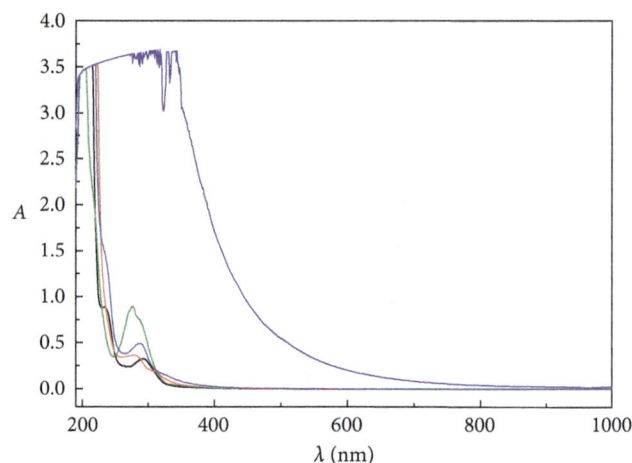

(a)

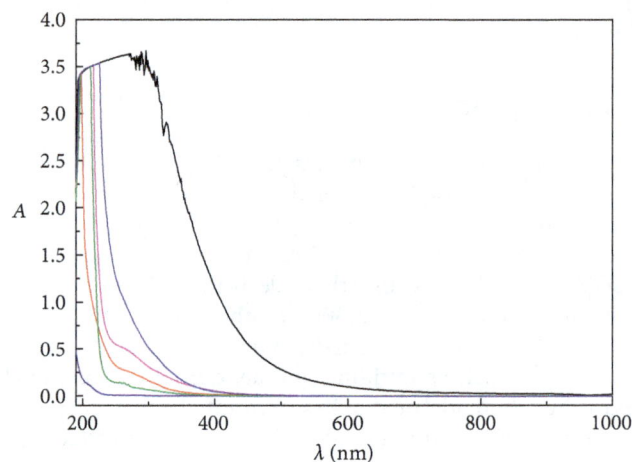

(b)

FIGURE 4: Ultraviolet-visible absorption spectroscopy of DOM components when the reaction time is 0 min (a) and 120 min (b).

generated a distinct absorbance peak at 270–290 nm. As shown in Figure 4(b), the absorbance at 270–290 nm disappeared after ozonation with corresponding R absorption band. This was a result of lone pairs of electrons on the heteroatoms connected with double bonds being transferred to the π^* antibonding orbital and indicated that double-bond heteroatoms were all oxidized during ozonation.

The spectrum is unable to provide much information about the properties of DOM. Moreover, the degree of humification maintains at a high level, and unsaturated organic components are concentrated in the ultraviolet area. The complexity of organic components also leads to the overlap of absorption peaks; therefore, more detailed information about components is difficult to achieve. E_{254}/E_{365}, E_{300}/E_{400}, and SR values of each component are shown in Table 2.

TABLE 2: UV-visible spectral parameters of the DOM components.

0 min	UV_{254}	E_{254}/E_{365}	E_{300}/E_{400}
HoA	3.609	1.34	2.10
HoB	0.407	15.65	35.77
HoN	0.403	17.52	42
HiA	0.441	7.47	11.82
HiB	0.283	17.69	39.86
HiN	0.354	11.42	16.69
120 min	UV_{254}	E_{254}/E_{365}	E_{300}/E_{400}
HoA	3.609	3.42	3.925
HoB	0.267	5.93	4.69
HoN	0.011	5.5	3
HiA	0.398	4.85	5.78
HiB	0.121	9.64	12.2
HiN	1.245	6.96	6.53

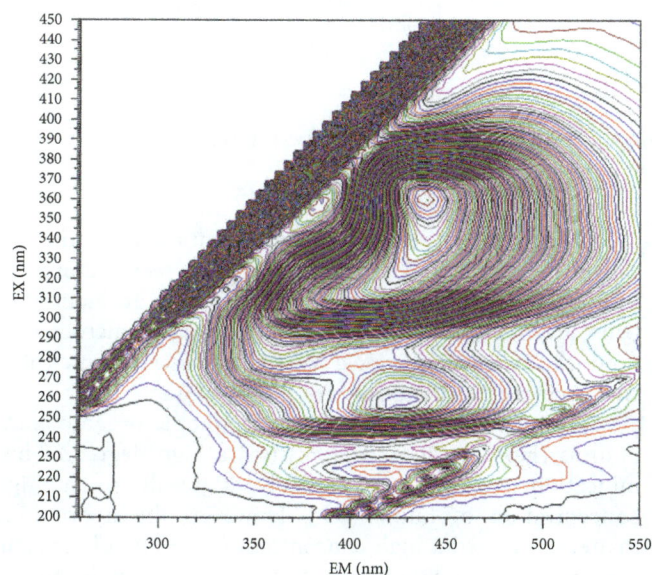

(a)

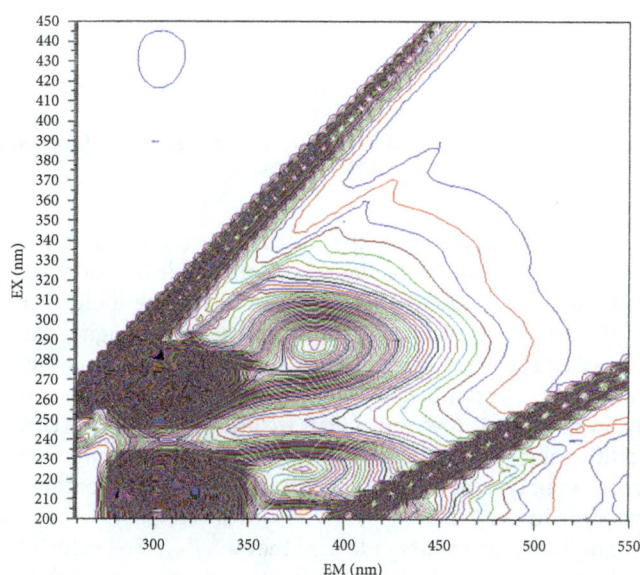

(b)

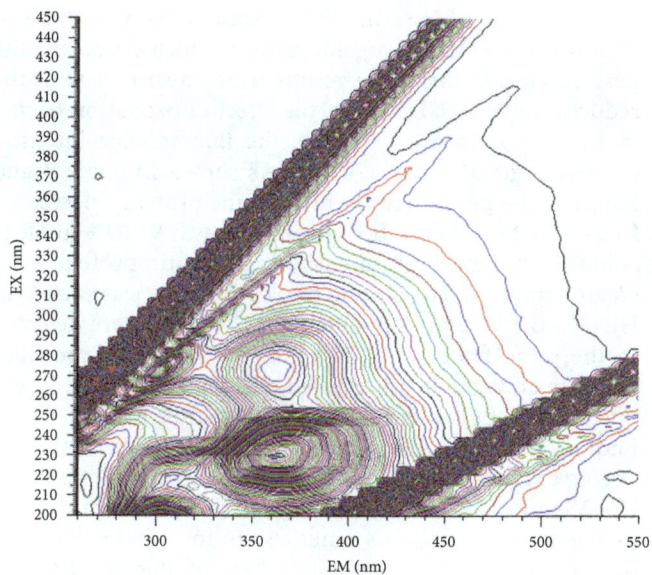

(c)

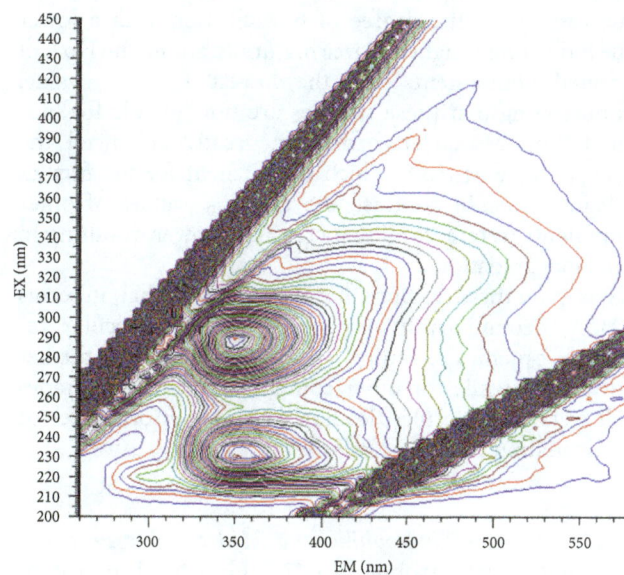

(d)

FIGURE 5: Continued.

(e)

(f)

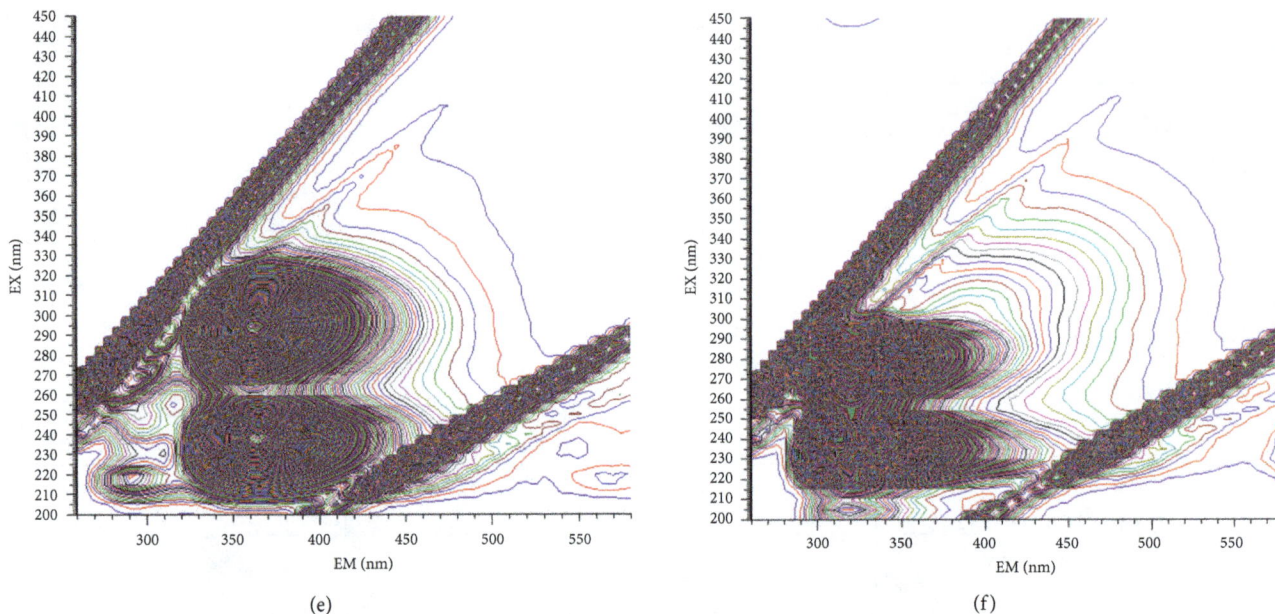

FIGURE 5: 3D EEM of fractions in the LCGW. (a) HoA. (b) HoB. (c) HoN. (d) HiA. (e) HiB. (f) HiN.

As shown in Table 2, the E_{254}/E_{365} value of HoA increased throughout the oxidation process while that of other components decreased. This value can reflect the molecular weight, with a smaller value indicating a larger weight. The increasing E_{254}/E_{365} value of HoA shows that ozone mainly reacts with macromolecular organic materials, resulting in the formation of low molecular weight organic materials. Additionally, the decreasing E_{254}/E_{365} values indicate that ozone mainly reacts with low molecular weight organic materials in the HoA.

Similarly, the results indicated that the E_{300}/E_{400} value of HoA was increasing throughout the oxidation process while that of other components was decreasing. The E_{254}/E_{365} value can reflect the degree of humification with a lower value indicating a higher degree of humification. The HoA of untreated components had the lowest E_{300}/E_{400} value, demonstrating that these samples are not suitable for biodegradation. Moreover, ozonation greatly enhanced the E_{300}/E_{400} value of HoA, which is of benefit for the further biodegradation. In contrast, the E_{300}/E_{400} values of other components were becoming increasingly lower, resulting in poor biodegradation.

Based on these findings, it can be inferred that, in terms of HoA, ozone mainly reacts with larger molecular recalcitrant organics to improve the performance for further degradation, while for other components, ozone degrades small organic molecules, leading to poor performance for biodegradation.

3.4. 3D EEM for Hydrophilic and Hydrophobic Fractions before and after Ozonation.

The 3D EEM has been extensively applied to investigate the different chemical fractions of dissolved organic matter by identifying certain classes of organic matter in wastewater. The changes in 3D EEM spectra of LCGW treated by ozonation at different reaction times are shown in Figure 5. The five main regions were commonly detected in different fractions, which contain aromatic protein I, aromatic protein II, fulvic acid III, soluble microbial by-product IV, and humic acid V. The fluorescence intensities corresponding to the peaks are shown in Table 3.

As shown in Figure 5, the hydrophobic organic compounds (HoA, HoB, and HoN) had a high degree of humification, which was probably a result of a high concentration of humic acid. However, the hydrophilic matter contained a high concentration of aromatic protein and dissolved organic matter. As a result, hydrophobic organic matter should be of more concern following ozonation. To address this issue, wastewater was bubbled with ozone to oxidize organic matter, which decreased the peak intensity. The above results were consistent with the reduction of TOC. However, the effect of ozonation occurs in two steps. In the first step, the fluorescence intensity decreased greatly after 120 min. As shown in Table 3 and Figure 6, the peak intensity of aromatic protein I decreased from 7218 to 643 and the removal rate was 91.09% in HoB. Additionally, the peak intensity of aromatic protein I decreased from 6700 to 226 and the removal rate was 96.63% in HiN. In the second step, different substances were detected in the same fractions at 120 min. Evaluation of the fluorescence at 360/440 nm revealed that the humic acid V was not present, but it was detected at 120 min in HiN. These findings indicated that parts of HoA were oxidized to HiN during the reaction, which was consistent with the results of UV-Vis and TOC analysis. Moreover, there were no changes in the types of organic materials in any of the fractions during ozonation, but all intensities of fluorescence decreased. In conclusion, during the 120 min ozonation of effluent after ammonia distillation and dephenolization, the macromolecular organics were oxidized to smaller

TABLE 3: Three-dimensional fluorescence spectroscopy table of DOM components.

Time (min)		$\lambda_{EX}/\lambda_{EM}$ (nm)					Fluorescence intensity (a.u.)				
		I	II	III	IV	V	I	II	III	IV	V
0	HoA			225/420		260/425			141		427
						360/440					1160
	HoB	215/300		205/380	270/300	290/385	7218		1058	5737	972
				225/380					774		
	HoN		230/360		270/360			719			346
	HiA		230/350		290/350			344		523	
	HiB	220/290		240/365	295/365		227		2509	2401	
	HiN	235/320			280/320		6700			4729	
120	HoA			245/415		310/410			612		682
	HoB	215/315		230/380		300/380	539		390		340
	HoN										
	HiA			225/400	270/300	310/405			197	216	131
	HiB	215/290	220/365		290/370		207	342		213	
	HiN	235/320		235/405		315/395	155		445		581

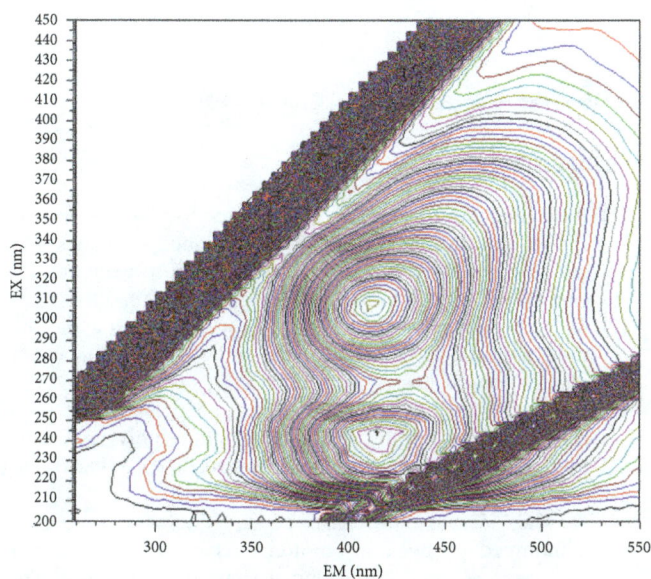

(a)

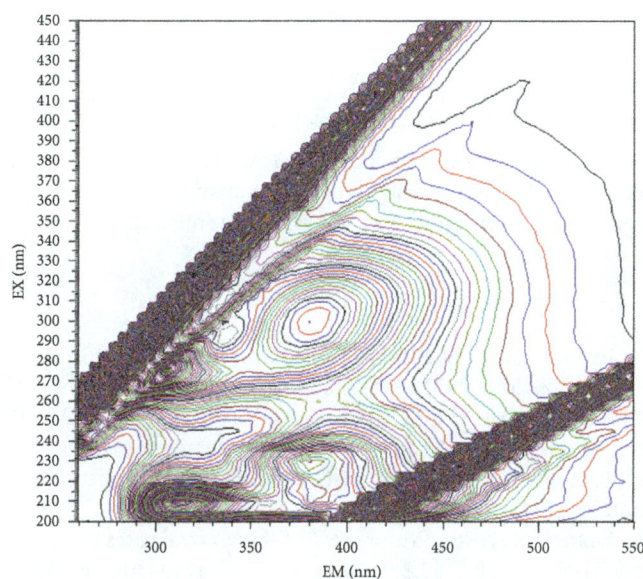

(b)

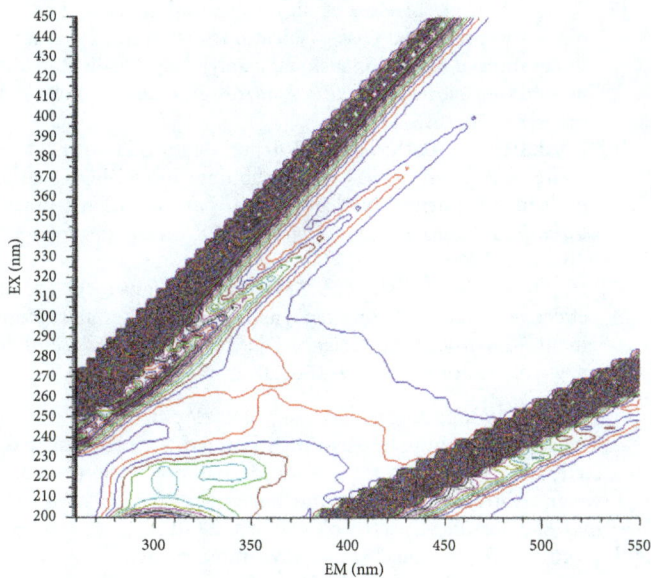

(c)

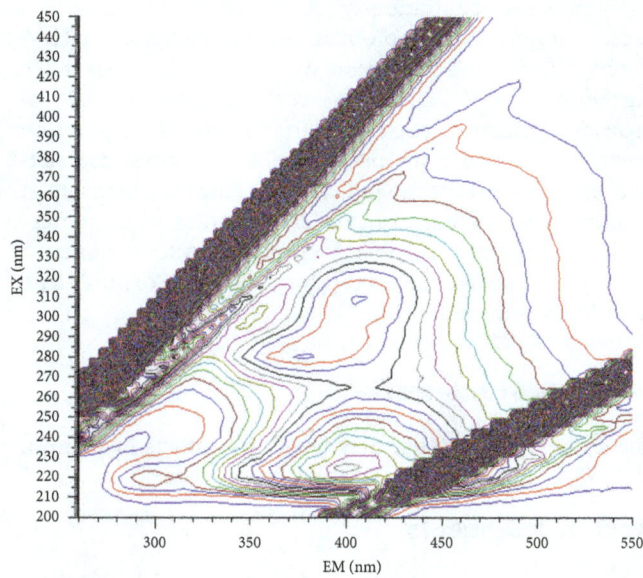

(d)

FIGURE 6: Continued.

(e)

(f)

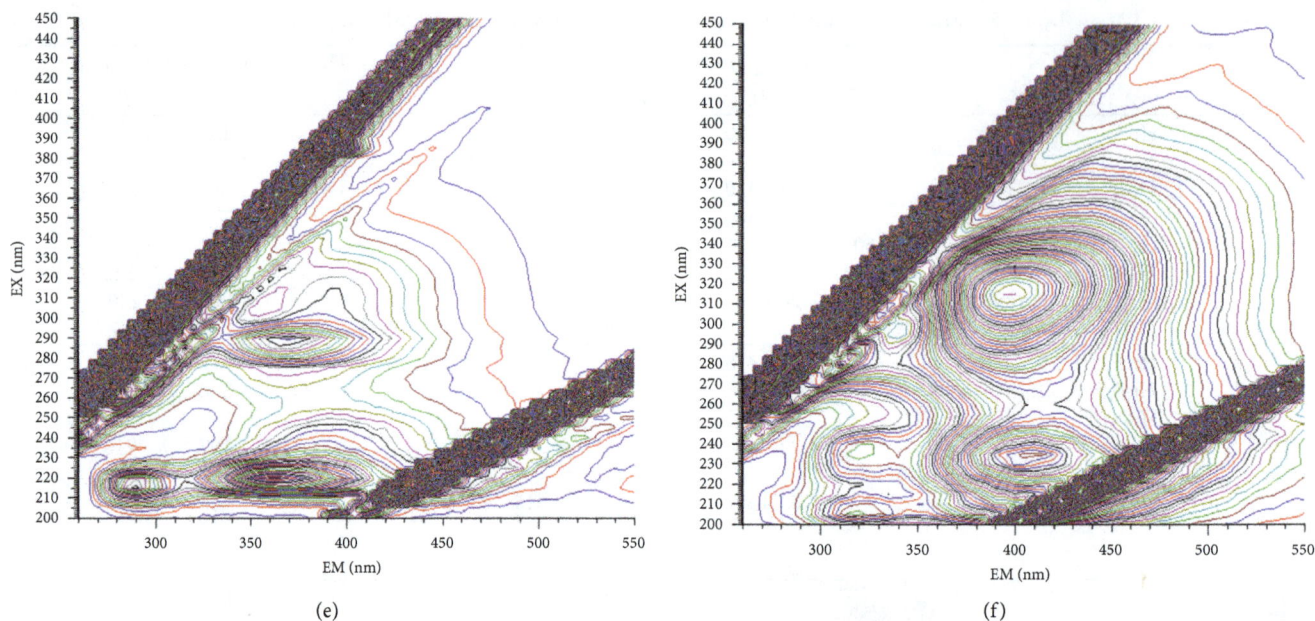

FIGURE 6: 3D EEM of fractions after 120 min of ozonation.(a) HoA. (b) HoB. (c) HoN. (d) HiA. (e) HiB. (f) HiN.

molecules with the degradation of soluble microbial by-products. After the organic types were changed, the residual organics were oxidized and completely removed.

4. Conclusions

During ozonation of effluent from ammonia distillation and dephenolization, DOM was fractionated into six classes by resin adsorption. The TOC, UV$_{254}$, UV-Vis, and 3D EEM of fractions were found to have different effects on ozonation. Additionally, the HoA and HiN fractions were found to be containing the majority TOC (about 380.21 mg·L^{-1} and 646.84 mg·L^{-1}, resp.). Moreover, TOC removal rates of HoA and HiN reached 42.85% and 67.13%, respectively. The UV$_{254}$ values of HoA remained unchanged, while that of HiN increased continuously before and after ozonation because humic macromolecular organic materials in HoA were oxidized, and a portion of their product was HiN. Furthermore, UV-Vis analysis revealed that the HoA of ozonation primarily reacts with larger molecular recalcitrant organics to improve the performance for further degradation. The 3D EEM spectra indicated that the macromolecular organics were oxidized to smaller molecules with the degradation of soluble microbial by-products. After the organic types were changed, the residual organics were oxidized and completely removed.

Conflicts of Interest

The authors declare that they have no conflicts of interest.

Acknowledgments

This research was funded by the 863 Project of the Ministry of Science and Technology from China (2015AA050501).

References

[1] A. K. Singh and J. Kumar, "Fugitive methane emissions from Indian coal mining and handling activities: estimates, mitigation and opportunities for its utilization to generate clean energy," *Energy Procedia*, vol. 90, pp. 336–348, 2016.

[2] C. He, X. Feng, and K. H. Chu, "Process modeling and thermodynamic analysis of Lurgi fixed-bed coal gasifier in an SNG plant," *Applied Energy*, vol. 111, no. 11, pp. 742–757, 2013.

[3] X. Hu, K. Chen, X. Lai et al., "Treatment of pretreated coal gasification wastewater (CGW) by magnetic polyacrylic anion exchange resin," *Journal of Environmental Chemical Engineering*, vol. 4, no. 2, pp. 2040–2044, 2016.

[4] H. Gai, H. Song, M. Xiao et al., "Conceptual design of a modified phenol and ammonia recovery process for the treatment of coal gasification wastewater," *Chemical Engineering Journal*, vol. 304, pp. 621–628, 2016.

[5] S. Jia, H. Han, B. Hou et al., "Advanced treatment of biologically pretreated coal gasification wastewater by a novel integration of three-dimensional catalytic electro-Fenton and membrane bioreactor," *Bioresource Technology*, vol. 198, pp. 918–921, 2015.

[6] J. Akhlas, A. Bertucco, F. Ruggeri et al., "Treatment of wastewater from syngas wet scrubbing: model-based comparison of phenol biodegradation basin configurations," *Canadian Journal of Chemical Engineering*, vol. 95, pp. 1652–1660, 2017.

[7] H. An, Z. Liu, X. Cao et al., "Mesoporous lignite-coke as an effective adsorbent for coal gasification wastewater treatment," *Environmental Science Water Research and Technology*, vol. 3, no. 1, pp. 169–174, 2017.

[8] L. Xu, J. Wang, X. Zhang et al., "Development of a novel integrated membrane system incorporated with an activated coke adsorption unit for advanced coal gasification wastewater treatment," *Colloids and Surfaces A: Physicochemical and Engineering Aspects*, vol. 484, pp. 99–107, 2015.

[9] P. Xu, W. Ma, B. Hou et al., "A novel integration of microwave catalytic oxidation and MBBR process and its application in advanced treatment of biologically pretreated Lurgi coal

gasification wastewater," *Separation and Purification Technology*, vol. 177, pp. 233–238, 2017.

[10] Z. Wang, X. Xu, Z. Gong et al., "Removal of COD, phenols and ammonium from Lurgi coal gasification wastewater using A^2O-MBR system," *Journal of Hazardous Materials*, vol. 235-236, pp. 78–84, 2012.

[11] Y. Li, S. Tabassum, and Z. Zhang, "An advanced anaerobic biofilter with effluent recirculation for phenol removal and methane production in treatment of coal gasification wastewater," *Journal of Environmental Sciences*, vol. 47, pp. 23–33, 2016.

[12] Y. Jeong and J. S. Chung, "Simultaneous removal of COD, thiocyanate, cyanide and nitrogen from coal process wastewater using fluidized biofilm process," *Process Biochemistry*, vol. 1, no. 41, pp. 1141–1147, 2006.

[13] Y. Li, G. Gu, J. Zhao et al., "Treatment of coke-plant wastewater by biofilm systems for removal of organic compounds and nitrogen," *Chemosphere*, vol. 52, no. 6, pp. 997–1005, 2003.

[14] F. Guan, G. Gao, and Q. Zhao, "Treatment of coal gasification wastewater by A/O biofilm process," *China Water and Wastewater*, vol. 25, pp. 74–76, 2009.

[15] H. Zhuang, S. Shan, C. Fang et al., "Advanced treatment of biologically pretreated coal gasification wastewater using a novel expansive flow biological intermittent aerated filter process with a ceramic filler from reused coal fly ash," *RSC Advances*, vol. 6, no. 46, pp. 39940–39946, 2016.

[16] Q. Liu, S. Vijay, P. Z. Fu et al., "An anoxic-aerobic system for simultaneous biodegradation of phenol and ammonia in a sequencing batch reactor," *Environmental Science and Pollution Research*, vol. 24, no. 12, pp. 11789–11799, 2017.

[17] Q. Zhao and Y. Liu, "State of the art of biological processes for coal gasification wastewater treatment," *Biotechnology Advances*, vol. 34, no. 5, pp. 1064–1072, 2016.

[18] C. Yang, Y. Qian, L. Zhang et al., "Solvent extraction process development and on-site trial-plant for phenol removal from industrial coal-gasification wastewater," *Chemical Engineering Journal*, vol. 117, no. 2, pp. 179–185, 2006.

[19] C. Lin, X. Zhuo, X. Yu et al., "Identification of disinfection by-product precursors from the discharge of a coking wastewater treatment plant," *RSC Advances*, vol. 54, no. 5, pp. 43786–43797, 2015.

[20] J. Swietlik, A. Dabrowska, and U. Raczykstanisławiak, "Reactivity of natural organic matter fractions with chlorine dioxide and ozone," *Water Research*, vol. 38, no. 3, pp. 547–558, 2004.

[21] T. Zhang, J. Liu, and J. Ma, "Comparative study of ozonation and synthetic goethite-catalyzed ozonation of individual NOM fractions isolated and fractionated from a filtered river water," *Water Research*, vol. 42, no. 6-7, pp. 1563–1570, 2008.

[22] F. Qi, W. Chu, and B. Xu, "Comparison of phenacetin degradation in aqueous solutions by catalytic ozonation with $CuFe_2O_4$, and its precursor: Surface properties, intermediates and reaction mechanisms," *Chemical Engineering Journal*, vol. 284, pp. 28–36, 2016.

2

Application of Chitosan Composite Flocculant in Tap Water Treatment

Awa Kangama (ID), **Defang Zeng** (ID), **Xu Tian, and Jinfu Fang**

School of Resource and Environmental Engineering, Wuhan University of Technology, 122 Luoshi Road, Wuhan 430070, China

Correspondence should be addressed to Defang Zeng; df5152@163.com

Academic Editor: Wenshan Guo

The chitosan is a good flocculant for tap water treatment because of its properties such as faster deposition rate and higher removal efficiency for COD (organic matter), SS (suspended solids), and metal ions. However, its high price limits the use in tap water treatment. In this paper, in order to reduce costs, chitosan (CTS), polyaluminum chloride (CF-PAC), and modified rectorite (Al(OH)$_3$ + HCl) were combined to prepare the flocculant for tap water treatment. In order to get the optimal composite flocculant formula, first, we combined these flocculants in two-by-two schema and then we combined all the three flocculants together with various dosing amounts. Through comparison between different combination schemas, the best formula of the composite chitosan flocculant was found to be CTS (ml) : CF-PAC (ml) : modified rectorite (Al(OH)$_3$ + HCl) (ml) = 1 : 30 : 5, with a turbidity removal rate of 96.38% and a removal rate of aluminum up to 80.1%, while the treatment cost is the lowest. In addition, we have designed a cost-effective method for the treatment cost evaluation. As raw water, we used water from the Han River, which is used as raw water at Zonguan Waterworks. In order to show the effectiveness of our optimal composite chitosan formula, we have compared our treatment results to those of the aluminum polyaluminum chloride flocculant currently used in Zonguan's water treatment plants.

1. Introduction

Water is an irreplaceable resource that drives our lives. An adult human should drink 2 to 5 liters of water a day [1–3], and the quality of drinking water is essential for public health. The treatment of drinking water or tap water by flocculation is an important research direction. The flocculation is a solid-liquid separation process, which consists of adding a flocculant into the raw water, after which finely divided or dispersed particles are aggregated or agglomerated together to form large particles of such a size (flocs) which settle and cause clarification of the system [4]. According to the literature [5–7], the main flocculants used in tap water treatment are chitosan (CTS) and polyaluminum chloride (PAC). The chitosan is one of the most promising biopolymers for extensive application due to its cationic behaviour. It is a partially deacetylated polymer obtained from the alkaline deacetylation of chitin, a biopolymer extracted from shellfish sources. And it is a linear hydrophilic aminopolysaccharide with a rigid structure containing both glucosamine and acetyl

glucosamine units, and each glucosamine unit is composed of a free amino group (-NH$_2$). As the active amino groups (-NH$_2$) in the chitosan molecule can be protonated with H$^+$ in water into a cationic polyelectrolyte [8], chitosan has characteristics of static attraction and adsorption and is widely used in the treatment of wastewater and the elimination of dyes [9]. The chitosan is insoluble in either water or organic solvents but soluble in dilute organic acids such as acetic acid and formic acid and inorganic acids where the free amino groups are protonated and the biopolymer becomes fully soluble [10, 11]. For more detailed information about the chitosan description, the interested readers can refer to Reference [8]. The flocculation efficiency of the chitosan flocculant in tap water treatment highly depends on its degree of deacetylation (DD) and the molecular weight (MW), which are two very important structural factors for chitosan. In addition, the effects of DD and MW on the final flocculation performance are usually complementary and synergistic [12]. The chemical structures of chitin and chitosan are shown in Figure 1. Compared to traditional chemical flocculants,

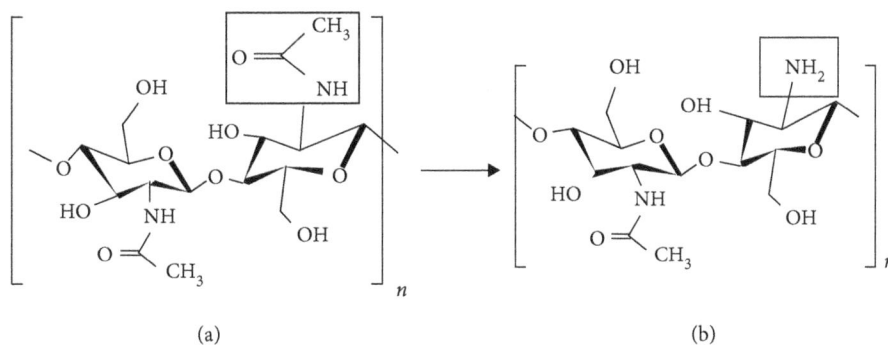

FIGURE 1: Chemical structures of chitin (a) and chitosan (b).

chitosan has a lower dosage, a faster deposition rate, better removal of COD (organic matter) and suspended solids, easier sludge treatment, no pollution, and presence of metal ions amino-D-glucose. However, when we use only the chitosan as a flocculant to treat the tap water, the treatment cost is too high compared to that of traditional chemical flocculants. The unit price of chitosan is much higher than that of traditional chemical flocculants. Therefore, there is a need to prepare a low-cost and efficient composite chitosan flocculant for the tap water treatment. In this paper, we prepare a low-cost and efficient composite chitosan flocculant for the tap water treatment by combining chitosan (CTS), polyaluminum chloride (CF-PAC), and modified rectorite (Al(OH)$_3$ + HCl).

The remainder of the paper is organized as follows: the experiment method and procedure are presented in Section 2. The experimental results are in detail presented and discussed in Section 3, and Section 4 presents the conclusion.

2. Experiment

2.1. Raw Water. In this experiment, raw water from Zonguan Waterworks was used, and raw water from Zonguan Waterworks came from the Han River. The Han River is the largest tributary of the Yangtze River. The choice of the Han River as the source of our water sample is that, in recent years, the pollution of the Han River has increased and the state departments concerned by the various measures to be taken have invested a lot of money to improve the quality of the water from the Han River and got results.

2.2. Main Reagents. Chitosan (CTS) with a mass concentration of chitosan 0.5% diluted 5 times, polyaluminum chloride (CF-PAC) with a mass concentration of chitosan 2%, and modified rectorite (Al(OH)$_3$ + HCl) diluted 5 times are used.

2.3. Main Equipments. MY3000-6 combination mixer (Qianjiang Meiyu Instrument Co., Ltd.) was used to fully control time and stirring speed. Turbidity was measured by using the automatic turbidimeter (2100 PTURBIMETER HACH), electronic scale, 100 ml volumetric flask, pipette, beaker, and so on. The flocculant solution was added dropwise with a dropper.

2.4. Experimental Methods and Procedures. We used the water sample by measuring its initial value. With a pipette, 1 L of water was taken from the water sample, by adding a flocculant to it and stirring it rapidly (200 stirs for 1 minute). After that, stir slowly (90 stirs for 6 minutes first and then 50 stirs for 8 minutes), stop the precipitation for 10 minutes, and then measure its turbidity. The experiment of flocculation and coagulation was performed using MY3000-6. The turbidity is determined by an automatic turbidimeter (2100 PTURBIMETER HACH). The flocculant solution was added dropwise with a dropper. In our experiments, the used degree of deacetylation of chitosan is 88% and the used molecular weight of chitosan is 500000.

3. Results and Discussion

3.1. Experimental Results of Tap Water Treatment with Simple Flocculants: Chitosan, Polyaluminum Chloride, and Modified Rectorite. As it can be seen from Figure 2, when different quantities of chitosan are added to the raw water, the turbidity of the treated water is between 7 and 9. According to the requirements of the current tap water treatment process, the turbidity of the water after flocculation and sedimentation should be less than 5 NTU before filtering in order to make the final effluent turbidity below 1 NTU. That is, if only chitosan is used to treat the source water of tap water, the treatment effect is not ideal, and also if the used quantity of chitosan is 3 drops per liter of water, the cost is 0.0162 yuan per ton of water, which is nearly 44.6% higher than the cost of flocculation used at Zonguan Waterworks, 0.0112 yuan per ton of water. Therefore, to sum up in one sentence, the only use of chitosan to treat the source water of tap water cannot reduce costs and improve the treatment effect. Therefore, in this work, we decided to use the composite flocculant composed of polyaluminum chloride, modified rectorite, and chitosan to deal with the treatment of raw water of tap water. The specific circumstances can be found below.

In Figure 3, it can be seen that, with the gradually increasing dosage of polyaluminum chloride, the turbidity of tap water gradually reduces; however, there is not a linear relationship between the amount of polyaluminum chloride and treatment effect. From the treatment effect point of view, it is much better than the chitosan, and tap water turbidity is

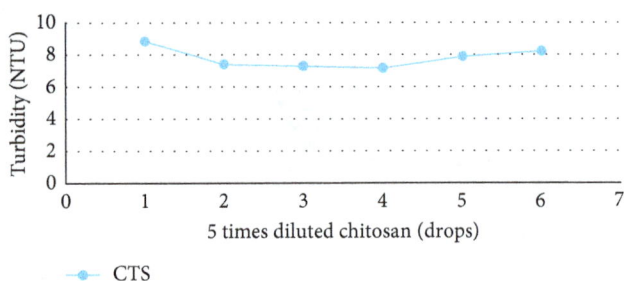

FIGURE 2: Turbidity of raw water of tap water treated with chitosan.

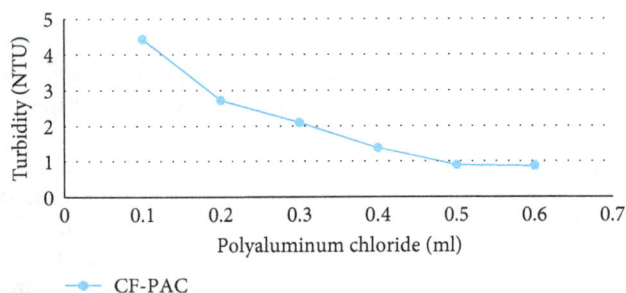

FIGURE 3: Turbidity of raw water of tap water treated with poly-aluminum chloride.

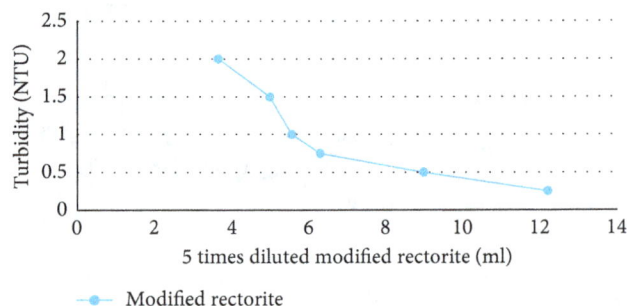

FIGURE 4: Turbidity of raw water of tap water treated with modified rectorite.

Polyaluminum Chloride, and Modified Rectorite. In this section, based on the above experimental data and results, we made the experiments of treating the source water of Zonguan Waterworks by combining chitosan, poly-aluminum chloride, and modified rectorite, in two-by-two schema. The specific circumstances of the experiments are as follows.

below 5 NTU, but according to Zonguan Waterworks, the added quantity of the flocculant is 20 kg per ton of water and the percentage of the polyaluminum chloride in 1 kg of flocculant is 10%; this value exceeds when CF-PAC is added at 0.2 ml, which is bound to bring in a large quantity of aluminum, which is not consistent with the goal of reducing the tap water content in aluminum. If the other non-aluminum and nontoxic and harmless flocculants are combined with PAC, the same or better result or effect can be obtained, and also the content of the tap water in aluminum can be reduced.

It can be seen from Figure 4 that, with the increase of the dosage of the modified rectorite, the turbidity of the effluent also decreases, but there is no linear relationship between the dosage and the treatment effect. However, when the dosage is 0.75 ml, the treatment effect is better than the best treatment effect of chitosan, and it costs 0.010699 yuan per ton of water. In addition, when the dosage of the modified rectorite is 0.3 ml, the effluent turbidity is only 4.98 NTU, which is less than 5 NTU, and the corresponding cost is also 58.2% higher than the current cost of flocculation at Zon-guan Waterworks.

Based on the above situations (Figures 2–4), the separate use of chitosan, polyaluminum chloride, and modified rectorite in various situations does not have high treatment costs and may cause secondary pollution to tap water. However, if they are combined, it is possible to improve the treatment effect, reduce costs, and reduce or not produce secondary pollution to tap water. Therefore, it is of great value and significance to compound them.

3.2. Experimental Results of Tap Water Treatment with Pair Composite Flocculants from Single Flocculants: Chitosan,

3.2.1. Cost-Effective Calculation Method. The cost-effective calculation method consists, first, of the calculation of the turbidity removal rate and then the calculation of the cost. To obtain the turbidity removal rate, we subtract the tur-bidity of the treated effluent from the turbidity of the raw water of the tap water and divide this result by the turbidity of the raw water of the tap water. And by dividing the costs of the treated effluent and the raw water of the tap water by the obtained turbidity removal rate, respectively, we will get a value: the greater this value, the higher the cost ratio of the corresponding formula, and vice versa.

3.2.2. Flocculation Experiment with Pair Composite Floccu-lant of Chitosan and Modified Rectorite (CTS + Modified Rectorite). It can be seen from Figure 5 that when the dosage of modified rectorite is 0.25 ml and 0.75 ml, the turbidity of the effluent does not decrease gradually with the increase of the dosage of chitosan. When the quantity of added chitosan is 2 drops, the effect becomes worst. While when the quantity of modified rectorite is 1.0 ml, the turbidity decreased with the increase of the dosage chitosan.

Using the above cost-effective calculation method with 1 drop of chitosan and 0.25 ml of modified rectorite, we can get the highest cost ratio with a value of 9695.13. At this time, the cost is 0.008966 yuan per ton of water and the turbidity removal rate is 86.93%. The added amount of chitosan and modified rectorite was 1 drop and 0.50 ml, respectively.

3.2.3. Flocculation Experiment with Pair Composite Floccu-lant of Chitosan and Polyaluminum Chloride (CTS + PAC). From Figure 6, it can be seen that, with the increase of the dosage of chitosan and polyaluminum chloride, the tur-bidity of the effluent decreases gradually. When the amount of chitosan and polyaluminum chloride is 3 drops and is 0.4 ml, respectively, the best treatment effect is achieved,

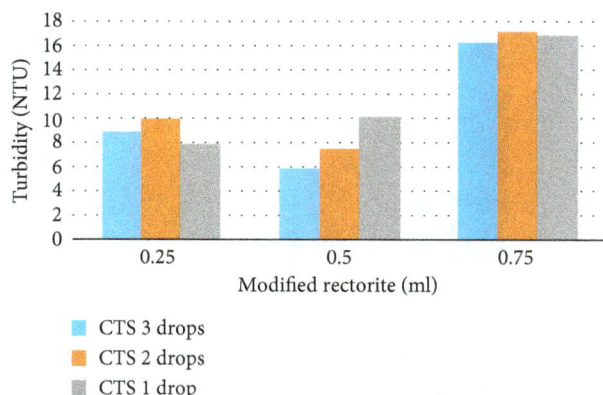

FIGURE 5: Turbidity of the effluent with chitosan and modified rectorite.

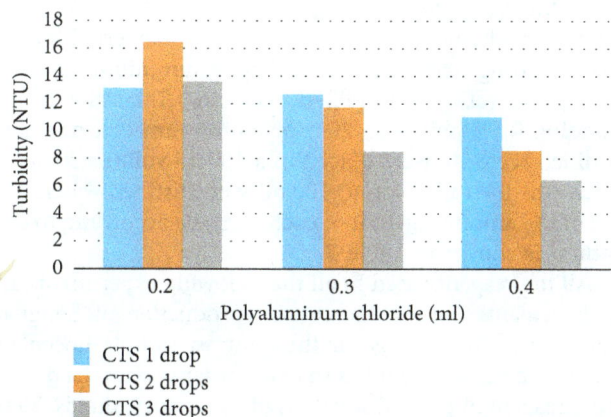

FIGURE 6: The effluent turbidity of raw water with chitosan and polyaluminum chloride.

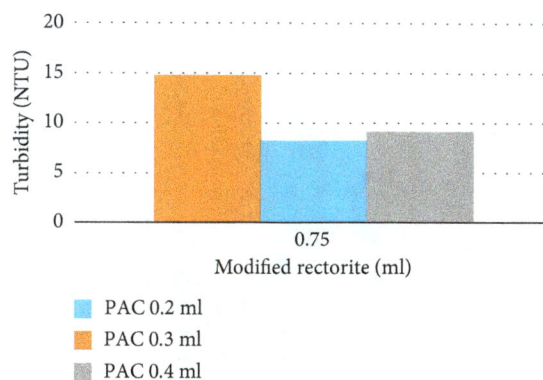

FIGURE 7: Turbidity of water treated with polyaluminum chloride and modified rectorite.

and the effluent turbidity is 6.52 NTU. At this time, the effluent turbidity has not reached the standard below 5 NTU. However, its cost has reached 0.0322 yuan per ton of water. Continuing to increase the dosage of chitosan and polyaluminum chloride is of no practical significance; however, according to the abovementioned cost-effective calculation method, we can get some of the best cost-effective formulas. Through the calculation, when the added amount of chitosan and polyaluminum chloride is 1 drop and 0.2 ml, respectively, the cost performance ratio is the highest, its value is 6964.93, the cost is 0.016099 yuan per ton of water, and the turbidity removal rate is 93.33%. While in the added amount of 1 drop of chitosan and 0.3 ml of polyaluminum chloride, the cost is 0.0174 yuan per ton of water.

3.2.4. Flocculation Experiment with Pair Composite Flocculant of Polyaluminum Chloride and Modified Rectorite (PAC + Modified Rectorite).
From Figure 7, it can be seen that when polyaluminum chloride is combined with modified rectorite and the amount of rectorite is constant while the amount of polyaluminum chloride increases, the treatment effect is better; however, the reverse is not the case. When

using certain amount of polyaluminum chloride with the added amount of rectorite increasing from 0.50 ml to 0.75 ml, the turbidity of the effluent will be increased. According to the above cost-effective method of calculation, when the added amount of polyaluminum chloride and modified rectorite is 0.2 ml and 0.25 ml, respectively, the cost performance ratio is the highest, its value is 8066.50, the cost is 0.011566 yuan per ton of water, and the turbidity removal rate is 93.30%.

3.3. Experimental Results of Water Treatment with Composite Flocculation of Chitosan, Polyaluminum Chloride, and Modified Rectorite.
In this section, based on the above experimental data and results, we designed experiments of treatment of the raw water of the tap water with the combination of chitosan, polyaluminum chloride, and modified rectorite as the composite flocculant. According to the cost-effective calculation method introduced in Section 3.2.1, we have got several experimental formulas with high cost performance. The following are the specific experimental cases.

From Figure 8, it can be seen that when the dosage of chitosan is 1 drop per liter of raw water, not only the turbidity of the effluent decreases with the increase of polyaluminum chloride and modified rectorite, but also the turbidity of the effluent begins to appear below 5 NTU when the dosage of polyaluminum chloride is 0.2 ml in addition; with the increase of dosage of modified rectorite and polyaluminum chloride, the turbidity of the effluent is getting lower and lower. The turbidity of the effluent reaches the lowest 0.67 NTU when the amount of polyaluminum chloride and modified rectorite is 0.6 ml and 2.0 ml, respectively. Therefore, from the experimental results, the effect of the combination of the flocculants is obvious; however, it is expensive if we calculate the cost. According to the cost-effective calculation method, when the added amount of chitosan, polyaluminum chloride, and modified rectorite was 1 drop, 0.10 ml, and 0.50 ml, respectively, the cost-effective ratio is the highest, its value is 5807.55, the cost is 0.01653272 yuan per ton of water, and the turbidity removal rate is 96.01%. In addition, when the added amount of chitosan, polyaluminum chloride, and modified rectorite

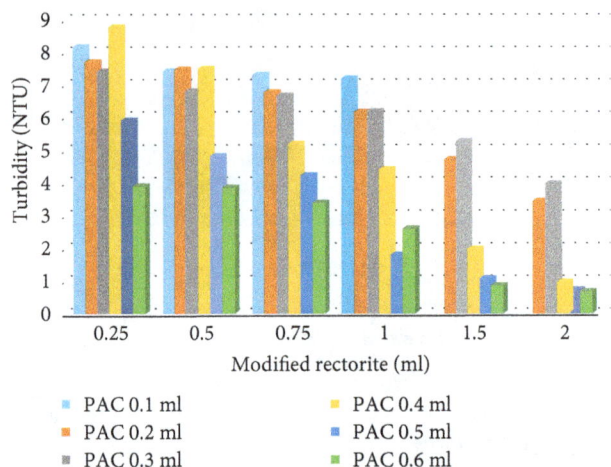

FIGURE 8: Turbidity of tap water treated with 1 drop of chitosan, polyaluminum chloride, and modified rectorite.

was 1 drop, 0.20 ml, and 0.25 ml, the cost of water treatment is 0.016966 yuan per ton; when the added amount of chitosan, polyaluminum chloride, and modified rectorite was 1 drop, 0.3 ml, and 0.25 ml, the cost of water treatment is 0.020966 yuan per ton.

As it can be seen in Figure 9, when the added amount of chitosan is 2 drops per liter of the water sample, from the result of the treatment, the effluent turbidity is also the same as that in the case of adding 1 drop of chitosan, and also the effluent turbidity decreases with the increase of the amount of polyaluminum chloride and modified rectorite. However, there is no linear relationship between them, but the greater the added amount of polyaluminum chloride and modified rectorite, the slower the decrease of turbidity of the effluent. With the application of the above cost-effective calculation method, the best formula is 2 drops of chitosan, 0.1 ml of polyaluminum chloride, and 0.25 ml of modified rectorite, the cost performance value is the value of 4957.38, the cost of the treatment of the raw water is 0.018966 yuan per ton of water, and the turbidity removal rate is 94.02%. In the formula, 2 drops of chitosan, 0.1 ml of polyaluminum chloride, and 0.50 ml of modified rectorite, the cost-effective value is 4363.83, the cost of processing the raw water is 0.021933 yuan per ton of water, and the turbidity removal rate is 95.71%.

3.4. Experimental Results and Discussion of a Formula with High Cost Performance for Each Composite Scheme. Based on the above experimental data and results, the most cost-effective formulas were obtained from each composite solution, as shown in Table 1. The conversion values in Table 1 is obtained by converting the diluted amount of chitosan and modified rectorite into the corresponding amount of 0.5% chitosan without dilution and modified rectorite without dilution, and then converting the newly obtained 0.5% chitosan: 2% polyaluminium chloride: modified rectorite ratio into ml. The value of the example is multiplied by 100 times, and the conversion value in the table is obtained.

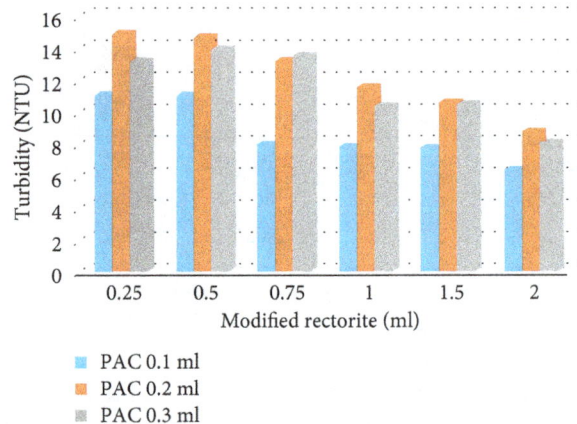

FIGURE 9: Turbidity of the effluent with 2 drops of chitosan, polyaluminum aluminum chloride, and modified rectorite.

3.4.1. The Configuration Method of Each Formula. Each composite scheme formula in Table 1 was converted into its corresponding 0.5% chitosan: 2% polyaluminium chloride: double the ratio of modified rectorite. Then the corresponding 0.5% chitosan, 2% polyaluminium chloride, and modified rectorite were placed in a 100 ml volumetric flask, which was then filled up to 100 ml with distilled water. The cost of the amount of 1 ml of each formula according to this method is shown in Table 2.

All the reagents used in all the following experiments are the flocculants used in the current production of Zonguan Waterworks. The dosage of the contrast sample flocculant added in the experiment is also in accordance with the dosage of the reagent in general situation of the factory, that is, 20 kg for thousand tons of water. The specific approach is described as follows: first, with the electronic balance we measure 2.00 g of flocculant used in the current production of Zonguan waterworks, and then we place this 2.00 g of flocculant in a 100 ml volumetric flask, which is then filled up to 100 ml with water. Second, we take 1.0 ml of the prepared solution, which contains 0.0200 g of flocculant. Finally, we add that 1.0 ml of the prepared solution to 1 liter of the source water sample, and this is equivalent to adding 20 kg of the flocculant into one thousand tons of water tons of source water.

3.4.2. The Experimental Results and Summary of Tap Water Source Water Treatment of Water Source by Each Formula. From Table 3, we can see that when the adding amount in formula no. 1 is 1.6 ml, the effluent turbidity is 15.7 NTU which is higher than the turbidity of the comparative sample equal to 10.2 NTU, which in turn is obviously larger than 5.5 NTU, apparently in order to achieve the same treatment effect as the comparative sample but also to continue to increase the dosing amount. However, when the dosage (the added amount) of the formula is 1.6 ml per liter, the cost has reached 0.022594 yuan per ton of water, which is more than double the cost of the contrast sample.

From Table 4, it can be seen that when the dosing amount per liter of water in formula no. 2 is 0.8 ml and 0.6 ml, the effluent turbidity is 16.1 NTU and 18.1 NTU, respectively, and the water turbidity of the water sample after the contrast sample treatment is 16.48 NTU. Therefore, to

TABLE 1: The correlation of the most cost-effective formulations in each of the composite schemes.

Number	CTS (drop) : CF-PAC (ml) : modified rectorite (ml) ratio	After conversion, CTS (ml) : CF-PAC (ml) : modified rectorite (ml) ratio	Turbidity removal rate (%)	Cost-effective value
1	1 : 0 : 0.5	1 : 0 : 10	83.25	6642.60
2	1 : 0.3 : 0	1 : 30 : 0	92.35	5307.49
3	1 : 0.2 : 0	1 : 20 : 0	93.33	6964.93
4	0 : 0.2 : 0.25	0 : 20 : 5	93.30	8066.50
5	1 : 0.1 : 0.25	1 : 10 : 5	75.09	5790.87
6	1 : 0.2 : 0.25	1 : 20 : 5	96.25	5673.04
7	1 : 0.3 : 0.25	1 : 30 : 5	96.38	4596.97
8	2 : 0.1 : 0.5	2 : 10 : 10	95.71	4363.83

TABLE 2: Cost of 1 ml solution when each formula is configured into 100 ml solution.

Number	1	2	3	4	5	6	7	8
The cost of 1 ml solution	0.01253	0.01740	0.01340	0.01157	0.01270	0.01670	0.020966	0.021933

TABLE 3: Formula no. 1 treatment of tap water source water-related case.

Turbidity of raw water	43.9 NTU			Raw water temperature		24.5°C	
Number			1			Sample of comparison	
Added amount (ml)	0.8	1.0	1.2	1.4	1.6	1.0	
Water turbidity (NTU)	18.5	18.4	17.2	15.4	15.7	10.2	

TABLE 4: Formula no. 2 treatment of tap water source water-related case.

Turbidity of raw water	65.9 NTU			Raw water temperature		24.0°C	
Number			2			Sample of comparison	
Added amount (ml)	0.6	0.8	1.0	1.2	1.5	1.0	
Water turbidity (NTU)	18.1	16.1	11.9	10.3	8.67	16.48	

achieve a treatment effect close to the contrast sample treatment effect, the dosing amount of the formula should be between 0.6 and 0.8 ml per liter of the sample water, with the corresponding treatment cost of 0.01044–0.01392 yuan per ton of water, which is close to the treatment cost of the contrast sample of 0.0112 yuan per ton of water.

From Table 5, it can be seen that when the dosing amount per liter of water in formula no. 3 is 1.2 and 1.5 ml, the effluent turbidity is 22.3 NTU and 15.1 NTU, respectively, and the water turbidity of the water sample after the contrast sample treatment is 21.9 NTU. Therefore, to achieve a treatment effect close to the contrast sample treatment effect, the dosing amount of the formula should be between 1.2 and 1.5 ml per liter of the sample water, with corresponding treatment cost of 0.01608–0.0201 yuan per ton of water, which is much higher than the treatment cost of the contrast sample of 0.0112 yuan per ton of water.

From Table 6, it can be seen that when the dosing amount per liter of water in formula no. 4 is 1.2 ml, the effluent turbidity reaches 14.23 NTU, which is less than the effluent turbidity of the contrast sample (the comparative effluent turbidity) equal to 14.7 NTU, and when the dosing amount is 1.0 ml, the effluent turbidity is 15.2 NTU, which is slightly larger than 14.7 NTU. Therefore, formula no. 4 can achieve the creatment effect close to the contrast sample treatment effect when the dosing amount is between 1 and

1.2 ml, and the corresponding treatment costs between 0.011566 and 0.0138792 yuan per ton of water, which is higher than the treatment cost of the contrast sample only between 0.0027 and 0.00366 yuan per ton of water.

From Table 7, it can be seen that when the dosing amount per liter of water in formula no. 5 is 1.0 ml and 1.2 ml, the effluent turbidity is 9.93 NTU and 6.03 NTU, respectively, while the effluent turbidity of the comparative sample is 8.32 NTU. The dosage of 1.0 ml is lower, while the dosage of 1.2 ml is higher, and the corresponding cases are shown in Table 8. It can be concluded that formula no. 5 can achieve the treatment effect close to that of the comparative sample, when the dosing amount is between 1 and 1.2 ml with the corresponding water treatment cost of 0.01270 and 0.01524 yuan per ton of water, which is slightly higher than that of the comparative sample equal to 0.0112 yuan per ton of water.

From Table 8, it can be seen that when the dosing amount per liter of water in formula no. 6 is 0.8 and 1.0 ml, the effluent turbidity is 15.1 NTU and 13 NTU, respectively, and the effluent turbidity of the comparative sample is 13.3 NTU. Therefore, formula no. 6 can achieve the treatment effect close to that of the contrast sample when the dosing amount per liter of water is between 0.8 and 1.0 ml and closer to 1.0 ml, with the corresponding water treatment costs of 0.01357– 0.01697 yuan per ton water, which are of 0.00237–0.00577

TABLE 5: Formula no. 3 treatment of tap water source water-related case.

Turbidity of raw water	134 NTU		Raw water temperature			23.1°C
Number			3			Sample of comparison
Adding amount (ml)	0.8	1.0	1.2	1.5	2.0	1.0
Water turbidity (NTU)	27.1	23.8	22.3	15.1	8.36	21.9

TABLE 6: Formula no. 4 treatment of tap water source water-related case.

Turbidity of raw water	99.8 NTU		Raw water temperature			24.0°C
Number			4			Sample of comparison
Adding amount (ml)	0.8	1.0	1.2	1.5	2.0	1.0
Water turbidity (NTU)	18.1	17.0	15.2	8.33	5.03	14.7

TABLE 7: Formula no. 5 treatment of tap water source water-related case.

Turbidity of raw water	23.1 NTU		Raw water temperature			24.5°C
Number			5			Sample of comparison
Adding amount (ml)	0.8	1.0	1.2	1.4	1.6	1.0
Water turbidity (NTU)	11.4	9.93	6.03	7.31	6.14	8.32

TABLE 8: Formula no. 6 treatment of tap water source water-related case.

Turbidity of raw water	99.8 NTU		Raw water temperature			24.0°C
Number			6			Sample of comparison
Adding amount (ml)	0.6	0.8	1.0	1.2	1.5	1.0
Water turbidity (NTU)	18.5	15.1	13.0	11.4	8.07	13.3

yuan per ton, respectively, higher than the comparative sample treatment cost of 0.0112 yuan per ton of water.

From Table 9, it can be seen that when the dosing amount per liter of water of formula no. 7 is 0.5 ml, the turbidity of the treated sample water is 13.8 NTU, which is higher than the effluent turbidity of the contrast sample (the comparative effluent turbidity) equal to 14.3 NTU. The water treatment cost of the dosing amount of 0.5 ml is only 0.010483 yuan per ton which is lower (of 0.000717 yuan per ton of water) than the water treatment cost of the comparative sample equal to 0.0112 yuan per ton of water.

From Table 10, it can be seen that when the dosing amount of formula no. 8 is 0.8 and 1.0 ml, the turbidity of the treated sample water is 11.6 NTU and 8.22 NTU, respectively, and effluent turbidity of the comparative sample is 11.1 NTU. Therefore, in formula no. 8, to achieve a treatment effect close to that of the comparative sample, the dosing amount of the formula should be between 0.8 and 1.0 ml and closer to 0.8 ml, with the corresponding cost of 0.017546–0.02133 yuan per ton water, which is more than 50% higher (0.0063–0101 yuan per ton of water) than the treatment cost of the comparative sample of 0.0112 yuan per ton of water.

From the comparison between the above 8 formulas and the contrast samples, the effect of formula no. 1, namely, chitosan and modified rectorite composite scheme, is relatively poor, and to achieve the same treatment effect as the comparative sample, the cost will be more than 1 time higher. The chitosan and polyaluminum chloride composite scheme, namely, formula nos. 2 and 3, has a better treatment

effect than the chitosan and modified rectorite composite scheme. In particular, formula no. 2 is the exact formula, and the cost of achieving the same treatment effect as the comparative sample is between 0.01044 and 0.01392 yuan per ton of water. The composite solution of polyaluminum chloride and modified rectorite is also better than that of chitosan and modified rectorite. The cost of formula no. 4 is close to 0.112 yuan per ton of water while achieving the same treatment as that of the comparative sample. The combination of chitosan, polyaluminum chloride, and modified rectorite is the best combination of the three. The treatment effect of the application of formula no. 7 to treat the water samples is better than that of the contrast sample when the dosing amount is close to 0.5 ml, and the corresponding cost is only 0.010483 yuan per ton of water, which is 0.000717 yuan per ton of water lower than the treatment cost of the comparative sample of 0.0112 yuan per ton of water.

3.5. The Determination of the Best Formula. According to the above summary, three best formula nos. 2, 4, and 7 have been obtained. In order to further determine their performance, they are further compared with the comparative sample. The specific information is shown in Tables 10–12.

Tables 11–13 show formula nos. 2, 4, and 7, configured according to the configuration method described above. The dosing amount of formula no. 2 per liter of the water sample is 0.65 ml, whose treatment effect is better than that of the comparative sample. The corresponding treatment

TABLE 9: Formula no. 7 treatment of tap water source water-related case.

Turbidity of raw water	61.6 NTU		Raw water temperature			21.8°C
Number			7			Sample of comparison
Adding amount (ml)	0.5	0.6	0.7	0.8	1.0	1.0
Water turbidity (NTU)	13.8	13.0	11.9	11.44	12.3	14.3

TABLE 10: Formula no. 8 treatment of tap water source water-related case.

Turbidity of raw water	123 NTU		Raw water temperature			21.8°C
Number			8			Sample of comparison
Adding amount (ml)	0.5	0.6	0.7	0.8	1.0	1.0
Water turbidity (NTU)	17.4	16.1	15.5	11.6	8.22	14.3

TABLE 11: Comparison experiment of formula nos. 2, 4, and 7 and contrast sample.

Number	Dosing amount (ml)	Turbidity of treated water (NTU)	Turbidity of raw water	Raw water temperature
4	1.1	4.45		
4	1.2	3.40		
2	0.8	3.38		
2	0.7	5.00	75.1 NTU	25.6°C
7	0.5	2.59		
Comparative sample	1.0	6.14		

TABLE 12: Comparison experiment of formula nos. 2, 4, and 7 and contrast sample.

Number	Dosing amount (ml)	Turbidity of treated water (NTU)	Turbidity of raw water	Raw water temperature
4	1.0	4.02		
2	0.65	6.08		
2	0.6	6.38		
7	0.4	6.86	89.3 NTU	25.6°C
7	0.5	3.67		
Comparative sample	1.0	6.33		

TABLE 13: Comparison experiment of formula nos. 2, 4, and 7 and contrast sample.

Number	Dosing amount (ml)	Turbidity of treated water (NTU)	Turbidity of raw water	Raw water temperature
4	1.1	22.7		
4	1.2	19.2		
2	0.8	16.3		
2	0.7	16.5	174 NTU	21.8°C
7	0.4	13.3		
Comparative sample	1.0	23.6		

cost is 0.01131 yuan per ton of water, which is 0.00011 yuan per ton of water higher than the cost of the contrast sample which is 0.01120 yuan per ton of water. The dosing amount of formula no. 4 per liter of the water sample is 1.0 ml, whose treatment effect is better than that of the comparative sample. The corresponding treatment cost is 0.01157 yuan per ton of water, which is only 0.00037 yuan per ton of water higher than the cost of the contrast sample which is 0.01120 yuan per ton of water. And the dosing amount of formula no. 7 per liter of sample water is 0.5 ml, whose treatment effect is better than that of the comparative sample. The corresponding treatment cost is 0.01048 yuan per ton of water, which is 0.00072 yuan per ton of weater lower than that of the contrast sample. In addition, when

the dosing amount of formula no. 2, no. 4, and no. 7 is 0.65 ml, 1.0 ml, and 0.5 ml, respectively, the treatment effect of formula no. 7 is the best and the corresponding turbidity of effluent is the lowest.

3.6. Comparison of the Effect of Aluminum Removal. In order to show the effectiveness of the solution of formula no. 7 in terms of aluminum removal effect, we treat the raw water of tap water of Zonguan waterworks with the solution of formula no. 7, and we compare the aluminum removal effect of the treated water with that of the flocculant used in Zonguan Waterworks, which is the contrast sample. The used dosing amount of formula no.7 is 0.5 ml, and the used dosing amount of the contrast sample is 1.0 ml.

TABLE 14: Aluminum concentration in water after treatment by formula no. 7 of the chitosan composite flocculant and contrast sample.

Project	Aluminum concentration (mg/L)	Turbidity (NTU)
Raw water	0.493	280
Effluent after treatment of raw water with formula no. 7	0.098	4.58
Effluent after treatment of raw water with the contrast sample	0.151	12.9

As can be seen from Table 14, the use of chitosan compound flocculant no. 7 to treat water supply has a good effect on the removal of aluminum in water, and the removal rate of aluminum is up to 80.1%. The aluminum polyaluminum chloride flocculant used in Zonguan Waterworks now has a removal rate of only 69.4%, which is 10.7% lower than the optimal formula in this paper. Therefore, the use of the chitosan composite flocculant to treat the tap water raw water can not only improve the effect and reduce the cost but also reduce the content of aluminum ion in the effluent.

4. Conclusion

In this work, a low-cost and efficient composite chitosan flocculant for the tap water treatment has been prepared by combining chitosan (CTS), polyaluminum chloride (CF-PAC), and modified rectorite ($Al(OH)_3$ + HCl). The formula of the prepared composite flocculant is CTS (ml) : CF-PAC (ml) : modified rectorite (ml) = 1 : 30 : 5. Compared with PAC as the flocculant used at Zonguan Waterworks for tap water treatment, the prepared flocculant has a turbidity removal rate of 96.38% and a removal rate of aluminum up to 80.1%, while the PAC has a removal rate of aluminum of only 69.4%, which is 10.7% lower than the optimal formula in this paper. In addition, the treatment cost was also reduced by about 0.0007 yuan per ton of water. The prepared composite chitosan flocculant is of better environmental and economic benefits than the conventional PAC flocculant in tap water treatment. In order to show the effectiveness of our optimal composite formula as the future work, we will apply the optimal composite formula to other tap water raw water.

Conflicts of Interest

The authors declare that they have no conflicts of interest.

Supplementary Materials

The supplementary material file contains our experimental data and results. First, we use the single flocculants (chitosan, polyaluminum chloride, and modified rectorite) to treat 1 liter of our sample water and then we measure the turbidity of the treated water. Second, we repeat the experiments using the pair composite flocculant: chitosan and modified rectorite, chitosan and polyaluminum chloride, and polyaluminum chloride and modified rectorite. Finally, we repeat the water treatment experiments by combining chitosan, polyaluminum chloride, and modified rectorite. Through comparison between different combination schemas of three flocculants (Tables 1–14), the best formula of the composite chitosan flocculant was found to be CTS (ml) : CF-PAC (ml) : modified rectorite ($Al(OH)_3$ + HCl) (ml) = 1 : 30 : 5 with a turbidity removal rate of 96.38% and a removal rate of aluminum up to 80.1%, while the treatment cost is the lowest. (*Supplementary Materials*)

References

[1] G. Howard and J. Bartram, *Domestic Water Quantity. Service Level and Health*, World Health Organization Document Production Services, Geneva, Switzerland, 2003.

[2] N. Ayni, M. Eben-Chaime, and G. Oron, "Optimizing desalinated sea water blending with other sources to meet magnesium requirements for potable and irrigation waters," *Water Research*, vol. 47, no. 7, pp. 2164–2167, 2013.

[3] I. Rosborg, B. Nihlgard, and M. Ferrante, "Mineral composition of drinking water and daily uptake," in *Drinking Water Minerals and Mineral Balance Importance, Health Significance, Safety Precautions*, I. Rosborg, Ed., pp. 25–32, Springer International Publishing, Basel, Switzerland, 2015.

[4] B. R. Sharma, N. C. Dhuldhoya, and U. C. Merchant, "Flocculants—an ecofriendly approach," *Journal of Polymers and the Environment*, vol. 14, no. 2, pp. 195–202, 2006.

[5] W. Chun Ming, L. Sha, and Y. Huiling, "Composite flocculant preparation and application in manure wastewater treatment," *Journal of Residuals Science & Technology*, vol. 14, no. 2, pp. 137–142, 2017.

[6] Z. Defang and Y. Jizu, "Preparation and application of a new composite chitosan flocculant," *International Journal of Environment and Pollution*, vol. 21, no. 5, pp. 417–424, 2004.

[7] D. Zeng, P. Zhang, and Z. Feng, "Production of chitosan used for flocculant in medium scale," *Chinese Journal of Environmental Science*, vol. 254-255, no. 1, pp. 62–65, 2002.

[8] C. S. Lee, J. Robinson, and M. F. Chong, "A review on application of flocculants in wastewater treatment," *Process Safety and Environmental Protection*, vol. 92, no. 6, pp. 489–508, 2014.

[9] C. G. Li, F. Wang, W. G. Peng, and Y. H. He, "Preparation of chitosan and epichlorohydrin cross-linked adsorbents and adsorption property of dyes," *Applied Mechanics and Materials*, vol. 423–426, pp. 584–587, 2013.

[10] F. Renault, B. Sancey, P. M. Badot, and G. Crini, "Chitosan for coagulation/flocculation processes—an eco-friendly approach," *European Polymer Journal*, vol. 45, no. 5, pp. 1337–1348, 2009.

[11] A. Szyguła, E. Guibal, M. A. Palacín, M. Ruiz, and A. M. Sastre, "Removal of an anionic dye (Acid Blue 92) by coagulation-flocculation using chitosan," *Journal of Environmental Management*, vol. 90, no. 10, pp. 2979–2986, 2009.

[12] R. Yang, H. Li, M. Huang, Y. Hu, and A. Li, "A review on chitosan-based flocculants and their applications in water treatment," *Water Research*, vol. 95, pp. 59–89, 2016.

3

Adsorption of Cr(VI) onto Hybrid Membrane of Carboxymethyl Chitosan and Silicon Dioxide

Yanling Deng,[1] Naoki Kano,[2] and Hiroshi Imaizumi[2]

[1]*Graduate School of Science and Technology, Niigata University, Niigata 950-2181, Japan*
[2]*Department of Chemistry and Chemical Engineering, Faculty of Engineering, Niigata University, Niigata 950-2181, Japan*

Correspondence should be addressed to Naoki Kano; kano@eng.niigata-u.ac.jp

Academic Editor: Khalid Z. Elwakeel

In this study, a new adsorbent material was synthesized by using carboxymethyl chitosan and silicon dioxide. The hybrid membrane was used as an adsorbent for the removal of Cr(VI) from aqueous solutions. The adsorption potential of Cr(VI) by the hybrid materials was investigated by varying experimental conditions such as pH, contact time, and the dosage of the hybrid membrane. Adsorption isotherms of Cr(VI) onto the hybrid membrane were studied with varying initial concentrations under optimum experiment conditions. The surface property of the hybrid membrane was characterized by SEM (scanning electron microscope) and Fourier transform infrared spectrometer (FTIR). The concentrations of Cr(VI) in solution are determined by ICP-AES (inductively coupled plasma atomic emission spectrometry). The present study investigates the adsorption mechanisms of Cr(VI) onto the hybrid membrane. The results provide new insight, demonstrating that the modified hybrid membrane can be an efficient adsorbent for Cr(VI) from the aqueous solution.

1. Introduction

The environmental conservation is of increasing social and economic importance. One of the intractable environmental problems is water pollution by heavy metals [1]. Heavy metals in the environmental water have been a major preoccupation for many years because of their toxicity towards aquatic life, human beings, and the environment [2].

Among the toxic metals, chromium has been reported to be toxic to animals and humans, and it is known to be carcinogenic [3, 4]. Chromium consists of two stable oxidation states such as trivalent state Cr(III) and hexavalent state Cr(VI) in natural environment. Cr(VI) is more toxic, carcinogenic, and mutagenic. It is highly toxic as it can diffuse as CrO_4^{2-} or $HCrO_4^-$ through cell membranes [5]. The effluents are discharged onto the open land or into the sewage system. These industries are major sources of chromium pollution in the environment [6].

Due to serious hazardous effects of heavy metal ions on human health and toxicity in the environment [7], it is important to develop a simple and highly effective removal method as well as sensitive analytical method for environmental pollutants to improve the quality of the environment and the human life.

Various treatment technologies such as ion exchange, precipitation, ultrafiltration, reverse osmosis, and electrodialysis have been used for the removal of heavy metal ions from aqueous solution [8]. However, these processes have some disadvantages, such as high consumption of reagent and energy, low selectivity, and high operational cost [9].

Many works for the removal of heavy metals by adsorption have been reported [10–12]. Particularly, the development of high efficiency and low cost adsorbents has aroused general interest in recent years.

Chitosan, whose full chemical name is known as (1,4)-2-amino-2-deoxy-β-D-glucose, can be environmentally friendly adsorbent due to the low price and no second pollution. Chitosan has free amino groups and hydroxyl groups, which can remove the heavy metal ions by forming stable metal chelates.

However, chitosan had some defects such as notable swelling in aqueous media and nonporous structure resulting in a very low surface area [10]. Therefore, many types of chemical modification can be undertaken to produce some chitosan derivatives for improving the removal efficiency of the heavy metal [13]. For example, silicon dioxide can be one of the materials for offsetting the defects of chitosan because it has many characteristics such as rigid structure, porosity, and high surface area.

Silicon dioxide is a synthetic amorphous polymer with silanol groups on the surface allowing metal adsorption [14, 15]. In case of silicon dioxide, the modified silicon dioxide through the graft between silanol groups and ligands has been developed [16–18].

Therefore, we have investigated the efficiency of the hybrid membrane of chitosan and silicon dioxide as sorbent for Cr(VI) [19]. In this study, carboxymethyl chitosan has been prepared by using chloroacetic acid (and chitosan) under alkaline conditions to improve the removal efficiency of Cr(VI) by the hybrid membrane.

Then, novel adsorption materials were designed to combine the beneficial properties of carboxymethyl chitosan and silicon dioxide. Moreover, the novel adsorbent can be recycled for adsorption of heavy metal ions compared with the disposable adsorbent.

From the abovementioned, the hybrid membrane of carboxymethyl chitosan and silicon dioxide was synthesized in this work to enhance the adsorption potential of Cr(VI). In present studies, the adsorption capacity of the hybrid membrane was investigated for the removal of Cr(VI) from aqueous solution under varying experimental conditions.

2. Experimental

2.1. Materials, Reagent, and Apparatus. 3-Aminopropyltriethoxysilane was purchased from Nacalal Tesque, Inc. (Tokyo, Japan), and chitosan was from Tokyo Chemical Industry Co. (Tokyo, Japan). Cr(VI) standard solutions were prepared by diluting a standard solution ($1{,}005\,mg{\cdot}dm^{-3}$ $K_2Cr_2O_7$ solution) purchased from Kanto Chemical Co., Inc. All other chemical reagents were also bought from Kanto Chemical Co., Inc. All reagents used were of analytical grade, and water ($>18.2\,M\Omega$ in electrical resistance), which was treated by an ultrapure water system (Advantec aquarius: RFU 424TA, Advantec, Toyo, Japan), was employed throughout the work.

The pH of Cr(VI) aqueous solution was measured by the pH meter (HORIBA UJXT 06T8, Japan). The surface property of the hybrid membrane of carboxymethyl chitosan and silicon dioxide was characterized by a SEM (JEOL, JSM-5800, Japan) and Fourier transform infrared spectroscopy in pressed KBr pellets (FTIR-4200, Jasco, Japan). The concentrations of Cr(VI) in the solution were determined by ICP-AES (inductively coupled plasma atomic emission spectrometry).

2.2. Prepared Carboxymethyl Chitosan. Under alkaline conditions, chitosan can react with chloroacetic acid to obtain the carboxymethyl chitosan. 5 g of chitosan was accurately weighed into a round-bottomed flask containing 75 ml isopropanol and 25 ml ultrapure water, and then 6.75 g of

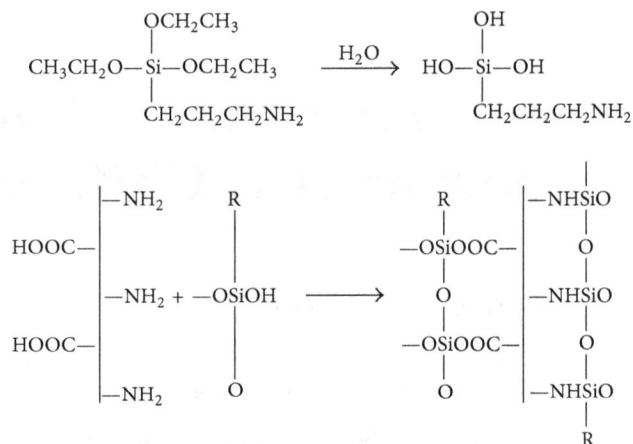

FIGURE 1: The reaction principle of the carboxymethyl chitosan crosslinking with 3-Aminopropyltriethoxysilane.

sodium hydroxide was added for alkalization. The mixed solution was stirred in a water bath at 50°C for 2 h and was cooled to room temperature after continuing stirring for 4 h. In addition, chloroacetic acid solution was prepared by dissolving 6 g of chloroacetic acid in 25 ml isopropanol solution and slowly dropped into the round-bottomed flask under stirring for 4 h. The solution was adjusted to neutral using hydrochloric acid, washed three times with 70% isopropanol, and then filtered. After washing completely with 90% isopropanol again, the solution was filtered. Then, carboxymethyl chitosan was dried at 50°C and used for the preparation of hybrid membrane.

2.3. The Prepared Hybrid Membrane of Carboxymethyl Chitosan and Silicon Dioxide. The reaction process of hybrid membrane synthesized from carboxymethyl chitosan and silicon dioxide is shown in Figure 1. The solution of carboxymethyl chitosan (3%, w/v) was prepared by dissolving 3 g of carboxymethyl chitosan in 100 ml ultrapure water. Silica sols (prepared by dissolving 5 ml of 3-Aminopropyltriethoxysilane in 100 ml ethanol) was add to the solution of carboxymethyl chitosan (3%, w/v) at 25°C, and the solution was stirred for 24 h. The hybrid membrane was dried at 25°C.

2.4. Adsorption Experiment of Cr(VI) Using Hybrid Membrane of Carboxymethyl Chitosan and Silicon Dioxide. The adsorption of Cr(VI) using hybrid membrane was studied by a batch method with various pH range of 1–7, contact time range of 20–120 min, sorbent dosage range from 0.05 to $0.3\,g{\cdot}dm^{-3}$, and initial concentrations range of $10–50\,mg{\cdot}dm^{-3}$.

Firstly, 50 ml solution of Cr(VI) was placed into a 200 ml conical flask. During the uptake experiment, the pH of the Cr(VI) solution was adjusted using $0.1\,mol{\cdot}dm^{-3}$ NaOH or $0.1\,mol{\cdot}dm^{-3}$ HCl. The concentration of Cr(VI) was determined by ICP-AES. Adsorption isotherms of Cr(VI) using the hybrid membrane were studied with varying initial concentrations under optimum experiment conditions.

The adsorption capacity of hybrid membrane for Cr(VI) was calculated using the mass balance equation:

$$q_e = \frac{(C_i - C_e)}{m} \cdot V,\qquad(1)$$

where q_e is the adsorption capacity (mg·g^{-1}) of Cr(VI) by the hybrid membrane at equilibrium, C_i and C_e are the concentrations of Cr(VI) at initial and equilibrium in a batch system, respectively (mg·dm^{-3}), V (dm^{-3}) is the volume of the heavy metal solution, and m (g) is the mass of the hybrid membrane.

2.5. Langmuir and Freundlich Isotherm Model. To understand the adsorption process of Cr(VI) using the hybrid membrane, adsorption isotherms of Langmuir and Freundlich were investigated under the optimal conditions.

Langmuir isotherm equation is defined as follows:

$$\frac{C_e}{q_e} = \frac{C_e}{q_{max}} + \frac{1}{K_L q_{max}},\qquad(2)$$

where C_e is the concentration of Cr(VI) at equilibrium (mg·dm^{-3}), q_e and q_{max} are the amount of adsorption of Cr(VI) at equilibrium (mg·g^{-1}) and the maximum adsorption capacity by the hybrid membrane (mg·g^{-1}), respectively, K_L (dm^{-3}·mg^{-1}) is the adsorption constant of Langmuir isotherm.

The linearized Freundlich isotherm equation is defined as follows:

$$\log_{10} q_e = \log_{10} K_F + \left(\frac{1}{n}\right)\log_{10} C_e.\qquad(3)$$

In this equation, K_F is the adsorption capacity ((mg·g^{-1})· (dm^{-3}·mg^{-1})$^{1/n}$) and $1/n$ is the adsorption intensity. The values of $1/n$ and K_F were determined on the basis of the plots of q_e versus C_e in log scale.

2.6. Kinetic Models. Kinetic models have been proposed to determine the rate of adsorption of the adsorbent. In addition, the process of kinetic study is very important for understanding the reaction process and the rate of the adsorption reactions.

The pseudo-first-order model is given by the following equation:

$$\ln\left(q_e - q_t\right) = \ln\left(q_e\right) - k_1 t,\qquad(4)$$

where q_e and q_t are the adsorption capacity of Cr(VI) using the hybrid membrane at equilibrium and time t, respectively (mg·g^{-1}), and k_1 is the rate constant of the pseudo-first-order adsorption (h^{-1}).

The pseudo-second-order rate equation is expressed as follows:

$$\frac{t}{q_t} = \frac{1}{k q_e^2} + \frac{t}{q_e},\qquad(5)$$

where k (g·mg^{-1}·h^{-1}) is the rate constant of the second-order model and q_e and q_t are the adsorption capacities of Cr(VI) using the hybrid membrane at equilibrium and time t, respectively (mg·g^{-1}).

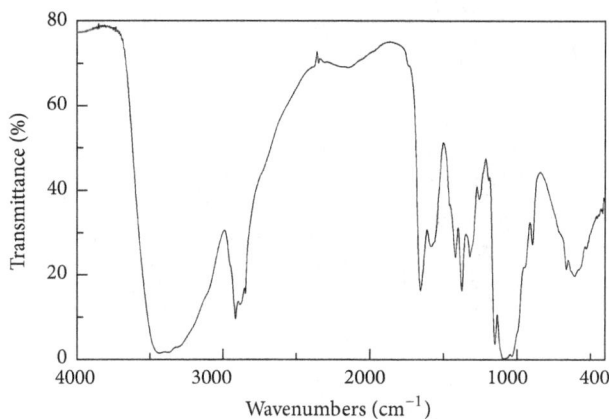

FIGURE 2: FTIR spectra of hybrid membrane of carboxymethyl chitosan and silicon dioxide.

3. Results and Discussion

3.1. Characteristics of the Hybrid Membrane of Carboxymethyl Chitosan and Silicon Dioxide. The FTIR spectroscopy is an important technique of characterization used to explain the changes in the chemical structures (i.e., the functional group on the surface of the samples). FTIR spectra of the hybrid membrane of carboxymethyl chitosan and silicon dioxide are presented in Figure 2. The strong broadband at the wave number region of 3300–3500 cm^{-1} is the characteristic of –NH$_2$ stretching vibration, and the bands at 3400 cm^{-1} are related to the symmetrical valent vibration of free NH$_2$ and –OH groups [27]. The –CH stretching vibration in –CH and –CH$_2$ was observed at 2916 cm^{-1} and 1376 cm^{-1}. The –NH$_2$ bending vibration was observed at 1652 cm^{-1} shifted to lower frequencies (the lower frequencies observed in the hybrid membrane may be explained by the presence of primary amine salt –NH$_3{}^+$ [28]). A strong C=O stretching band at 1655 cm^{-1} may be related to the carboxymethyl group. Other bands at 1090 cm^{-1} are related to Si–O–Si valent vibrations [10]. The results of FTIR analysis show that the hybrid membrane of carboxymethyl chitosan and silicon dioxide was prepared successfully in this study.

The surface property of the hybrid membrane of carboxymethyl chitosan and silicon dioxide was investigated by SEM, and SEM images are shown in Figure 3. The surface morphology of the hybrid membrane showed the form of grain coalescence, which may be due to the crosslinking among adjacent carboxymethyl chitosan groups. Moreover, there was the porous structure in the surface of hybrid membrane. It indicates that silicon dioxide was incorporated into the carboxymethyl chitosan definitely, and thereby the porous structure increased.

Carboxymethyl has a high chelating ability for metal ions to form stable metal chelates. The lone pair electrons on the nitrogen atom can also constitute coordination bonds with the metal ions to form the complex precipitation. The molecule also may contain free amino groups and hydroxyl groups, which can remove the heavy metal ions by chelation mechanisms.

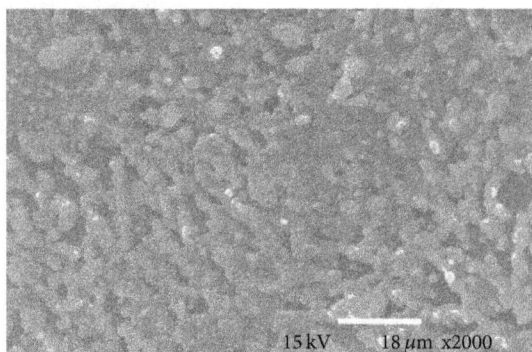

FIGURE 3: SEM pictures of hybrid membrane of carboxymethyl chitosan and silicon dioxide.

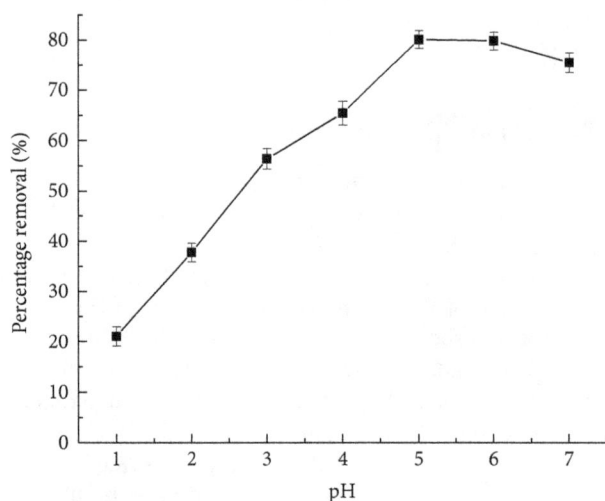

FIGURE 4: Effect of pH for Cr(VI) adsorption using hybrid membrane of carboxymethyl chitosan and silicon dioxide.

FIGURE 5: Effect of contact time for Cr(VI) adsorption using hybrid membrane of carboxymethyl chitosan and silicon dioxide.

3.2. Effect of pH. The effects of pH on the removal of Cr(VI) were investigated under the following condition: initial concentration of Cr(VI) for $50\,mg{\cdot}dm^{-3}$, the contact time of 120 min, and the dosage of the adsorbent for $0.2\,g{\cdot}dm^{-3}$.

The effect of pH on the removal of Cr(VI) using the hybrid membrane is shown in Figure 4.

It is well known that pH influences significantly the adsorption processes by affecting both the protonation of the surface groups and the chemical form of Cr(VI).

Cr(VI) exists in a variety of forms with different pH, and it exists in the form of H_2CrO_4 at pH 1.0 [23] and different forms such as $Cr_2O_7^-$, $HCrO_4^-$, $Cr_3O_{10}^{2-}$, and $Cr_4O_{13}^{2-}$, while $HCrO_4^-$ predominates at the pH range from 2.0 to 6.0. Furthermore, this form shifts to CrO_4^{2-} and $Cr_2O_7^{2-}$ when pH increases [29, 30]. The process of shifts is given by the following equations:

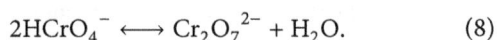

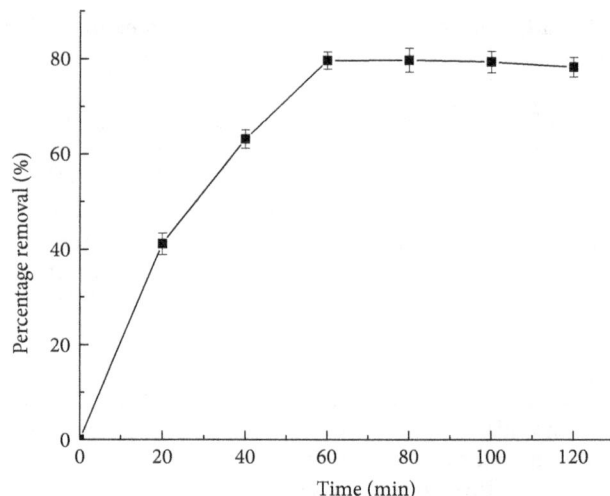

$$H_2CrO_4 \longleftrightarrow H^+ + HCrO_4^- \qquad (6)$$

$$HCrO_4^{2-} \longleftrightarrow H^+ + CrO_4^{2-} \qquad (7)$$

$$2HCrO_4^- \longleftrightarrow Cr_2O_7^{2-} + H_2O. \qquad (8)$$

It is found that the adsorption capacity was relatively low at pH 1. It may be attributable to the strong competition between H_2CrO_4 and protons for adsorption sites. The adsorption efficiency of Cr(VI) increased with the increase of pH and reached the maximum at pH 5 (80%). It is considered that the $(-NH_2)$ in the adsorbent may be protonated to form $(-NH_3^+)$ at pH 2–6. The surface of the hybrid membrane became positively charged due to the strong protonation at these pH ranges, which leads to a stronger attraction between the positively charged surface and the negatively charged $Cr_2O_7^{2-}$ and $HCrO_4^-$. Then, the protonation will enhance the Cr(VI) adsorption at pH 5-6. However, at higher pH, Cr may precipitate from the solution as its hydroxides. Hence, pH 5 was considered as the optimum pH for further work.

3.3. Effect of Contact Time. Adsorption experiments were performed in order to determine the optimum contact time at pH 5 under the condition of the concentration of Cr(VI) for $50\,mg{\cdot}dm^{-3}$ and the dosage of the adsorbent for $0.2\,g{\cdot}dm^{-3}$. The experimental results are shown in Figure 5. It can be observed that the adsorption capacity of Cr(VI) increases with increasing time within 60 min. The removal rate for Cr(VI) reached approximately 80% at 60 min, and after that there is no appreciable increase. Then, 60 min was selected as the optimized contact time.

3.4. Effect of Hybrid Membrane Dosage. In order to estimate the optimal dosage of the hybrid membrane, the adsorption experiments were carried out with the range of 0.05–$0.3\,g{\cdot}dm^{-3}$ for the adsorbent under the optimum conditions of pH (pH 5) and contact time (60 min) and the concentration of Cr(VI) for $50\,mg{\cdot}dm^{-3}$. The results are shown in Figure 6. The results indicate that the adsorption capacity of hybrid membrane for Cr(VI) reached the adsorption equilibrium at the dosage of $0.25\,g{\cdot}dm^{-3}$ and that no significant change is observed at a dosage from

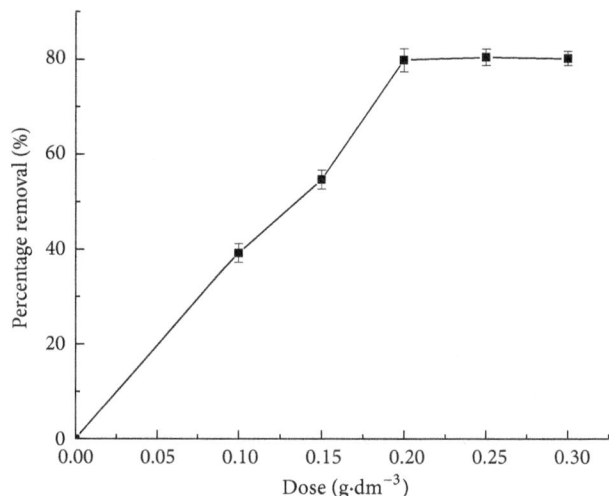

FIGURE 6: Effect of dosage of adsorbent for Cr(VI) adsorption using hybrid membrane of carboxymethyl chitosan and silicon dioxide.

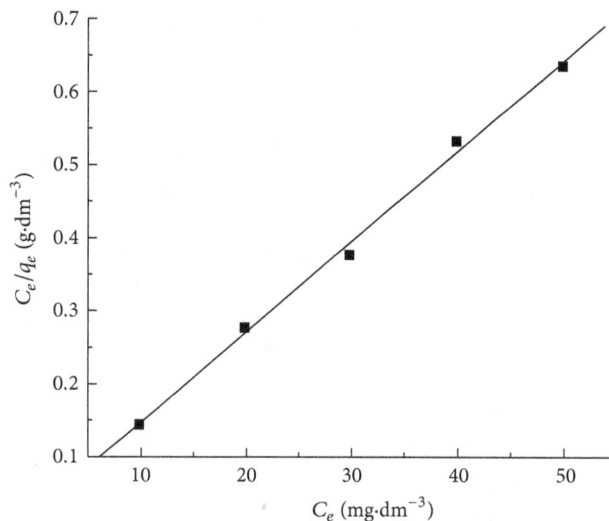

FIGURE 8: Langmuir isotherm of Cr(VI) adsorption using hybrid membrane of carboxymethyl chitosan and silicon dioxide.

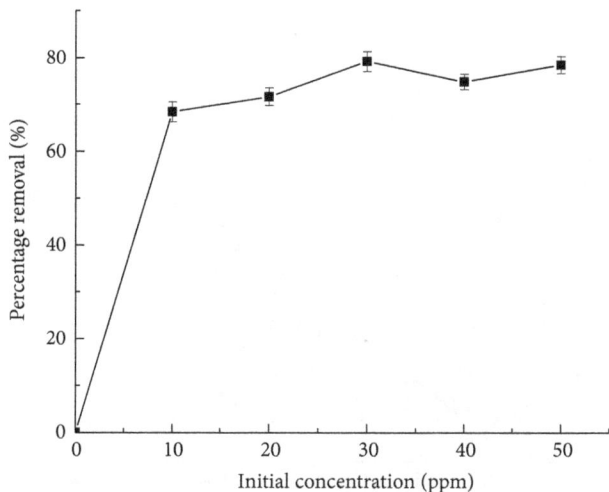

FIGURE 7: Effect of initial concentration for Cr(VI) adsorption using hybrid membrane of carboxymethyl chitosan and silicon dioxide.

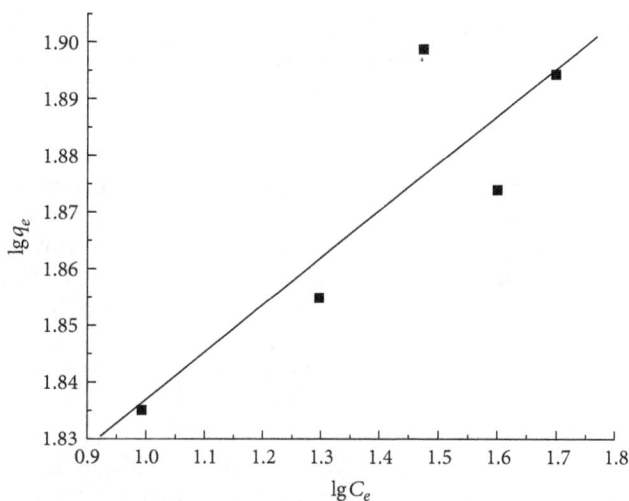

FIGURE 9: Freundlich isotherm of Cr(VI) adsorption using hybrid membrane of carboxymethyl chitosan and silicon dioxide.

0.2 to $0.3 \, \text{g·dm}^{-3}$. The removal rate reached about 80% at $0.25 \, \text{g·dm}^{-3}$, and $0.25 \, \text{g·dm}^{-3}$ was selected as the optimal dosage.

3.5. Effect of Initial Concentration. The experiments were performed by varying concentrations from 10 to $50 \, \text{mg·dm}^{-3}$ under optimized condition of pH (pH 5) and contact time (60 min) with adsorbent dosage ($0.25 \, \text{g·dm}^{-3}$). The results are shown in Figure 7. There was a slight increase from 20 to $50 \, \text{mg·dm}^{-3}$ except at the concentrations of $30 \, \text{mg·dm}^{-3}$. The initial concentrations were taken as $40 \, \text{mg·dm}^{-3}$. The data from these studies were fitted to the Langmuir and Freundlich isotherm equations.

3.6. Adsorption Isotherms. In this work, Langmuir and Freundlich isotherms were investigated in order to evaluate the performance of the adsorbents in the adsorption processes.

The adsorption data obtained for Cr(VI) using the hybrid membrane were analyzed by Langmuir (Figure 8) and Freundlich (Figure 9) equations. The correlation coefficient (R^2) of Langmuir and Freundlich isotherms for Cr(VI) using the hybrid membrane is shown in Table 1 along with other relevant parameters.

The maximum adsorption capacity ($q_{\max}$) calculated from Langmuir model was $80.7 \, \text{mg·g}^{-1}$. Based on Table 1, it is found that R^2 value of Langmuir isotherm is larger than that of Freundlich isotherm. This result suggests that the adsorption of Cr(VI) on the hybrid membrane of carboxymethyl chitosan and silicon dioxide mainly occurred by monolayer reaction.

The comparison of the maximum adsorption capacity of the hybrid membrane of carboxymethyl chitosan and silicon dioxide for Cr(VI) in the present study with that of other

TABLE 1: Coefficient of Langmuir and Freundlich isotherms for Cr(VI) using hybrid membrane of carboxymethyl chitosan and silicon dioxide.

	Hybrid membrane					
	Langmuir isotherm			Freundlich isotherm		
q_{max} [mg·g^{-1}]	K_L [dm^{-3}·mg^{-1}]	R^2	K_F [(mg·g^{-1})·(dm^{-3}·mg^{-1})$^{1/n}$]	$1/n$		R^2
80.7	0.531	0.998	56.7	0.0834		0.867

TABLE 2: Comparison of adsorption capacities for Cr(VI) by different adsorbents.

Adsorbent	Adsorption capacity (mg·g^{-1})	References
Crosslinked chitosan bentonite composite	89.1	[20]
Chitosan/polyvinyl alcohol/containing cerium(III)	52.9	[21]
STAC-modified rectorite	21.0	[22]
Ethylenediamine-modified crosslinked magnetic chitosan	51.8	[23]
Clarified sludge	26.3	[24]
A novel modified graphene oxide/chitosan	86.2	[25]
Chitosan-g-poly/silica gel nanocomposite	55.7	[26]
Hybrid membrane of carboxymethyl chitosan and silicon dioxide	80.7	This study

adsorbents in literatures [20–26] is presented in Table 2. As seen in Table 2, the adsorption capacity of the hybrid membrane for Cr(VI) in this work is on a level with that of another adsorbents in previous works.

3.7. Kinetic Studies. Kinetic models were tested in this study for the adsorption of Cr(VI) onto the hybrid membrane under the optimized experimental conditions. Adsorption time is one of the important factors which help us to predict the kinetics as well as the mechanism of the uptake of heavy metals on material surface [31].

The results for the rate constant (k) and the amount of the adsorbed Cr(VI) (q_e) are shown in Table 3 along with the regression coefficients (R^2). From Table 3, it is found that R^2 value of the pseudo-second-order model is larger than that of the pseudo-first-order model; therefore, the rate of adsorption is more suitable to the pseudo-second-order model than to the pseudo-first-order model.

Then, the linear plot of t/qt versus t for Cr(VI) adsorption system is shown in Figure 10. It implies that the adsorption kinetics based on the experimental values is in good agreement with the pseudo-second-order kinetic model and that the rate constant of the second-order equation (k) is 3.4×10^{-2} g·mg^{-1}·h^{-1} in this work.

3.8. Adsorption Mechanism of Cr(VI) by the Hybrid Membrane. The novel adsorbent for Cr(VI) was synthesized by using carboxymethyl chitosan and silicon dioxide. The hybrid membrane has carboxymethyl, free amino group, and hydroxyl groups on its surface as the adsorption site. It can remove Cr(VI) by forming stable metal chelates, and the porous structure of the hybrid membrane enhances the adsorption capacity of Cr(VI).

From the kinetic studies, the adsorption reaction was found to conform to the pseudo-second-order kinetics, suggesting the importance of porous structure. This implies that the adsorption process for Cr(VI) was mainly chemical

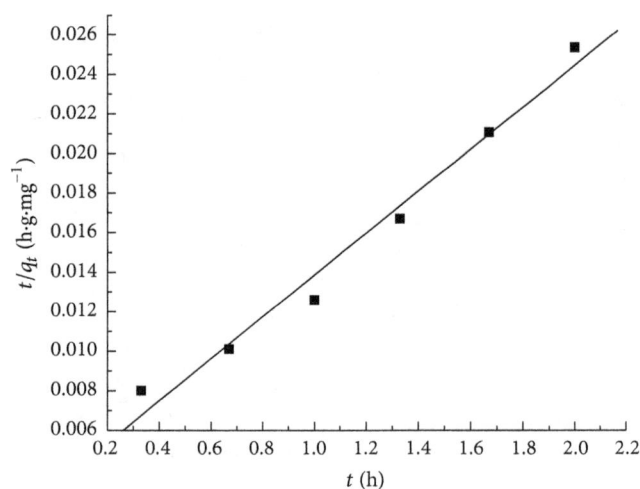

FIGURE 10: The pseudo-second-order kinetic model of Cr(VI) adsorption using hybrid membrane of carboxymethyl chitosan and silicon dioxide.

and that the adsorption process involves the valency forces through sharing electrons between the metal ions and the adsorbent.

Moreover, the adsorption isotherm of Cr(VI) by the hybrid membrane was more suitably described by Langmuir model, indicating that monolayer adsorption of Cr(VI) on the hybrid membrane is more dominant.

3.9. Regeneration Studies. From industrial and technological point of view, it is desirable to recover and reuse the adsorbed material. Then, regeneration experiments were conducted using the hybrid membrane after adsorption of Cr(VI) at pH 13.5. In each desorption experiment, 75 mg of the spent adsorbent after adsorption was treated with 200 ml of 0.5 mol·dm^{-3} NaOH and 2 mol·dm^{-3} NaCl solution as desorption agent

TABLE 3: The kinetic coefficient for Cr(VI) using hybrid membrane of carboxymethyl chitosan and silicon dioxide.

	Cr(VI)				
Pseudo-first-order			Pseudo-second-order		
q_e (mg·g^{-1})	K_1 (h^{-1})	R^2	q_e (mg·g^{-1})	K_2	R^2
79.7	8.91	0.924	94.4	3.42×10^{-2}	0.990

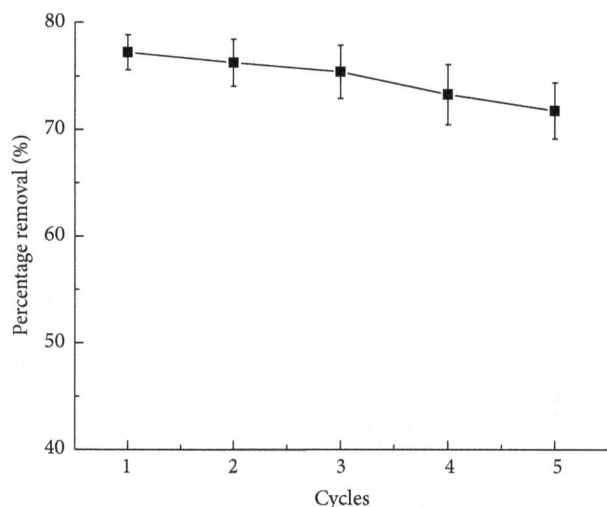

FIGURE 11: The adsorption capacity after desorption using hybrid membrane of carboxymethyl chitosan and silicon dioxide for Cr(VI).

and then filtered. Cr(VI) content in the filtrate was determined by ICP-AES. Adsorption and desorption studies have been continued during five cycles at room temperature for 4 hours as eluent. The adsorption capacity after desorption using the above leaching agents is shown in Figure 11. From this figure, it is found that the hybrid membrane still presents the high adsorption capacity (74.6%) towards Cr(VI) within 3 cycles.

4. Conclusions

The present study investigated the efficiency of the hybrid membrane synthesized from carboxymethyl chitosan and silicon dioxide as an adsorbent for Cr(VI) by batch techniques. The following conclusions can be drawn considering the results of this work:

(1) The hybrid membrane exhibited high adsorption capacity for Cr(VI). The removal of Cr(VI) was more than 80% under the optimal experimental conditions (at pH 5, contact time of 60 min, dosage of 0.25 g·dm^{-3}, and initial Cr(VI) concentration of 40 mg·dm^{-3}).

(2) The adsorption isotherm of Cr(VI) by the hybrid membrane of carboxymethyl chitosan and silicon dioxide was more suitably described by Langmuir model, and the correlation coefficients were more than 0.998. This suggests that monolayer chemical adsorption of Cr(VI) on the hybrid membrane is more dominant. The maximum adsorption capacity was estimated as 80.7 mg·g^{-1} for Cr(VI) under the optimum conditions. The adsorption capacity of the hybrid membrane for Cr(VI) in this work is on a level with that of other adsorbents in previous works.

(3) The best fit was obtained with a pseudo-second-order kinetic model while investigating the adsorption kinetics of Cr(VI) adsorption on the hybrid membrane, and the correlation coefficient was more than 0.9. The rate constant (k) is 3.4×10^{-2} g·mg^{-1}·h^{-1}.

(4) From this work, it was quantitatively found that the hybrid membrane of carboxymethyl chitosan and silicon dioxide could be an efficient adsorbent for Cr(VI). It is very significant information from the viewpoint of environmental protection and can be used for treating industrial wastewaters including pollutants.

Conflicts of Interest

The authors declare that there are no conflicts of interest regarding the publication of this paper.

Acknowledgments

The present work was partially supported by a Grant-in-Aid for Scientific Research from the Japan Society for the Promotion of Science (Research Program (C), no. 16K00599) and a fund for the promotion of Niigata University KAAB Projects from the Ministry of Education, Culture, Sports, Science and Technology, Japan. The authors are also grateful to Mr. M. Ohizumi of Office for Environment and Safety in Niigata University and to Dr. M. Teraguchi, Mr. T. Nomoto, and Professor T. Tanaka of Faculty of Engineering in Niigata University for permitting the use of ICP-AES, FTIR, and SEM and for giving helpful advice in measurement.

References

[1] Y. Sağ and Y. Aktay, "Kinetic studies on sorption of Cr(VI) and Cu(II) ions by chitin, chitosan and *Rhizopus arrhizus*," *Biochemical Engineering Journal*, vol. 12, no. 2, pp. 143–153, 2002.

[2] N. Chiron, R. Guilet, and E. Deydier, "Adsorption of Cu(II) and Pb(II) onto a grafted silica: isotherms and kinetic models," *Water Research*, vol. 37, no. 13, pp. 3079–3086, 2003.

[3] C. Raji and T. S. Anirudhan, "Batch Cr(VI) removal by polyacrylamide-grafted sawdust: kinetics and thermodynamics," *Water Research*, vol. 32, no. 12, pp. 3772–3780, 1998.

[4] E. Demirbas, M. Kobya, E. Senturk, and T. Ozkan, "Adsorption kinetics for the removal of chromium (VI) from aqueous

solutions on the activated carbons prepared from agricultural wastes," *Water SA*, vol. 30, no. 4, pp. 533–539, 2004.

[5] N. Sankararamakrishnan, A. Dixit, L. Iyengar, and R. Sanghi, "Removal of hexavalent chromium using a novel cross linked xanthated chitosan," *Bioresource Technology*, vol. 97, no. 18, pp. 2377–2382, 2006.

[6] S. A. Khan, Riaz-ur-Rehman, and M. A. Khan, "Adsorption of chromium (III), chromium (VI) and silver (I) on bentonite," *Waste Management*, vol. 15, no. 4, pp. 271–282, 1995.

[7] Z. Guo, D. Li, X. Luo et al., "Simultaneous determination of trace Cd(II), Pb(II) and Cu(II) by differential pulse anodic stripping voltammetry using a reduced graphene oxide-chitosan/poly-l-lysine nanocomposite modified glassy carbon electrode," *Journal of Colloid and Interface Science*, vol. 490, no. 5, pp. 11–22, 2017.

[8] F. Fu and Q. Wang, "Removal of heavy metal ions from wastewaters: a review," *Journal of Environmental Management*, vol. 92, no. 3, pp. 407–418, 2011.

[9] M. M. Montazer-Rahmati, P. Rabbani, A. Abdolali, and A. R. Keshtkar, "Kinetics and equilibrium studies on biosorption of cadmium, lead, and nickel ions from aqueous solutions by intact and chemically modified brown algae," *Journal of Hazardous Materials*, vol. 185, no. 1, pp. 401–407, 2011.

[10] E. Repo, J. K. Warchoł, A. Bhatnagar, and M. Sillanpää, "Heavy metals adsorption by novel EDTA-modified chitosan-silica hybrid materials," *Journal of Colloid and Interface Science*, vol. 358, no. 1, pp. 261–267, 2011.

[11] R. M. Ali, H. A. Hamad, M. M. Hussein, and G. F. Malash, "Potential of using green adsorbent of heavy metal removal from aqueous solutions: adsorption kinetics, isotherm, thermodynamic, mechanism and economic analysis," *Ecological Engineering*, vol. 91, pp. 317–332, 2016.

[12] J.-H. Park, Y. S. Ok, S.-H. Kim et al., "Competitive adsorption of heavy metals onto sesame straw biochar in aqueous solutions," *Chemosphere*, vol. 142, pp. 77–83, 2016.

[13] A. J. Varma, S. V. Deshpande, and J. F. Kennedy, "Metal complexation by chitosan and its derivatives: a review," *Carbohydrate Polymers*, vol. 55, no. 1, pp. 77–93, 2004.

[14] P. Michard, E. Guibal, T. Vincent, and P. Le Cloirec, "Sorption and desorption of uranyl ions by silica gel: pH, particle size and porosity effects," *Microporous Materials*, vol. 5, no. 5, pp. 309–324, 1996.

[15] H. H. Tran, F. A. Roddick, and J. A. O'Donnell, "Comparison of chromatography and desiccant silica gels for the adsorption of metal ions—I. Adsorption and kinetics," *Water Research*, vol. 33, no. 13, pp. 2992–3000, 1999.

[16] L. Bois, A. Bonhommé, A. Ribes, B. Pais, G. Raffin, and F. Tessier, "Functionalized silica for heavy metal ions adsorption," *Colloids and Surfaces A: Physicochemical and Engineering Aspects*, vol. 221, no. 1–3, pp. 221–230, 2003.

[17] P. K. Jal, S. Patel, and B. K. Mishra, "Chemical modification of silica surface by immobilization of functional groups for extractive concentration of metal ions," *Talanta*, vol. 62, no. 5, pp. 1005–1028, 2004.

[18] M. Rajiv Gandhi and S. Meenakshi, "Preparation and characterization of silica gel/chitosan composite for the removal of Cu(II) and Pb(II)," *International Journal of Biological Macromolecules*, vol. 50, no. 3, pp. 650–657, 2012.

[19] Y. Deng, N. Kano, and H. Imaizumi, "Removal of chromium from aqueous solution using hybrid membrane of chitosan and silicon dioxide," *Journal of Chemistry and Chemical Engineering*, vol. 10, no. 5, pp. 199–206, 2016.

[20] Q. Liu, B. Yang, L. Zhang, and R. Huang, "Adsorptive removal of Cr(VI) from aqueous solutions by cross-linked chitosan/bentonite composite," *Korean Journal of Chemical Engineering*, vol. 32, no. 7, pp. 1314–1322, 2015.

[21] F. F. Wang and M. Q. Ge, "Fibrous mat of chitosan/polyvinyl alcohol/containing cerium(III) for the removal of chromium(VI) from aqueous solution," *Textile Research Journal*, vol. 83, no. 6, pp. 628–637, 2013.

[22] H. Hong, W.-T. Jiang, X. Zhang, L. Tie, and Z. Li, "Adsorption of Cr(VI) on STAC-modified rectorite," *Applied Clay Science*, vol. 42, no. 1-2, pp. 292–299, 2008.

[23] X.-J. Hu, J.-S. Wang, Y.-G. Liu et al., "Adsorption of chromium (VI) by ethylenediamine-modified cross-linked magnetic chitosan resin: isotherms, kinetics and thermodynamics," *Journal of Hazardous Materials*, vol. 185, no. 1, pp. 306–314, 2011.

[24] A. K. Bhattacharya, T. K. Naiya, S. N. Mandal, and S. K. Das, "Adsorption, kinetics and equilibrium studies on removal of Cr(VI) from aqueous solutions using different low-cost adsorbents," *Chemical Engineering Journal*, vol. 137, no. 3, pp. 529–541, 2008.

[25] L. Zhang, H. Luo, P. Liu, W. Fang, and J. Geng, "A novel modified graphene oxide/chitosan composite used as an adsorbent for Cr(VI) in aqueous solutions," *International Journal of Biological Macromolecules*, vol. 87, pp. 586–596, 2016.

[26] R. Nithya, T. Gomathi, P. N. Sudha, J. Venkatesan, S. Anil, and S.-K. Kim, "Removal of Cr(VI) from aqueous solution using chitosan-g-poly(butyl acrylate)/silica gel nanocomposite," *International Journal of Biological Macromolecules*, vol. 87, pp. 545–554, 2016.

[27] G. N. Kousalya, M. Rajiv Gandhi, and S. Meenakshi, "Sorption of chromium(VI) using modified forms of chitosan beads," *International Journal of Biological Macromolecules*, vol. 47, no. 2, pp. 308–315, 2010.

[28] D. Lin-Vien, N. B. Colthup, W. G. Fateley, and J. G. Grasselli, *The Handbook of Infrared and Raman Characteristic Frequencies of Organic Molecules*, Elsevier, Amsterdam, Netherlands, 1991.

[29] T. Karthikeyan, S. Rajgopal, and L. R. Miranda, "Chromium(VI) adsorption from aqueous solution by Hevea Brasilinesis sawdust activated carbon," *Journal of Hazardous Materials*, vol. 124, no. 1–3, pp. 192–199, 2005.

[30] M. Akram, H. N. Bhatti, M. Iqbal, S. Noreen, and S. Sadaf, "Biocomposite efficiency for Cr(VI) adsorption: kinetic, equilibrium and thermodynamics studies," *Journal of Environmental Chemical Engineering*, vol. 5, no. 1, pp. 400–411, 2017.

[31] V.-P. Dinh, N.-C. Le, T.-P. Nguyen, and N.-T. Nguyen, "Synthesis of α-MnO$_2$ nanomaterial from a precursor γ-MnO$_2$: characterization and comparative adsorption of Pb(II) and Fe(III)," *Journal of Chemistry*, vol. 2016, Article ID 8285717, 9 pages, 2016.

4

Green and Efficient Extraction of Resveratrol from Peanut Roots Using Deep Eutectic Solvents

Jingnan Chen (ID)**, Xingxing Jiang, Guolong Yang, Yanlan Bi, and Wei Liu** (ID)

Provincal Key Laboratory for Transformation and Utilization of Cereal Resource, College of Food Science and Technology, Henan University of Technology, Zhengzhou 450001, China

Correspondence should be addressed to Jingnan Chen; chenjingnan813@126.com and Wei Liu; liuwei307@hotmail.com

Guest Editor: Mariusz Korczyński

Deep eutectic solvents (DESs), a new group of ecofriendly solvent combined with the ultrasonic-assisted extraction (UAE) technique, were first successfully used for extraction of resveratrol from peanut roots. Resveratrol in the extract was analyzed and quantified using a HPLC-UV method. A series of DESs consisting of choline chloride (ChCl) and 1,4-butanediol, citric acid, and ethylene glycol were formulated, finding ChCl/1,4-butanediol was a most proper extraction system. The optimal extraction parameters were obtained using a Box–Behnken design (BBD) test combined with response surface methodology as follows: 40% of water in ChCl/1,4-butanediol (1/3, g/g) at 55°C for 40 min and solid/liquid ratio of 1:30 g/mL. The total extraction content and extraction yield of resveratrol from peanut roots could reach 38.91 mg/kg and 88.19%, respectively, under such optimal conditions. The present study will provide a typical example for using DESs to extract natural bioactive compounds from plants.

1. Introduction

Peanut (*Arachis hypogaea* Linn.) belongs to the family of Rosales and is widely cultivated around the world as an important economical crop. This species is capable of producing stilbene derivatives, including resveratrol (3,5,4'-trihydroxy-stilbene) and other stilbenoids [1]. Resveratrol has attracted tremendous interest ascribing to its strong biological activity, such as antioxidant, anti-inflammatory, anticancer, cardiovascular protection, and cardioprotection functions [2–4]. In addition, it has been widely used as an active ingredient in cosmetic, medicine, and health products. Naturally, resveratrol is extracted from peanut [5], Japanese knotweed [6], grape [7], and other plants. Peanut roots are the waste products in the field after peanut harvesting, which are the cheapest sources of resveratrol [8].

Extraction of resveratrol from peanut roots should be based on green and sustainable technology. Conventionally, organic solvents, such as alcohols and acetone, are widely used in extraction of bioactive components from natural products resources in the fields of cosmetic, food, and pharmaceutical industries [9, 10]. However, the organic solvents pollution is a serious environmental issue. Therefore, the green extraction by using the new and environmentally friendly solvents was needed urgently [11]. In this connection, the deep eutectic solvents (DESs) have emerged as a better alternative to conventional solvents [12]. DESs can be easily synthesized with quaternary ammonium salt and hydrogen bond donor (HBD) with a gentle heating temperature range (70°C–90°C), and no further purification is required [13]. Numerous DESs consisting of different components, such as choline chloride, urea, organic acids, polyols, and sugars, have been developed [14–16]. In addition to as an green extraction solvent, DESs have other advantages, including the simple synthesis, low cost, biodegradable, chemical inertness, and no toxic factors [17, 18]. In fact, DESs have been applied in the extraction of a wide variety of natural compounds including phenolic compounds [19, 20], flavonoids [21, 22], sugars [23], and proteins [24, 25].

With the aim of development of green extraction technology for separation bioactive compounds from oil processing by-products, the present study was to establish a highly efficient and green extraction technology for the extraction of resveratrol from peanut roots using deep eutectic solvents assisted by ultrasonic extraction methods (DESs-UAE). First, a series of DESs were prepared by mixing the varying ratios of polyols, organic acid, and carbamide with ChCl. Second, various parameters of DESs-UAE in extracting resveratrol were optimized and systematically evaluated using a Box–Behnken design.

2. Materials and Methods

2.1. Chemicals and Reagents. 1,4-Butanediol (>98%) and citric acid (>99%) were obtained from Kwangfu Fine Chemical Industry Research Institute (Tianjin, China). Ethylene glycol (>98%) was purchased from Fuyu Fine Chemical Co., Ltd. (Tianjin, China). Lactic acid (>95%) was purchased from Sinopharm Chemical Reagent Co., Ltd. (Shanghai, China). Glycerol (>99%), 1,2-propylene glycol (>99%), DL-malic acid (>99%), and carbamide (>99%) were purchased from Kermel Chemical Reagent Co., Ltd. (Tianjin, China). Choline chloride (>98%) was obtained from Macklin Biochemical Co., Ltd. (Shanghai, China). Resveratrol (3,5,4′-trihydroxy stilbene, ≥98%) was obtained from Sigma-Aldrich Co. (St. Louis, MO, USA). Methanol (the chromatographic grade) was purchased from VBS biologic Co. (USA). All samples and solutions prepared for HPLC analysis were filtered through the 0.45 μm nylon membranes prior to use.

Standard stock solutions: resveratrol compound was directly prepared in methanol. The concentration of resveratrol in the standard stock solutions was all 500 μg/mL. Working standard solutions: resveratrol compound was obtained by diluting the stock solutions with methanol to a series of proper concentrations. The standard stock solutions and working standard solutions were all stored at 4°C in a dark place.

2.2. Plant Materials. Peanut roots were purchased from Hebei province in China. The plant material was cleaned and dried at 40°C. The dry plant material was ground into powder with a blender and passed through 50 mesh and then stored in a dry place at room temperature until used.

2.3. Preparation of DESs. All of the chemicals used in DESs preparations were dried at 60°C for 24h. The DESs were prepared at specific ratios of choline chloride to the hydrogen donor (i.e., 1,4-butanediol, glycerol, and lactic acid; Table 1). The varying ratios of choline chloride with the hydrogen donor were stirred in a flask at 80°C for 5–10 min, until a homogeneous transparent colourless liquid was formed. These DESs samples were treated by vacuum drying prior to use.

2.4. Deep Eutectic Solvents Ultrasonic-Assisted Extraction of Resveratrol. About 0.50 g peanut roots powder was weighed into an 50 mL centrifuge tube and followed by addition of 10 mL of extraction solvent. The mixture was then ultrasonically treated. Ultrasound-assisted extraction (UAE) was

TABLE 1: Different systems of natural deep eutectic solvents (DESs).

Abbreviation	Component 1	Component 2	Molar ratio
DES-1	Choline chloride	1,4-Butanediol	1:2
DES-2	Choline chloride	1,2-Propylene glycol	1:2
DES-3	Choline chloride	Glycerol	1:2
DES-4	Choline chloride	Ethylene glycol	1:2
DES-5	Choline chloride	DL-Malic acid	1:2
DES-6	Choline chloride	Lactic acid	1:2
DES-7	Choline chloride	Carbamide	1:2
DES-8	Choline chloride	Citric acid	1:2

performed using an ultrasonicator under 40 kHz and 400 W (SCQ-7201B, Shengyan Ultrasonic Instrument Co., Ltd., Shanghai, China). The extraction was carried out under different conditions. After extraction, the mixture was centrifuged (10 min, 2500 rpm) with a bench-scale centrifuge. The supernate was separated and filtered through a 0.45 μm membrane prior to HPLC analysis.

The extraction content was calculated according to the following equation:

$$\text{extraction content (mg/kg)} = \frac{\text{mass of resveratrol (mg)}}{\text{mass of weighed peanut roots powder (kg)}}. \tag{1}$$

The extraction yield was defined as follows [26]:

$$\text{extraction yield (\%)} = \frac{\text{mass of the resveratrol in extraction solution}}{\text{sum of the mass of resveratrol in sample}} \times 100\%. \tag{2}$$

The mass of resveratrol in the extraction solution (one-step extraction) was determined by HPLC-UV. The sum mass of resveratrol in the sample was calculated by analysis of the total mass of resveratrol in the combined extraction solutions afforded by continuously extracting three times with methanol according to the standard method.

2.5. Experimental Design and Statistical Analysis. Firstly, the mole ratio of hydrogen bond donors and ChCl (1,4-butanediol:ChCl = 1:1, 2:1, 3:1, 4:1, and 5:1), the percentage of water (10%, 20%, 30%, 40%, 50%, and 100%), solid-liquid ratio (1:10, 1:20, 1:30, 1:40, and 1:50), extraction temperature (20, 30, 40, 50, 60°C, and 70°C), and extraction time (20, 30, 40, 50, and 60 min) were optimized by single-factor experiments, respectively. Furthermore, Box–Behnken experimental design (BBD) with response surface methodology (RSM) was used to estimate the most effective combination of extraction parameters according to the single-factors experiments. A three-level (−1, 0, and +1) four-factor Box–Behnken design (BBD) was applied to evaluate the interaction effect of the factors: the percentage of water (A), solid/liquid ratio (B), extraction temperature

(C), and extraction time (D) on the extraction content of resveratrol (Y). 29 experiments running with different combinations of four factors were carried out totally. The second-order polynomial given in the equation was applied to correlate the interaction of each factor to the response. For the four factors, this equation is

$$Y = \beta_0 + \sum_{i=1}^{4} \beta_i X_i + \sum_{i=1}^{4} \beta_{ii} X_i^2 + \sum_{i=1}^{4} \sum_{j=i+1}^{4} \beta_{ij} X_i X_j, \quad (3)$$

where X_i and X_j are the independent coded variables which influence the response Y. Y is the predicted response; β_i is the parameter estimated for the variable; β_{ii} and β_{ij} are the parameters estimated for the interaction between variables i and i and i and j; and β_0, β_i, β_{ii}, and β_{ij} are the regression coefficients for intercept, linearity, square, and interaction, respectively. The variables of each factor were changed in the range of between −1 and 1 for the appraisals, while the dependent variable was the extraction content of resveratrol.

All above experimental statistical analyses were analyzed using the software Design-Expert 8.0.6 (Stat-Ease, Minneapolis, MN, USA). Analysis of variance (ANOVA) was carried out to evaluate the optimal conditions for the resveratrol extraction from peanut roots using the DESs-UAE method. A significance level of $p < 0.05$ was performed for each influential factor.

2.6. HPLC Analysis of Resveratrol.

The determination of resveratrol content was carried out on an HPLC system (Waters e2695, USA). The chromatographic separation of resveratrol was performed on Waters Symmetry C18 reversed-phase column (250 mm × 5 mm × 4.6 mm, 5 μm). The mobile phase consisting of 0.5% formic acid aqueous solution (A) and methanol (B) was filtered through a 0.45 μm membrane filter prior to use. The gradient elution was performed as follows: 0–6 min, 28% B; 6–15 min, 28–60% B; 15–20 min, 60–100% B; and 20–25 min, 100% B. The flow rate and injection volume were 1.0 mL/min and 10 μL, respectively, and the column temperature was set at 30°C. The resveratrol was identified by comparing the retention time with the standard, and the quantification of resveratrol was carried out at 306 nm. The HPLC analysis of the resveratrol standard and peanut roots sample was shown in Figure 1.

2.7. Statistical Analysis.

Experimental results were obtained as the mean value ± standard deviation (SD) ($n = 3$). The significance of difference was assessed using ANOVA. Differences were considered significant when the p value was <0.05.

3. Results and Discussion

3.1. Effect of DESs on Resveratrol Extraction.

The components of DESs have significant influence on their physicochemical properties, such as polarity, viscosity, and dissolving capacity, which will directly influence their extraction efficiency. In the present study, ChCl-based DESs were

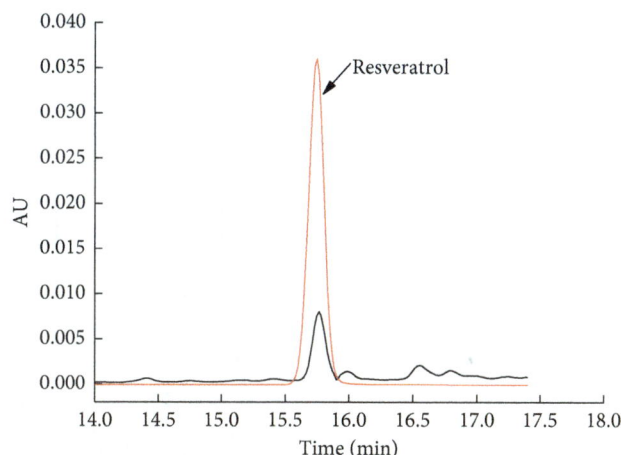

FIGURE 1: HPLC chromatogram of resveratrol extracted from peanut roots (the red and black lines mean the standard of resveratrol and sample, respectively).

synthesized by ChCl combining with different hydrogen bond donors (HBDs) including 1,4-butanediol, citric acid, ethylene glycol, lactic acid, glycerol, 1,2-propylene glycol, DL-malic acid and carbamide. The obtained eight DESs with different physicochemical properties were used for extracting resveratrol from peanut roots. The extraction contents were shown in Figure 2. The results indicated that the DESs type indeed strongly influenced the resveratrol extraction efficiency. The sequence of DESs for the extraction contents of resveratrol was as follows: DES-1 (ChCl/1,4-Buta) > DES-2 (ChCl/1,2-PG) > DES-4 (ChCl/EG) > DES-6 (ChCl/LA) > DES-7 (ChCl/Ca) > DES-3 ((ChCl/Gly) > DES-5 (ChCl/MA) > DES-8 (ChCl/CA). The optimal DES which provided the highest extraction content (26.44 ± 0.06 mg/kg) of resveratrol was DES-1 (composed of ChCl/1,4-butanediol). These data indicated that the polyalcohol-based DESs had a better extraction efficiency than organic acid-based DESs except for DES-3 ((ChCl/glycerol). This was because the strength of H-bonding interactions of the organic acid-based DESs would be the most efficient or the enthalpy of hole formation of polyalcohol-based DESs would be better for the resveratrol extraction [27]. Moreover, resveratrol belong to polyhydroxyphenols, and the polyalcohol-based DESs had a more suitable polarity for the resveratrol extraction. In addition, the much higher viscosity of glycerol-based DES also limited its extraction efficiency of resveratrol. Therefore, the following experiments were aimed at optimizing the extraction processing of resveratrol using choline chloride-1,4-butanediol (ChCl/1,4-Buta) (DES-1) as an extraction solvent.

3.2. Effect of Choline Chloride/1,4-Butanediol Molar Ratio on Resveratrol Extraction.

The effect of the choline chloride/1,4-butanediol molar ratio (1:1, 1:2, 1:3, 1:4, and 1:5) on the resveratrol extraction was examined (Figure 3). The results showed that the maximum resveratrol content (28.79 ± 0.12 mg/kg) could be achieved at choline

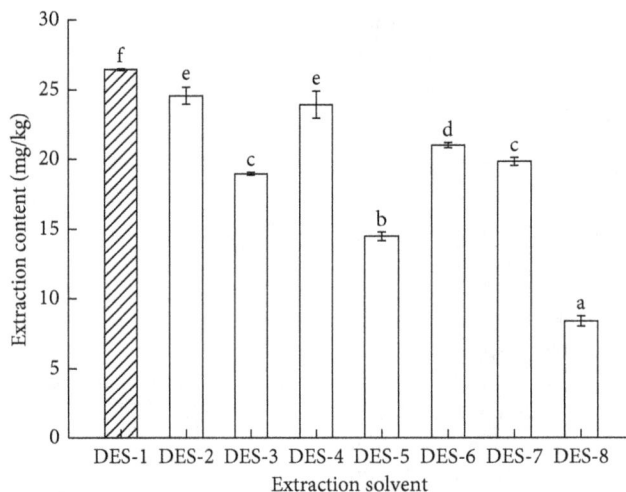

FIGURE 2: Effect of DESs on the extraction content of resveratrol from peanut roots. UAE conditions: solid/liquid ratio of 1: 20 (g/mL), 25°C, 30 min, ultrasonic power of 40 kHz, and choline chloride/1,4-butanediol molar ratio of 1 : 2, 20% of water (v/v) (DES-1: ChCl/1,4-Buta; DES-2: ChCl/1,2-PG; DES-3: ChCl/Gly; DES-4: ChCl/EG; DES-5: ChCl/MA; DES-6: ChCl/LA; DES-7: ChCl/Ca; and DES-8: ChCl/CA) (data are expressed as the mean ± SD; means in the group with different letters differ significantly at $p < 0.05$).

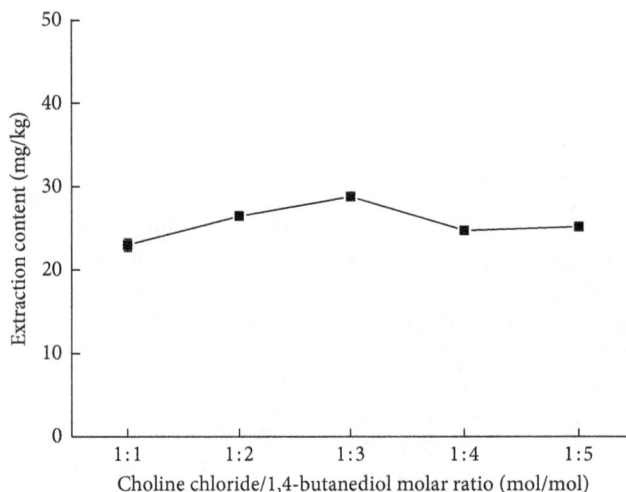

FIGURE 3: Effect of choline chloride/1,4-butanediol molar ratio on the extraction content of resveratrol from peanut roots. UAE conditions: solid/liquid ratio of 1: 20 (g/mL), 25°C, 30 min, ultrasonic power of 40 kHz, extraction solvent of choline chloride/1,4-butanediol (DES-1), and 20% of water (v/v).

chloride/1,4-butanediol molar ratio of 1 : 3. Further increasing of the proportion of 1,4-butanediol (>1 : 3) would cause the decline of resveratrol content (1 : 4, 24.76 ± 0.47 mg/kg, and 1 : 5, 25.21 ± 0.30 mg/kg). The increase of the 1,4-butanediol molar ratio in DES would result in the viscosity of DES decreasing and the polarity increasing, which might affect the effectiveness of mass transport and diffusion of resveratrol from peanut roots [28–30]. Therefore, the ChCl/1,4-butanediol molar ratio of 1 : 3 (mol/mol) was selected for the next experiments.

3.3. Effect of Water Content in DESs on Resveratrol Extraction. The high viscosity of DESs not only hinders the mass transport from plant matrices to solution but also leads to handling difficulties. Polarity is another important property

of DESs since it affects the solubilizing ability of DESs. The addition of water to DES can decrease the viscosity of the DESs, adjust the polarity, and increase the solubility of the target compounds. In this study, ChCl/1,4-Buta- (DES-1-) water mixture with water fraction ranging from 10% to 50% (v/v) was evaluated for the extraction of resveratrol from peanut roots (Figure 4).

As shown in Figure 4, the extraction power of the resveratrol was significantly improved with the increasing proportion of water, up to maximum (28.26 ± 0.77 mg/kg) at 30% (v/v). This was because that the addition of water led to a decrease in the viscosity of the reaction media, improving the mass transfer from peanut roots to solution, therefore, enhancing the extraction efficiency. However, higher concentration of water in ChCl/1,4-Buta (DES-1) (40%–100%) led to the decrease in the extraction amount of resveratrol

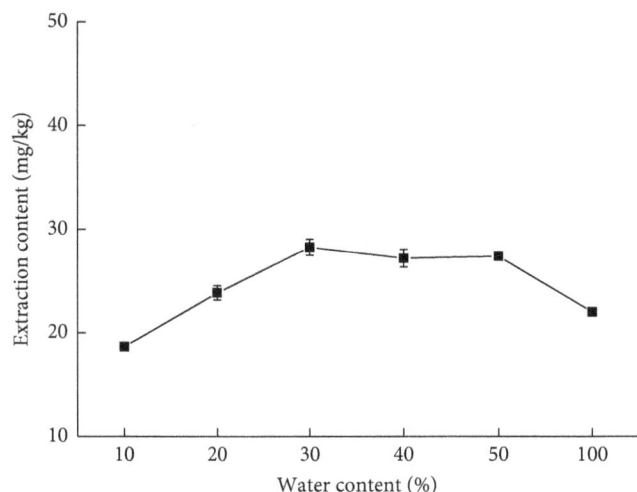

FIGURE 4: Effect of water content on the extraction content of resveratrol from peanut roots. UAE conditions: solid/liquid ratio of 1: 20 (g/mL), 25°C, 30 min, ultrasonic power of 40 kHz, extraction solvent of choline chloride/1,4-butanediol (DES-1), and choline chloride/1,4-butanediol molar ratio of 1 : 3.

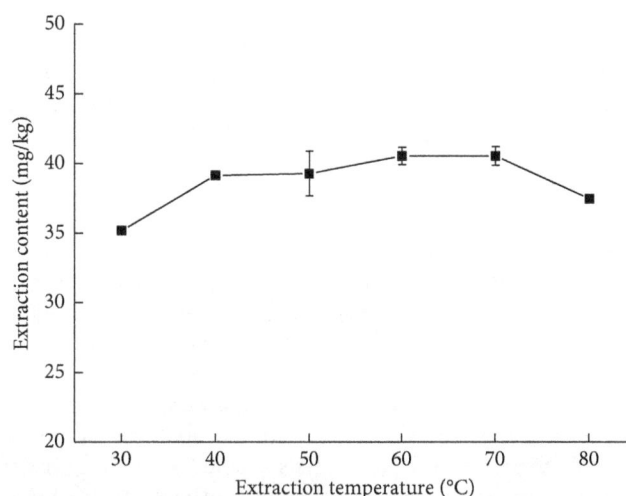

FIGURE 5: Effect of solid/liquid ratio on the extraction content of resveratrol from peanut roots. UAE conditions: 25°C, 30 min, ultrasonic power of 40 kHz, extraction solvent of choline chloride/1,4-butanediol (DES-1), choline chloride/1,4-butanediol molar ratio of 1 : 3, and 30% of water (v/v).

(40%, 27.20 ± 0.84; 50%, 27.39 ± 0.14; and 100%, 22.02 ± 0.21) (mg/kg). This was probably because higher concentration of water weakened the interactions between resveratrol and ChCl/1,4-Buta (DES-1) and also increased the polarity of extraction solution. Furthermore, the excess water made the ChCl/1,4-Buta (DES-1) diluted, which might result in the disruption of hydrogen bonds of DESs components and the loss of the supermolecular structure consequently [31]. Hence, water content of 30% in the ChCl/1,4-Buta was considered as the optimal ratio.

3.4. Effect of Solid/Liquid Ratio on Resveratrol Extraction. The solid/liquid ratio was evaluated (Figure 5). From the results, we could find that the extraction efficiency of resveratrol increased from 13.17 ± 0.32 to 38.34 ± 0.54 mg/kg with the increase of the solid/liquid ratio from 1 : 10 to 1 : 30 (g/mL). But further increase of the solid/liquid ratio had no obvious effect on the extraction content of resveratrol (1 : 40, 36.88 ± 0.18, and 1 : 50, 38.04 ± 0.21) (mg/kg), indicating that the target compound could be fully extracted at 1 : 30 g/mL of the solid/liquid ratio. Therefore, 1 : 30 g/mL of solid/liquid ratio was selected for the further experiments.

3.5. Effect of Extraction Temperature on Resveratrol Extraction. The temperature affects the viscosity and solubility of solvents and therefore affects the extraction efficiency of resveratrol. As shown in Figure 6, the extraction content of the resveratrol increased continually with the increasing extraction temperatures from 20°C to 60°C (20°C, 35.66 ± 0.15, and 60°C, 40.53 ± 0.67) (mg/kg). The elevated temperature might decrease the viscosity of the DESs, inducing the full contact of the material with the extraction solvent. The extraction efficiency at 70°C (40.52 ± 0.67 mg/kg) had no change compared to 60°C. However, with further increase of

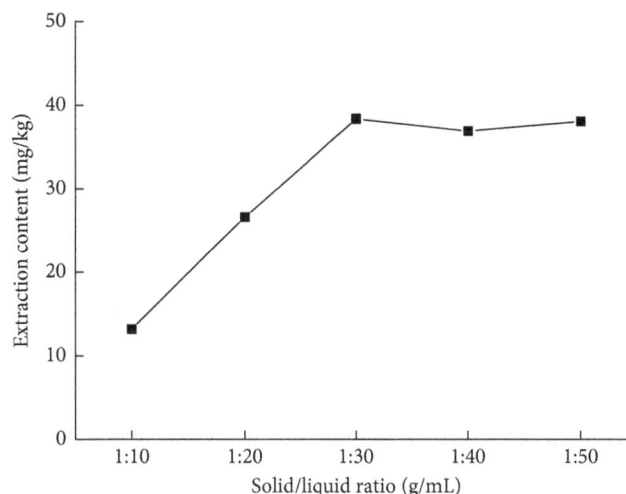

FIGURE 6: Effect of extraction temperatures on the extraction content of resveratrol from peanut roots. UAE conditions: solid/liquid ratio of 1: 30 (g/mL), 30 min, ultrasonic power of 40 kHz, extraction solvent of choline chloride/1,4-butanediol (DES-1), choline chloride/1,4-butanediol molar ratio of 1:3, and 30% of water (v/v).

temperature, the extraction content of resveratrol decreased slightly (80°C, 37.46 ± 0.002 mg/kg), probably because the higher extraction temperature would make the resveratrol oxidized or decomposed. Comprehensively considering the extraction efficiency and energy saving, 60°C was selected as the optimal extraction temperature.

3.6. Effect of Extraction Time on Resveratrol Extraction. The extraction time was also investigated, and the results were shown in Figure 7. The highest extraction content of the resveratrol was obtained at 30 min (39.15 ± 0.07 mg/kg).

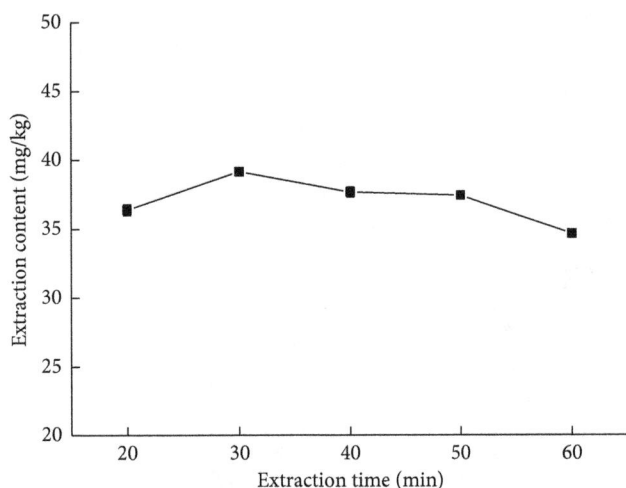

FIGURE 7: Effect of extraction time on the extraction content of resveratrol from peanut roots. UAE conditions: solid/liquid ratio of 1 : 30 (g/mL), 60°C, ultrasonic power of 40 kHz, extraction solvent of choline chloride/1,4-butanediol (DES-1), choline chloride/1,4-butanediol molar ratio of 1 : 3, and 30% of water (v/v).

With the prolonging of extraction time from 30 to 60 min, the resveratrol content decreased slightly (40 min, 37.65 ± 0.37; 50 min, 37.41 ± 0.05; and 60 min, 34.64 ± 0.06) (mg/kg). This trend might be because resveratrol took place oxidation or decomposition during the long time extraction process. Thus, 30 min was chosen as the optimal extraction time.

3.7. Optimization of the Extraction Process by the BBD Assay.

Further optimization of DESs-UAE resveratrol extraction conditions (water content, liquid/solid ratio, extraction temperature, and extraction time) was carried out by a Box-Behnken design (BBD) method. The data were analyzed using Design-Expert 8.0.6 software for statistical analysis of variance (ANOVA) and regression analysis (Table 2). The regression equation model for resveratrol extraction was obtained and shown in the following equation:

$$\begin{aligned} Y = {}& 37.58 + 2.52A + 4.81B + 1.47C + 1.74D + 0.85AB \\ & - 1.55AC - 0.8AD - 1.53BC + 0.57BD - 0.93CD \\ & - 6.21A^2 - 9.48B^2 - 1.82C^2 - 3.27D^2, \end{aligned}$$

(4)

where Y is the extraction content of resveratrol (mg/kg) and A, B, C, and D represented water content, solid/liquid ratio, extraction temperature, and extraction time, respectively.

The analysis of variance (ANOVA) was performed to evaluate the optimal extraction conditions of resveratrol (Table 3). The F-value of the model was 31.03 ($p < 0.0001$), indicating that the afforded model was significant. "Lack of fit F-value" was 1.74 ($p = 0.3117$), demonstrating that the lack of fit of the quadratic models was not significant, and the experiment data fitted well to the model. The regression analysis of the data showed the coefficient of the

TABLE 2: Experimental data and the obtained response values with different combinations of water content (A), solid/liquid ratio (B, g/mL), extraction temperature (C, °C), and extraction time (D, min) used in BBD.

Run	Factor				Extraction content Y
	A	B	C	D	
1	70	30	30	40	32.40
2	40	30	55	40	38.33
3	40	10	30	40	18.17
4	10	30	55	10	22.44
5	70	50	55	40	29.08
6	40	50	55	70	30.66
7	10	30	30	40	23.54
8	40	30	30	10	26.86
9	40	30	55	40	38.12
10	40	50	55	10	27.88
11	40	30	55	40	38.19
12	10	30	55	70	28.26
13	10	50	55	40	2.93
14	70	30	55	10	28.95
15	10	30	80	40	30.76
16	70	30	55	70	31.56
17	70	30	80	40	33.41
18	40	10	55	10	21.07
19	40	30	55	40	38.03
20	40	30	80	10	33.14
21	40	10	55	70	21.57
22	40	30	30	70	33.34
23	70	10	55	40	18.77
24	40	30	55	40	35.22
25	40	10	80	40	21.49
26	40	30	80	70	35.88
27	40	50	80	40	30.73
28	10	10	55	40	16.03
29	40	50	30	40	33.53

determination ($R^2 = 0.9688$) value for resveratrol was significant, implying that this quadratic model was suitable to describe the response of the experiment regarding to the resveratrol.

The effect of these factors affecting the resveratrol extraction was in an order of B (liquid/solid ratio) > A (water content) > D (extraction time) > C (extraction temperature), which was determined by the absolute value of the liner term coefficient of the regression equation. The p value of the quadratic term of A^2 and B^2 was both <0.0001, respectively, implying that water content (A) and solid/liquid (B) ratio both had significant effects on the extraction content of resveratrol.

The effect and interaction of four factors on the extraction yields of resveratrol were examined by the three-dimensional response surface (Figure 8). Figure 8(a) showed the effects of water content, solid/liquid ratio, and their interaction on the extraction content of resveratrol. It was observed that the highest extraction content was afforded with the water content range of 15%–55% and solid/liquid ratio of 25–38 mL/g. When solid/liquid ratio was a certain value, the extraction content of resveratrol had the trend of increasing first and then decreasing with the increase of water content. When the water content was fixed, the yield of

TABLE 3: ANOVA statistics analysis of the model for the extraction of resveratrol.

Source	Sum of squares	df	Mean square	F value	p value probability > F	Significance
Model	1168.03	14	83.43	31.03	<0.0001	**
A (water content; %)	76.10	1	76.10	28.31	0.0001	**
B (solid/liquid ratio; g/mL)	277.52	1	277.52	103.22	<0.0001	**
C (extraction temperature; °C)	25.78	1	25.78	9.59	0.0079	*
D (extraction time; min)	36.45	1	36.45	13.56	0.0025	*
AB	2.9	1	2.9	1.08	0.3164	
AC	9.64	1	9.64	3.58	0.0792	
AD	2.56	1	2.56	0.95	0.3458	
BC	9.38	1	9.38	3.49	0.0829	
BD	1.31	1	1.31	0.49	0.4969	
CD	3.49	1	3.49	1.30	0.2738	
A^2	249.87	1	249.87	92.24	<0.0001	**
B^2	583.45	1	583.45	217.01	<0.0001	**
C^2	21.44	1	21.44	7.97	0.0135	
D^2	69.44	1	69.44	25.83	0.0002	
Residual	37.64	14	2.69			
Lack of fit	30.62	10	3.06	1.74	0.3117	Not significant
Pure error	7.02	4	1.76			
Correlation total	1205.67	28				

$^* p < 0.01$; $^{**} p < 0.001$.

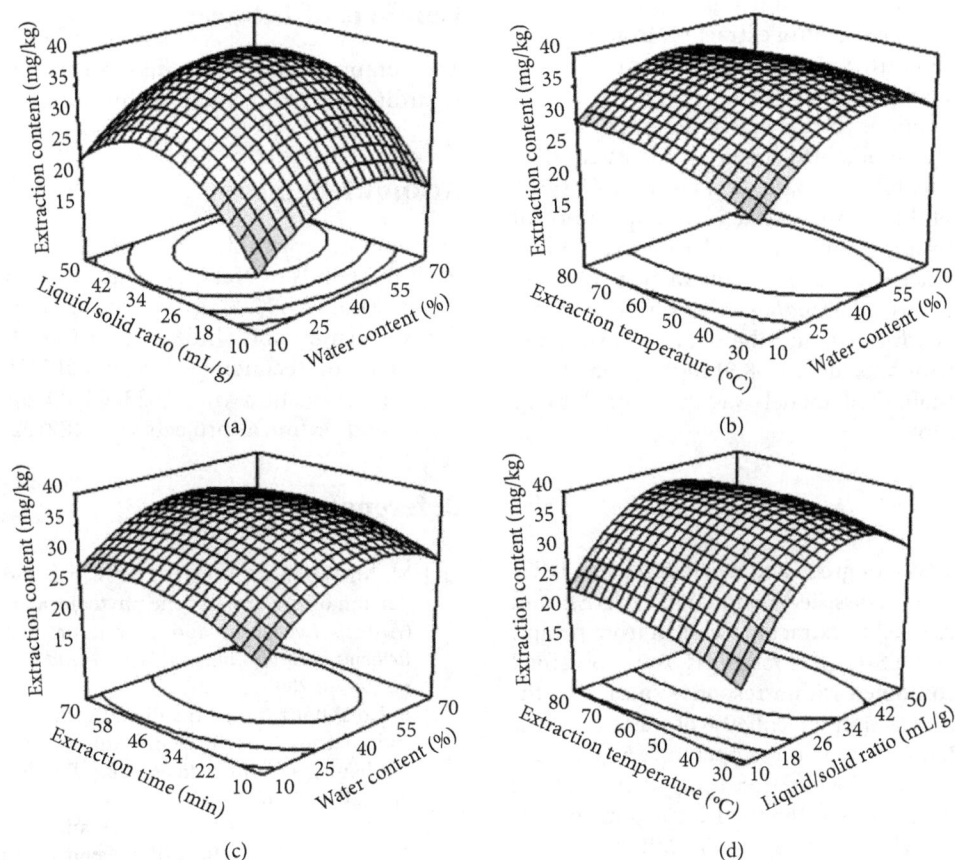

(a)

(b)

(c)

(d)

FIGURE 8: Continued.

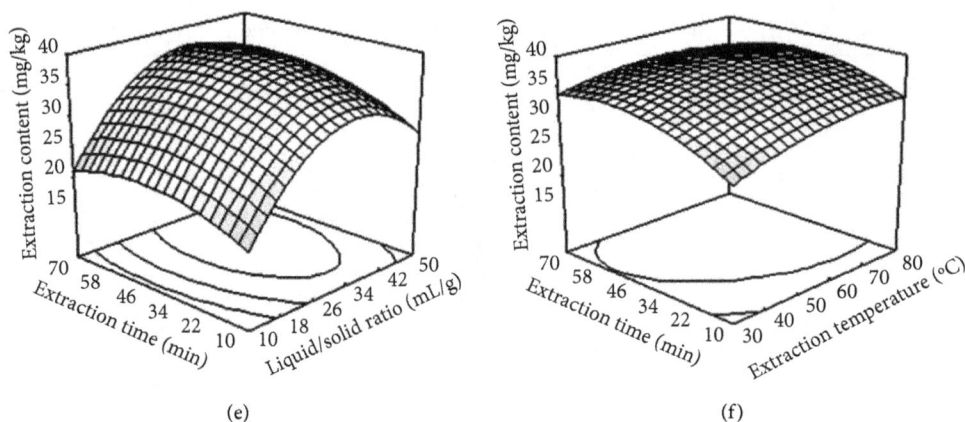

FIGURE 8: (a–f) refers to the target compounds response values under the two variables. Response surfaces representations for resveratrol from peanut roots: (a) varying water content and liquid/solid ratio; (b) varying water content and temperature; (c) varying water content and time; (d) varying temperature and liquid/solid ratio; (e) varying liquid/solid ratio and time; (f) varying temperature and time.

resveratrol was increased and then kept at the stable level with the increase of solid/liquid ratio. The interaction effect of water content and solid/liquid ratio on the resveratrol content was not significant. The same trend was observed in Figures 8(b)–8(f) (the interactive effects of water content, extraction time, and solid/liquid ratio on the extraction content of resveratrol). In summary, the interactions in water content, extraction time, extraction temperature, and solid/liquid ratio on the resveratrol extraction from peanuts roots were no significant, which was consistent with the analysis result of the regression model.

The optimal conditions for the extraction of resveratrol from peanut roots were obtained based on the established model, which were as follows: water content of 40% (v/v), solid/liquid ratio of 1:30 g/mL, extraction temperature of 55°C, and extraction time of 40 min. Under these optimum conditions, the predicted extraction content of resveratrol from peanut roots was 38.39 mg/kg. The verification experiment was also performed, and the obtained extraction content of resveratrol was up to 38.91 mg/kg, which indicated that the established model was considered to be reliable and reasonable.

4. Conclusion

In this study, a new type of green and efficient solvent DESs coupled with ultrasonic-assisted extraction (UAE) and HPLC-UV was developed to extract resveratrol from peanut roots. The optimal DES-UAE conditions were obtained using a BBD test combined with a response surface methodology as follows: extraction solvent 40% of water in ChCl/1,4-butanediol (1/3, g/g), extraction temperature 55°C, solid/liquid ratio 1:30 g/mL, and extraction time 40 min. Under the above optimum conditions, the total extraction content of resveratrol from peanut roots was up to 38.91 mg/kg, and the extraction yield was 89.46%. It was concluded that this DES-UAE-HPLC method was a fast, safe, and efficient extraction method for the preparation and determination of resveratrol from peanut roots.

Abbreviations

DESs: Deep eutectic solvents
UAE: Ultrasonic-assisted extraction
BBD: Box–Behnken design
HBD: Hydrogen bond donor.

Conflicts of Interest

The authors declare that there are no conflicts of interest regarding the publication of this article.

Acknowledgments

The authors gratefully acknowledge the financial support from the Funds of National Natural Science Foundation of China (no. 31772003); Provincal Key Laboratory for Transformation and Utilization of Cereal Resource, Henan University of Technology (no. PL2017001); Henan Natural Science Foundation (no. 162300410046); and Henan scientific and technical projects (no. 182102110024).

References

[1] M. M. Hasan, M. Cha, V. K. Bajpai, and K. H. Baek, "Production of a major stilbene phytoalexin, resveratrol in peanut (*Arachis hypogaea*) and peanut products: a mini review," *Reviews in Environmental Science and Bio/Technology*, vol. 12, no. 3, pp. 209–221, 2012.

[2] R. Kolahdouz Mohammadi and T. Arablou, "Resveratrol and endometriosis: in vitro and animal studies and underlying mechanisms," *Biomedicine and Pharmacotherapy*, vol. 91, pp. 220–228, 2017.

[3] L. M. Hung, J. K. Chen, S. S. Huang, R. S. Lee, and M. J. Su, "Cardioprotective effect of resveratrol, a natural antioxidant derived from grapes," *Cardiovascular Research*, vol. 47, no. 3, p. 549, 2000.

[4] D. G. Wanga, W. Y. Liub, and G. T. Chena, "A simple method for the isolation and purification of resveratrol from Polyg-

onum cuspidatum," *Journal of Pharmaceutical Analysis*, vol. 3, no. 4, pp. 241–247, 2013.

[5] Q. Xiong, Q. Zhang, D. Zhang, Y. Shi, C. Jiang, and X. Shi, "Preliminary separation and purification of resveratrol from extract of peanut (*Arachis hypogaea*) sprouts by macroporous adsorption resins," *Food Chemistry*, vol. 145, pp. 1–7, 2014.

[6] D. Zhang, X. Li, D. Hao et al., "Systematic purification of polydatin, resveratrol and anthraglycoside B from Polygonum cuspidatum Sieb, et Zucc," *Separation and Purification Technology*, vol. 66, no. 2, pp. 329–339, 2009.

[7] M. C. Pascual-Martí, A. Salvador, A. Chafer, and A. Berna, "Supercritical fluid extraction of resveratrol from grape skin of Vitis vinifera and determination by HPLC," *Talanta*, vol. 54, no. 4, pp. 735–740, 2001.

[8] C. D. Liu, Y. Y. Wen, and J. M. Chiou, "Comparative characterization of peanuts grown by aquatic floating cultivation and field cultivation for seed and resveratrol production," *Journal of Agricultural and Food Chemistry*, vol. 51, no. 6, pp. 1582–1585, 2003.

[9] S. Jiang, Q. Liu, Y. Xie et al., "Separation of five flavonoids from tartary buckwheat (Fagopyrum tataricum (L.) Gaertn) grains via off-line two dimensional high-speed countercurrent chromatography," *Food Chemistry*, vol. 186, pp. 153–159, 2015.

[10] X. H. Yao, D. Y. Zhang, M. H. Duan et al., "Preparation and determination of phenolic compounds from Pyrola incarnata, Fisch. with a green polyols based-deep eutectic solvent," *Separation and Purification Technology*, vol. 149, pp. 116–123, 2015.

[11] M. Cvjetko Bubalo, S. Vidović, I. Radojčić Redovniković, and S. Jokić, "Green solvents for green technologies," *Journal of Chemical Technology and Biotechnology*, vol. 90, no. 9, pp. 1631–1639, 2015.

[12] E. L. Smith, A. P. Abbott, and K. S. Ryder, "Deep eutectic solvents (DESs) and their applications," *Chemical Reviews*, vol. 114, no. 21, pp. 11060–11082, 2014.

[13] A. P. Abbott, D. Boothby, G. Capper, D. L. Davies, and R. K. Rasheed, "Deep eutectic solvents formed between choline chloride and carboxylic acids: versatile alternatives to ionic liquids," *Journal of the American Chemical Society*, vol. 126, no. 29, pp. 9142–9147, 2004.

[14] A. Paiva, R. Craveiro, I. Aroso, M. Martins, R. L. Reis, and A. R. C. Duarte, "Natural deep eutectic solvents-solvents for the 21st century," *ACS Sustainable Chemistry and Engineering*, vol. 2, no. 5, pp. 1063–1071, 2014.

[15] K. Radosevic, N. Curko, V. G. Srcek et al., "Natural deep eutectic solvents as beneficial extractants for enhancement of plant extracts bioactivity," *LWT-Food Science and Technology*, vol. 73, pp. 45–51, 2016.

[16] Y. T. Dai, J. V. Spronsenb, and G. J. Witkamp, "Natural deep eutectic solvents as new potential media for green technology," *Analytica Chimica Acta*, vol. 766, no. 5, pp. 61–68, 2013.

[17] Z. F. Wei, X. Q. Wang, X. Peng et al., "Fast and green extraction and separation of main bioactive flavonoids from Radix Scutellariae," *Industrial Crops and Products*, vol. 63, no. 4, pp. 175–181, 2015.

[18] B. M. Cvjetko, N. Ćurko, M. Tomašević, K. K. Ganić, and R. I. Redovniković, "Green extraction of grape skin phenolics by using deep eutectic solvents," *Food Chemistry*, vol. 200, pp. 159–166, 2016.

[19] K. Pang, Y. C. Hou, W. Z. Wu, W. J. Guo, W. Peng, and K. N. Marsh, "Efficient separation of phenols from oils via forming deep eutectic solvents," *Green Chemistry*, vol. 14, no. 9, pp. 2398–2401, 2012.

[20] Y. T. Dai, G. J. Witkamp, R. Verpoorte, and Y. H. Choi, "Natural deep eutectic solvents as a new extraction media for phenolic metabolites in Carthamus tinctorius L.," *Analytical Chemistry*, vol. 85, no. 13, pp. 6272–6278, 2013.

[21] M. W. Nam, J. Zhao, M. S. Lee, J. H. Jeong, and J. Lee, "Enhanced extraction of bioactive natural products using tailor-made deep eutectic solvents: application to flavonoid extraction from Flos sophorae," *Green Chemistry*, vol. 17, no. 3, pp. 1718–1727, 2015.

[22] W. T. Bi, M. Tian, and K. H. Row, "Evaluation of alcohol-based deep eutectic solvent in extraction and determination of flavonoids with response surface methodology optimization," *Journal of Chromatography A*, vol. 1285, pp. 22–30, 2013.

[23] A. K. Das, M. Sharma, D. Mondal, and K. Prasad, "Deep eutectic solvents as efficient solvent system for the extraction of k-carrageenan from Kappaphycus alvarezii," *Carbohydrate Polymers*, vol. 136, pp. 930–935, 2016.

[24] N. Li, Y. Z. Wang, K. J. Xu, Y. H. Huang, Q. Wen, and X. Q. Ding, "Development of green betaine-based deep eutectic solvent aqueous two-phase system for the extraction of protein," *Talanta*, vol. 152, pp. 23–32, 2016.

[25] K. Xu, Y. Z. Wang, Y. H. Huang, N. Li, and Q. Wen, "A green deep eutectic solvent-based aqueous two-phase system for protein extracting," *Analytica Chimica Acta*, vol. 864, pp. 9–20, 2015.

[26] F. Y. Du, X. H. Xiao, and G. K. Li, "Application of ionic liquids in the microwave-assisted extraction of trans -resveratrol from Rhizma Polygoni Cuspidati," *Journal of Chromatography A*, vol. 1140, no. 1-2, pp. 56–62, 2007.

[27] A. P. Abbott, A. Y. M. Al-Murshedi, O. A. O. Alshammari et al., "Thermodynamics of phase transfer for polar molecules from alkanes to deep eutectic solvents," *Fluid Phase Equilibria*, vol. 448, pp. 99–104, 2017.

[28] M. Hayyan, C. Y. Looi, A. Hayyan, and W. F. Wong, "In vitro and in vivo toxicity profiling of ammonium-based deep eutectic solvents," *Plos One*, vol. 10, no. 2, Article ID e0117934, 2015.

[29] A. P. Abbott, R. C. Harris, and K. S. Ryder, "Application of hole theory to define ionic liquids by their transport properties," *Journal of Physical Chemistry B*, vol. 111, no. 18, pp. 4910–4913, 2007.

[30] C. D'Agostino, R. C. Harris, A. P. Abbott, L. F. Gladden, and M. D. Mantle, "Molecular motion and ion diffusion in choline chloride based deep eutectic solvents studied by 1H pulsed field gradient NMR spectroscopy," *Physical Chemistry Chemical Physics*, vol. 13, no. 48, pp. 21383–21391, 2011.

[31] Y. T. Dai, G. J. Witkamp, R. Verpoorte, and Y. H. Choi, "Tailoring properties of natural deep eutectic solvents with water to facilitate their applications," *Food Chemistry*, vol. 187, pp. 14–19, 2015.

5

Green Synthesis and Characterization of Biosilica Produced from Sugarcane Waste Ash

Rodrigo Heleno Alves,[1] **Thais Vitória da Silva Reis,**[1,2]
Suzimara Rovani,[1] **and Denise Alves Fungaro**[1]

[1]*Instituto de Pesquisas Energéticas e Nucleares (IPEN-CNEN/SP), Av. Prof. Lineu Prestes, No. 2242, Cidade Universitária, 05508-000 São Paulo, SP, Brazil*
[2]*Faculdades Oswaldo Cruz, Rua Brigadeiro Galvão, 540 Barra Funda, 01151-000 São Paulo, SP, Brazil*

Correspondence should be addressed to Denise Alves Fungaro; dafungaro@gmail.com

Academic Editor: Barbara Gawdzik

In this study, ash from sugarcane waste was used in the synthesis of biosilica using alkaline extraction followed by acid precipitation. Different parameters that could influence the silica particle synthesis were evaluated. The ash and synthesized biosilica were characterized by a combination of spectroscopic and chemical techniques such as XRD, XRF, SEM, particle size analyser, N_2 adsorption analysis, TGA, and FTIR. The best condition for biosilica production was achieved with fusion method and aging temperature of 80°C for 1 h during gel formation. X-ray powder diffraction pattern confirms the amorphous nature of synthesized silica. The purity of the prepared silica was 99% silica which was confirmed by means of XRF. The experimental data suggest that the sugarcane waste ash could be converted into a value-added product, minimizing the environmental impact of disposal problems.

1. Introduction

Brazil is the global leader in the production of sugarcane, harvesting more than 600 million tonnes of it every year. The state of São Paulo, in the South-Central region, is the largest producer in the country and is responsible for 61% of Brazil's sugarcane production [1].

The processing of sugarcane to produce sugar and ethanol generates various agricultural wastes, especially straw and bagasse. Each ton of sugarcane generates between 250 and 270 kg of bagasse and 200 kg of straw and tips [2].

About 50% of these residues are used in distillery plants as a source of energy; the remainder is stockpiled. The burning of the waste generates 1–4% ash. Considering the amount of sugarcane produced in this harvest, between 3 and 12 million tons of ash are produced each year.

Generally, the destination of biomass ash generated in the sugar industry is the disposal in landfills or used as fertilizer in the plantations. These practices have caused problems to public health and unwanted environmental impacts mainly associated with soil and water [3]. Moreover, sugarcane waste ash can be used as a substitute for cement or sand in civil construction [4, 5].

The ash of the sugarcane waste contains high levels of silica [3]. Silica is considered a value-added product for presenting numerous applications in various industries like rubber industry as a reinforcing agent, in tooth pastes as a cleansing agent, as an anticaking agent in salts, in cosmetics, and so forth [6]. Thus, efforts have been done in order to find economical way to extract silica of waste ash materials.

The objective of this study was to evaluate the effect of different synthesis parameters to optimize conditions for obtaining biosilica from sugarcane biomass ash. This not only provides value addition but also solves the problem of large amount of ash disposal.

2. Experimental

2.1. Materials. All the regents used were of analytical grade. Sodium hydroxide and acetic acid (analytical grade) were

TABLE 1: Parameters of the biosilica synthesis process.

| Samples | Alkaline fusion | | Hydrothermal treatment | | Silica gel formation | Silica gel | Synthesis product |
	Time (h)	Agitation time (h)	Time (h)	Temperature (°C)		m^a (g)	
BS1	1	24	1	100	No	—	—
BS2	1	24	2	100	Yes	0.232	Amorphous
BS3	1	24	3	100	Yes	0.248	Amorphous
BS4	1	24	6	100	Yes	0.366	Amorphous
BS5	1	24	20	100	Yes	0.431	Amorphous
BS6	1	24	24	100	Yes	0.475	n.a[b]
BS7	0	0	3	100	No	—	—
BS8	1	24	2	90	Yes	0.388	n.a[b]
BS9	1	24	3	90	Yes	0.402	Amorphous
BS10	1	24	6	90	Yes	0.530	Amorphous
BS11	1	24	20	90	Yes	0.689	Amorphous
BS12[c]	1	24	20	90	Yes	0.742	Amorphous

[a]m = mass of product formed; [b]n.a = not totally amorphous; [c]BS12 = aging temperature of 80°C.

obtained from Merck (Darmstadt, Germany). The sample of ash derived from burning of the sugarcane waste was provided by the company COSAN S.A. (São Paulo, Brazil). Quantitative filter paper Whatman Number 41 was purchased from Whatman Plc (Kent, England). Oven (Fanen Orion 515 Model), muffle furnace (Quimis, Q-318 M24 Model), and mechanical shaker (Ethik Technology 430 Model) were used.

2.2. Methods. The synthesis of biosilica was carried out by different methods. A schematic flow chart for preparation of silica gel from sugarcane waste ash (SWA) is proposed as shown in Figure 1. The experimental conditions are summarized in Table 1.

2.2.1. Preparation of Sodium Silicate Solution by Two-Step Process. Typically, 1.0 g of SWA was mixed with 1.3 g of NaOH in capsule and triturated to obtain a homogeneous mixture. This mixture was then calcined at 550°C for 1 h in a muffle furnace. After cooling to room temperature, the fusion products were ground and placed in beaker with 100 mL of distilled water. The suspension was stirred for 24 h (120 rpm) at room temperature. The resulting slurry of first step was submitted to hydrothermal treatment in an oven at different temperatures (90 and 100°C) and reaction times (1–24 h). The reaction products were filtered with a quantitative filter paper Whatman Number 41 and the filtered liquid (sodium silicate solution) was stored for biosilica preparation.

2.2.2. Preparation of Sodium Silicate Solution by One-Step Process. In a Teflon beaker, 1.0 g of SWA was mixed with 10 mL of 3 mol L^{-1} aqueous NaOH solution. The mixture was incubated at 100°C in oven for 3 h. At the end of the treatment, the mixture was filtered and the filtered liquid was allowed to cool to room temperature and was stored.

2.2.3. Synthesis of Biosilica. In the as-obtained solution of sodium silicate, 3 mol L^{-1} acetic acid was added dropwise under constant stirring until pH 7. The silica gel formed was

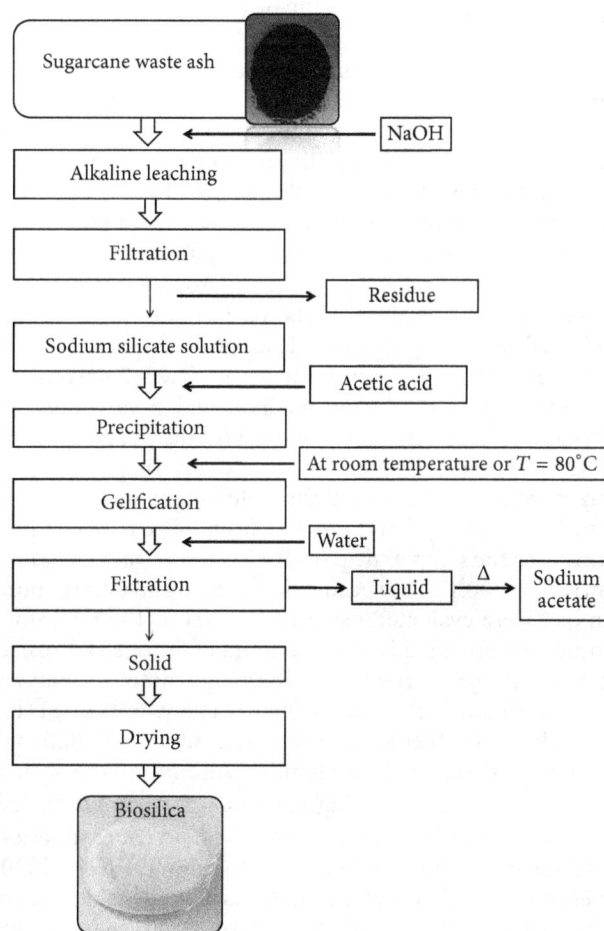

FIGURE 1: Flow diagram of the procedure used to produce biosilica from sugarcane biomass ash.

aged for at least 20 h at room temperature. After this time, the silica gel was washed with distilled water and filtered. The filtrate was stocked and the silica retained on the filter was dried for 24 h at 80°C [7–10].

2.2.4. Synthesis of Biosilica with Aging Process at 80°C. The sodium silicate solution was obtained using the two-step process described in Section 2.2.1. The conditions used in the second step were 90°C for 20 h. In the sodium silicate solution 3 mol L^{-1} acetic acid up was added to pH 7. The resulting gel mixture was aged at 80°C for 1 h (aging step). The gel was washed twice with 100 mL of distilled water, filtered (sodium acetate solution), and separated from solid (silica gel). The silica gel was dried at 135°C for 24 h [11]. The sodium acetate solution was dried at 100°C to obtain sodium acetate as salt.

2.3. Characterization of Materials. The chemical composition of materials in the form of oxides was analysed by energy dispersive X-ray fluorescence spectrometry (XRF) (RX Axios Advanced, PANalytical, Phillips spectrometer) and wavelength dispersive X-ray fluorescence spectroscopy (WDXRF) (Rigaku Co. spectrometer, model RIX 3000). The mineralogical compositions of the samples were determined by X-ray diffraction analyses (XRD) with an automated Rigaku multiflex diffractometer with Cu anode using Co Kα radiation at 40 kV and 20 mA over the range (2θ) of 5–80° with a scan time of 0.5°/min. The crystalline phases present in the samples were identified with the help of ICDD (International Centre for Diffraction Data). The Fourier transform infrared spectroscopy (FTIR) spectra were recorded on ALPHA FTIR Spectrometer from Bruker, operating in attenuated total reflectance (ATR). The spectra were obtained using 200 cumulative scans, range 375 to 4000 cm^{-1}. Scanning electron micrograph was obtained by a JEOL JSM-7401 scanning electron microscope, at a typical acceleration voltage of 3 KV. The particle size analysis was conducted by laser particle analyser Malvern MSS Mastersizer 2000 Ver. 5.54 (United Kingdom) (0.02–2000 μm) in isopropyl alcohol dispersing medium and pump speed of 2500 rpm. Adsorption-desorption isotherms were measured with Micrometrics TriStar volumetric adsorption analyser using nitrogen of 99.999% purity. Measurements were performed in range of relative pressure from 10E-6 to 0.99 liquid nitrogen on the samples degased for 2 h, under vacuum about 50 mTorr, at 200°C. The specific surface area and average pore diameter were evaluated using BET method. The total pore volume was estimated from the amount of nitrogen adsorbed at the relative pressure of 0.99. Thermogravimetric analyses were recorded in a thermogravimetric analyser TGA/SDTA 851 produced by Mettler Toledo. Dried samples (~10.0 mg) were analysed under dynamic nitrogen atmosphere with a flow of 100.0 mL min^{-1}, using a alumina-port sample heated 1000°C with a heating rate of 5° or 10°C min^{-1}. Determination of carbon and sulfur was done by LECO model TCHEN600. Other physical-chemical properties (bulk density, cation exchange capacity, pH, and pH of point of zero charge) have been described in a previous paper [12].

3. Results and Discussion

3.1. Characterization of Sugarcane Waste Ash. The chemical composition of sugarcane waste ash (SWA) is shown in Table 2. The major component was SiO$_2$ content with 81.60 wt%., indicating that this material is suitable to use

TABLE 2: Chemical composition of sugarcane waste ash and synthesized biosilica.

Oxides	SWA (wt.%)	Biosilica (wt.%)
SiO$_2$	81.60	99.1
Al$_2$O$_3$	7.94	0.33
Fe$_2$O$_3$	2.31	0.08
K$_2$O	2.10	0.07
MgO	1.26	—
P$_2$O$_5$	1.07	—
CaO	0.98	0.03
TiO$_2$	0.67	—
SO$_3$	0.52	—
Cl	0.36	0.13
Na$_2$O	0.26	0.22
MnO	0.06	—
Ta$_2$O$_5$	0.04	—
NiO	0.03	0.03
As$_2$O$_3$	0.03	—
CuO	—	<0.01
ZnO	—	<0.01
LI[a]	0.79	—

[a]LI = loss on ignition at 1050°C for one hour.

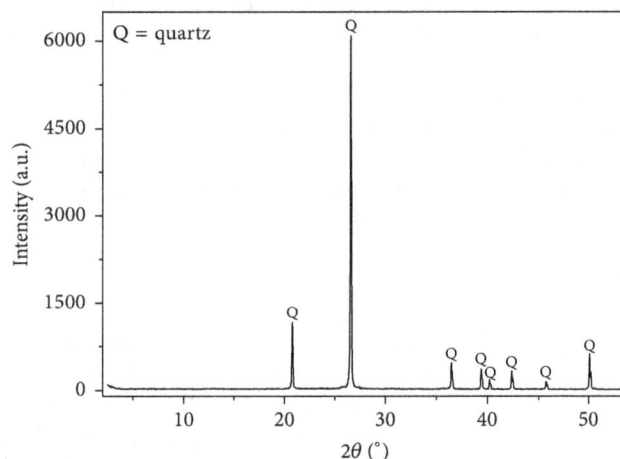

FIGURE 2: X-ray diffractogram of sugarcane biomass ash.

as a source of silica for producing sodium silicate solution. The compounds of aluminum, iron, potassium, and magnesium were found in amounts ≥1.3%. The other oxides were considered impurities for presenting content ≤1.00%. The percentage of carbon and sulfur was 0.16 and 0.04%, respectively.

The sugarcane absorbs more silicon than any other cultivated plant. Sugarcane can remove from 500 to 700 kg Si ha^{-1} during their development [13]. Monosilicic acid (H$_4$SiO$_4$) is the form of silicon used by plants, which is found in both liquid and adsorbed phases of silicon in soils [14–17].

The SWA diffractogram (Figure 2) indicated the presence of silica only in the crystalline phase (SiO$_2$, ICDD01-085-0794). The peaks observed at 20.9°, 26.6°, 36.5°, 39.4°,

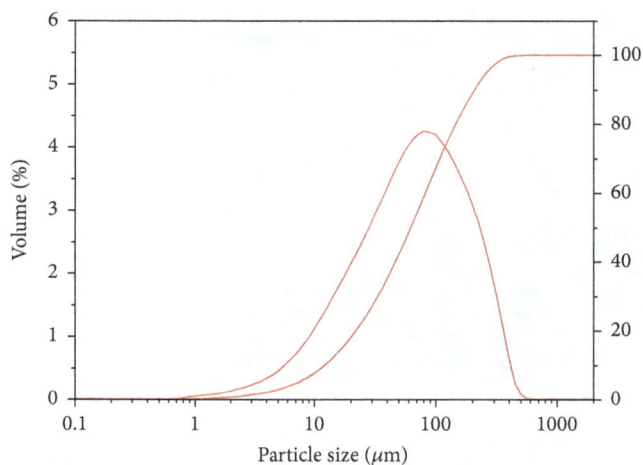

FIGURE 3: Particle size distribution curves of sugarcane biomass ash.

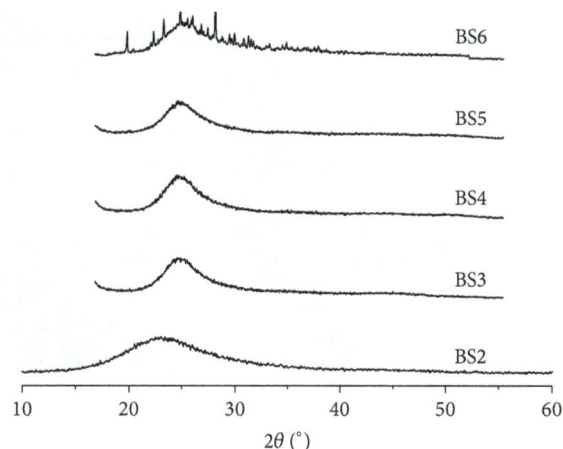

FIGURE 4: XRD patterns of the synthesis products obtained by hydrothermal treatment of sugarcane waste ash at 100°C.

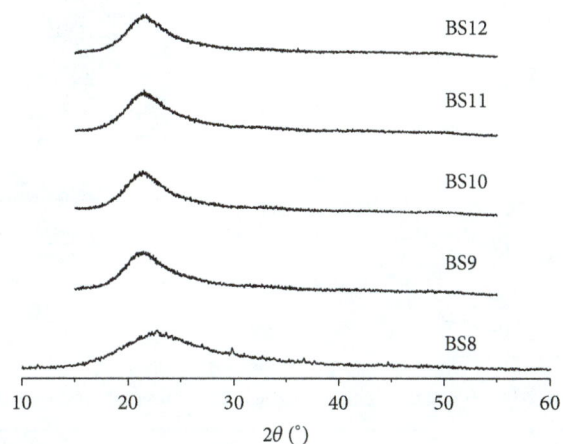

FIGURE 5: XRD patterns of the synthesis products obtained by hydrothermal treatment of sugarcane waste ash at 90°C.

40.2°, 42.4°, 45.8°, and 50.2° are typical of quartz. The crystalline phase of silica in the sugarcane ash is related to the conditions of combustion (mainly time and temperature). At temperatures of 500–800°C or when the exposure to high temperature is small, the silica contained in the ash is predominantly amorphous. At temperatures greater than 800°C, the amorphous silica present in sugarcane is converted in crystalline silica polymorphs, such as quartz, cristobalite, and tridymite [18].

The discrete particle diameter distribution and the cumulative distribution of SWA were shown in Figure 3. The diameters corresponding to the cumulative distribution in the contents of 10.0%, 50.0%, and 90.0% ($d_{0.1}$, $d_{0.5}$, $d_{0.9}$, resp.) and Sauter mean diameter ($d[3, 2]$) presented the following values: 12.170, 62.528, 208.140, and 26.084 μm, respectively. The range of the estimated particle size was between 0.893 and 563.677 μm.

3.2. X-Ray Diffraction Analysis of Biomass Ash-Based Products.

Tests were performed using different experimental conditions in order to determine the optimal conditions of synthesis for obtaining the biosilica. The broad X-ray diffraction peak at theta equal to 22 degrees confirms the formation of amorphous silica, in general [7]. The results are summarized in Figures 4 and 5. The yield of the synthesis was estimated using the mass of the actual product obtained (Table 1).

In the set of experiments at 100°C (Table 1, BS1–BS7 samples, Figure 4), XRD showed that the extracted silica is predominantly amorphous for BS2 to BS5 samples. For shorter reaction time (1 h) there was no gel formation (BS1 sample). A difference in the patterns of diffraction was recognized after 24 h of hydrothermal treatment due to the formation of only crystalline material (BS6 sample). There was no gel formation in BS7 sample, and therefore it did not form silica. This fact suggests that the alkaline fusion method should be applied to transform the high crystalline silica content of SWA in sodium silicate solution prior to the hydrothermal step.

In the set of experiments at 90°C (Table 1, BS8–BS12, Figure 5), essentially noncrystalline (amorphous) biosilica was formed (except BS8 sample) with yield directly proportional to the time of hydrothermal treatment. XRD of BS8 sample show amorphous silica along with some crystalline silica.

The greatest silica yield (91%) was obtained for the synthesis with aging process at 80°C (BS12 sample). Furthermore, the gelation time presented a drastic reduction (at least 20 h for 1 h) and the synthesis was reproducible. Aside from biosilica xerogel, 1.663 g of sodium acetate can also be derived from synthesis through evaporation of the aqueous solution (Figure 1). Thus, besides the mineralogical analysis, other characterizations were carried out for the silica gel obtained by this experimental procedure.

3.3. Characterization of Biosilica Gel

3.3.1. Physical and Chemical Characteristics.
X-Ray fluorescence showed that the produced biosilica had high purity (99.1%), free of MgO, P_2O_5, TiO_2, SO_3, MnO, Ta_2O_5, and

(a)

(b)

(c)

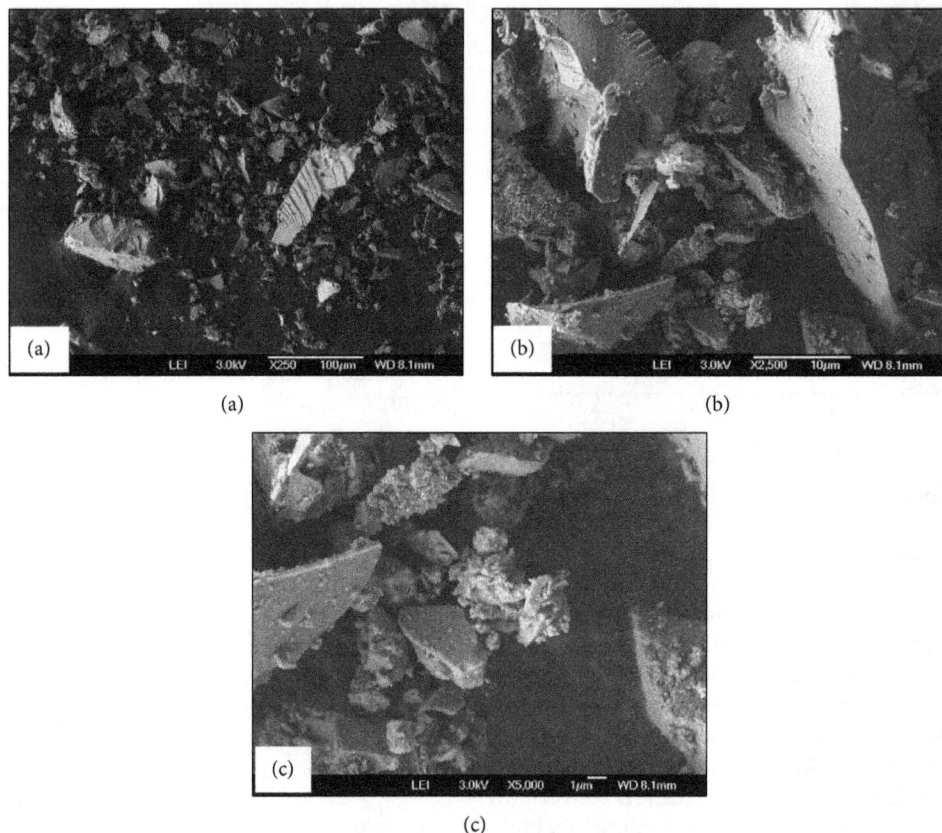

FIGURE 6: Scanning electron microscopy of biosilica. Magnification of 250x (a), 2500x (b), and 5000x (c).

TABLE 3: Physical-chemical properties of biomass ash and biosilica.

	SWA	Biosilica
pH	5.70	7.80
pH_{PZC}[b]	—	5.29
CEC^a (meq 100 g)	—	4.25
Bulk density (g cm^{-3})	1.41	0.708

[a]CEC = cationic exchange capacity; [b]pH_{PZC} = point of zero charge.

As_2O_3 that were present in the sugarcane waste ash (Table 2). Also, minor amounts of Na_2O, Al_2O_3, and Cl and traces of Fe, K, Ca, Ni Cu, and Zn were found in synthesized material. Some impurities were resulting from the sodium hydroxide used in the synthesis. The high purity was obtained after washing the silica gel (B12 sample) repeatedly with deionized water to remove sodium salt [8–10].

The other physical-chemical properties of biosilica and the ash that served as raw material are shown in Table 3. The pH of the leached of the ash is slightly acid and the pH of silica gel is slightly alkaline. To produce silica, it is necessary to add acid to silicate solution. When acid was added to silicate solution (pH > 11), silica gel started to form rapidly as soon as pH 10 was reached and formed a rigid gel at pH ~ 7.0. The incomplete neutralization or the presence of residual OH^- may be responsible for slightly alkaline pH

of the leached biosilica. The surface of the biosilica has a negative charge in aqueous solution because the pH value is above the pH in the point of zero charge (pH_{PZC}) indicating that is a material with cationic exchange capacity. The bulk density of the biosilica was lower than 1.3 g cm^{-3}, showing that this product is classified as lightweight material. This property may be beneficial for the use of the prepared gel in many environmental and industrial applications [19].

3.3.2. Morphology Analysis of Silica Gel. The surface morphology of the biosilica was obtained using SEM images with different magnifications (Figure 6). The silica particles with size between <1 to 100 μm were grouped into clusters, forming an irregular and cohesive surface. In addition, smooth surfaces were also observed. The agglomeration of particles forming clusters is a result of alkaline treatment and the lack of uniformity is a result of the laboratory grinding process.

3.3.3. Thermogravimetric Analysis. The thermogravimetry profiles of sugarcane waste ash (SWA), biosilica (BS), and biosilica repeatedly washed with water (BSW) are shown in Figure 7. For the BS and BSW sample, the first mass loss (33–90°C) is ascribed to moisture loss [20]. The second mass loss (90–180°C) corresponds to the loss of physically adsorbed water from the surface [21]. The third step (180–530°C) was related to the loss of the organic structure (CH_3COONa)

TABLE 4: Assignments of the IR vibrations of studied samples.

Frequency (cm^{-1})	Assignment	Spectra of samples*	References
400	Si-O-M (metal impurities bonding)	SWA	[26]
450–455	Si-O-Si symmetric stretching (siloxane group)	All samples	[23–25]
520	Al-O-Si bending vibrations	SWA	[22]
550–560	(Si)-Al-O stretching	BS and BSW	[29]
615	CO$_2$ rocking (acetate group)	BS	[30]
645	CO$_2$ deformation torsion about C-C bond (acetate group)	BS	[30]
690	Si-O-Al deformation	SWA	[21]
780	Si-O stretching showed presence of quartz	SWA	[22]
790–797	Si-O-Si symmetric stretching (siloxane group)	BS and BSW	[24, 25, 27]
925	C-C stretching (acetate group)	BS	[30]
957	Si-OH bending vibrational absorption (silanol group)	BSW	[25, 27]
1050–1070	Si-O-Si asymmetric stretching	All samples	[23, 25, 31]
1415	C-O stretching (acetate group)	BS	[30]
1560	C-O stretching (acetate group)	BS	[30]

*SWA = sugarcane waste ash; BS = biosilica; BSW = biosilica washed with water.

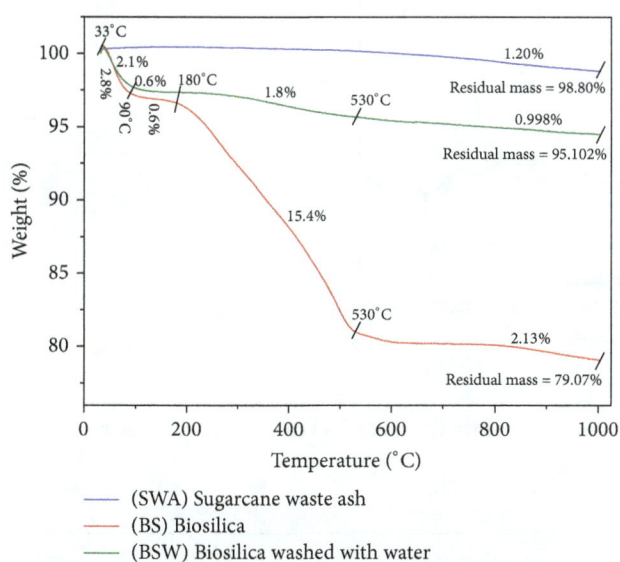

FIGURE 7: TG curves of SWA (blue), BS (red), and BSW (green) under an inert atmosphere of nitrogen.

and the loss of the chemically adsorbed water bonded to Si-OH (silanol) through the hydrogen bond [21, 22]. The fourth weight loss (530–1000°C) is due to the surface dehydroxylation reaction [22]. For the SWA, only a mass loss of 1.20% probably related to moisture loss was observed.

3.3.4. FTIR Analysis. The FTIR spectra of sugarcane waste ash (SWA), biosilica (BS), and biosilica after washing repeatedly with deionized water (BSW) are shown in Figures 8(a), 8(b), and 8(c), respectively. For all samples, the bands observed at about 450–455 cm^{-1} and 1050–1070 cm^{-1} were attributed to the Si-O-Si symmetric and asymmetric stretching, respectively [8, 23–25].

According to the literature, the bands observed at 400, 520, 690, and 780 cm^{-1} for SWA sample were due to Si-O-M (metal impurities bonding), Al-O-Si bending vibrations, Si-O-Al deformation, and Si-O stretching indicating the presence of crystalline quartz, respectively [21, 22, 26, 27]. These results are in agreement with chemical analysis (Table 2) and XRD patterns (Figure 2).

The appearance of absorption bands at 550 and 560 cm^{-1} assigned to the (Si)-Al-O stretching indicated the presence of small content of aluminum in BS and BSW samples [28]. Also, this is in agreement with chemical analysis.

The presence of band about 790–797 cm^{-1} for BS and BSW samples was due to Si-O-Si symmetric stretching [21, 29]. The bands at 615, 645, 925, 1415, and 1560 cm^{-1}, for BS sample, were due to the presence of sodium acetate (CH$_3$COONa), which is impurity of the synthesis process [30]. The absence of these peaks in BSW sample confirmed once again that the repeated washing step removes completely soluble salts of synthesized silica gel.

The band at 957 cm^{-1} for BSW sample is due to the presence of silanol group Si-OH bending vibration absorption [27, 29]. According to the TGA thermogram (Figure 7), BS sample also has the silanol group in its structure; however the silanol band was probably overlaid by the C-C stretching band at 925 cm^{-1}. Table 4 shows FTIR assignments of the peaks observed in Figures 8(a), 8(b), and 8(c).

3.3.5. N$_2$ Adsorption-Desorption Isotherms. Figure 9 presents the nitrogen adsorption/desorption isotherms measured at 77 K for the sugarcane waste ash and biosilica. The sugarcane waste ash exhibits a typical type II isotherm of nonporous materials according to the IUPAC classification (Figure 9(a)) [31, 32].

The isotherm obtained for BS (Figure 9(b)) was attributed to type IV, which is associated with a complex mesoporous structure where the distribution of pore size and shape were not well defined. In addition, the multilayer adsorption might

(a)

(b)

(c)

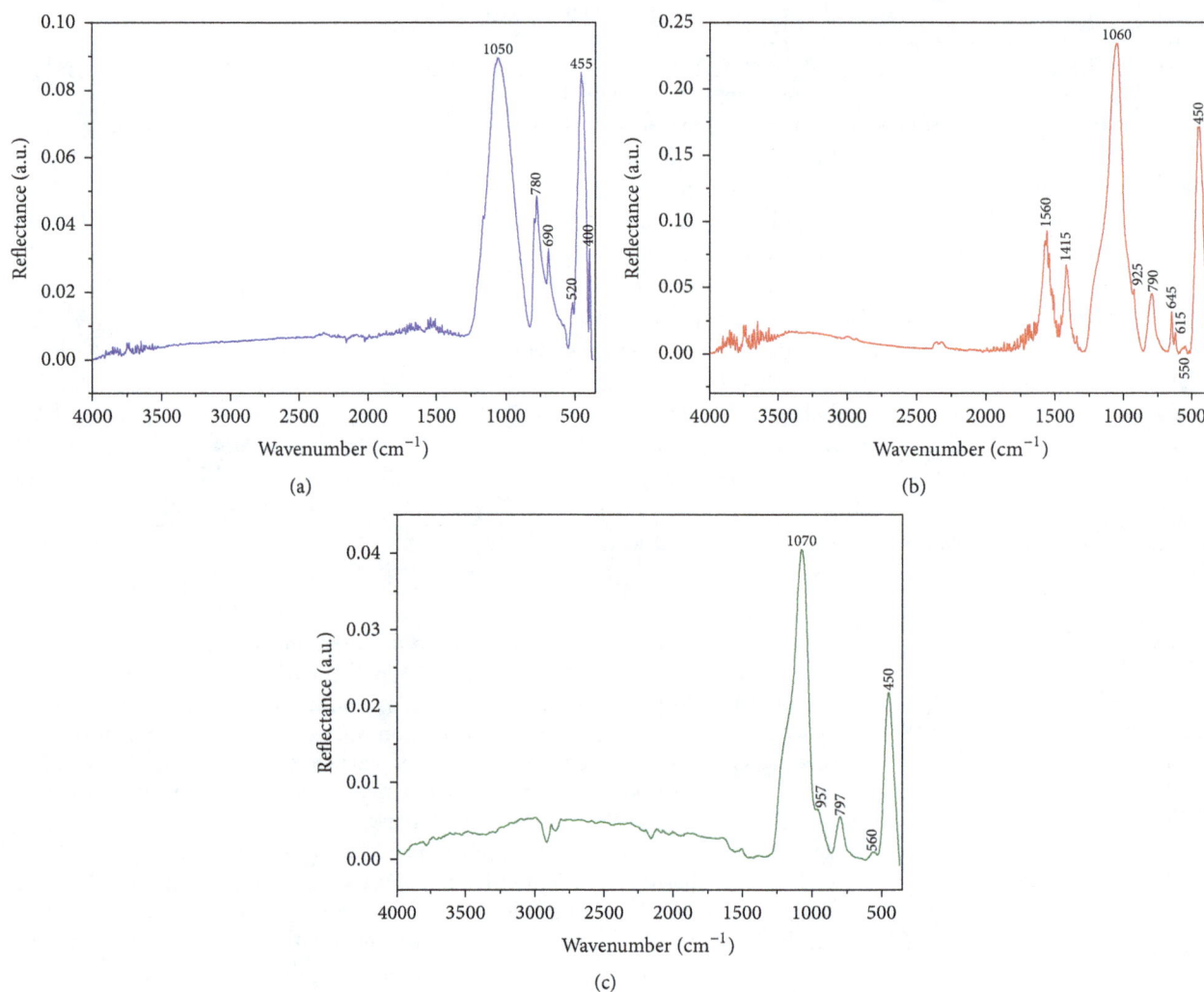

FIGURE 8: FTIR vibrational spectra for the samples: sugarcane waste ash (a), biosilica (b), and biosilica washed with water (c).

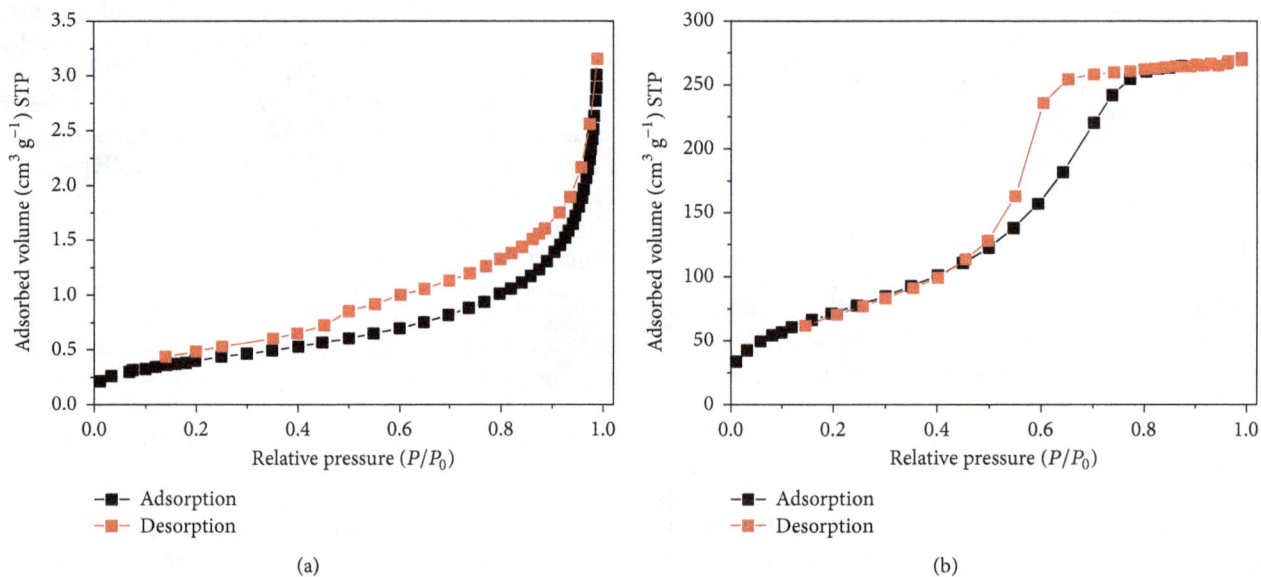

(a)

(b)

FIGURE 9: Nitrogen adsorption-desorption isotherms for the samples: ash (a) and biosilica (b).

TABLE 5: Textural properties of biosilica and ash that serve as raw material.

Samples	S_{BET} [a] $(m^2\,g^{-1})$	V_p [b] $(cm^3\,g^{-1})$	D_p [c] (nm)
Biosilica	265	0.425	6.250
Ash	1.5	0.0049	10.790

[a] S_{BET} = specific area; [b] V_p = total pore volume; [c] D_p = average pore diameter.

occur in the middle relative pressure range. The desorption cycle of the isotherm displays a H2 type of the hysteresis loop, which is often associated with a complex mesoporous structure where the distribution of pore size and shape is not well defined [33, 34].

The surface area of all the materials was evaluated using BET method and the pore size by BJH method (isotherms with hysteresis), and their textural properties are summarized in Table 5. It is seen that the sugarcane waste ash shows low specific surface area and porosities. The synthesized silica showed high specific area with dimensions and pore volume within the specifications of mesoporous materials, which can also be proved by type IV isotherms, characterizing its mesoporosity in Figure 9.

4. Conclusions

Pure amorphous silica was successfully extracted at a 99.1% yield from sugarcane waste ash generated by a two-stage heating process followed by precipitation by acidification of the sodium silicate solution, so obtained. The XFR, FTIR, and XRD data gave clear evidence to the high purity of amorphous silica extracted from sugarcane waste ash. The use of the one-step process did not result in the formation of biosilica. Experimental results show that optimum conditions for producing biosilica were 90°C for 20 h in the hydrothermal treatment of two-step method, aging temperature of 80°C, and an aging time of 1 h. The mesoporous structure of silica makes it a potential catalytic (or catalyst support) or adsorbent material.

Conflicts of Interest

The authors declare that there are no conflicts of interest regarding the publication of this paper.

Acknowledgments

The authors are grateful to Conselho Nacional de Desenvolvimento Científico e Tecnológico (CNPq) and Coordenação de Aperfeiçoamento de Pessoal de Nível Superior (CAPES) for financial support, Dr. Paola Corio and Dr. Jonnatan J. Santos (Institute of Chemistry, USP) for SEM and FTIR-ATR analysis, Dr. M. A. Logli for nitrogen adsorption desorption isotherms analysis, and COSAN S. A. (São Paulo, Brazil) for supplying the sugarcane waste ash.

References

[1] Conab—Companhia Nacional de Abastecimento/Acompanhamento da safra brasileira: Cana-de-açúcar, Terceiro levantamento, V. 3. Safra - no. 3, 2016.

[2] UNICA—União da Indústria de Cana-de-Açúcar (Bioeletricidade) http://www.unica.com.br/colunas/470156692036979688/bioeletricidade-por-cento3A-o-que-falta-para-esta-alternativa/.

[3] A. Sales and S. A. Lima, "Use of Brazilian sugarcane bagasse ash in concrete as sand replacement," Waste Management, vol. 30, no. 6, pp. 1114–1122, 2010.

[4] G. C. Cordeiro, R. D. Toledo Filho, L. M. Tavares, and E. M. R. Fairbairn, "Pozzolanic activity and filler effect of sugar cane bagasse ash in Portland cement and lime mortars," Cement and Concrete Composites, vol. 30, no. 5, pp. 410–418, 2008.

[5] G. C. Cordeiro, R. D. Toledo Filho, L. M. Tavares, and E. D. M. R. Fairbairn, "Ultrafine grinding of sugar cane bagasse ash for application as pozzolanic admixture in concrete," Cement and Concrete Research, vol. 39, no. 2, pp. 110–115, 2009.

[6] Y. Wang, A. Kalinina, T. Sun, and B. Nowack, "Probabilistic modeling of the flows and environmental risks of nano-silica," Science of the Total Environment, vol. 545-546, pp. 67–76, 2016.

[7] S. R. Kamath and A. Proctor, "Silica gel from rice hull ash: preparation and characterization," Cereal Chemistry, vol. 75, no. 4, pp. 484–487, 1998.

[8] U. Kalapathy, A. Proctor, and J. Shultz, "A simple method for production of pure silica from rice hull ash," Bioresource Technology, vol. 73, no. 3, pp. 257–262, 2000.

[9] U. Kalapathy, A. Proctor, and J. Shultz, "Production and properties of flexible sodium silicate films from rice hull ash silica," Bioresource Technology, vol. 72, no. 2, pp. 99–106, 2000.

[10] U. Kalapathy, A. Proctor, and J. Shultz, "An improved method for production of silica from rice hull ash," Bioresource Technology, vol. 85, no. 3, pp. 285–289, 2002.

[11] L. S. Ferret, "Obtenção de zeólitas e sílica de cinzas de carvão," Anais do V Congresso Brasileiro de Carvão Mineral, ISBN: 978-85-66380-02-6, Editora SATC (Educação & Tecnologia), Gramado, RS, Brazil, pp. 404–409, 2013.

[12] T. C. R. Bertolini, J. C. Izidoro, R. R. Alcântara, L. C. Grosche, and D. A. Fungaro, "Surfactant-modified zeolites from coal fly and bottom ash as adsorbents for removal of crystal violet from aqueous solution," ActaVilet, vol. 1, no. 4, pp. 78–94, 2015.

[13] D. L. Anderson, "Soil and leaf nutrient interactions following application of calcium silicate slag to sugarcane," Fertilizer Research, vol. 30, no. 1, pp. 9–18, 1991.

[14] A. G. Sangster, M. J. Hodson, and H. J. Tubb, "Silicon deposition in higher plants," in Silicon in Agriculture, G. H. Datnoff and G. H. Snyder, Eds., vol. 8, pp. 85–113, Elsevier, 2001.

[15] V. V. Matichenkov and D. V. Calvert, "Silicon as a beneficial element for sugarcane," Journal of the American Society of Sugar Cane Technologists, vol. 22, pp. 21–30, 2002.

[16] R. J. Schaetzl and S. Anderson New York: Cambridge University Press, New York, NY, USA, 2005.

[17] B. T. Tubana and J. R. Heckman, "Silicon in soils and plants," in Silicon and Plant Diseases, F. A. Rodrigues and L. E. Datnoff, Eds., chapter 2, pp. 7–51, Springer International Publishing, Switzerland, Europe, 2015.

[18] J. S. Le Blond, C. J. Horwell, B. J. Williamson, and C. Oppenheimer, "Generation of crystalline silica from sugarcane burning," Journal of Environmental Monitoring, vol. 12, no. 7, pp. 1459–1470, 2010.

[19] A. Sdiri, T. Higashi, S. Bouaziz, and M. Benzina, "Synthesis and characterization of silica gel from siliceous sands of southern Tunisia," Arabian Journal of Chemistry, vol. 7, no. 4, pp. 486–493, 2014.

[20] L. Franken, L. S. Santos, E. B. Caramão, T. M. H. Costa, and E. V. Benvenutti, "Xerogel p-anisidinapropilsílica: estudo da estabilidade térmica e da resistência à lixiviação com solventes," *Química Nova*, vol. 25, no. 4, pp. 563–566, 2002.

[21] M. A. Girsova, G. F. Golovina, I. A. Drozdova, I. G. Polyakova, and T. V. Antropova, "Infrared studies and spectral properties of photochromic high silica glasses," *Optica Applicata*, vol. 44, no. 2, pp. 337–344, 2014.

[22] H. Wanyika, E. Maina, A. Gachanja, and D. Marika, "Instrumental characterization of montmorillonite clays by X-ray fluorescence spectroscopy, fourier transform infrared spectroscopy, X-ray diffraction and uv/visible spectrophotometry," *Jomo Kenyatta University of Agriculture and Technology*, vol. 17, no. 1, pp. 224–239, 2016.

[23] S. Hu and Y.-L. Hsieh, "Preparation of activated carbon and silica particles from rice straw," *ACS Sustainable Chemistry and Engineering*, vol. 2, no. 4, pp. 726–734, 2014.

[24] A. Mourhly, M. Khachani, A. E. Hamidi, M. Kacimi, M. Halim, and S. Arsalane, "The Synthesis and characterization of low-cost mesoporous silica SiO_2 from local pumice rock," *Nanomaterials and Nanotechnology*, vol. 5, no. 35, pp. 1–7, 2015.

[25] A. F. Boza, V. L. Kupfer, A. R. Oliveira et al., "Synthesis of α-aminophosphonates using a mesoporous silica catalyst produced from sugarcane bagasse ash," *RSC Advances*, vol. 6, no. 29, pp. 23981–23986, 2016.

[26] A. F. Hassan, A. M. Abdelghny, H. Elhadidy, and A. M. Youssef, "Synthesis and characterization of high surface area nanosilica from rice husk ash by surfactant-free sol—gel method," *Journal of Sol-Gel Science and Technology*, vol. 69, no. 3, pp. 465–472, 2014.

[27] P. Su, R. Wang, Y. Yu, and Y. Yang, "Microwave-assisted synthesis of ionic liquid-modified silica as a sorbent for the solid-phase extraction of phenolic compounds from water," *Analytical Methods*, vol. 6, no. 3, pp. 704–709, 2014.

[28] P. L. King, P. F. McMillan, and G. M. Moore, "Infrared spectroscopy of silicate glasses with application to natural systems," *Infrared Spectroscopy of Silicate Glasses with Application to Natural Systems*, Chapter 4, pp. 93–134, 2004.

[29] R. Yuvakkumar, V. Elango, V. Rajendran, and N. Kannan, "High-purity nano silica powder from rice husk using a simple chemical method," *Journal of Experimental Nanoscience*, vol. 9, no. 3, pp. 272–281, 2014.

[30] L. H. Jones, "Infrared spectra and structure of the crystalline sodium acetate complexes of U(VI), Np(VI), Pu(VI), and Am(VI). A comparison of metal-oxygen bond distance and bond force constant in this series," *The Journal of Chemical Physics*, vol. 23, no. 11, pp. 2105–2107, 1955.

[31] K. S. Sing, "Reporting physisorption data for gas/solid systems with special reference to the determination of surface area and porosity (Recommendations 1984)," *Pure and Applied Chemistry*, vol. 57, no. 4, pp. 603–619, 1985.

[32] J. Rouquerol and F. Rouquerol, "Adsorption at the liquid-solid interface: thermodynamics and methodology," in *Adsorption by Powders and Porous Solids (Second Edition)*, F. R. R. S. W. S. L. Maurin, Ed., pp. 105–158, Academic Press: Oxford, 2014.

[33] K. S. W. Sing, "Reporting physisorption data for gas/solid systems with special reference to the determination of surface area and porosity (Provisional)," *Pure and Applied Chemistry, Pergamon Press*, vol. 54, no. 11, pp. 2201–2218, 1982.

[34] S. D. Bhagat, Y.-H. Kim, M.-J. Moon, Y.-S. Ahn, and J.-G. Yeo, "A cost-effective and fast synthesis of nanoporous SiO_2 aerogel powders using water-glass via ambient pressure drying route," *Solid State Sciences*, vol. 9, no. 7, pp. 628–635, 2007.

6

New Sustainable Biosorbent Based on Recycled Deoiled Carob Seeds: Optimization of Heavy Metals Remediation

M. Farnane,[1] A. Machrouhi,[1] A. Elhalil,[1] M. Abdennouri,[1] S. Qourzal,[2] H. Tounsadi (ID),[1,3] and N. Barka (ID)[1]

[1]Laboratoire des Sciences des Matériaux, des Milieux et de la Modélisation (LS3M), FPK, Univ Hassan 1, B.P. 145, 25000 Khouribga, Morocco
[2]Equipe de Catalyse et Environnement, Département de Chimie, Faculté des Sciences, Université Ibn Zohr, B.P. 8106 Cité Dakhla, Agadir, Morocco
[3]Université Sidi Mohamed Ben Abdellah, Faculté des Sciences Dhar Elmehraz, Laboratoire d'Ingénierie, d'Electrochimie, de Modélisation et d'Environnement, Fès, Morocco

Correspondence should be addressed to N. Barka; barkanoureddine@yahoo.fr

Academic Editor: Adina Negrea

In this study, an efficient biosorbent was developed from deoiled carob seeds, a agroindustrial waste. The biosorption efficiency was evaluated for cadmium and cobalt ions removal from aqueous solution under various parameters such as treating agent, solution pH, biosorbent dosage, contact time, initial metal ions concentration, and temperature. The effect of some major inorganic ions including Na^+, K^+, Ca^{2+}, Mg^{2+}, and Al^{3+} on the biosorption was also established. Based on this preliminary study, four independent variables including solution pH, biosorbents dosage, initial metal concentration, and treating agent were chosen for the optimization of the process using full-factorial experimental design. It was found that chemical pretreatment of the raw deoiled carob seeds with NaOH strongly enhances its biosorption potential. Thus, the optimal conditions for high biosorption of cadmium(II) and cobalt(II) were achieved at pH of 6, biosorbent dosage of 1 g/L, and initial metal concentration of 50 mg/L. The biosorbents were characterized by Fourier transform infrared spectroscopy (FT-IR), scanning electron microscopy (SEM), energy dispersive X-ray (EDX), Boehm titration, and the point of zero charge (pH_{PZC}).

1. Introduction

Heavy metal pollution imposes ecological and public health problems according to hazardous and irrecoverable effects of metal ions on the environment and aquatic ecosystems [1]. Besides, the toxic and harmful effects to organisms living in water, heavy metals also accumulate throughout the food chain and may affect human beings [2]. In this way, the removal and recovery of heavy metals from aqueous effluents before being disposed in the environment is required [3].

Various methods have been used to remove heavy metal ions such as chemical precipitation [4], membrane filtration [5], ion exchange [6], solvent extraction [7], flotation [8], and electrochemical treatment [9]. Among all these mentioned methods, sorption is an effective and eco-friendly method for the removal of heavy metal ions from wastewaters due to its simple design, easy operation handling, and availability of different sorbents with large efficacity to remove a wide range of heavy metals [10–12]. Activated carbons are widely used as adsorbents in wastewater treatment because of their high surface areas and active functional groups, but their high cost inhibits sometimes their use [13]. Therefore, there is a need to develop other biosorbents from alternative low-cost raw materials for the same role as activated carbon. Recently, low-cost precursors have become the focus of researchers for example, chitosans [14], bark of *Pinus elliottii* [15], *Jatropha curcas* [16], sugarcane bagasse [17], *Eichornia crassipes* [18], coconut [19], agricultural peels [20–22], sunflower stalks [23], raw carob shells and chemically treated carob shells [24], *Diplotaxis harra*, and *Glebionis coronaria* L [25].

The aim of this study is to assess the applicability of chemically treated deoiled carob seeds for the removal of cadmium(II) and cobalt(II) ions from aqueous solution. Biosorption studies were carried out under various parameters such as solution pH, biosorbents dosage, contact time, initial metal ions concentration, and temperature. The biosorption kinetic data were tested by pseudo-first-order and pseudo-second-order kinetic models. The equilibrium data were analyzed using Langmuir and Freundlich models. This paper also discussed the combined effect of the most influencing parameters, which are solution pH, biosorbent dosage, concentration of the solution, and the treating agent. Full-factorial experimental design with two levels (2^4) and surface response methodology were used to acquire the optimal parameters for high removal of Cd(II) and Co(II) ions.

2. Experimental

2.1. Materials. All the chemicals used in the preparation and the biosorption studies were of analytical grade. Cd$(NO_3)_2 \cdot 4H_2O$ (98%), Co$(NO_3)_2 \cdot 6H_2O$ (98%), NaCl (99.5%), Al$(NO_3)_3 \cdot 9H_2O$ (≥98%), HCl (37%), Na_2CO_3, and $NaHCO_3$ were obtained from Sigma-Aldrich (Germany). Mg$(NO_3)_2 \cdot 6H_2O$ (97%) was provided from SDS (France), Ca$(NO_3)_2 \cdot 4H_2O$ and HNO_3 (65%) from Scharlau (Spain), and NaOH (98%) was provided from Merck (Germany).

2.2. Preparation of the Biosorbents. The deoiled residue was obtained as a by-product from the hydrodistillation process of carob seeds from the region of Khenifra in Morocco. The biomass was repeatedly washed with demineralized water and then oven dried at about 120°C for 24 h min in order to remove excess moisture. The dried biomass was then grounded using mortar and pestle and sieved to get a size fraction lower than 160 µm, referred as Raw-seeds. For the chemical treatment, 10 g of the Raw-seeds was treated with 100 mL of 1 M solution of HCl or NaOH for 2 h. The biosorbents were then filtered and washed with distilled water until neutral pH. The pretreated biosorbents were then dried in an oven at 120°C for 24 h and stored in glass bottles under following names HCl-seeds and NaOH-seeds for further use.

2.3. Characterisation. FTIR transmittance spectra of the biosorbents were recorded in the region of 4000–400 cm^{-1} using a Scotech-SP-1 spectrophotometer. Basic and oxygenated acidic surface groups were assessed by Boehm titrations [26]. About 0.1 g of each sample was mixed with 50 mL of 0.01 M aqueous reactant solution (NaOH, Na_2CO_3, $NaHCO_3$, or HCl). The mixtures were stirred at 500 rpm for 24 h at room temperature. Then, the suspensions were filtrated by a 0.45 µm membrane filter. To determine the oxygenated groups content, back titrations of the filtrate (10 mL) were achieved with standard 0.01 M·HCl solution.

Basic groups contents were also determined by back titration of the filtrate with 0.01 M·NaOH solution. The morphological characteristics were analyzed by scanning electron microscopy (SEM). Small amount of each sample was finely powdered and mounted directly onto aluminum sample holder using the two-sided adhesive carbon model. Energy dispersive X-ray (EDX) was also performed to determine the elemental composition of the raw carob seeds and the both treated samples. The point of zero charge (pH$_{PZC}$) was determined by the pH drift method according to the method proposed by Noh and Schwarz [27]. The pH of NaCl aqueous solution (50 mL at 0.01 mol/L) was adjusted to successive initial values in the range from 2.0 to 12.0 by addition of HNO_3 (0.1 N) and/or NaOH (0.1 N). Furthermore, 0.05 g of each biosorbent was added in 50 mL of solution and stirred for 6 h. The final pH was measured and plotted against the initial pH. The pH$_{PZC}$ was determined at the value for which pH$_{final}$ = pH$_{initial}$.

2.4. Batch Biosorption Procedure. Stock solutions were prepared by dissolving desired weight of each metal ion in distilled water, and necessary concentrations were obtained by dilution. Biosorption experiments were investigated in a series of beakers containing 100 mL of the metal ion solution at desired concentration and desired weight of the biosorbent. The mixtures were stirred for 2 h at 500 rpm using a magnetic stirrer. The influence of pH was performed by varying the pH from 2.0 to 7.0 at an initial metal concentration of 100 mg/L. The pH of the solutions was adjusted with either 0.1 M of HCl or 0.1 M of NaOH and using a SensION + PH31 pH meter. The biosorbent dosage was varied from 0.5 to 5 g/L. The contact time was varied between 5 and 210 min at room temperature with initial pH solution. Biosorption equilibrium was established for different metal ion concentration between 20 and 200 mg/L. The effect of temperature was tested from 10 to 50°C using a thermostatically controlled incubator. The effect of some major inorganic ions including Na$^+$, K$^+$, Mg^{2+}, Ca^{2+}, and Al^{3+} on the biosorption was studied for each heavy metal at a constant initial metal concentration of 100 mg/L. The concentration of each ion was varied from 10 to 100 mg/L.

After each biosorption experiment completed, the solid phase was separated from the liquid phase by centrifugation at 3000 rpm for 10 min. Metal ions concentration was determined using a PerkinElmer atomic absorption spectrophotometer (Analyst 200).

The biosorption capacity and biosorption removal efficiency were calculated using the following equations:

$$q_{t,e} = \frac{\left(C_0 - C_{t,e}\right)}{R}, \tag{1}$$

$$\% \text{ removal} = \frac{\left(C_0 - C\right)}{C_0} * 100, \tag{2}$$

where $q_{t,e}$ (mg/g) is the biosorbed quantity at any time or at equilibrium, C_0 (mg/L) is the initial metal ion concentration, $C_{t,e}$ (mg/L) is the metal ion concentration at a time t or at equilibrium, and R (g/L) is the mass adsorbents per liter of solution.

Kinetic and equilibrium parameters were estimated with the aid of the nonlinear regression method using Origin 6.0 software.

2.5. Design of Experiment for the Optimization of Cd(II) and Co(II) Biosorption. The methodology of experimental design was used for modeling and optimization of the biosorption processes of Cd(II) and Co(II) ions from aqueous solutions. The four most influencing factors are solution pH (A), biosorbents dosage (B), initial metals concentration (C), and treating agent (D). The values of variable levels are presented in Table 1. The experiments were made according to a full-factorial design at two levels (2^4), with 16 experiments.

In addition, a first-order polynomial model was also used for modeling sorption of Cd(II) or Co(II) ions. The general equation of the first-order polynomial model is presented in

$$Y = b_0 + b_1 A + b_2 B + b_3 C + b_4 D + b_{12} AB + b_{13} AC$$
$$+ b_{14} AD + b_{23} BC + b_{24} BD + b_{123} ABC + b_{124} ABD$$
$$+ b_{134} ACD + b_{234} BCD + b_{1234} ABCD,$$

$$(3)$$

where Y (mg/g) is the responses of interest (adsorption capacity of Cd(II) ($Y1$) and adsorption capacity of Co(II) ($Y2$)).

The results were analyzed using the Trial software Design Expert 10.0.0.

3. Results and Discussion

3.1. Characterization

3.1.1. FT-IR Analysis of the Biosorbents. The infrared spectra of Raw-seeds, NaOH-seeds, and HCl-seeds are illustrated in Figure 1. The figure shows broad absorption band for Raw-seeds at 3200–3600 cm^{-1} due to the stretching of the N–H bond of amino groups and indicative of bonded hydroxyl group [11], and this band was separated into more resolute three bands after the chemical pretreatments with NaOH and HCl. Two bands at around 3200–3400 cm^{-1} indicate the presence of carboxylic acid and amino groups, and the other one near 3500 cm^{-1} is related to the OH stretching vibration mode in alcohol and phenol groups. The band at 2930 cm^{-1} corresponds to the symmetrical and asymmetrical–CH– vibrations in lipids. The peaks located at 1620 cm^{-1} are characteristics of C=O stretching for aldehydes and ketones, which can be conjugated or nonconjugated to aromatic rings [28]. The–C–O, C–C, and–C–OH stretching vibrations can be attributed to peaks in the region of 1180–1048 cm^{-1}. The spectra showed bands located at 630 cm^{-1}, assigned to OH− ions.

3.1.2. Boehm Titration of the Biosorbents. The oxygen functional groups are very important characteristics of the biosorbents because they determine the surface properties

TABLE 1: Process factors and their levels.

Factors	Levels	
	Low (−)	High (+)
A. pH	4	6
B. Biosorbent dose (g/l)	1	3
C. Initial concentration (mg/l)	50	100
D. Treating agent	HCl	NaOH

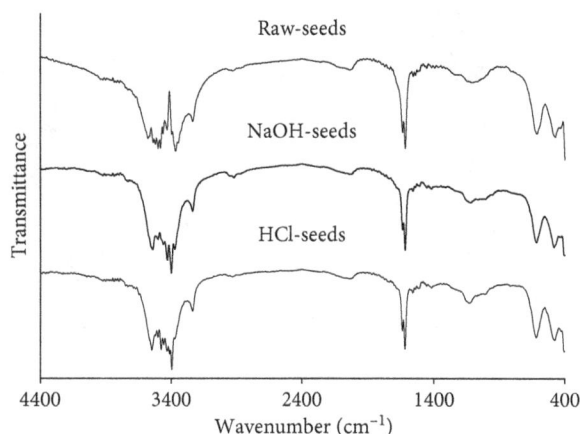

FIGURE 1: FT-IR spectra of Raw-seeds, HCl-seeds, and NaOH-seeds biosorbents.

and hence their quality as biosorbents. These functional groups are mainly divided as acidic or basic, which affect the surface charge and consequently the biosorption capability of the biosorbents. The Boehm's titration method provides qualitative and quantitative information regarding the total amount of basic groups and the amounts of acidic functional groups such as carboxylic, lactonic, and phenolic. From Table 2, it can be seen that Raw-seeds, HCl-seeds, and NaOH-seeds' surface constituted mainly of acidic groups, which are due to phenolic, lactonic, and carboxylic groups and a less quantity of basic groups. So, the surface of these biosorbents is acidic. The biosorbents having greater surface acid groups have higher cation exchange properties. According to the experimental data, the HCl-seeds had an important amount of acidic groups than Raw-seeds followed by NaOH-seeds. The use of chemical reagents acid in the treatment process produces an increase in the amount of acid groups present in the biosorbent surface. It was observed that the concentration of lactonic and phenolic groups in HCl-seeds is higher than those of carboxylic groups.

3.1.3. Morphology of the Biosorbents. The surface texture and morphology of biosorbents were analyzed by SEM in order to compare the morphology of raw and chemically treated carob seeds. The SEM images of these biosorbents are depicted in Figure 2. As it is clearly shown, there is a significant difference among the tree samples. In fact, no obvious pores can be seen for Raw-seeds. Then, for the HCl-seeds, the surface morphology does not have well-defined pores. However, the NaOH-seeds' surface indicates some

TABLE 2: Chemical groups on the surface of the biosorbents.

Biosorbent	Carboxylic groups (meq/g)	Lactonic groups (meq/g)	Phenolic groups (meq/g)	Total acid groups (meq/g)	Total basic groups (meq/g)
Raw-seeds	0.4070	0.4880	0.4900	1.3850	0.3750
HCl-seeds	0.3810	0.5150	0.5240	1.4200	0.3650
NaOH-seeds	0.4090	0.4830	0.4990	1.3910	0.3870

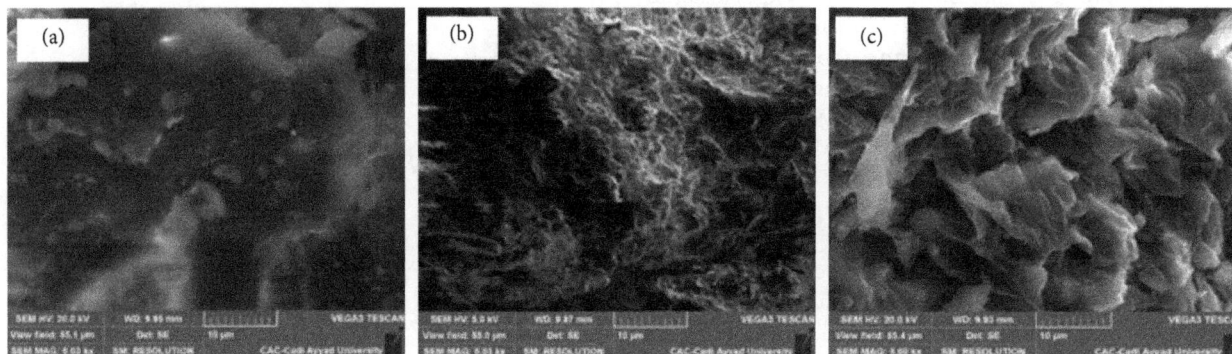

FIGURE 2: SEM images of (a) Raw-seeds, (b) HCl-seeds, and (c) NaOH-seeds.

irregular cavities and a changing in the external texture as a result of the reaction between the raw material and the treating agent. These different characteristics allow NaOH-seeds to contribute a high adsorption performance of heavy metals. These results suggest that the NaOH is an effective agent for creating well-developed pores on the surface of the raw material, which is already shown for the alkaline-treated carob shells [24].

3.1.4. EDX Analysis.

3.1.4. EDX Analysis. EDX is an analytical technique to identify the element presence on the material surface based on its characteristic X-ray energy. This technique is normally coupled with SEM analysis to gain more complete result. The elemental compositions of Raw-seeds, NaOH-seeds, and HCl-seeds are tabulated in Table 3. In addition, the elemental compositions are presented under peaks in Figure 3. The major components of the raw carob seeds were set up to be carbon (56.13 weight%) and oxygen (35.31 weight%). However, these Raw-seeds also contain a small percentage of phosphorus, sulfur, chlorine, potassium, and calcium. After the HCl treatment, a small reduction in the percentage of carbon (54.71 weight%) and a significant rise in the quantity of oxygen (45.10 weight%) was seen. An extinction of potassium and calcium was noticed. However, for NaOH-seeds, it can be also seen that the percentage of carbon decreases (48.32 weight%), and the oxygen content increased to acquired 40.35 weight%. A change in the elemental composition was noticed after the NaOH treatment which mainly falls within an appearance of new elements, including sodium, magnesium, and aluminum and a disappearance of other elements such as phosphorus, sulfur, and chlorine. As a result, it might be concluded that the alkaline treatment increases the percentage of oxygen in HCl-seeds followed by NaOH-seeds and reduced the amount of carbon for both biosorbents as the same order.

3.1.5. pH of Zero Charge. The pH_{PZC} is an important characteristic for the biosorbent as it indicates its acidity-basicity and the net surface charge of the biosorbent in solution. The pH_{PZC} was 5.9, 2.3, and 6.9, respectively, for Raw-seeds, HCl-seeds, and NaOH-seeds. We can see that the Raw-seeds have an acidic surface. After the treatment by HCl, we found an increase in the acidity of the biomaterial at 2.3. But, after NaOH treatment, an increase in the pH_{PZC} appears. The low pH_{PZC} is in agreement with the predominance of surface acid groups. This result indicates that, for pH lower than 5.9, 2.3, and 6.9, the surfaces of the Raw-seeds, HCl-seeds, and NaOH-seeds are positively charged. Then at these pH, the biosorption of the studied metals was inhibited, due to the electronic repulsion between metal ions and positively charged functional groups. Inversely, for pH superior of 5.9, 2.3, and 6.9, the number of negatively charged sites on the Raw-seeds, HCl-seeds, and NaOH-seeds' surface increases, and metal biosorption becomes more important.

3.2. Biosorption Performance

3.2.1. Effect of Solution pH on the Biosorption. The pH of the solution has a significant impact on the uptake of heavy metals, since it indicates the surface charge of the adsorbent and the degree of ionization and speciation of the adsorbate [29]. The pH of the solution controls the electrostatic interactions between the sorbent and the sorbate [30]. However, the dependence of heavy metal biosorption on pH was different for each metal. Figure 4 shows the effect of pH on biosorption of Cd(II) and Co(II) onto Raw-seeds, HCl-seeds, and NaOH-seeds. It can be seen that the metal biosorption increases with increasing solution pH, and it is strongly dependent on pH solution. It is known generally that the percent removal of the heavy

TABLE 3: Elemental composition of Raw-seeds, HCl-seeds, and NaOH-seeds.

Element	Raw-seeds		HCl-seeds		NaOH-seeds	
	Weight %	Atomic %	Weight %	Atomic %	Weight %	Atomic %
C	56.13	65.81	54.71	61.72	48.32	58.11
O	35.31	31.09	45.10	38.19	40.35	36.43
Na	—	—	—	—	4.13	2.59
Mg	—	—	—	—	0.90	0.53
Al	—	—	—	—	0.34	0.18
P	0.23	0.11	0.10	0.05	—	—
S	0.23	0.10	0.07	0.03	—	—
Cl	0.21	0.08	0.02	0.01	—	—
K	4.94	1.78	—	—	0.17	0.17
Ca	2.95	1.04	—	—	1.98	1.98

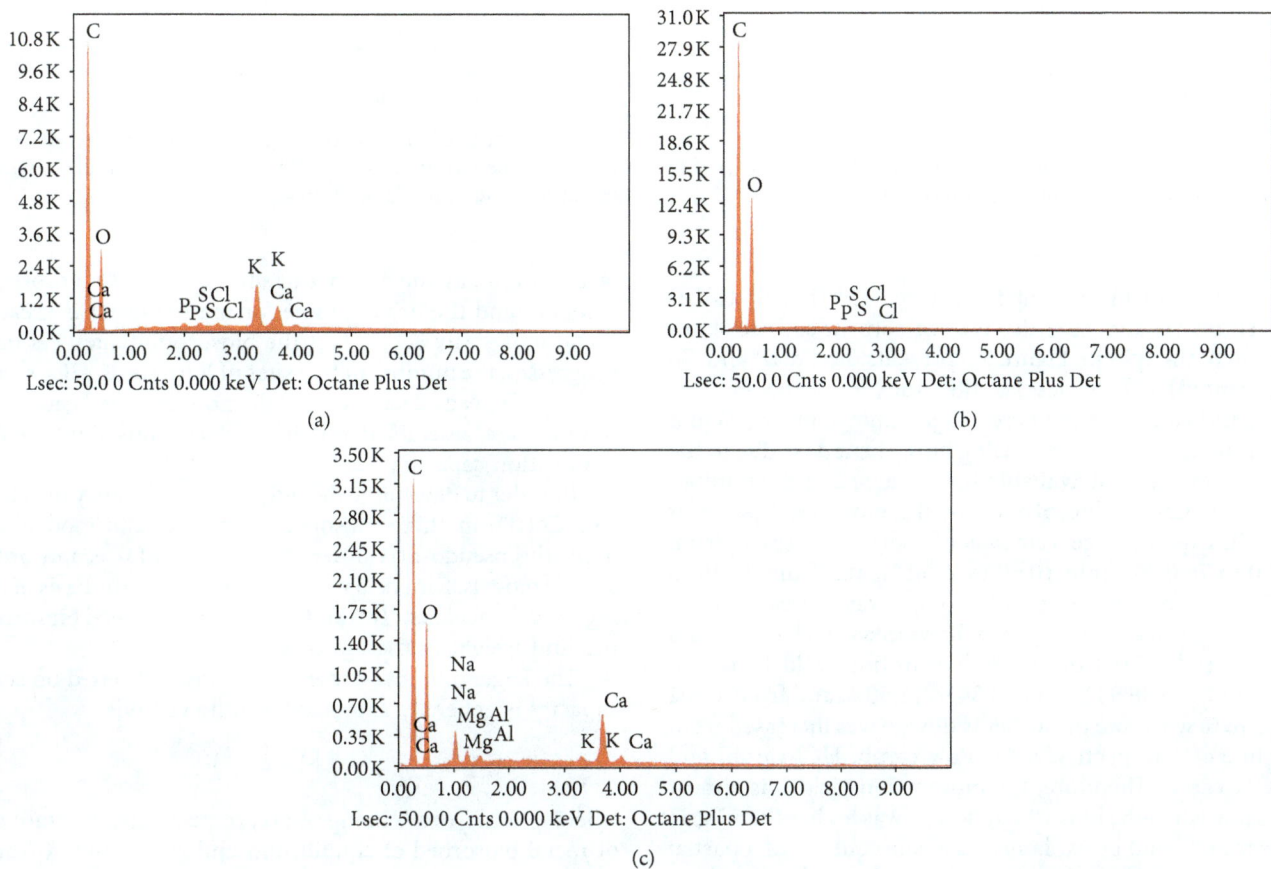

(a)

(b)

(c)

FIGURE 3: EDX analysis of (a) Raw-seeds, (b) HCl-seeds, and (c) NaOH-seeds.

metal ions increases with pH. At low pH, the cations compete with the H^+ ions in the solution for the sorption sites and therefore biosorption declines. In contrast, as pH increased, the competition between proton and metal cation decreases which means that there are more negative groups available for the binding of metal ions which results a greater metal uptake. On the other hand, at higher pH, metal cations start to form hydroxide complexes or precipitate as their hydroxides, which decrease the biosorption of metal ions [31, 32]. The pH_{PZC} values indicate that the biosorbent acquires a positive charge below a pH of 5.9, 2.3, and 6.9, respectively, for Raw-seeds, HCl-seeds,

and NaOH-seeds. Above these values, the biosorbents' surface becomes negatively charged [33]. Therefore, the ionic sorbent-sorbate interaction becomes progressively significant for pH higher than 5.9, 2.3, and 6.9. As shown in the figure, the sorption capacities for Cd(II) and Co(II) by NaOH-seeds are greater than those biosorbed by Raw-seeds and HCl-seeds. This may be related to the properties of biosorbent and metal sorbate.

3.2.2. Effect of Biosorbents Dosage. The biosorbent dosage is an important parameter because this parameter determines

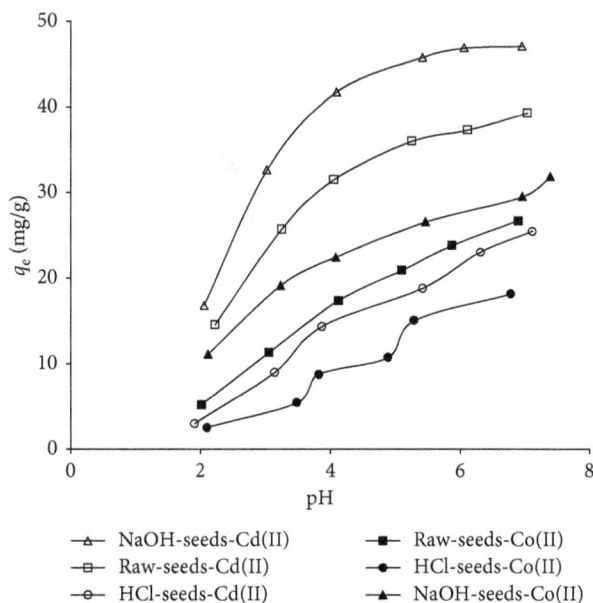

FIGURE 4: Effect of pH on the biosorption of Cd(II) and Co(II) onto the biosorbents: $C_0 = 100$ mg/L, contact time = 120 min, $R = 2$ g/L, and T = 25°C.

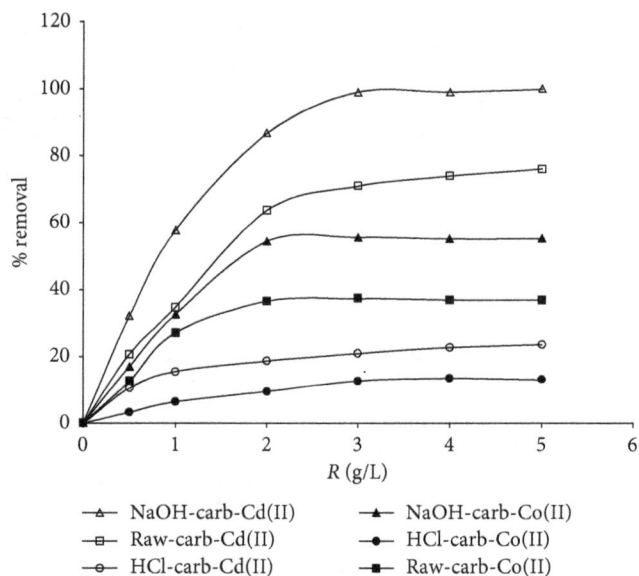

FIGURE 5: Effect of biosorbents dosage on the removal of Cd(II) and Co(II) by the biosorbents: $C_0 = 100$ mg/L, contact time = 120 min, initial pH, and temperature = 25°C.

the capacity of biosorbent for a given Cd(II) and Co(II) concentration and also determines sorbent-sorbate equilibrium of the system. Figure 5 represents the Cd(II) and Co (II) removal efficiencies for the study biosorbents. This figure indicates that the percentage removal of Cd(II) and Co(II) increased with increasing biosorbent dose due to the increase in the total available surface area and the number of active sites for biosorption of the biosorbent particles [32]. The percentage removal of Cd(II) increased from 20.70 to 76.06%, from 10.60 to 23.65%, and from 32.20 to 99.91% when the biosorbents dosage was increased from 0.5 to 5 g/L, respectively, for Raw-seeds, Hcl-seeds, and NaOH-seeds. For Co(II), the biosorption yield increased from 12.73 to 36.94%, from 3.34 to 13.50%, and from 17.02 to 55.69% when the biosorbents dosage was increased from 0.5 to 5 g/L, respectively, for Raw-carob, HCl-carob, and NaOH-carob. Therefore, the biosorption yield was almost the same when the biosorbent dosage was higher than 3 g/L. This trend could be explained as a consequence of a partial aggregation of biosorbent at higher sorbent dosage, which results in the decrease in effective surface area for the [34] biosorption [35].

3.2.3. Biosorption Kinetics. The effect of contact time on the biosorption of Cd(II) and Co(II) metal ions is reported in Figure 6. The evolution of the biosorbed amount of metal ions with the contact time indicates that the equilibrium was relatively fast and was totally reached in about 45 min for the biosorption of Cd(II) by both the biosorbents and was 60 min in the case of Co(II). This equilibrium time is very short in comparison with other literature results [36–38], which is one of the advantages of our biosorbents. In Figure 4, two kinetic regions can be observed: the first one is characterized by a high biosorption rate, which is due to the

fact that initially the number of sites of available biosorbent is higher and the driving force for mass transfer is greater. Metal ions easily access first the biosorption sites. As time progresses, the number of free sites of Raw-seeds, HCl-seeds, and NaOH-seeds decreases and the nonbiosorbed cations in solution are assembled on the surface, thus limiting the biosorption capacity.

In order to determine the biosorption efficiency of Cd(II) and Co(II) on three biosorbents, two kinetic models are used; the pseudo-first-order and the pseudo-second-order kinetic models. Kinetic data were analyzed on the basis of the regression coefficient (r^2) and the amount of metal biosorbed per unit weight of the biosorbent.

The Lagergren first-order rate expression based on solid capacity is generally expressed as follows [39]:

$$q = q_e\left(1 - e^{-k_1 t}\right), \tag{4}$$

where q_e and q (both in mg/g) are, respectively, the amounts of metal biosorbed at equilibrium and at any time t (min) and k_1 (1/min) is the rate constant of biosorption.

The pseudo-second-order equation is expressed as [40]

$$q = \frac{k_2 q_e^2 t}{1 + k_2 q_e t}, \tag{5}$$

where k (g/mg min) is the pseudo-second order rate constant.

The obtained data and the correlation coefficients, r^2, are listed in Table 4. The results show that the calculated equilibrium values using pseudo-first-order model kinetic were very close to the experimental ones (q_{exp}) than the others calculated from the pseudo-second-order model and also showed the best fit to the experimental data with the highest correlation coefficients ($r^2 = 0.999$) for the both metal ions. From these results, it was concluded that

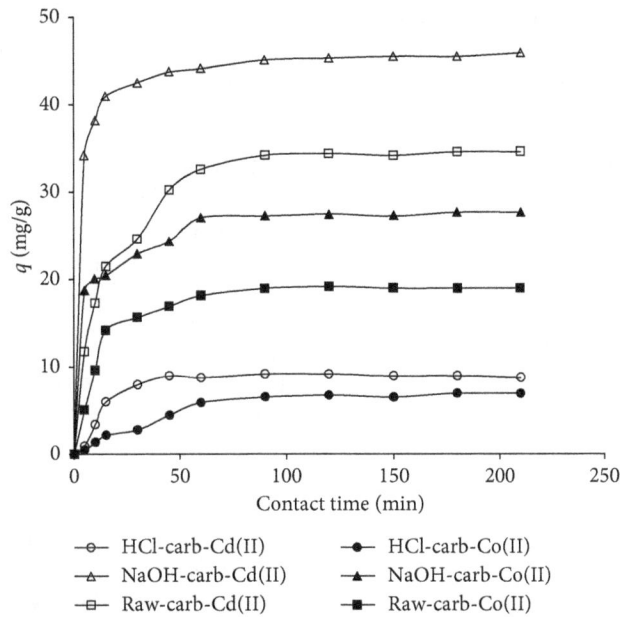

FIGURE 6: Kinetics of Cd(II) and Co(II) biosorption by the biosorbents: $C_0 = 100$ mg/L, $R = 2$ g/L, initial pH, and temperature $= 25°$C.

TABLE 4: Pseudo-first-order and pseudo-second-order kinetic parameters for the biosorption of Co(II) and Cd(II).

Metal	Biosorbent	q_e (mg/g)	Pseudo-first-order			Pseudo-second-order		
			q_e (mg/g)	k_1 (1/min)	r^2	q_e (mg/g)	k_2 (g/mg·min)	r^2
	Raw-seeds	34.578	34.159	0.071	0.996	37.230	0.002	0.992
Cd(II)	HCl-seeds	8.942	9.076	0.043	0.991	10.380	0.005	0.972
	NaOH-seeds	45.905	45.863	0.296	0.992	42.649	0.009	0.994
	Raw-seeds	18.963	18.964	0.064	0.998	20.697	0.004	0.993
Co(II)	HCl-seeds	6.929	7.090	0.022	0.984	8.907	0.002	0.973
	NaOH-seeds	27.676	27.605	0.051	0.999	28.741	0.005	0.997

the biosorption of Cd(II) and Co(II) onto Raw-seeds, HCl-seeds, and NaOH-seeds could be better described by the pseudo-first-order model. This may be due to rapidity transfer speed of Cd and Co molecules to the surface of the biosorbent and the availability of active sites.

3.2.4. Biosorption Isotherms. The biosorption isotherms describe how the sorbate molecules are distributed between the liquid phase and solid phase when the system reaches the equilibrium. The analysis of isotherm data by fitting them to different models is important to find a sustainable model that can be used [41]. The biosorption isotherms are illustrated in Figure 7. It is obvious that the amount of metal biosorbed increases as its equilibrium concentration increased. This trend may be due to the high driving force for mass transfer at a high initial heavy metal concentration. In addition, if the heavy metal concentration in solution is higher, the active sites of biosorbent are surrounded by much more ions, and the biosorption phenomenon occurs more efficiently. Thus, biosorption amount increases with

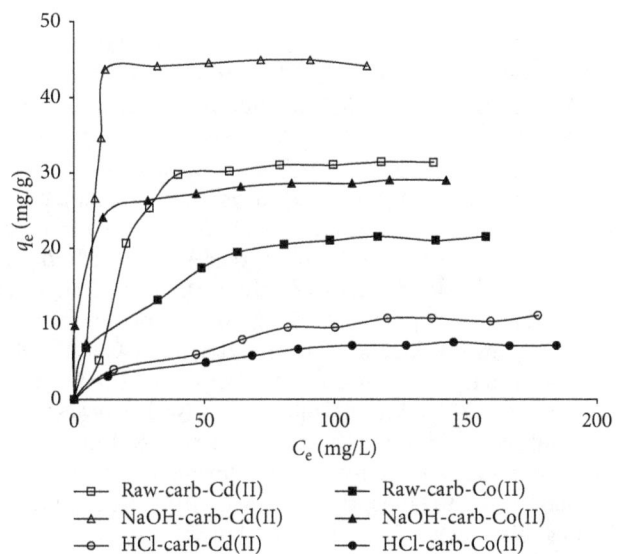

FIGURE 7: Adsorption isotherms of Cd(II) and Co(II) biosorption by the biosorbents: $R = 2$ g/L, initial pH, contact time $= 120$ min, and temperature $= 25°$C.

the increase of initial ion concentration [42]. The isotherms' form was type L for Raw-seeds, HCl-seeds, and NaOH-seeds according to Giles classification [43].

Several biosorption isotherms can be used to correlate the biosorption equilibrium in heavy metals biosorption on several biosorbents. Some well-known isotherms are Langmuir and Freundlich models.

(1) Langmuir Model. Langmuir isotherm assumes two main points in the biosorption process. First, the biosorption happens at specific homogeneous biosorption sites in the biosorbent. Second, the monolayer biosorption and maximum biosorption occurs when biosorbed molecules form a saturated layer on the surface of adsorbent. All biosorption sites involved are energetically identical, and the intermolecular force decreases as the distance from the biosorption surface increases [44].

$$q_e = \frac{q_m K_L C_e}{1 + K_L C_e}, \tag{6}$$

where q_m (mg/g) is the maximum monolayer biosorption capacity, K_L (L/mg) is the Langmuir equilibrium constant related to the biosorption affinity, and C_e is the equilibrium concentration.

(2) Freundlich Model. Freundlich isotherm is an empirical equation assuming that the biosorption process takes place on heterogeneous surfaces, and biosorption capacity is related to the concentration of biosorbed metal ions at equilibrium. Freundlich isotherm is suitable in treating metal ions' biosorption at higher concentrations. However, this isotherm is not suitable for low concentration range.

Freundlich isotherm model is represented by the following equation [45]:

$$q_e = K_F C_e^{1/n}, \tag{7}$$

where K_F ($mg^{1-1/n}/g/L^n$) is the Freundlich constant and n is the heterogeneity factor. The K_F value is related to the biosorption capacity, while $1/n$ value is related to the biosorption intensity.

(3) Analysis of Adsorption Isotherms. The calculated isotherm parameters for each model and correlation coefficients analyzed by nonlinear regression method are presented in Table 5. This table shows that the Langmuir model indicates higher values of correlation coefficients ($r^2 > 0.993$) in the biosorption of Co(II) than Cd(III) biosorption onto the three biosorbents. For these reasons, it can be approved that Langmuir equilibrium isotherm describes the metal biosorption process using the studied biosorbents well. This process is occurred by the formation of metal ion monolayer onto the biosorbent surface with finite number of identical sites, which are homogeneously distributed over the biosorbent surface [46]. The q_{max} values were 34.85, 14.08, and 49.11 mg/g for Cd(II) and 24.74, 7.65, and 28.18 mg/g in the case of Co(II) respectively, for Raw-seeds, HCl-seeds, and NaOH-seeds. We can also conclude that NaOH-seeds' biosorption capacity is superior than that of

Raw-seeds and HCl-seeds. On the other hand, the biosorption of Cd(II) on the three biosorbents is greater in comparison to that of the Co(II) biosorption. This may be due to the nature of the interaction between each sorbate and biosorbent. However, the Freundlich model provides two parameters: k_f and n. k_f is related to the biosorption capacity and biosorption intensity of the metal ions on the different biosorbents and represents the quantity of metal ions biosorbed onto biosorbent at equilibrium concentration. n represents the strength of metal ions' biosorption on the biosorbent. n value from 1 to 10 indicates relatively strong biosorption. The obtained n values in the studied biosorbents were more than 1, which indicates relatively a strong biosorption of the metal ions on the biosorbents [47].

Table 6 presents a comparison of the maximum adsorption capacity of Raw-seeds, HCl-seeds, and NaOH-seeds with various adsorbents reported in the literature for the adsorptive removal of Cd(II) and Co(II). Though there were variations in some experimental conditions, the table shows that the biosorption capacity of Raw-seeds, HCl-seeds, and NaOH-seeds for the both heavy metals is most higher compared to other adsorbents used in previous studies. This can be explained by the nature of functional groups present in the surface of Raw-seeds, HCl-seeds, and NaOH-seeds.

3.2.5. Effect of Temperature. The variation of sorption efficiencies of Cd(II) and Co(II) on Raw-seeds, HCl-seeds, and NaOH-seeds as function of solution temperature is shown in Figure 8. It was observed that the temperature does not have significant influence on the biosorption capacity in the studied range. The variation in temperature had two major effects on the sorption process. An increase in the temperature is known to increase the rate of diffusion of the adsorbate molecules across the external boundary layer and in the internal pores of the adsorbent particles as a result of reduced viscosity of the solution [61]. In addition, several authors have shown that further increases in the temperature may lead to a decrease in the metal removal percentage. This may be attributed to an increase in the relative desorption of the metal from the solid phase to the liquid phase, deactivation of the biosorbent surface, destruction of the active sites on the biosorbent surface owing to bond disruption [62], or weakness of the sorbent active site binding forces and the sorbate species and also between the adjacent molecules of the sorbed phase [63].

3.2.6. Effect of Inorganic Ions. Industrial effluents often contain more than one metal ion. Consequently, biosorption becomes competitive, in which several metal ions compete for a limited number of binding sites [64]. Figure 9 shows the effect of Na^+, K^+, Mg^{2+}, Ca^{2+}, and Al^{3+} ions on the biosorption of Cd(II) and Co(II). From this figure, we can conclude that the monovalent cations K^+ and Na^+ have less influence on the biosorption. On the other hand, divalent and trivalent ions inhibit the biosorption on the both biosorbents. The degree of inhibition by the inorganic ions followed the sequence: $K^+ < Na^+ < Mg^{2+} < Ca^{2+} < Al^{3+}$ for the both metals. The monovalent cations Na^+ and K^+ are

TABLE 5: Model isotherm constants for the biosorption of Cd(II) and Co(II) biosorption onto Raw-seeds, HCl-seeds, and NaOH-seeds.

Isotherms	Parameters	Raw-seeds		HCl-seeds		NaOH-seeds	
		Cd(II)	Co(II)	Cd(II)	Co(II)	Cd(II)	Co(II)
Langmuir	q_m (mg/g)	34.85	24.74	14.08	7.65	49.11	28.18
	K_L (L/mg)	0.082	0.054	0.021	0.086	0.150	1.100
	r^2	0.996	0.997	0.992	0.999	0.999	0.993
Freundlich	N	5.176	5.890	3.352	12.082	4.564	6.091
	K_F (mg1-1/n/g/Ln)	12.61	9.47	2.40	13.97	17.53	4.68
	r^2	0.988	0.994	0.988	0.999	0.991	0.973

TABLE 6: Comparison of biosorption capacity (qm) of Raw-seeds, HCl-seeds, and NaOH-seeds for Cd(II) and Co(II) with different other biosorbents.

Adsorbent	q_m (mg/g) Cd(II)	q_m (mg/g) Co(II)	References
Pine bark	28.00	—	[48]
Peat	22.50	—	[49]
Hazelnut shells	5.42	—	[50]
Waste tea leaves	31.48	—	[51]
Aquatic plant *Najas graminea*	28.00	20.6	[52]
Black carrot residues	—	5.35	[53]
Coir pith	—	12.82	[54]
J. rubens (red algae)	30.50	32.6	[55]
Saw dust	26.73	—	[56]
Neem bark	25.57	—	[57]
Mangosteen shell	3.15	0.34	[11]
Areca catechu	10.66	—	[58]
	—	27.15	[59]
Rose waste biomass	—	20.63	[60]
	31.35	17.41	[24]
Sargassum wightii	14.90	10.46	[24]
Raw-carob	49.63	30.04	[24]
HCl-carob			
NaOH-carob			
Raw-seeds	31.35	17.41	Present study
HCl-seeds	14.90	10.46	Present study
NaOH-seeds	49.63	30.04	Present study

bound by ionic attraction and therefore do not compete directly with the binding of heavy metals by the biosorbents. However, divalent and trivalent ions prevent the biosorption on the both biosorbents. The decreasing of Al^{3+} with high concentrations may be attributed to favorable electrostatic effects due to the increased number of positively charged surface binding sites arising from Al^{3+}. However, the effect of ionic strength is explained as the result of competition of ions with the heavy metals for electrostatic binding to the surface of Raw-seeds, HCl-seeds, and NaOH-seeds. Because, the functional groups are negatively charged, they will electrostatically attract any cation, be it inorganic or heavy metal ions of interest [65].

3.3. Optimization of Biosorption Conditions

3.3.1. Experimental Design. Table 7 shows the adsorption conditions and experimental results for the two responses of cadmium and cobalt removal. As can be seen, there was a considerable variation in the removal efficiency of cadmium and cobalt at different values of the already selected factors. Consequently, the maximum sorption capacities of cadmium and cobalt were 85.73 and 51.90 mg/g, respectively. These greater capacities were obtained for the initial concentration of 50 mg/l, pH = 6, biosorbent dose of 1 g/l with carob seeds treated by NaOH agent. The regression analysis was performed to adjust the response functions with the experimental data. The values of regression coefficient estimated are presented in Table 8. From the table, the biosorbent dose and initial concentration present a negative effect on cadmium and cobalt ions removal, while the pH and treating agent have a positive effect on elimination of two heavy metals. The analysis of the interaction effects shows a significant interaction between biosorbent dose and treating agent for cadmium removal ($b_{24} = -6.09$) with a negative impact and a significant interaction between biosorbent dose and initial concentration for cobalt removal with a positive effect ($b_{23} = 2.58$).

3.3.2. Analysis of Variance (ANOVA). Analysis of variance (ANOVA) was carried out to justify the adequacy of the models. After discarding the insignificant terms, the

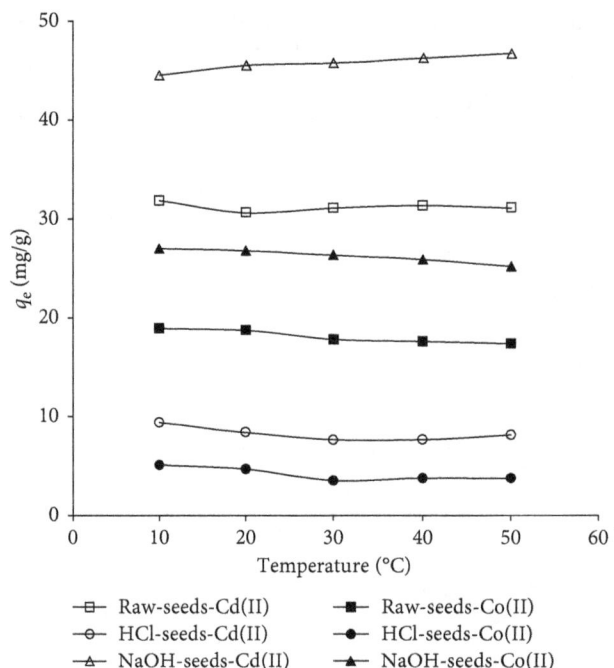

FIGURE 8: Effect of temperature on Cd(II) and Co(II) biosorption by the biosorbents: $R = 2$ g/L, initial pH, contact time = 120 min, and $C_0 = 100$ mg/L.

ANOVA data for the coded quadratic models for the two responses at a confidence level of 95% are reported in both Tables 9 and 10. The quality of the model developed was evaluated based on correlation coefficient, R^2, and standard deviation. Data given in Tables 9 and 10 demonstrate that the two models were significant at p values < 0.05. The closer the R^2 to unity and the smaller the standard deviation, the more accurate the response could be predicted by the model. The regression equations, in terms of their coded factors, are expressed by the following second-order polynomial equations:

$$Y1 = 38.93 + 5.23A - 12.86B - 4.42C + 11.89D - 2.41AB$$
$$- 1.57AD + 1.95BC - 6.09BD + 1.66BCD,$$

(8)

$$Y2 = 23.24 + 2.15A - 4.93B - 6.32C + 7.83D - 0,95AB$$
$$+ 2.58BC - 2.48BD.$$

(9)

According to these equations, it is revealed that the pH and treating agent have a positive effect on cadmium ions removal. Moreover, the removal of the cadmium decreased if biosorbents dosage and initial metals concentration increases. Additionally, the interactions pH (A)– biosorbents dosage (B), pH (A)–treating agent (D), biosorbents dosage (B)–initial concentration (C), biosorbents dosage (B)– treating agent (D), and biosorbents dosage (B)–initial concentration (C)–treating agent (D) also had a significant effect on cadmium sorption. The interactions AB, AD, and BD presented a negative effect on cadmium ions removal. Then, the interactions BC and BCD presented a positive

effect on cadmium removal. When this interaction is in the high level, the removal of cadmium increased. It is the same in the case of cobalt. Furthermore, the interactions pH de la solution (A)–biosorbents dosage (B), biosorbents dosage (B)–initial metals concentration (C), and biosorbents dosage (B)–treating agent (D) also had a significant effect on cobalt sorption. In fact, the interaction BC had a positive effect on cobalt sorption proving an increase in this response. On the other hand, the interaction AB and BD present a negative effect for cobalt sorption.

3.3.3. Response Surface Analysis. The three-dimensional (3D) response surface of the tested factors is presented for identifying the type of interaction between the studied factors. For biosorption of cadmium, there are significant interactions between pH and biosorbent dose and another one between biosorbent dose and initial concentration. Figure 10 presents response surfaces plots for these significant interactions. This figure indicates that the sorption of cadmium increased with increasing of pH and and decrease of biosorbent dose with a fixed initial concentration at 50 mg/l with NaOH treating carob (Figure 10(a)). Furthermore, the sorption of Cd(II) increased with decreasing of biosorbent dose and initial concentration (Figure 10(b)). A maximal sorption cadmium response is observed at pH = 6 and initial concentration of 50 mg/l at a biosorbent dose of 1 g/l with NaOH treating carob.

Figure 11 presents the significant interaction between pH and biosorbent dose and another one between biosorbent dose and initial concentration, for the sorption of cobalt. Figure 11(a) shows that the biosorption of Co(II) increased with increasing of pH and decreasing of biosorbent dose with a fixed initial concentration at 50 mg/l

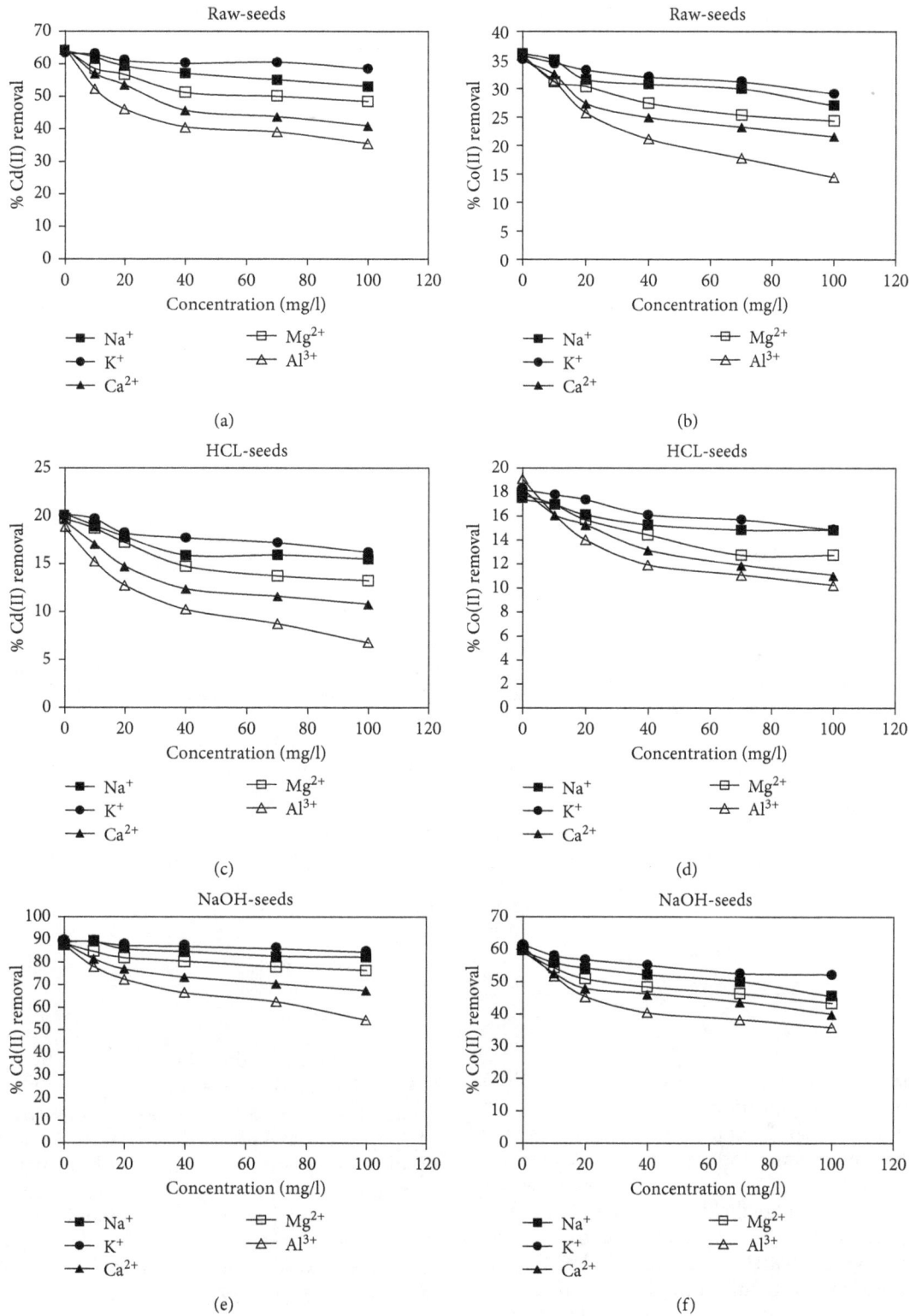

FIGURE 9: Effect of Na$^+$, K$^+$, Ca^{2+}, Mg^{2+}, and Al^{3+} concentration on the inhibition of the biosorption of Cd(II) and Co(II) onto biosorbents.

with NaOH treating carob. Moreover, when the initial concentration and the biosorbent dose decrease, the sorption of cobalt increases at pH = 6 with NaOH treating carob (Figure 11(b)). So, the best cobalt biosorption was obtained with pH = 6, biosorbent dose of 1 g/l, concentration of 50 mg/l, and NaOH treating carob.

3.3.5. Optimization. The two responses were optimized simultaneously by using the desirability function approach. The response variable was 1.00. Hence, "R^2" is in consistent enough with the "R^2_{Adj}." Thus, $R^2 = 0.9979$, $R^2 = 0.9955$ and $R^2_{Adj} = 0.9947$, $R^2_{Adj} = 0.9916$ for Cd(II) and Co(II) sorption responses. The model F-value of the

TABLE 7: Factorial experimental design matrix coded, real values, and experimental results of the responses.

Run	Coded values				Actual values				Responses	
	A	B	C	D	A	B	C	D	Cd(II)	Co(II)
1	−1	−1	−1	−1	4	1	50	HCl	30.59	24.22
2	+1	−1	−1	−1	6	1	50	HCl	46.21	29.30
3	−1	+1	−1	−1	4	3	50	HCl	19.23	14.48
4	+1	+1	−1	−1	6	3	50	HCl	29.34	18.05
5	−1	−1	+1	−1	4	1	100	HCl	19.56	5.58
6	+1	−1	+1	−1	6	1	100	HCl	38.86	12.36
7	−1	+1	+1	−1	4	3	100	HCl	11.57	8.64
8	+1	+1	+1	−1	6	3	100	HCl	20.92	10.71
9	−1	−1	−1	+1	4	1	50	NaOH	70.11	42.86
10	+1	−1	−1	+1	6	1	50	NaOH	85.73	51.90
11	−1	+1	−1	+1	4	3	50	NaOH	32.10	26.53
12	+1	+1	−1	+1	6	3	50	NaOH	33.48	29.17
13	−1	−1	+1	+1	4	1	100	NaOH	56.32	27.61
14	+1	−1	+1	+1	6	1	100	NaOH	66.89	31.56
15	−1	+1	+1	+1	4	3	100	NaOH	30.11	18.81
16	+1	+1	+1	+1	6	3	100	NaOH	31.80	20.13

TABLE 8: Values of model coefficients of the two responses.

Main coefficients	Y1	Y2
b_0	38.93	23.24
b_1	5.23	2.15
b_2	−12.86	−4.93
b_3	−4.42	−6.32
b_4	11.89	7.82
b_{12}	−2.41	−0.95
b_{13}	−0.11	−0.39
b_{14}	−1.57	−0.03
b_{23}	1.95	2.58
b_{24}	−6.09	−2.48
b_{34}	−0.11	−0.22
b_{123}	0.06	0.03
b_{124}	−0.49	−0.18
b_{134}	−0.49	−0.41
b_{234}	1.66	−0.22
b_{1234}	0.61	0.43

TABLE 9: Analysis of variance for sorption of cadmium ions.

Source	Sum of squares	df	Mean square	F value	p value; prob > F	
Model	6488.97	9	720.99	313.12	<0.0001	Significant
A	437.22	1	437.22	189.88	<0.0001	
B	2645.07	1	2645.072	1148.73	<0.0001	
C	313.043	1	313.04	135.95	<0.0001	
D	2262.37	1	2262.37	982.524	<0.0001	
AB	93.14	1	93.14	40.445	0.0007	
AD	39.45	1	39.45	17.13	0.0061	
BC	61.03	1	61.03	26.51	0.0021	
BD	593.24	1	593.24	257.64	<0.0001	
BCD	44.40	1	44.40	19.28	0.0046	
Residual	13.81	6	2.30			
Cor. total	6502.796	15				

$R^2 = 0.9979$; $R^2_{adj} = 0.9947$.

both responses is greater, in order of 313.12 and 255.37 for Cd(II) and Co(II) sorption, respectively. Further, these results explain that the models are suitable. Then, it was found that there was good agreement between experimentally and model predicted response factor which confirms the adequacy and the significance of the proposed model. The optimal conditions for high sorption of cadmium(II) and cobalt(II) were achieved at pH = 6, biosorbent dose of 1 g/l, and initial concentration = 50 mg/l with carob seeds treated by NaOH. The greater sorption capacities were 85.73 mg/g for Cd(II) and 51.90 mg/g for Co(II).

4. Conclusion

During this study, raw carob seeds and chemically treated carob seeds were used as low-cost natural biosorbents for the removal of Cd(II) and Co(II) from aqueous solutions.

The biosorption productivity was tested by using different biosorption conditions According to these studies, it was found that the biosorption yield increases with the increase of biosorbent dosage with an optimum at 0.1 g/L. The optimum sorption was obtained at basic pH medium. The sorption process was very rapid, since the equilibrium time was obtained at 90 min for Cd(II) and 60 min for Co(II). Biosorption kinetics data were properly fitted with the pseudo-first-order kinetic model. The equilibrium biosorption was increased with an increase in the initial ions concentration in solution. The biosorption isotherm could be well fitted by the Langmuir equation. The temperature does not have much influence on the biosorption performance. Other tests show that the presence of inorganic ions had specific effects on Cd(II) and Co(II) biosorption onto biosorbents, with the inhibition effect observing the following sequence: $K^+ < Na^+ < Mg^{2+} < Ca^{2+} < Al^{3+}$. From these studies, it can be also seen that chemical pretreatment the raw carob shell with NaOH strongly enhances its

TABLE 10: Analysis of variance for sorption of cobalt ions.

Source	Sum of squares	df	Mean square	F value	p value; prob > F	
Model	2302.19	7	328.88	255.37	<0.0001	Significant
A	74.23	1	74.23	57.64	<0.0001	
B	388.97	1	388.97	302.03	<0.0001	
C	639.204	1	639.20	496.33	<0.0001	
D	980.24	1	980.24	761.14	<0.0001	
AB	14.54	1	14.54	11.29	0.0099	
BC	106.31	1	106.31	82.55	<0.0001	
BD	98.69	1	98.69	76.63	<0.0001	
Residual	10.30	8	1.29			
Cor. total	2312.50	15				

$R^2 = 0.9955$; $R^2_{adj} = 0.9916$.

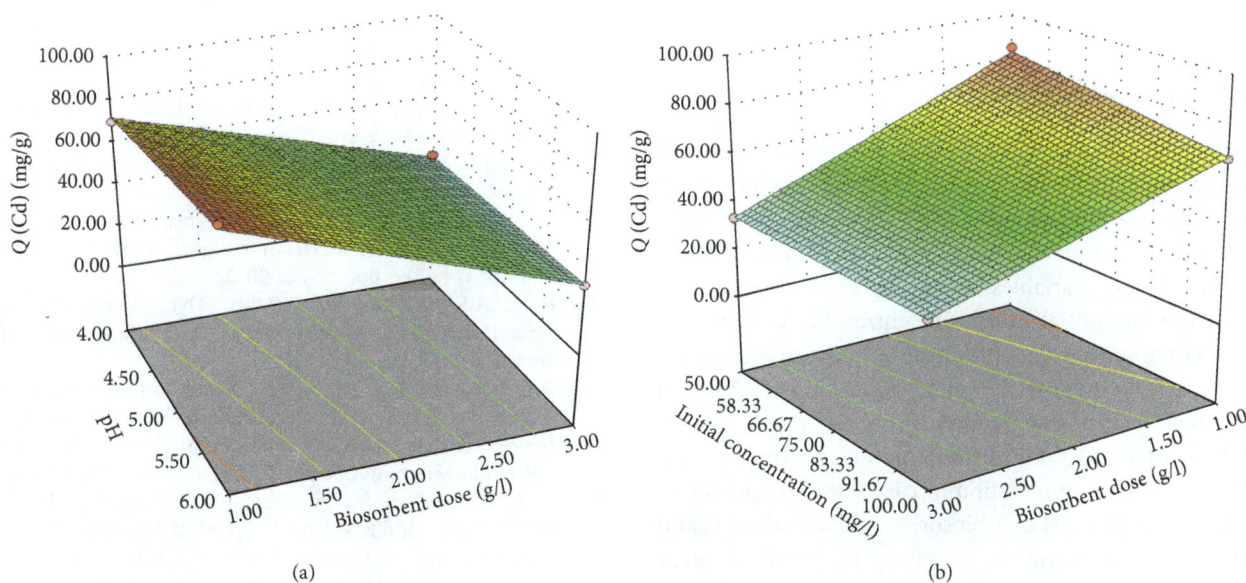

(a)   (b)

FIGURE 10: Surface response plots for the cadmium ions removal.

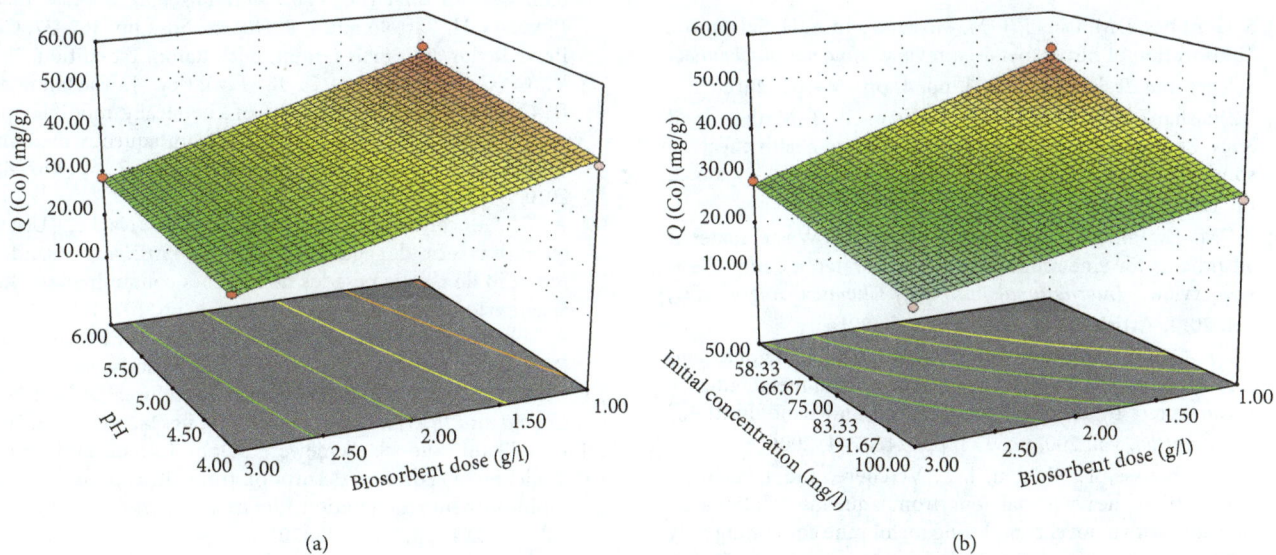

(a)   (b)

FIGURE 11: Surface response plot for the cobalt ions removal.

biosorption potential for the cadmium(II) and cobalt(II) ions. Full-factorial experimental design was used to determine the optimum conditions of different variables which affect the removal of Cd(II) and Co(II) ions. In order to achieve approximately the greater values of biosorption capacities of treated carob shell for Cd(II) and Co(II) ions, the optimum values of different process parameters were found to be 85.73 mg/g for Cd(II) and 51.90 mg/g for Co (II). These higher sorption efficiencies are obtained at pH = 6, biosorbent dose of 1 g/l, and initial concentration = 50 mg/l with carob shell treated by NaOH.

Conflicts of Interest

The authors declare that they have no conflicts of interest.

Supplementary Materials

In summary, efficient biosorbents were developed from deoiled carob seeds. The biosorption efficiency was evaluated for cadmium and cobalt ions removal from aqueous solution under various parameters. Based on this preliminary study, four independent variables including solution pH, biosorbents dosage, initial metal concentration, and treating agent were chosen for the optimization of the process using full-factorial experimental design. It was found that chemical pretreatment of the raw deoiled carob seeds with NaOH strongly enhances its biosorption potential. The optimal conditions for high biosorption of cadmium(II) and cobalt (II) were achieved at pH of 6, biosorbent dosage of 1 g/L, and initial metal concentration of 50 mg/L. (*Supplementary Materials*)

References

[1] S. H. Abbas, I. M. Ismail, T. M. Mostafa, and A. H. Sulaymon, "Biosorption of heavy metals: a review," *Journal of Chemical Science and Technology*, vol. 3, no. 4, pp. 74–102, 2014.

[2] M. Jaishankar, T. Tseten, N. Anbalagan, B. B. Mathew, and K. N. Beeregowda, "Toxicity, mechanism and health effects of some heavy metals," *Interdisciplinary Toxicology*, vol. 7, no. 2, pp. 60–72, 2014.

[3] H. M. Zwain, M. Vakili, and I. Dahlan, "Waste material adsorbents for zinc removal from wastewater: a comprehensive review," *International Journal of Chemical Engineering*, vol. 2014, Article ID 347912, 13 pages, 2014.

[4] M. J. González-Muñoz, M. A. Rodríguez, S. Luque, and J. R. Álvareza, "Recovery of heavy metals from metal industry waste waters by chemical precipitation and nanofiltration," *Desalination*, vol. 200, no. 1–3, pp. 742–744, 2006.

[5] H. Bessbousse, T. Rhlalou, J. F. Verchère, and L. Lebrun, "Removal of heavy metal ions from aqueous solutions by filtration with a novel complexing membrane containing poly (ethyleneimine) in a poly(vinyl alcohol) matrix," *Journal of Membrane Science*, vol. 307, no. 2, pp. 249–259, 2008.

[6] R. Kiefer, A. I. Kalinitchev, and W. H. Höll, "Column performance of ion exchange resins with aminophosphonate functional groups for elimination of heavy metals," *Reactive and Functional Polymers*, vol. 67, no. 12, pp. 1421–1432, 2007.

[7] J. Konczyk, C. Kozlowski, and W. Walkowiak, "Lead(II) removal from aqueous solutions by solvent extraction with tetracarboxyl- resorcin [4] arene," *Physicochemical Problems of Mineral Processing*, vol. 49, no. 1, pp. 213–222, 2013.

[8] H. Polat and D. Erdogan, "Heavy metal removal from waste waters by ion flotation," *Journal of Hazardous Materials*, vol. 148, no. 1-2, pp. 267–273, 2007.

[9] A. Hamid Sulaymon, A. Obaid Sharif, and T. K. Al-Shalch, "Removal of cadmium from simulated wastewaters by electrodeposition on stainless steel tubes bundle electrode," *Desalination and Water Treatment*, vol. 29, no. 1–3, pp. 218–226, 2011.

[10] K. Vijayaraghavan, K. Palanivelu, and M. Velan, "Biosorption of copper(II) and cobalt(II) from aqueous solutions by crab shell particles," *Bioresource Technology*, vol. 97, no. 12, pp. 1411–1419, 2006.

[11] P. Chakravarty, N. Sen Sarma, and H. P. Sarma, "Biosorption of cadmium(II) from aqueous solution using heartwood powder of Areca catechu," *Chemical Engineering Journal*, vol. 162, no. 3, pp. 949–955, 2010.

[12] I. Anastopoulos and Z. George Kyzas, "Progress in batch biosorption of heavy metals onto algae," *Journal of Molecular Liquids*, vol. 209, pp. 77–86, 2015.

[13] K. A. Adegoke and O. S. Bello, "Dye sequestration using agricultural wastes as adsorbents," *Water Resources and Industry*, vol. 12, pp. 8–24, 2015.

[14] S. Y. Bratskaya, A. V. Pestov, Y. G. Yatluk, and V. A. Avramenko, "Heavy metals removal by flocculation/precipitation using N-(2-carboxyethyl)chitosans," *Colloids and Surfaces A: Physicochemical and Engineering Aspects*, vol. 339, no. 1–3, pp. 140–144, 2009.

[15] A. C. Gonçalves Jr., L. Strey, C. A. Lindino, H. Nacke, D. Schwantes, and E. P. Seidel, "Applicability of the Pinus bark (*Pinus elliottii*) for the adsorption of toxic heavy metals from aqueous solutions," *Acta Scientiarum Technology*, vol. 34, no. 1, pp. 79–87, 2012.

[16] H. Nacke, A. C. Gonçalves Jr., G. F. Coelho, L. Strey, and A. Laufer, "Renewable energy technologies: removal of cadmium from aqueous solutions by adsorption on Jatropha biomass," in *Green Design, Materials and Manufacturing Processes*, H. Bártolo and J. P. Duarte, Eds., pp. 367–37, CRC Press Taylor & Francis Group, Boca Raton, 1st edition, 2013.

[17] V. C. G. Dos Santos, C. R. T. Tarley, J. Caetano, and D. C. Dragunski, "Assessment of chemically modified sugarcane bagasse for lead adsorption from aqueous medium," *Water Science and Technology*, vol. 62, no. 2, pp. 457–465, 2010.

[18] A. C. Gonçalves Jr., C. Selzlein, and H. Nacke, "Uso de biomassa seca de aguapé (*Eichornia crassipes*) visando à remoção de metais pesados de soluções contaminadas," *Acta Scientiarum Technology*, vol. 31, no. 1, pp. 103–108, 2009.

[19] A. Bhatnagar, V. J. P. Vilar, C. M. S. Botelho, and R. A. R. Boaventura, "Coconut-based biosorbents for water treatment a review of the recent literature," *Advances in Colloid and Interface Science*, vol. 160, no. 1-2, pp. 1–15, 2010.

[20] E. Njikam and S. Schiewer, "Optimization and kinetic modeling of cadmium desorption from citrus peels: a process for biosorbent regeneration," *Journal of Hazardous Materials*, vol. 213-214, pp. 242–248, 2012.

[21] S. Liang, X. Guo, and Q. Tian, "Adsorption of Pb^{2+}, Cu^{2+} and Ni^{2+} from aqueous solutions by novel garlic peel adsorbent,"

Desalination and Water Treatment, vol. 51, no. 37–39, pp. 7166–7171, 2013.

[22] F. A. Ugbe, A. A. Pam, and A. V. Ikudayisi, "Thermodynamic properties of chromium(III) ion adsorption by sweet orange (*Citrus sinensis*) peels," *American Journal of Analytical Chemistry*, vol. 5, no. 10, pp. 666–673, 2014.

[23] B. I. Hussein, "Removal of copper ions from waste water by adsorption with modified and unmodified sunflower stalks," *Journal of Engineering*, vol. 16, no. 3, pp. 5411–5421, 2010.

[24] M. Farnane, H. Tounsadi, R. Elmoubarki et al., "Alkaline treated carob shells as sustainable biosorbent for clean recovery of heavy metals: Kinetics, equilibrium, ions interference and process optimisation," *Ecological Engineering*, vol. 101, pp. 9–20, 2017.

[25] H. Tounsadi, A. Khalidi, M. Abdennouri, and N. Barka, "Biosorption potential of *Diplotaxis harra* and *Glebionis coronaria* L. biomasses for the removal of Cd(II) and Co(II) from aqueous solutions," *Journal of Environmental Chemical Engineering*, vol. 3, no. 2, pp. 822–830, 2015.

[26] H. P Boehm, E. Diehl, W. Heck, and R. Sappok, "Surface oxides of carbon," *Angewandte Chemie International Edition in English*, vol. 3, no. 10, pp. 669–677, 1964.

[27] J. S. Noh and J. A. Schwarz, "Estimation of the point of zero charge of simple oxides by mass titration," *Journal of Colloid and Interface Science*, vol. 130, no. 1, pp. 157–164, 1989.

[28] C. R. Teixeira Tarley and M. A. Zezzi Arruda, "Biosorption of heavy metals using rice milling by-products. Characterisation and application for removal of metals from aqueous effluents," *Chemosphere*, vol. 54, no. 7, pp. 987–995, 2004.

[29] O. E. Abdel Salam, N. A. Reiad, and M. M. ElShafei, "A study of the removal characteristics of heavy metals from wastewater by low-cost adsorbents," *Journal of Advanced Research*, vol. 2, no. 4, pp. 297–303, 2011.

[30] D. Özcimen and A. Ersoy-Meriçboyu, "Removal of copper from aqueous solutions by adsorption onto chestnut shell and grapeseed activated carbons," *Journal of Hazardous Materials*, vol. 168, no. 2-3, pp. 1118–1125, 2009.

[31] J. P. Chen and X. Wang, "Removing copper, zinc, and lead ion by granular activated carbon in pretreated fixed-bed columns," *Separation and Purification Technology*, vol. 19, no. 3, pp. 157–167, 2000.

[32] E.-Z. El-Ashtoukhy, N. K. Amin, and O. Abdelwahab, "Removal of lead(II) and copper(II) from aqueous solution using pomegranate peel as a new adsorbent," *Desalination*, vol. 223, no. 1–3, pp. 162–173, 2008.

[33] C. A. Rozaini, K. Jain, C. W. Oo et al., "Optimization of nickel and copper ions removal by modified mangrove barks," *International Journal of Chemical Engineering and Applications*, vol. 1, no. 1, pp. 84–89, 2010.

[34] N. Barka, S. Qourzal, A. Assabbane, A. Nounah, and Y. Ait-Ichou, "Removal of reactive yellow 84 from aqueous solutions by adsorption onto hydroxyapatite," *Journal of Saudi Chemical Society*, vol. 15, no. 3, pp. 263–267, 2011.

[35] A. Nasrullah, H. Khan, A. S. Khan et al., "Potential biosorbent derived from calligonum polygonoides for removal of methylene blue dye from aqueous solution," *Scientific World Journal*, vol. 2015, Article ID 562693, 11 pages, 2015.

[36] Y.-j. Gao, *Cadmium and Cobalt Removal from Heavy Metal Solution using Oyster Shells Adsorbent*, pp. 1098–1101, 2011.

[37] X. Li, Y. Tang, X. Cao, D. Lu, F. Luo, and W. Shao, "Preparation and evaluation of orange peel cellulose adsorbents for effective removal of cadmium, zinc, cobalt and nickel," *Colloids and Surfaces A: Physicochemical and Engineering Aspects*, vol. 317, no. 1–3, pp. 512–521, 2008.

[38] G. Ozdemir, N. Ceyhan, and E. Manav, "Utilization of an exopolysaccharide produced by *Chryseomonas luteola* TEM05 in alginate beads for adsorption of cadmium and cobalt ions," *Bioresource Technology*, vol. 96, no. 15, pp. 1677–1682, 2005.

[39] S. Lagergren, "About the theorie of so-called adsorption of soluble substance Seven Vetenskapsakad," *Handlingar*, vol. 24, pp. 1–39, 1898.

[40] Y. S. Ho and G. Mckay, "Pseudo-second order model for sorption processes," *Process Biochemistry*, vol. 34, no. 5, pp. 451–465, 1999.

[41] B. H. Hameed, I. A. W. Tan, and A. L. Ahmad, "Adsorption isotherm, kinetic modeling and mechanism of 2,4,6-trichlorophenol on coconut husk-based activated car- bon," *Chemical Engineering Journal*, vol. 144, no. 2, pp. 235–244, 2008.

[42] M. H. Jnr and A. I. Spiff, "Effect of metal ion concentration on the biosorption of Pb^{2+} and Cd^{2+} by *Caladium bicolor* (wild cocoyam)," *Electronic Journal of Biotechnology*, vol. 8, no. 2, pp. 162–169, 2005.

[43] C. H Giles, D. Smith, and A. Huitson, "A general treatment and classification of the solute adsorption isotherm. I. Theoretical," *Journal of Colloid and Interface Science*, vol. 47, no. 3, pp. 755–765, 1974.

[44] I. Langmuir, "The constitution and fundamental properties of solids and liquids," *Journal of the American Chemical Society*, vol. 38, no. 11, pp. 2221–2295, 1916.

[45] H. Freundlich and W. Heller, "The adsorption of cis- and trans-azobenzene," *Journal of the American Chemical Society*, vol. 61, no. 8, pp. 2228–2230, 1939.

[46] G. D. Halsey, "The role of surface heterogeneity in adsorption," *Advances in Catalysis*, vol. 4, pp. 259–269, 1952.

[47] B. Kiran and A. Kaushik, "Chromium binding capacity of Lyng by aputealisexo poly-saccharides," *Biochemical Engineering Journal*, vol. 38, pp. 47–54, 2008.

[48] S. Al-Asheh and Z. Duvnjak, "Binary metal sorption by pine barks: study of equilibria and mechanisms," *Separation Science and Technology*, vol. 33, no. 9, pp. 1303–1329, 1998.

[49] T Gosset, J. L Transcart, and D. R Thevenot, "Batch metal removal by peat: kinetics and thermodynamics," *Water Research*, vol. 20, no. 1, pp. 21–26, 1986.

[50] G. Cimino, A. Passerini, and G. Toscano, "Removal of toxic cations and Cr(VI) from aqueous solution by hazelnut shell," *Water Research*, vol. 34, no. 11, pp. 2955–2962, 2000.

[51] T. W. Tee and A. R. M. Khan, "Removal of lead, cadmium and zinc by waste tea leaves," *Environmental Technology Letters*, vol. 9, no. 11, pp. 1223–1232, 1988.

[52] C. L. Lee, T. C. Wang, C. K. Lin, and H. K. Mok, "Heavy metals removal by a promising locally available aquatic plant, *Najas graminea* del., in Taiwan," *Water Science and Technology*, vol. 39, no. 10-11, pp. 177–181, 1999.

[53] F. Guzel, H. Yakut, and G. Topal, "Determination of kinetic and equilibrium parameters of the batch adsorption of Mn (II), Co(II), Ni(II) and Cu(II) from aqueous solution by black carrot (*Daucus carota* L.) residues," *Journal of Hazardous Materials*, vol. 153, no. 3, pp. 1275–1287, 2008.

[54] H. Parab, S. Joshi, N. Shenoy, A. Lali, U. S. Sarma, and M. Sudersanan, "Determination of kinetic and equilibrium of Co(II), Cr(III), and Ni(II) onto coir pith," *Process Biochemistry*, vol. 41, no. 3, pp. 609–615, 2006.

[55] M. Wael Ibrahim, "Biosorption of heavy metal ions from aqueous solution by red macroalgae," *Journal of Hazardous Materials*, vol. 192, no. 3, pp. 1827–1835, 2011.

[56] T. Kumar Naiya, P. Chowdhury, A. K. Bhattacharya, and S. K. Das, "Saw dust and neem bark as low-cost natural

biosorbent for adsorptive removal of Zn(II) and Cd(II) ions from aqueous solutions," *Chemical Engineering Journal*, vol. 148, no. 1, pp. 68–79, 2009.

[57] R. Zein, R. Suhaili, F. Earnestly, Indrawati, and E. Munaf, "Removal of Pb(II), Cd(II) and Co(II) from aqueous solution using *Garcinia mangostana* L. fruit shell," *Journal of Hazardous Materials*, vol. 181, no. 1–3, pp. 52–56, 2010.

[58] M. A. Javeda, H. N. Bhattia, M. A. Hanifa, and R. Nadeema, "Kinetic and equilibrium modeling of Pb(II) and Co(II) sorption onto rose waste biomass," *Separation Science and Technology*, vol. 42, no. 16, pp. 3641–3656, 2007.

[59] K. Vijayaraghavan, J. Jegan, K. Palanivelu, and M. Velan, "Biosorption of cobalt(II) and nickel(II) by seaweeds: batch and column studies," *Separation and Purification Technology*, vol. 44, no. 1, pp. 53–59, 2005.

[60] N. Barka, M. Abdennouri, M. El Makhfouk, and S. Qourzal, "Biosorption characteristics of cadmium and lead onto eco-friendly dried cactus (*Opuntia ficus-indica*) Cladodes," *Journal of Environmental Chemical Engineering*, vol. 1, no. 3, pp. 144–149, 2013.

[61] M. Saleem, T. Pirzada, and R. Qadeer, "Sorption of acid violet 17 and direct red 80 dyes on cotton fiber from aqueous solutions," *Colloids and Surfaces A: Physicochemical and Engineering Aspects*, vol. 292, no. 2-3, pp. 246–250, 2007.

[62] A. Sari and M. Tuzen, "Biosorption of cadmium(II) from aqueous solution by red algae (*Ceramium virgatum*): equilibrium, kinetic and thermodynamic studies," *Journal of Hazardous Materials*, vol. 157, no. 2-3, pp. 448–454, 2008.

[63] D. Kratochvil and B. Volesky, "Advances in the biosorption of heavy metals," *Trends in Biotechnology*, vol. 16, no. 7, pp. 291–300, 1998.

[64] S. Schiewer and B. Volesky, "Ionic strength and electrostatic effects in biosorption of divalent metal ions and protons," *Environmental Science & Technology*, vol. 31, no. 9, pp. 2478–2485, 1997.

[65] V. Murphy, "An investigation into the mechanisms of heavy metal binding by selected seaweed species," Ph.D. Thesis, Waterford Institute of Technology, Waterford, Ireland, 2007.

7

Adsorption of Phthalates on Municipal Activated Sludge

Hongbo Wang,[1,2] Haining Li,[1] Qicheng Song,[1] Lili Gao,[3] and Ning Wang[1]

[1]*School of Municipal and Environmental Engineering, Shandong Jianzhu University, Jinan 250101, China*
[2]*Shandong Co-Innovation Center of Green Building, Jinan, China*
[3]*Shandong Urban and Rural Planning Design Institute, Jinan, Shandong 250013, China*

Correspondence should be addressed to Ning Wang; wangning@sdjzu.edu.cn

Academic Editor: Athanasios Katsoyiannis

Phthalates (PAEs) are commonly detected in discharge of municipal wastewater treatment plants. This study investigated the removal of six typical PAEs with activated sludge and the results revealed that concentrations of aqueous PAEs decreased rapidly during the beginning 15 min and reached equilibrium within 2 hours due to the adsorption of activated sludge. The process followed first-order kinetic equation, except for dioctyl phthalate (DOP). The factors influencing the adsorption were also evaluated and it was found that higher initial concentrations of PAEs enhanced the removal but affected little the adsorption equilibrium time. The adsorption of PAEs favored lower operating temperature (the optimum temperature was approximately 25°C in this research), which could be an exothermic process. Additionally, lower aqueous pH could also benefit the adsorption.

1. Introduction

Phthalates (PAEs), a group of artificial chemicals, are characterized with low water solubility, low volatility, and low temperature resistance, which are widely used in agriculture, cosmetics, coatings, and so forth [1]. Multiple studies have reported that PAEs enter human body through breathing, eating, and skin contacting and possibly cause mutagens, teratogens, and carcinogens [2]. Major developed countries recognized PAEs as a group of serious pollutants. The compounds were found in atmosphere [3], water [4], soil [5], and biological agents, including human body [6]. Owing to the long half-life time in environment and strong bioaccumulation tendency, PAEs are thought to be a group of persistent organic pollutants (POPs). The US EPA listed six PAEs as priority toxic pollutants, including dimethyl phthalate (DMP), diethyl phthalate (DEP), dibutyl phthalate (DBP), benzyl butyl phthalate (BBP), di-2-ethylhexyl phthalate (DEHP), and dioctyl phthalate (DOP), and China also regulated three PAEs (DEP, DMP, and DOP) as environmental priority control pollutants. Previous researchers have reported that municipal sewage treatment plants are major pathway of PAEs into the environment [7–9].

To control the pollutants, adsorption is a popular and effective method to separate PAEs from aqueous phase and many sorbents have been considered, such as activated carbon, polymeric adsorbent, carbon nanotube, chitosan, and seaweed. The adsorption of DBP on activated carbon is spontaneous and endothermic and favors higher temperatures [10]. Polymer resins were described to eliminate aqueous DEP and the adsorption processes are limited by both film and intraparticle diffusions [11]. Nanomaterials, such as carbon nanotube, were also found to adsorb DMP and DEP, and the performance is influenced by nanotube size, surface modification, and adsorption temperature [12]. Chen et al. [13] conducted a pilot column test of six PAEs removal by chitosan and the breakthrough times vary from 4.5 to 8 hrs. Seaweed, a biomass adsorbent, could remove DEHP with the adsorption capacity of 5.68–6.54 mg/g in a batch test [14]. Although the mentioned adsorbents could effectively eliminate aqueous PAEs, the economic cost must be considered in practical application and cheaper alternative is preferred.

One substitution would be activated sludge, yellowish-brown flocculants and particulates, which was confirmed to adsorb and accumulate PAEs from municipal wastewater [9, 15, 16], due to the composition of microbial communities characterized with porous structure and extracellular polymers resulting in large surface area, strong adsorption, and plentiful biological activities. Previous research [17] showed DEP (0.73 mg/g) and DBP (17.6 mg/g) are substantially adsorbed by activated sludge and higher hydrophobicity of PAEs promotes the process. Additionally, the high octanol/water partition coefficient of PAEs assists the accumulation in biomass [18], resulting in greater tendency to attach on activated sludge. Besides, as a solid waste, sludge was an economic method to control the aqueous PAEs.

This research symmetrically investigated the adsorption of six PAEs on activated sludge and the influencing factors to improve the removal rate. The partitions between sludge and water were also studied.

2. Method and Materials

2.1. Chemicals and Sample Preparation. The regents used in this research were listed as follows: 2000 μg/L of six PAE standard solutions, including DMP, DEP, DBP, DEHP, DOP, and BBP (Sigma-Aldrich, USA); methanol (HPLC grade, America world); dichloromethane (HPLC grade, Tianjin Kermel Chemical Reagent Co., Ltd., China); acetone and hydrochloric acid (GR, Tianjin Kermel Chemical Reagent Co., Ltd., China); and ultra-pure water (resistivity = 18.6 M omega, pH = 2.5).

Experimental water and activated sludge: The experimental water was made with glucose as carbon source, NH_4Cl as nitrogen source, certain amount of Mg, P, Fe, Ca, and Zn ions as trace nutrients, and sodium azide as inhibitor, following the instructions of pervious research [19]. The water was adjusted to COD of 300 mg/L and total nitrogen of 50 mg/L to mimic municipal wastewater. The activated sludge sample was taken from a secondary sediment tank of Jinan water treatment factory. Before experiments, the sewage was triply rinsed and washed with distilled water to avoid the disturbance of background PAEs.

2.2. Methodology

Adsorption Experiment. All experiments were carried out in flasks stirred with a thermostatic oscillator (SHA-CA, Jinan Jingke Corporation, China) at 130 rpm. The sludge concentration (MLSS) was adjusted to 4000 mg/L (typical concentration in activated sludge tank of a municipal wastewater treatment plant) with prepared experimental water and a certain amount of mixed PAE standard solution. Subsequently, at expected time spans, certain amounts of mixture of sludge and water were separated by centrifuging at high speed rpm and filtrating with 0.45 μm glass fiber. The filtrate (water) was adjusted to pH equal to 2.5 with hydrochloric acid and stored in sealed glass bottles at 4°C and the remaining activated sludge was frozen and dried for extraction.

Extraction of PAEs from Water. Solid phase extraction was applied with C18 column (Supelco, USA), which was sequentially cleaned and activated by dichloromethane, methanol,

and ultra-pure water before usage. The water sample passed the C18 column, which was dried by vacuum to remove residual moisture after the extraction. The target PAEs were eluted into a small test tube with dichloromethane, dehydrated with anhydrous sodium sulfate, and concentrated to 1 ml with nitrogen gas blower (KL-512J, Beijing Kang Lin Science & Technology Corporation, China) for further analysis.

Extraction of PAEs from Sludge. Soxhlet extraction method, modified from EPA method 8061A, was employed for the dry sludge sample. After being added with anhydrous sodium sulfate and ground until passing through a 1 mm sieve, the sample was packed with filter paper and placed in Soxhlet extraction tube, which was rinsed and extracted with acetone and dichloromethane (volume ratio 1 : 1) mixture at 60°C for 24 hours. The extracts were concentrated to 5 ml with rotary evaporator before being purified with anhydrous Na_2SO_4-silica gel-Al_2O_3 chromatography column, which was eluted with dichloromethane for three times. Then, the volume of eluent was reduced to 1 ml by rotary evaporating and nitrogen blowing.

Instrument Analysis. PAEs were quantitatively determined with GC/MS (QP2010 Plus, Shimadzu Corporation, Japan). The column temperature was programed as follows: the initial temperature was 120°C, immediately raised to 300°C at 10°C/min, and held for 2 min. Inlet temperature was 300°C. The sample inject volume was 1 μL and the split ratio was 20 : 1. The carrier gas was helium with the flow rate of 1 mL/min. The pressure of the inlet column was 106 KPa. The bombardment source was EI+ with the ion source temperature of 250°C. SIM mode was used and the details of mass spectrum characteristic fragment ions and retention time of the peaks were shown in Table 1.

For the purpose of QA/QC, all the experimental processes were triplicated to minimize operational error. The detection limits were concentrations when signal/noise was three. The spiked recovery of PAEs was 83.20%–111.78% with standard deviation of 2.29%–8.99%.

3. Results and Discussion

3.1. Effect of Initial Concentration of PAEs. Different initial concentrations of adsorption experiments were chosen as 40 μg/L and 80 μg/L, and the results are shown in Figure 1. Both concentration profiles were similar: at the beginning 15 mins, all PAEs decreased sharply; during 15 min–2 h, most PAEs kept going down slowly; after 2 hr, PAEs fluctuated in a small concentration range. Then, it is likely considered that the adsorption equilibrium was reached in approximately 2 h. For the initial concentration of 40 μg/L, DMP showed an ascending trend after 15 min, while other PAEs were either relatively stable or slightly decreased as time elapsed; for initial concentration of 80 μg/L, DMP, DBP, and DOP showed slightly increasing trend, while DEP and DEHP were stable and BBP cannot be detected after 15 mins. Slow desorption of PAEs from sludge might contribute to the observation in a longer time experiment [20].

TABLE 1: Quantitation information of PAEs for GC-MS.

PAEs	Retention time (min)	Qualitative ion (m/z)	Quantitative ion (m/z)	Auxiliary quantitative ion (m/z)
DMP	5.27	163 : 164 : 92 : 194	163	164, 92
DEP	6.72	149 : 177 : 150 : 105	149	177
DBP	10.53	149 : 150 : 223 : 205	149	150
BBP	14.09	149 : 206 : 91 : 123	149	91
DEHP	15.58	149 : 167 : 279 : 113	149	167
DOP	17.07	149 : 279 : 150 : 390	149	150

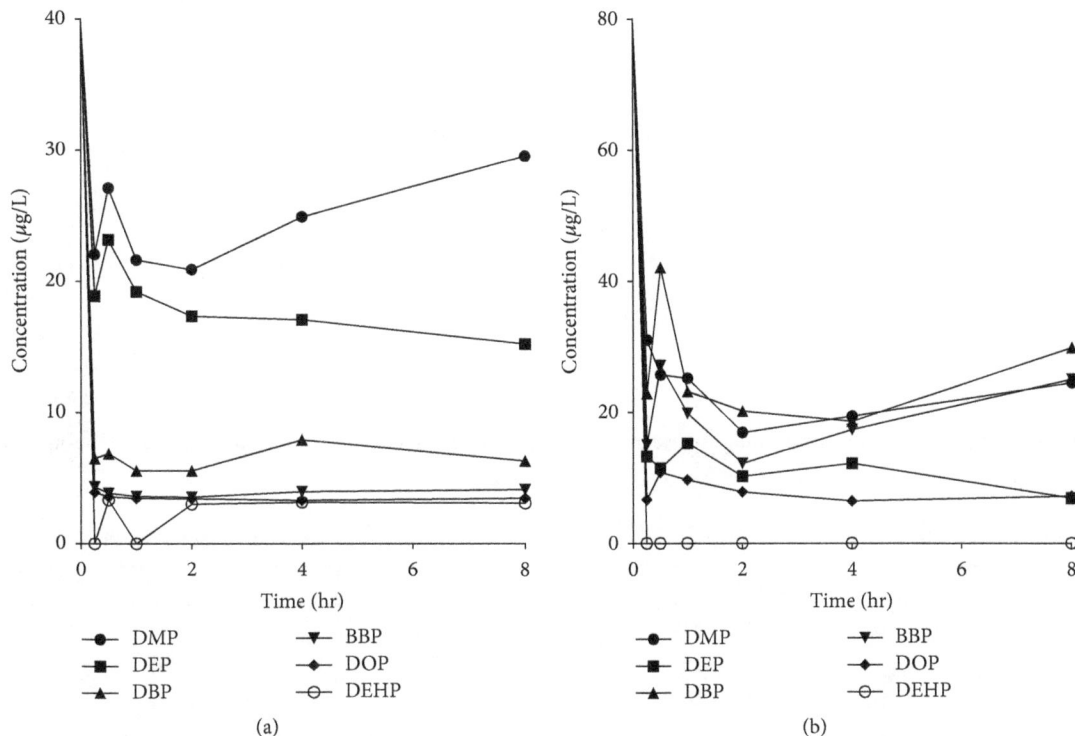

FIGURE 1: The change of PAEs in water phase over time with initial concentration of 40 μg/L (a) or 80 μg/L (b).

Figure 2 presents the adsorption capacities of activated sludge in two hours. Figure 2(a) shows that the adsorption capacity of activated sludge of 6 PAEs reached maximum in 0.5 hr when initial concentration was 40 μg/kg; adsorption capacity did not vary a lot during 0.5 hr to 1 hr but DEHP was not detected; during 1 hr to 4 hr, PAEs in mud phase presented a downward trend, while, during 4 hr to 8 hr, DEP, DBP, and DOP increased slightly and BBP were not detected. Figure 2(b) shows that the adsorption maximum occurred in 0.25 hr when the initial concentration was 80 μg/kg and then showed a downward trend except that DEP kept stable. The decrease was mainly due to the sludge biological degradation [21]. The higher the initial concentration of PAEs, the stronger the adsorption of activated sludge, because of the diffusion force of greater concentration gradient in water. When reaching the equilibrium, the removal rate of DEHP was the highest, while the removal rate of DMP was lowest. Sludge adsorbed much less DMP and DEP than DOP and BBP.

In this research, the process of activated sludge adsorbing PAEs was fitted to first-order kinetics equation as follows:

$$C = C_0 \cdot e^{-kt}, \tag{1}$$

$$t_{1/2} = \frac{\ln 2}{k}, \tag{2}$$

where C (μg/L) is the residual concentration of PAEs at time t, C_0 (μg/L) is the initial concentration of PAEs, k is degradation rate constant, and $t_{1/2}$ (h) is the half-life of the degradation.

Equation (1) was transferred into (3) to obtain rate constants and half-life, shown in Table 2, by linearly fitting above data.

$$\ln C = \ln C_0 - kt. \tag{3}$$

Except for DOP, the removal process of other five PAEs basically was in accord with the first-order kinetics equation,

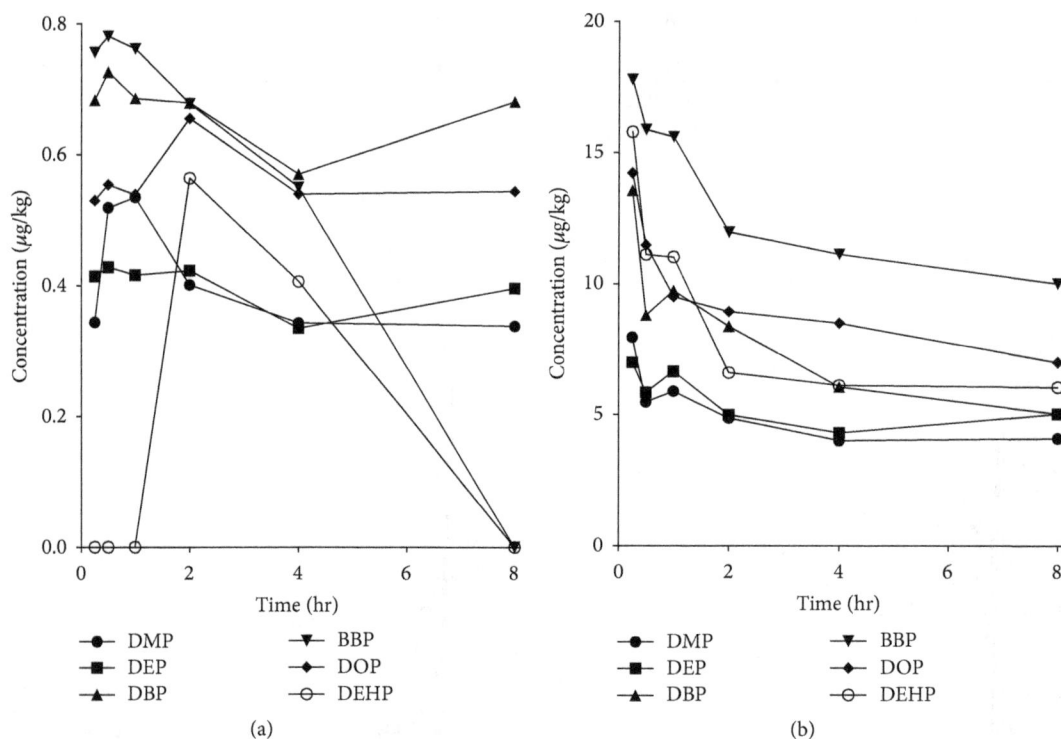

FIGURE 2: The change of PAEs in sludge phase over time with initial concentration of 40 μg/kg (a) or 80 μg/kg (b).

TABLE 2: The rate constants and half-life of activated sludge adsorbing PAEs fitting first-order reaction (initial PAE concentration was 80 μg/L).

Phthalate ester	Kinetic equations	K (h^{-1})	$t_{1/2}$ (h)	R^2
DMP	$\ln C = -2.486t + 4.846$	2.486	0.279	0.9657
DEP	$\ln C = -1.872t + 4.106$	1.872	0.370	0.9380
DBP	$\ln C = -1.599t + 4.499$	1.599	0.433	0.9958
DEHP	$\ln C = -0.697t + 3.011$	0.697	0.994	0.9754
DOP	$\ln C = -1.023t + 3.815$	1.023	0.677	0.5657
BBP	$\ln C = 4.629t + 4.361$	4.629	0.150	0.9630

while the half-life and degradation rate constant were significantly different. The degradation rate constants were sorted in order: $k_{BBP} > k_{DMP} > k_{DEP} > k_{DBP} > k_{DOP} > k_{DEHP}$, while the order of molecular weight was DMP < DEP < DBP < BBP < DOP = DEHP. It implied that the molecular weight and structure of PAEs influenced the adsorption and degradation property of activated sludge. With the increase of molecular weight, carbon chain, and volume of PAEs, steric effect of biological reaction was increased. At the same time, the increase of molecular weight reduced the water solubility of PAEs, had an effect on utilizing microorganism in water, and caused the decline of biological degradation rate. Then the PAEs compete with each other for the adsorption sites on sludge, resulting in the different rate constants.

The mass balance analysis after 8 hr adsorption is shown in Table 3. Except that DEHP, the most commercially applied phthalate [22], was majorly trapped on sludge, other PAEs lost more than one-third, which might result from the

biological degrading though NaN$_3$ was used to limit the activities of micrograms. It has been reported that about 36–42% removal of some PAE attributed to biodegradation of sludge in a nitrogen-removing sequencing batch reactor [21].

3.2. Effect of Temperature. The temperature effect is shown in Figure 3. Less aqueous PAEs remained when water temperature was 25°C, while the concentration of PAEs in sludge decreased with temperature increasing. In water, the temperature affected more DEP and DMP, which were 34.35 μg/L and 30.39 μg/L at 25°C, respectively, both significantly lower than those at 15°C and 35°C. This may be due to the fact that, at the low temperature, the viscosity of PAEs solution and the diffusion rate of molecules were low, resulting in less PAEs adsorbed on the surface of activated sludge; when the temperature raised, the adhesion of the solution decreases became lower and the diffusion was accelerated;

TABLE 3: Mass balance analysis in one-liter mixture solution.

Phthalate ester	Initial mass (μg)	Remains in sludge (μg)	Remains in water (μg)	Loss (μg)
DMP	80	16.4	24.49	39.11
DEP	80	20.12	6.85	53.03
DBP	80	20.12	29.82	30.06
DEHP	80	40.04	25.04	14.92
DOP	80	28	7.16	44.84
BBP	80	24.2	0	55.8

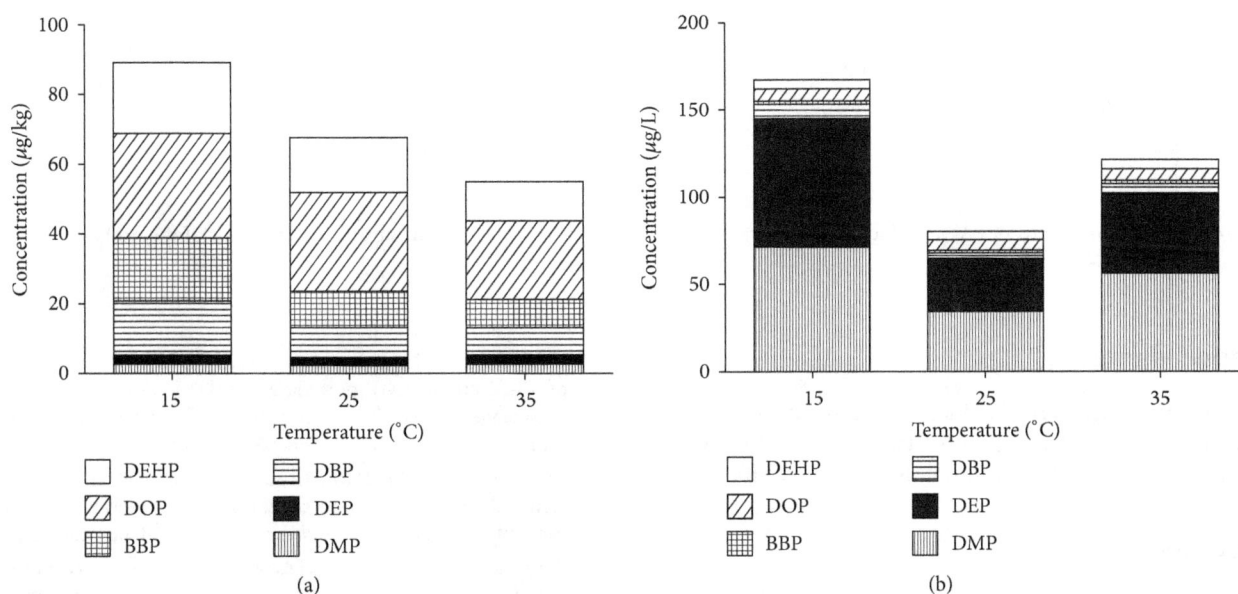

FIGURE 3: The concentrations of PAEs in sludge (a) and water (b) at different temperatures (initial concentration = 80 μg/L, pH = 7.0, and experiment duration = 2 hrs).

but, when temperature went higher, desorption was strengthened since the activated sludge adsorption of PAEs is an exothermic process. For sludge, higher temperature would increase the bacterial activities, enhancing the degradation of PAEs.

3.3. Effect of pH.
Different pH values might inhibit the activity of activated sludge adsorption of PAEs. The results are shown in Figure 4.

As shown in Figure 4(a), except BBP, PAEs in water showed a sharp decreasing as pH increased in the range of 5–9 but no significant change at pH of 9–11. The aqueous pH implies the concentrations of the H^+ and OH^-, and the microbial floc surface of activated sludge is full of protein molecules, containing carbonyl groups (-COOH). When the OH- content was high, -COOH would turn into -COO$^-$, resulting in high negative potential on surface. The hydrolysis of PAEs created electropositive potential, meaning they were easy to be adsorbed on the activated sludge at higher pH. As Figure 4(b) shows, the PAEs in mud concentration did not vary much along the pH range and showed no regular trend for all species.

4. Conclusions

The adsorption capacity of activated sludge was higher with the initial PAEs concentration of 80 μg/L than that of 40 μg/L. Both adsorption isotherm profiles were similar: fast adsorption rate in the first 30 min and adsorption equilibrium during 2 hrs. The adsorption followed first-order kinetic rate.

The adsorption of PAEs on activated sludge favored 25°C in this study, and only DEP and DMP out of six PAEs were influenced significantly by temperature.

The aqueous PAEs concentrations decreased significantly in the pH range of 5–9 and kept relatively constant when the pH was in the range of 9–11.

Competing Interests

The authors declare that they have no competing interests.

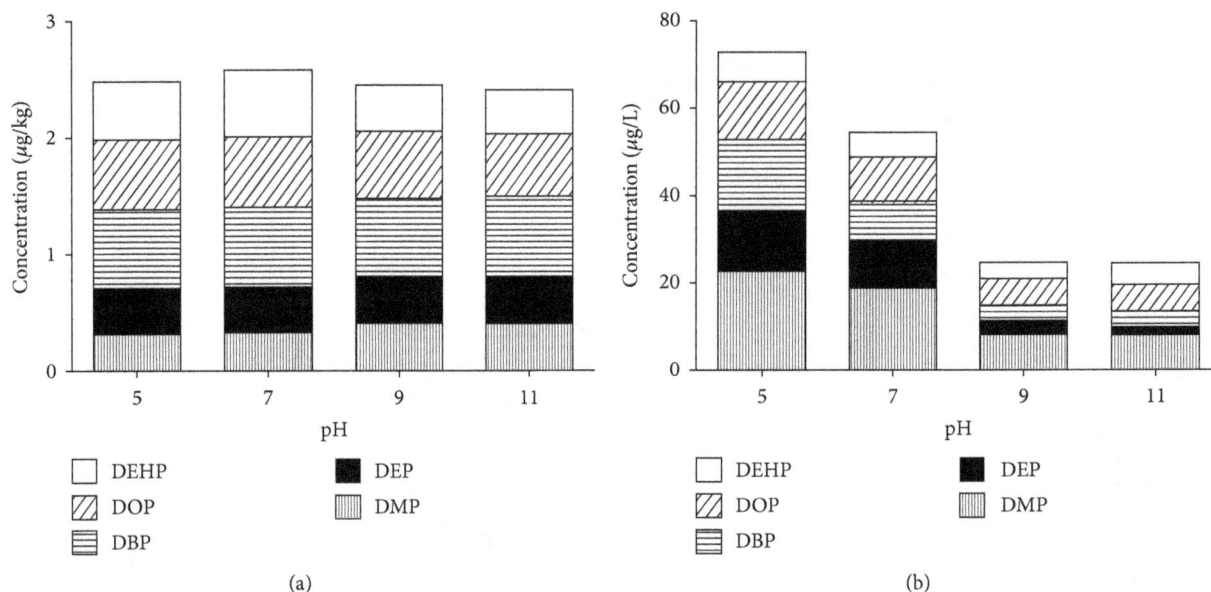

FIGURE 4: The concentrations of PAEs in sludge (a) and water (b) at different pH (initial concentration = 80 μg/L, temperatures = 25°C, and experiment duration = 2 hrs).

Acknowledgments

Funds for this work were provided by the National Natural Science Foundation of China (no. 21407097) and the Shandong Natural and Science Foundation (ZR2014EEM009).

References

[1] F. Alatriste-Mondragon, R. Iranpour, and B. K. Ahring, "Toxicity of di-(2-ethylhexyl) phthalate on the anaerobic digestion of wastewater sludge," *Water Research*, vol. 37, no. 6, pp. 1260–1269, 2003.

[2] M. Matsumoto, M. Hirata-Koizumi, and M. Ema, "Potential adverse effects of phthalic acid esters on human health: a review of recent studies on reproduction," *Regulatory Toxicology and Pharmacology*, vol. 50, no. 1, pp. 37–49, 2008.

[3] M. Wensing, E. Uhde, and T. Salthammer, "Plastics additives in the indoor environment—flame retardants and plasticizers," *Science of the Total Environment*, vol. 339, no. 1–3, pp. 19–40, 2005.

[4] D. Gao, Z. Li, Z. Wen, and N. Ren, "Occurrence and fate of phthalate esters in full-scale domestic wastewater treatment plants and their impact on receiving waters along the Songhua River in China," *Chemosphere*, vol. 95, pp. 24–32, 2014.

[5] L. He, G. Gielen, N. S. Bolan et al., "Contamination and remediation of phthalic acid esters in agricultural soils in China: a review," *Agronomy for Sustainable Development*, vol. 35, no. 2, pp. 519–534, 2015.

[6] E. Bruns-Weller and J. Pfordt, "Determination of phthalic acid esters in foodstuffs, mother's milk, dust, and textiles," *Umweltwissenschaften und Schadstoff-Forschung*, vol. 12, no. 3, pp. 125–130, 2000.

[7] S. Amir, M. Hafidi, G. Merlina et al., "Fate of phthalic acid esters during composting of both lagooning and activated sludges," *Process Biochemistry*, vol. 40, no. 6, pp. 2183–2190, 2005.

[8] H.-F. Cheng, M. Kumar, and J.-G. Lin, "Degradation kinetics of di-(2-ethylhexyl) phthalate (DEHP) and organic matter of sewage sludge during composting," *Journal of Hazardous Materials*, vol. 154, no. 1-3, pp. 55–62, 2008.

[9] M. Barret, L. Delgadillo-Mirquez, E. Trably et al., "Anaerobic removal of trace organic contaminants in sewage sludge: 15 years of experience," *Pedosphere*, vol. 22, no. 4, pp. 508–517, 2012.

[10] Z. Q. Fang and H. J. Huang, "Adsorption of di-n-butyl phthalate onto nutshell-based activated carbon. Equilibrium, kinetics and thermodynamics," *Adsorption Science & Technology*, vol. 27, no. 7, pp. 685–700, 2009.

[11] Z. Xu, W. Zhang, B. Pan, L. Lv, and Z. Jiang, "Treatment of aqueous diethyl phthalate by adsorption using a functional polymer resin," *Environmental Technology*, vol. 32, no. 2, pp. 145–153, 2011.

[12] W. Den, H. C. Liu, S. F. Chan, K. T. Kin, and C. Huang, "Adsorption of phthalate esters with multiwalled carbon nanotubes and its applications," *Journal of Environmental Engineering and Management*, vol. 16, no. 4, pp. 275–282, 2006.

[13] C.-Y. Chen, C.-C. Chen, and Y.-C. Chung, "Removal of phthalate esters by α-cyclodextrin-linked chitosan bead," *Bioresource Technology*, vol. 98, no. 13, pp. 2578–2583, 2007.

[14] H. W. Chan, T. C. Lau, P. O. Ang, M. Wu, and P. K. Wong, "Biosorption of di(2-ethylhexyl)phthalate by seaweed biomass," *Journal of Applied Phycology*, vol. 16, no. 4, pp. 263–274, 2004.

[15] Q.-Y. Cai, C.-H. Mo, Q.-T. Wu, Q.-Y. Zeng, and A. Katsoyiannis, "Occurrence of organic contaminants in sewage sludges from eleven wastewater treatment plants, China," *Chemosphere*, vol. 68, no. 9, pp. 1751–1762, 2007.

[16] X.-Z. Meng, Y. Wang, N. Xiang et al., "Flow of sewage sludge-borne phthalate esters (PAEs) from human release to human intake: implication for risk assessment of sludge applied to soil," *Science of the Total Environment*, vol. 476-477, pp. 242–249, 2014.

[17] H. H. P. Fang and H. Zheng, "Adsorption of phthalates by activated sludge and its biopolymers," *Environmental Technology*, vol. 25, no. 7, pp. 757–761, 2004.

[18] P. Serôdio and J. M. F. Nogueira, "Considerations on ultra-trace analysis of phthalates in drinking water," *Water Research*, vol. 40, no. 13, pp. 2572–2582, 2006.

[19] J. Zhao, Y. Li, C. Zhang, Q. Zeng, and Q. Zhou, "Sorption and degradation of bisphenol A by aerobic activated sludge," *Journal of Hazardous Materials*, vol. 155, no. 1-2, pp. 305–311, 2008.

[20] X. Vecino, L. Rodríguez-López, J. M. Cruz, and A. B. Moldes, "Sewage sludge polycyclic aromatic hydrocarbon (PAH) decontamination technique based on the utilization of a lipopeptide biosurfactant extracted from corn steep liquor," *Journal of Agricultural and Food Chemistry*, vol. 63, no. 32, pp. 7143–7150, 2015.

[21] S. K. Marttinen, M. Ruissalo, and J. A. Rintala, "Removal of bis (2-ethylhexyl) phthalate from reject water in a nitrogen-removing sequencing batch reactor," *Journal of Environmental Management*, vol. 73, no. 2, pp. 103–109, 2004.

[22] M. Julinová and R. Slavík, "Removal of phthalates from aqueous solution by different adsorbents: a short review," *Journal of Environmental Management*, vol. 94, no. 1, pp. 13–24, 2012.

8

Review of Upflow Anaerobic Sludge Blanket Reactor Technology: Effect of Different Parameters and Developments for Domestic Wastewater Treatment

M. K. Daud [ID],[1,2] Hina Rizvi [ID],[3] Muhammad Farhan Akram,[3] Shafaqat Ali [ID],[3,4] Muhammad Rizwan,[3] Muhammad Nafees,[5] and Zhu Shui Jin [ID][1]

[1]Institute of Crop Science, College of Agriculture and Biotechnology, Zhejiang University, Zijingang Campus, Hangzhou, China
[2]Department of Biotechnology and Genetic Engineering, Kohat University of Science and Technology, Kohat 26000, Pakistan
[3]Department of Environmental Sciences and Engineering, Government College University, Allama Iqbal Road, Faisalabad 38000, Pakistan
[4]Key Laboratory of Soil Environment and Pollution Remediation, Institute of Soil Science, Chinese Academy of Sciences, Nanjing 210008, China
[5]Institute of Soil & Environmental Sciences, University of Agriculture, Faisalabad 38000, Pakistan

Correspondence should be addressed to Hina Rizvi; hinarizvi1@gmail.com and Zhu Shui Jin; shjzhu@zju.edu.cn

Academic Editor: Claudio Di Iaconi

The upflow anaerobic sludge blanket (UASB) reactor has been recognized as an important wastewater treatment technology among anaerobic treatment methods. The objective of this study was to perform literature review on the treatment of domestic sewage using the UASB reactor as the core component and identifying future areas of research. The merits of anaerobic and aerobic bioreactors are highlighted and other sewage treatment technologies are compared with UASB on the basis of performance, resource recovery potential, and cost. The comparison supports UASB as a suitable option on the basis of performance, green energy generation, minimal space requirement, and low capital, operation, and maintenance costs. The main process parameters such as temperature, hydraulic retention time (HRT), organic loading rate (OLR), pH, granulation, and mixing and their effects on the performance of UASB reactor and hydrogen production are presented for achieving optimal results. Feasible posttreatment steps are also identified for effective discharge and/or reuse of treated water.

1. Introduction

The wastewater generated by a community is called "sewage," which is a mixture of domestic wastewater, industrial wastewater (where the industry is discharging its wastewater in the same sewage system), and rain water, where a single sewer system exists for wastewater and storm water [1]. In the developing/underdeveloped countries of the world, more than 90% of the sewage is discharged untreated in the environment due to lack of proper wastewater collection and treatment facilities. The quantity and strength of wastewater are governed by the size and socioeconomic condition of the population of the area [2]. The composition of sewage varies greatly and its characterization is important for determining the size and designing of treatment plant [3]. Table 1 provides an overview of characteristics of municipal sewage in various cities of the world.

Anaerobic treatment is preferred to treat municipal wastewater because of its merits over conventional treatment methods [1]. These advantages are (i) its ability to treat high COD loads and withstand fluctuation in the influent, (ii) biogas formation, and (iii) effective treatment of wastewater in a short period of time [4]. Anaerobic reactors reduce pollution load and provide good stabilization of solids. Furthermore, depending on the design of a UASB reactor, a high sludge hold-up time can be obtained so that the excess sludge needs to be discharged only once every three to four years [5]. The comparison of aerobic and anaerobic technologies is given in Table 2.

TABLE 1: Characteristics of sewage in different cities of the world.

Parameters	Pakistan [82] Karachi	Pakistan [82] Lahore	Palestine Al Bireh [3]	Brazil Pedregal [1]	Columbia Cali [1]	Netherlands Bennekom [1]
BOD	220–475	200–215	-	368	95	231
COD	200–1400	580–803	1586	727	267	520
Chlorides	300–1200	32–72	-	110	-	-
Sulfates	50–200	-	-	18	-	15
NH_4^+-N	-	-	80	34	17	-
Nkj-N	-	-	104	44	24	-
Total P	-	-	13	11	1.3	18
PO_4^{-3}-P	-	-	12.9	8	-	14
TDS	1000–1800	486–598	-	-	-	-
TSS	250–900	106–176	736	492	215	-
VSS	-	-	617	252	107	-
Temperature	-	-	-	24–26	24–27	8–20

TABLE 2: Comparison of aerobic and anaerobic treatment methods.

Parameter	Aerobic treatment	Anaerobic	References
Process	Degradation of organic matter occurs in the presence of oxygen	Degradation of organic matter occurs in the absence of oxygen	[83]
By-products	The process generates carbon dioxide, water, and excess biomass	The process generates carbon dioxide, methane, and excess biomass	[44, 84]
Applicability	Highest removal efficiency for wastewater having low to medium organic content (COD < 1000 ppm) that is complex to biodegrade, for example, municipal sewage and refinery wastewater	Highest removal efficiency for wastewater having medium to high organic content (COD > 1000 ppm) and easy biodegradability, for example, food and beverage industry's wastewater rich in organic content. But also applicable to low strength wastewater (COD > 300 and <1000 mg/L)	[25, 85]
Reaction kinetic	Decay rate $k_d = 0.06$ d^{-1}	Decay rate $k_d = 0.03$ d^{-1}	[44, 83, 84]
Sludge yield coefficient (kgVSS/Kg COD)	0.35–0.45 (relatively high) Biomass yield is fairly constant irrespective of the type of substrate metabolized	0.05–0.15 (relatively low) biomass yield is not constant and varies with the type of substrate metabolized	[86]
Posttreatment	Direct discharge, followed by filtration/disinfection	Generally done by aerobic methods	[21, 83, 87, 88]
Foot print	1.0 to 2.4 kgCO$_2$/kgCOD removed (depending on the wastewater strength)	0.5 to 1.0 KgCO$_2$/kgCOD removed (depending on the wastewater strength)	[44, 45, 85, 89]
Capital cost	12–40 US$/inhab.	40–65 US$/inhab.	[49, 90–92]
Typical technologies	Activated sludge, trickling filters, extended aeration, oxidation ditch, downflow hanging sponge (DHS), Membrane Bioreactor (MBR), Moving Bed Bioreactor (MBBR)	UASB, continuously stirred tank reactor/digester/Upflow Anaerobic Filters, ultrahigh rate fluidized bed reactors, hybrid high rate reactors, two-stage UASB reactor	[15, 24, 60, 67]

The major achievement in the development of anaerobic treatment was the introduction of high rate reactors in which biomass retention and liquid retention were not interlinked [5]. Among the various anaerobic wastewater treatment technologies, upflow anaerobic sludge blanket (UASB) reactors have achieved considerable success and these reactors have been applied to treat a wide range of effluents such as sugar, pulp and paper, dairy, chemical, potato starch, bean balancing, soft drinks, fish processing, noodle processing, yeast production, slaughterhouse, and coffee processing industries [6–8]. The UASB process was developed by Lettinga and coworkers in the late 1970s [6]. It is primarily used for the treatment of highly concentrated industrial wastewaters [7–9]; however, it can also be used for the treatment of low strength wastewater such as municipal wastewater with relatively lower contaminant strength [10–14]. As compared to aerobic technologies anaerobic treatment systems such as UASB are being encouraged because of several advantages, including plain design, uncomplicated construction and maintenance, small land requirement, low construction and operating cost, low excess sludge production, robustness in terms of COD removal efficiency, ability to cope with fluctuations in temperature, pH and influent concentration, quick biomass recovery after shutdown, and energy generation in the form of biogas or hydrogen [1, 9, 15–19]. These characteristics make UASB a popular wastewater treatment option [20, 21] and a large number of researchers have recommended UASB technology for the treatment of sewage wastewater in tropical and subtropical regions [10–12, 14, 22–24].

It is particularly appealing in tropical countries where the comparatively high ambient temperature is nearly optimum for the mesophilic methanogenic bacteria [11, 13]. Significant efforts were made in the past twenty years to ascertain the mass transfer and kinetic processes going on in the reactor. In this study, the advantages and disadvantages of anaerobic and aerobic bioreactors are highlighted and comparison is made of UASB with other sewage treatment methods to state their feasibility and efficiency in domestic wastewater treatment and identify possible future areas of research. The effects of main process parameters, temperature, HRT, OLR, pH, granulation, and mixing on the performance of UASB reactors and hydrogen production are provided for optimal growth of bacteria and performance of the system. Appropriate post-treatment options are also identified to be potentially used in developing countries having appropriate climate conditions.

2. Upflow Anaerobic Sludge Blanket Reactor Process

Wastewater to be treated is introduced from the bottom of the reactor and it flows upward through a blanket of biologically activated sludge, which is generally in the form of granular aggregates. The sludge aggregates have very good stability and do not get washed out under practical conditions and therefore provide good treatment efficiency when the wastewater comes in contact with the granules [25]. The gases (methane and carbon dioxide) produced under anaerobic conditions cause internal mixing, which helps in the formation and maintenance of biological granules. However, some of the gas produced in the sludge blanket is attached to the granules, and a gas-liquid-solid separator (GLSS) is added on the top of the reactor for the effective segregation of gas, liquid, and granules. In GLSS, the gas surrounded particles strike with the bottom of degassing baffles and fall back into the sludge blanket and the treated water flows out of the reactor [21, 26, 27].

There is lower gas production in sewage as compared to high strength wastewater, which leads to less circulation of gas to support the formation of biological granules. Therefore, control of channeling is important for weaker wastewaters like sewage [6, 13].

3. Treatment Potential of UASB Process

UASB technology has been effectively used for the treatment of a wide range of wastewaters but is generally inhibited by incomplete biodegradability of complex wastewaters [10, 28].

The UASB technology has been found to be very effective for the treatment of wastewater with a high content of carbohydrates. Carbohydrate rich organic wastewater, such as starch or canning industry wastewater is easily digested by microbes and is thus a nutrient-rich starting material for anaerobic hydrogen production. Upflow anaerobic sludge blanket (UASB) reactor has therefore turned out to be one of the most popular designs for the treatment of wastewaters from food processing industries. Anaerobic reactors have the ability to withstand variations in wastewater quality and complete shutdown of reactor in off season [28, 29]. Fang analyzed the microscopic (SEM and TEM) examinations of biogranules sampled from different UASB reactors. The microbial distribution in granules was found strongly dependent on the thermodynamics of degradation and kinetics of particular substrates. Biogranules degrading carbohydrates had a layered distribution in which the surface layer was occupied by hydrolytic/fermentative acidogens; the mid-layer consisted of syntrophic colonies and interior layer was comprised of acetotrophic methanogens, whereas substrates having a rate-limiting hydrolytic/fermentative step did not have a layered structure and bacteria were distributed evenly. The results reveal that granules are developed through evolution process and not through random aggregation of suspended microbes [29]. Moreover, as supported by experimental evidence biogranules degrading carbohydrates are more resistant than suspended sludge to the toxicity of hydrogen sulfide, heavy metals, and aromatic pollutants in wastewater [30].

About half the organic matter content in domestic wastewater is attributable to black water having a major portion of nutrients. High rate UASB reactors are popular and have been applied widely due to their ability to manage high OLRs, short HTRs, and low energy requirements [31, 32]. A number of full-scale plants are operational and several others are at present under construction, especially in tropical or subtropical areas [33, 34]. Table 3 summarizes the advantages and disadvantages of UASB reactor. Basic concept of UASB technique is to build up anaerobic sludge that has good settling characteristics [1] and that can hold highly active bacterial aggregation without the requirement for immobilization on a support material [35]. About 60% of the full-scale anaerobic treatment plants operating worldwide are currently based on the UASB design model, treating a wide range of industrial wastewaters [7, 8, 36, 37]. About 793 UASB reactors are installed worldwide showing the widest application of this technology out of total of approximately 1229 other anaerobic applications which include anaerobic contact filter, Expanded Granular Sludge Bed (EGSB) reactor, hybrid reactor, and fluidized bed. About half of these installations (338 out of 793) are in tropical and subtropical regions [38]. Table 4 shows the performance of UASB reactors as compared to other technologies. On the basis of overall low investment and O and M costs, satisfactory COD, BOD, and TSS removal, and energy generation potential, this technology is suitable as compared to other technologies especially in developing countries. New research studies in this area have proven the successful operation of this system to treat low strength wastewater [12, 39]. The concentration of COD in domestic wastewater is usually low and suspended solids are high along

TABLE 3: Advantages and disadvantages of UASB reactor technology.

Sr. number		References
	Advantages of UASB technology	
(1)	Provides high removal efficiency even at high OLR and low temperature and therefore requires smaller reactor volume	[11, 70]
(2)	Simple construction and low operation and maintenance cost due to local availability of construction material and other parts	[84]
(3)	Robustness in treatment efficiency and wide applicability from very small to very large scale	[11, 22]
(4)	Energy is generated as methane/hydrogen gas. Energy generated can be used to heat the boilers to reduce operation costs. Less energy demand when external temperature control is not required	[17, 18]
(5)	Less CO_2 emissions due to less energy requirement and additional energy production in the form of biogas that can be used to run the system	[44]
(6)	Low sludge production as compared to aerobic processes. The sludge produced is stabilized having good dewatering characteristics can be stored for extended time periods and reused as an inoculum for seeding UASB reactors	[77]
(7)	Quick startup time (about one week) by means of granular anaerobic sludge as seed	[49]
(8)	Ability to withstand organic shock loads	[11, 25]
(9)	Ability to treat sewage due to availability of macro- and micronutrients and stability of pH without addition of any chemicals	[73, 76]
	Disadvantages of UASB technology	
(10)	Needs posttreatment as pathogens are not removed completely, except for helminthes eggs that are successfully entrapped in the sludge. Incomplete removal of nutrients and therefore needing posttreatment	[21, 85]
(11)	Long startup time is required due to the slow growth rate of microorganisms in case activated sludge is not amply available	[20]
(12)	Odor, toxicity, and corrosion problem: H_2S is generated in anaerobic digestion, particularly when there is high concentration of sulfate in the wastewater. The biogas needs further suitable handling to avoid bad odor. The top gas box and the deflector can be constructed from concrete with epoxy coating to prevent corrosion and gas leakage	[85, 93]
(13)	In cold regions temperature needs to be maintained within (15–35°C) to achieve steady state performance	[5]
(14)	A considerable portion of biogas produced may be dissolved in the effluent whose recovery is needed	[10, 94]

with low methane yield that requires initial hydrolysis to convert the suspended solids into soluble substrate. Hydrolysis is commonly the limiting step at low temperature setting. Therefore, the UASB reactors for domestic wastewater treatment having high level of suspended solids are practicable only at higher temperature which may require an outside heat source [32, 40]. Better process understanding and operational knowledge on granules structure have made it possible to apply high organic loads and resulted in reducing startup time and providing a more sustainable operation [40, 41]. Its performance in the world proves it to be a consistent and efficient system for wastewater treatment [1].

4. Resource Recovery Potential

Anaerobic technologies have the ability to recover the chemical energy of organic carbon in wastewater as methane, hydrogen, and electricity. In UASB reactor, solids are captured and organic matter is converted into biogas consisting largely of methane and carbon dioxide. Organically bound nitrogen is transformed to ammonium and sulfate is reduced to hydrogen sulfide. Moreover, sludge generation is low and sludge produced is highly stable with good dewaterability

characteristics [42]. Under aerobic conditions, there is a complete loss of biomass energy in low value heat during the oxidation of organic matter, whereas under anaerobic degradation process, the original energy of the biomass is not changed. About 13.5 MJ of energy is stored in one kilogram COD. Depending on the extent to which anaerobic conversion is permitted, this energy is captured in the form of methane, hydrogen, other gases, or liquid compounds such as alcohols. Energy recovery is an old process which exists as alcohol fermentation and is now renewed by the production of biofuels from organically rich substrates [43]. In terms of carbon foot print, anaerobic wastewater treatment is more advantageous, based on the relative efficiency of the aerobic system (Table 2). In case of effluent having low strength (BOD less than 300 mg/L), aerobic process is favorable, as it generates less greenhouse gas. But, in case of wastewater having higher strength (BOD more than 300 mg/L) anaerobic treatment is more beneficial. The recovery of dissolved methane in effluents in an economic way makes anaerobic wastewater treatment favorable in reducing carbon foot print for all wastewater strengths [44, 45]. CH_4 produced in the reactor should always be used for energy production and methane dissociation in the effluent and sulfate reduction

TABLE 4: Performance comparison of UASB with other wastewater treatment technologies.

Sr. number	Parameter	UASB	Activated Sludge Process (ASP)	Trickling Filters (TF)	Waste stabilization pond (WSP)	Moving Bed Bioreactor (MBBR)	References
(1)	BOD removal, %	75–83	85–90	80–90	75–85	85–95	[10, 62, 92, 95–98]
(2)	COD removal, %	70–80	80–95	85–90	70–85	85–90	[10, 44, 62, 92, 95–100]
(3)	TSS removal, %	70–80	85–90	75–85	70–85	85–95	[10, 62, 92, 95–97]
(4)	Overall HRT	4–10 hrs	12–14 hrs	13–14 hrs	8–15 days	8–12 hrs	[49]
(5)	Average applied OLR for sewage treatment	1.0–2.0 KgCOD/m^3·day	0.3–0.55 KgBOD/m^3·day	1.5 to 2.0 KgCOD/m^3·day	50–450 kgBOD/ha·d	4–30 gCOD/m^2·day	[11, 20, 31, 42, 49, 101–106]
(6)	Average area required (m^2/mld)	1450	1820	1620	8000	450	[62]
(7)	Biogas generation	0.05–0.25 (m^3/Kg COD removed)	Nil	Nil	0.05 to 1.5 m^3/Kg BOD$_5$ removed (not usually collected)	Nil	[49, 104–106]
(8)	Economic life in years	30	30	30	30	30	[62]
(9)	Total annual O&M costs	1.0–1.5 US$/inhab. year	4.0–8.0 US$/inhab. year	4.0–6.0 US$/inhab. year	1.5–2.5 US$/inhab. year	10.98 Rupees Lacs/MLD year[a]	[62, 92]
(10)	Total construction costs	12–20 US$/inhab.	40–65 US$/inhab.	50–60 US$/inhab.	20–40 US$/inhab.	42.04 Rupees Lacs/MLD[a]	[62, 92]

[a]O and M costs are based on Indian Rupees in year 2008 and are calculated considering constant annual expenditure over the life of the plant.

must be examined. About 20% of the CH_4 produced is dissolved in the effluent at ambient temperature of 20°C and biogas having methane content of 70%. As a result the benefits of energy production are reduced significantly. Nonetheless, diffusion of methane is high at low temperatures (less than 18°C) and it can be removed easily by a stripping chamber or air diffusion [46]. In case of high sulfate or sulfite content, sulfate reducing bacteria compete with the methanogens for methanogenic substrates and therefore the amount of methane generated is decreased [3]. In the United States and other developed countries, methane generated from anaerobic digestion is combusted for energy production, at large treatment plants, and is at least flared and converted to CO_2 at smaller plants [47]. Increased biogas production, efficient biogas usage, and in turn decreased addition of external fossil will lead to reduction of carbon footprint of a wastewater treatment plant [48].

Hydrogen produced in anaerobic fermentation is a clean, carbon-free energy carrier gas, a substitute for fossil fuel, that is considered to play a main role in the future hydrogen based economy due to its high energy yield (122 kJ/g). The major problem in hydrogen production from biological processes is the low production rates and hydrogen yield at a large scale. Present research on hydrogen fermentation is directed towards determining culture details, beneficial substrates, conditions affecting microbial conversions, sequential fermentation, combined fermentation, mixed culture fermentation, optimizing enrichment, and stability of acidogenic sludge and reaction kinetics [49, 50]. Acetate (HAc), butyrate (HBu), caproate (HCa), propionate (HPr), and lactate (HLa) make up the metabolic pathways involved in the acidogenesis process in the UASB reactor for hydrogen production. HAc, HBu, and HCa are the desirable metabolites by-products for hydrogen production pathways. On the other hand, HLa and HPr are undesirable and consuming hydrogen pathways [51]. A number of studies have been done on the generation of biohydrogen from food processing industries' wastewater; however very few studies have been carried out on hydrogen production from domestic wastewater. Potential of hydrogen production is now being realized by using domestic wastewater in anaerobic reactors [52, 53]. Fernandes et al. [53] reported hydrogen yield of 200 ml H_2/gCOD by using domestic sewage as substrate in anaerobic reactor. Similarly, Paudel et al. [52] reported hydrogen yields of 1.014 mol H_2/mol glucose at 3 g COD/L day loading rate using synthetic domestic wastewater. The operational conditions play a crucial role in hydrogen production. A pH of about 5.0–5.5 is reported to favor hydrogen production as the intermediates produced in the reactor can form appropriate ratios and stimulate the metabolism of particular microbes responsible for hydrogen fermentation. For both high rate hydrogen production and organics removal, organic acids generated in effluent can be methanized in a methanogenic reactor after hydrogenic reactor and energy generated from hydrogen may be utilized in posttreatment units where anaerobic reactor is the core technology for sewage treatment [53].

Factors such as inoculum, temperature, OLR, HRT, and pH must be taken into account for optimal hydrogen production. Use of inoculum previously adapted to hydrogen production greatly enhances hydrogen rate and yield as compared to nonadapted one. In mixed culture, hydrogen producing bacteria may consume the hydrogen produced and they need to be removed to curtail hydrogen consumers and enrich hydrogen producers [52–54]. Previous studies do not show an optimal range of HRT and OLR in fermentative hydrogen production due to varying environmental process parameters or ranges of the variables reported in literature. The OLR can be optimized only when the microbes are well acclimatized to the applied OLR. The optimum temperature for hydrogen production through dark fermentation depends on the type of hydrogen producing bacteria and carbon source used [55]. Hydrogen producing bacteria belong to mesophilic and thermophilic groups and generally rate of hydrogen production increases with increase in temperature through thermophilic bacteria compared to mesophilic bacteria [56].

Mu and Yu [57] demonstrated that hydrogen could be generated continuously and steadily from an acidogenic-granule-based UASB reactor at varying concentration of substrate (5.33–28.07 g-COD/L) and HRTs of 3–30 h, during treatment of sucrose-rich synthetic wastewater. They found that H_2 partial pressure in biogas decreased with increasing substrate concentration; however it was not affected by the change of HRT in a range of 6–22 h. The rate of hydrogen production increased with increasing substrate concentration but decreased with increasing HRT with a hydrogen yield in the range of 0.49–1.44 mol-H2/mol glucose [47].

Furthermore, research is underway on directly producing electric current from electrons released in anaerobic fermentation. These electrons are scavenged at an anode and with the use of cathode under oxidizing conditions electric current is generated which can be used in decentralized plants directly without any further conversion process [58, 59].

5. Posttreatment Requirements

The UASB reactors do not warrant the removal of remaining organic matter, nutrients, and pathogens. Therefore, posttreatment of anaerobically treated wastewater is usually required in order to improve the quality of effluent in accordance with the irrigation standards. This has been successfully done by conventional systems such as maturation ponds, waste stabilization ponds, polishing ponds, constructed wetlands, rotating biological contactors, moving bed biofilm reactor, downflow hanging sponge [60–62], and advanced oxidative processes (AOPs) [63, 64].

Polishing ponds are natural systems mainly for removal of solids which have been effectively used for the posttreatment of upflow anaerobic sludge blanket (UASB) effluents. They are mainly used as maturation ponds for the removal of pathogens, nitrogen, and remaining organic matter from the UASB reactor effluent [65]. The use of sequential anaerobic-aerobic technology such as UASB-activated sludge for municipal wastewater treatment combines the benefits of both systems in an economical arrangement. The combined system is energy efficient, less sludge producer, and relatively less complex for domestic wastewater treatment as compared to other options. An added advantage is the biological oxidization of the dissolved methane; however, the overall greenhouse gas reduction will depend on the energy consumption of aerobic step [44, 66]. The UASB and posttreatment system can be applied consecutively or in an integrated manner.

6. Effect of Different Parameters on the Efficiency of UASB Reactor

The efficiency of UASB reactors is regulated by a large number of factors including wastewater characteristics, acclimatization of seed sludge, pH, nutrients, presence of toxic compounds, loading rate, upflow velocity (V_{up}), hydraulic retention time (HRT), liquid mixing, and reactor design that affect the growth of sludge bed [16, 67–69].

7. Effect of Temperature

Temperature considerably influences the growth and survival of microorganisms. Anaerobic treatment is possible at all three temperature ranges (psychrophilic, mesophilic, and thermophilic); however, low temperature generally leads to a decline in the maximum specific growth rate and methanogenic activity [39, 70, 71]. Methanogenic activity at this temperature range is 10 to 20 times lower than the activity at 35°C, which requires an increase in the biomass in the reactor (10 to 20 times) or operating at higher sludge retention time (SRT) and hydraulic retention time (HRT) in order to achieve the same COD removal efficiency as that obtained at 35°C [3, 67].

It is argued that the decrease of temperature slows down the hydrolysis and decreases the maximum growth and substrate utilization rates [5]. Singh et al. [72] treated municipal wastewater using a UASB system under low temperature conditions and reported 70% COD removal at 11°C and 30 to 50% at 6°C. Lew et al. [40] found a gradual decrease in COD removal efficiency as the temperature was decreased. They reported 82% COD removal at 28°C, 72% at 20°C, 68% at 14°C, and 38% at 10°C. Kalogo and Verstraete [73] also found that COD removal efficiency at temperature in the range of 10–15°C was lower than that of efficiency at 35°C.

Van Lier and Lettinga [74] studied the effect of transient temperature rise on the performance of a UASB reactor containing mesophilic microorganisms. There was an increase in the methane production with an increase in the temperature due to the accelerated methanogenic activity. However, a sharp drop in the methane generation was noted at the reactor temperature exceeding 45°C because of a substantial decline in the activity of mesophilic granular sludge due to bacterial inactivation. Halalsheh et al. [75] treated high strength sewage (COD = 1531 mg/L) using a UASB pilot plant under subtropical conditions. The COD_{tot} removal efficiencies were 62% and 51% in summer and winter, respectively, when the plant was operated at ambient temperature (18–25°C) and hydraulic retention time of 24 hours.

8. Effect of Hydraulic Retention Time (HRT)

Hydraulic retention time (HRT) is one of the most important parameters affecting the performance of the reactor especially in case of municipal wastewater [76, 77]. The upflow velocity (V_{up}) is directly related to HRT and plays an important role in entrapping suspended solids. A decrease in V_{up} entails an increase in HRT, which boosts suspended solids' (SS) removal efficiency of the system [1, 78, 79]. The COD removal efficiency of a UASB reactor also decreases at elevated upflow velocity because higher V_{up} reduces the contact time between sludge and wastewater in addition to smashing of sludge granules and resultantly higher washout of solids [3, 11, 80]. However, some scientists reported no distinct effect of HRT on the treatment efficiency of UASB reactor [22, 77]. The difference of opinion in scientific community is attributable to the difference in the reactor design, operating procedures, and range of HRT. Flow rate is also a key operational parameter that maintains the hydraulic retention time. In UASB process, if diameter of reactor will be too large then it may cause liquid channeling in the reactor leading to insufficient contact between the substrate and biomass. Therefore, large reactor will result in decreased biogas production and sludge washout due to insufficient mixing within the reactor. In contrast, comparatively more height may promote substrate mixing leading to proper contact of substrate with microorganisms resulting in more organic matter degradation and formation of biogas [81].

The upflow velocity (V_{up}) is helpful in providing adequate mixing of the substrate and biomass without channeling and maintaining the hydraulic retention time. The allowable limit of upflow velocity is 0.5–1.5 m/h as reported by different researchers. For the treatment of municipal wastewater in UASB reactor, researchers have reported successful operation at 0.31–0.426 m/h V_{up} at 3-4 h of HRT [32, 75].

9. Effect of Organic Loading Rate (OLR)

OLR is the main parameter significantly affecting microbial ecology and functioning of UASB process. In case of sewage, the OLR usually is applied in the range of 1.0–2.0 KgCOD/m^3·day [42, 101]. UASB reactor is preferred for its potential to treat wastewater having low content of suspended solids and gives higher methane yield [27]. The reactors seeded with granular activated sludge can give high performance within a brief startup period and can also adapt quickly to increase of OLR [25]. The effect of OLR on the performance of a UASB reactor depends on a number of factors which sometimes have a dissimilar effect, mostly contradictory, on the performance of UASB reactor [11]. Researchers have reported an increase in the efficiency of high rate anaerobic reactors with increasing OLR [107]. However, that increase is up to a certain OLR, beyond which there occurs sludge bed flotation and excessive foaming in the gas-liquid-solids separator (GLSS); therefore a range of optimum OLR is usually recommended for a given temperature range and wastewater [108]. Seghezzo [10] operated pilot-scale UASB reactors with OLR of 0.6 kgCOD/m^3·day (HRT of 6 hours and COD influent = 153 mg/L) and achieved maximum 63% COD removal efficiency at a low temperature of 17°C. Farajzadehha et al. [108] carried out a study on a lab-scale UASB reactor for determining the optimal organic loading rate and hydraulic retention time. Substrate used was fortified municipal wastewater at volumetric organic loadings of 3.6, 7.2, 10.8, and 14.4 kgCOD/m^3·day and temperature of 30 and 20°C. The results indicated an optimal organic loading rate range of 7.2 to 10.8 kgCOD/m^3·day at both temperature conditions attaining COD removal efficiency of 85% and 73% in UASB reactor at 30°C and 20°C, respectively. Nitrate removal efficiency was about 80% at optimized organic loading rate and there was a 30% decrease in nitrate removal efficiency when OLR was increased to 14.4 kgCOD/m^3·day. Halalsheh [22] determined the performance of UASB reactors for treatment of strong domestic wastewater at OLR of approximately 1.5 kgCOD/m^3·day (COD influent = 1500 mg/L) with a high portion of suspended solids (about 80%) at HRT of 24 hours. Even at considerably long HRT, the reactor achieved a COD removal efficiency of only 62% in summer (25°C). The comparatively low efficiency was caused by the sludge washout. Higher than optimal OLR results in the accumulation of biogas in the sludge bed forming gas pockets that ultimately cause sludge flotation [25]. Leitao [11] determined the effect of influent concentration, HRT, and OLR on the COD removal efficiency of UASB reactors employed to treat sewage. Eleven (11) pilot-scale UASB reactors were divided into three sets. In first set, five reactors were operated at constant HRT of 6 hrs and with different COD$_{Inf}$ (92–816 mg/L) and corresponding OLR ranging from 0.2 to 3.3 kgCOD/m^3·day. In second set, four reactors were operated with approximately the same COD$_{Inf}$ (~800 mg/L) but different HRTs (1, 2, 4, and 6 hrs) and corresponding OLR ranging from 3.3 to 17.6 kgCOD/m^3·day. In third set, HRTs were same as second set but the COD$_{Inf}$ was modified (136–816 mg/L) to have approximately the same organic loading rate (OLR~ 3.3 kgCOD/m^3·day) in the four reactors. Results revealed that although the reactors of third set were operated with same OLR (3.3 kg COD/m^3·day) and different HRTs and COD$_{Inf}$, they resulted in completely different COD removal efficiency varying from 13 to 57%. According to Leitao [11], as OLR is dependent on wastewater strength, upflow velocity, and volume of reactor, it is thus also dependent on the applied HRT. Therefore, impact of OLR on reactor performance is not simple, as it is dependent on other parameters, which have opposing effects on the removal efficiency of UASB; that is, increasing the feed concentration from 98 to 818 mg/L (and hence OLR of 0.4 to 3.3 kgCOD/m^3·day) up to a certain limit (OLR of 1.2 kg COD/m^3·day) caused an increase in removal efficiency (50 to 64%), whereas increasing the flow rate (and therefore OLR from 3.3 to 17.6 kg COD/m^3·day) caused a decrease in total COD efficiency from 57 to 36%.

In the case of effluent having COD contents lower than 300 mg/L (OLR: 0.4 to 0.8 kgCOD/m^3·day), the COD removal efficiency of UASB reactors was low (50–53%). However, reactors showed maximum COD removal efficiency of 64% when the COD concentration of influent was around 300 mg/L (OLR 1.2 kgCOD/m^3·day). At higher influent COD concentration (300 to 816 mg/L) and corresponding OLR of 2.2 to 3.3 kgCOD/m^3·day, the COD removal remained in the range of 57 to 60%. A decrease in SS removal as COD efficiency from 97 to 90% of UASB reactors with an increase in influent COD concentration from 92 to 816 mg/L and in

turn OLR from 0.4 to 3.3 kgCOD/m^3·day was reported [11]. When the increase in OLR (0.4 to 3.3 kgCOD/m^3·day) is due to an increase in the influent COD contents, a sharp decrease in SS removal efficiency may occur. That decrease owes to SS washout caused by turbulence due to higher rate of gas production. When the increase in OLR is associated with decreasing HRT (increase of flow rate), a decline in SS removal efficiency may occur due to sludge washout and short contact time between sludge bed and substrate, which results in the malfunctioning of physical and biological processes taking place in the reactor. However, SS removal efficiency of the reactor decreases to less extent when increase in OLR is associated with increased SS contents of substrate [11]. Miron [109] reported an increase in SS removal with an increase in OLR associated with higher influent SS. These differences among the research workers are due to use of primary sludge by Miron to increase OLR. Primary sludge is mainly comprised of settleable suspended solids which increased the SS removal efficiency and therefore reported higher SS removal [11, 108, 109].

Thus, in order to describe the performance of UASB for treatment of sewage, the OLR has to be examined along with HRT and/or COD$_{Inf}$ as advised by Mahmoud [3]. It is however clear from these studies that increasing the OLR will cause a decrease in efficiency of suspended solids removal. However, at a certain OLR, the efficiency of UASB would be optimal in case of long HRT and high COD load of influent (up to a certain limit). When the solids removal efficiency in UASB is related to the OLR, it is important to differentiate between these parameters. It is therefore inferred that OLR is an insufficient design parameter to guarantee good performance of anaerobic reactors [42].

10. Effect of pH

The pH of an anaerobic reactor is especially important because methanogenesis process can proceed at a high rate only when the pH is maintained in the range of 6.3–7.8 [110]. In the case of domestic sewage, pH naturally remains in this range because of the buffering capacity of the acid-base system (carbonate system), and addition of chemical is not required [1]. The UASB reactors employed for sewage treatment in tropical and subtropical countries are reported to be extremely stable in terms of pH and buffering capacity [10–12, 14, 22, 24, 111]. Improvement in both hydrolysis and acidogenesis rates is achieved when treating domestic wastewater using anaerobic reactor and pH 7 provides an optimal working environment for anaerobic digestion resulting into more than 80% TOC and COD removal [111].

11. Effect of Granulation

In UASB reactor, long HRTs have been found disadvantageous for the development of granular sludge [107, 112]. In contrast, very short HRTs lead to washout of biomass. Both situations are unacceptable for achieving optimum results from UASB reactor. Even though granulation has been considered essential for successful treatment of domestic wastewater in UASB reactors, these reactors are found to be effective even without granules. The formation of granules in startup is helpful in shortening startup time [41, 113, 114]. The high performance of the UASB reactor is based on the formation of an active sludge in the lower part of the reactor. The development of sludge bed occurs by the accumulation of incoming suspended solids and bacterial growth under specific conditions, due to the natural aggregation of bacteria in flocs and evolution of granules in the form of layered structure [29, 115]. These granules are not washed out from the reactor during operation of UASB. The diameter of the granulated sludge particles has been found in the range of 1.0 to 3.0 mm [8, 116–118]. The granular suspensions have greater settling velocities (20–80 mh^{-1}) as compared to upflow velocities (V_{up} = 0.1–1.0 mh^{-1}). Therefore, substantial quantity of biomass can accumulate in the reactor. Consequently, a high sludge loading rate (SLR) could be applied (up to 5 gCODgVSS^{-1}day^{-1}) with a relatively short HRT, less than 4 hours [25]. Formation of activated sludge is important, in either granular or flocculent form, in the reactor which ensures adequate removal efficiency even at high OLR [27].

12. Effect of Mixing

Mixing offers effective contact time to microbes and wastewater, decreases hurdles of transfer of mass, lowers the growth of repressive by-products, and provides uniform environmental conditions. If mixing is not proper, the main process rate will be hampered by pockets of substrate at separate digestion stages, consequently leading to pH and temperature changes at every stage [119]. Mixing can be achieved mechanically or by recirculation of methane gas or slurry. A number of researchers have observed that significant mixing affects the working of anaerobic reactors. Mixing enhanced efficiency of anaerobic systems treating wastewater with high COD concentration; moreover the recirculation of slurry exhibited better results as compared to biogas recirculation and impeller mixing mode [120].

Mixing also provides increased biogas production as compared to unmixed digesters [120]. Discontinuous mixing is beneficial over energetic mixing when it is applied in large municipal and farm waste digesters [119]. Formation of sludge granules occurs due to fluidization. Vigorous mixing is not recommended as methanogens are less effective in these conditions [121]. Similarly, Karim et al. [120] reported that mixing in startup period lowers the digester pH, resulting in unstable performance and prolonged startup period.

13. Conclusions and Future Research Areas

UASB has been successfully applied for the treatment of industrial wastewater with high organic content and municipal wastewaters with high COD load in countries having temperate climatic conditions. Depending on the sewerage system, the sewage wastewater stream may contain toxic substances. Buildup of this matter can affect reactor performance. Overall UASB technology has been found successful for treating domestic wastewater. Therefore these reactors should be installed on priority basis in small communities and towns especially in developing countries with suitable climate conditions. In sewage treatment, the modeling of anaerobic digestion and obtaining higher yield of hydrogen from domestic wastewater have been active research areas in the last few years. The modification of UASB reactor, sequential use of UASB reactor with activated sludge/sequencing batch reactor/septic tank, flash aeration, two-stage UASB reactor/anaerobic filter/hybrid reactor, use of cosubstrate, posttreatment of UASB reactor for pathogen removal and reuse options, kinetics reactions and transfer of mass in anaerobic granular sludge, generation of high methane content, sludge profiling of UASB reactor at various OLRs, modeling for anaerobic granulation, and hydraulics and performance evaluation have been some of active research areas in this field. Modeling of UASB for performance evaluation will be very useful in directing future research on UASB system for direct treatment of wastewater.

Conflicts of Interest

The authors declare no conflicts of interest.

References

[1] A. C. Haandel and G. Lettinga, *Anaerobic Sewage Treatment: A Practical Guide for Regions with A Hot Climate*, John Wiley and Sons, Chichester, UK, 3rd edition, 1994.

[2] M. Henze and A. Ledin, "Types, characteristics and quantities of classic, combined domestic wastewaters," 2004.

[3] N. Mahmoud, *Anaerobic pre treatment of sewage under low temperature (15°C) conditions in an integrated UASB-Digester system*, Department of Environmental Technology and the Department of Agro Technology and Food Sciences Wageningen University, Wageningen, The Netherlands, 2002.

[4] P. S. James and S. Kamaraj, "Immobilized cell anaerobic bioreactors for energy production from agro-industrial waste waters-An introduction," *Bioenergy News*, vol. 6, no. 3, article 10, 2002.

[5] G. Lettinga, S. Rebac, and G. Zeeman, "Challenge of psychrophilic anaerobic wastewater treatment," *Trends in Biotechnology*, vol. 19, no. 9, pp. 363–370, 2001.

[6] E. Metcalf and I. Eddy, *Wastewater Engineering Treatment and Reuse*, McGraw-Hill Education, New York, NY, USA, 4th edition, 2003.

[7] A. Farghaly and A. Tawfik, "Simultaneous hydrogen and methane production through multi-phase anaerobic digestion of paperboard mill wastewater under different operating conditions," *Applied Biochemistry and Biotechnology*, vol. 181, no. 1, pp. 142–156, 2017.

[8] K. Yetilmezsoy and S. Sakar, "Development of empirical models for performance evaluation of UASB reactors treating poultry manure wastewater under different operational conditions," *Journal of Hazardous Materials*, vol. 153, no. 1-2, pp. 532–543, 2008.

[9] L. Singh, Z. A. Wahid, M. F. Siddiqui, A. Ahmad, M. H. A. Rahim, and M. Sakinah, "Application of immobilized upflow anaerobic sludge blanket reactor using Clostridium LS2 for enhanced biohydrogen production and treatment efficiency of palm oil mill effluent," *International Journal of Hydrogen Energy*, vol. 38, no. 5, pp. 2221–2229, 2013.

[10] L. Seghezzo, *Anaerobic Treatment of Domestic Wastewater in Subtropical Regions*, Wageningen University, Wageningen, The Netherlands, 2004.

[11] R. C. Leitao, *Robustness of UASB Reactors Treating Sewage under Tropical Conditions*, Wageningen University, Wageningen, The Netherlands, 2004.

[12] H. Rizvi, N. Ahmad, F. Abbas et al., "Start-up of UASB reactors treating municipal wastewater and effect of temperature/sludge age and hydraulic retention time (HRT) on its performance," *Arabian Journal of Chemistry*, vol. 8, no. 6, pp. 780–786, 2015.

[13] A. A. Khan, I. Mehrotra, and A. A. Kazmi, "Sludge profiling at varied organic loadings and performance evaluation of UASB reactor treating sewage," *Biosystems Engineering*, vol. 131, pp. 32–40, 2015.

[14] S. M. Mgana, *Towards Sustainable And Robust Community on Site Domestic Wastewater [Phd Thesis]*, Wageningen University, Wageningen, The Netherlands, 2003.

[15] T. A. Elmitwalli, K. L. T. Oahn, G. Zeeman, and G. Lettinga, "Treatment of domestic sewage in a two-step anaerobic filter/anaerobic hybrid system at low temperature," *Water Research*, vol. 36, no. 9, pp. 2225–2232, 2002.

[16] W. R. Abma, W. Driessen, R. Haarhuis, and M. C. M. Van Loosdrecht, "Upgrading of sewage treatment plant by sustainable and cost-effective separate treatment of industrial wastewater," *Water Science and Technology*, vol. 61, no. 7, pp. 1715–1722, 2010.

[17] P. Kongjan, S. O-Thong, and I. Angelidaki, "Hydrogen and methane production from desugared molasses using a two-stage thermophilic anaerobic process," *Engineering in Life Sciences*, vol. 13, no. 2, pp. 118–125, 2013.

[18] M. Hernández and M. Rodríguez, "Hydrogen production by anaerobic digestion of pig manure: Effect of operating conditions," *Journal of Renewable Energy*, vol. 53, pp. 187–192, 2013.

[19] H. Rizvi, S. Ali, A. Yasar, M. Ali, and M. Rizwan, "Applicability of upflow anaerobic sludge blanket (UASB) reactor for typical sewage of a small community: its biomass reactivation after shutdown," *International Journal of Environmental Science and Technology*, pp. 1–12, 2017.

[20] J. A. Álvarez, I. Ruiz, M. Gómez, J. Presas, and M. Soto, "Start-up alternatives and performance of an UASB pilot plant treating diluted municipal wastewater at low temperature," *Bioresource Technology*, vol. 97, no. 14, pp. 1640–1649, 2006.

[21] S. Chong, T. K. Sen, A. Kayaalp, and H. M. Ang, "The performance enhancements of upflow anaerobic sludge blanket (UASB) reactors for domestic sludge treatment—a State-of-the-art review," *Water Research*, vol. 46, no. 11, pp. 3434–3470, 2012.

[22] M. Halalsheh, *Anaerobic Pre Treatment of Strong Sewage. A Proper Solution for Jordan*, Environmental Technology Department Wageningen University: Wageningen Research Center, Wageningen, The Netherlands, 2002.

[23] Q. H. Banihani and J. A. Field, "Treatment of high-strength synthetic sewage in a laboratory-scale upflow anaerobic sludge bed (UASB) with aerobic activated sludge (AS) post-treatment," *Journal of Environmental Science and Health, Part A: Toxic/Hazardous Substances and Environmental Engineering*, vol. 48, no. 3, pp. 338–347, 2013.

[24] P. F. F. Cavalcanti, *Integrated Application of The UASB Reactor And Ponds for Domestic Sewage Treatment in Tropical Regions*, Wageningen University, Wageningen, The Netherlands, 2003.

[25] S. V. Kalyuzhnyi, V. I. Sklyar, M. A. Davlyatshina et al., "Organic removal and microbiological features of UASB-reactor under various organic loading rates," *Bioresource Technology*, vol. 55, no. 1, pp. 47–54, 1996.

[26] G. Zeeman, W. Sanders, and G. Lettinga, "Feasibility of the on-site treatment of sewage and swill in large buildings," *Water Science and Technology*, vol. 41, no. 1, pp. 9–16, 2000.

[27] R. Shahperi, M. F. M. Din, S. Chelliapan et al., "Optimization of methane production process from synthetic glucose feed in a multi-stage anaerobic bioreactor," *Desalination and Water Treatment*, vol. 57, no. 60, pp. 29168–29177, 2016.

[28] B. Wolmarans and G. H. De Villiers, "Start-up of a UASB effluent treatment plant on distillery wastewater," *Water SA*, vol. 28, no. 1, pp. 63–68, 2002.

[29] H. H. P. Fang, "Microbial distribution in UASB granules and its resulting effects," *Water Science and Technology*, vol. 42, no. 12, pp. 201–208, 2000.

[30] G. Kyazze, R. Dinsdale, A. J. Guwy, F. R. Hawkes, G. C. Premier, and D. L. Hawkes, "Performance characteristics of a two-stage dark fermentative system producing hydrogen and methane continuously," *Biotechnology and Bioengineering*, vol. 97, no. 4, pp. 759–770, 2007.

[31] I. Ruiz, M. Soto, M. C. Veiga, P. Ligero, A. Vega, and R. Blazquez, "Performance of and biomass characterisation in a UASB reactor treating domestic waste water at ambient temperature," 1998.

[32] S. Uemura and H. Harada, "Treatment of sewage by a UASB reactor under moderate to low temperature conditions," *Bioresource Technology*, vol. 72, no. 3, pp. 275–282, 2000.

[33] W. Zhou, B. Wu, Q. She, L. Chi, and Z. Zhang, "Investigation of soluble microbial products in a full-scale UASB reactor running at low organic loading rate," *Bioresource Technology*, vol. 100, no. 14, pp. 3471–3476, 2009.

[34] B. Wu and W. Zhou, "Investigation of soluble microbial products in anaerobic wastewater treatment effluents," *Journal of Chemical Technology and Biotechnology*, vol. 85, no. 12, pp. 1597–1603, 2010.

[35] J. E. Schmidt and B. K. Ahring, "Granular sludge formation in upflow anaerobic sludge blanket (UASB) reactors," *Biotechnology and Bioengineering*, vol. 49, no. 3, pp. 229–246, 1996.

[36] T. G. Jantsch, I. Angelidaki, J. E. Schmidt, B. E. Braa de Hvidsten, and B. K. Ahring, "Anaerobic biodegradation of spent sulphite liquor in a UASB reactor," *Bioresource Technology*, vol. 84, no. 1, pp. 15–20, 2002.

[37] K. Karim and S. K. Gupta, "Continuous biotransformation and removal of nitrophenols under denitrifying conditions," *Water Research*, vol. 37, no. 12, pp. 2953–2959, 2003.

[38] H. Fang and Y. Liu, "Anaerobic wastewater treatment in (sub-) tropical regions," in *Advances in Water and Wastewater Treatment Technology*, T. Matsuo, K. Hanaki, S. Takizawa, and H. Satoh, Eds., vol. 285, 2001.

[39] K. S. Singh, T. Viraraghavan, and D. Bhattacharyya, "Sludge blanket height and flow pattern in UASB reactors: Temperature effects," *Journal of Environmental Engineering*, vol. 132, no. 8, pp. 895–900, 2006.

[40] B. Lew, I. Lustig, M. Beliavski, S. Tarre, and M. Green, "An integrated UASB-sludge digester system for raw domestic wastewater treatment in temperate climates," *Bioresource Technology*, vol. 102, no. 7, pp. 4921–4924, 2011.

[41] Y. Liu, H.-L. Xu, S.-F. Yang, and J.-H. Tay, "Mechanisms and models for anaerobic granulation in upflow anaerobic sludge blanket reactor," *Water Research*, vol. 37, no. 3, pp. 661–673, 2003.

[42] A. Abdelgadir, X. Chen, J. Liu et al., "Characteristics, process parameters, and inner components of anaerobic bioreactors," *BioMed Research International*, vol. 2014, Article ID 841573, 10 pages, 2014.

[43] A. C. van Haandel, "Integrated energy production and reduction of the environmental impact at alcohol distillery plants," *Water Science and Technology*, vol. 52, no. 1-2, pp. 49–57, 2005.

[44] F. Y. Cakir and M. K. Stenstrom, "Greenhouse gas production: A comparison between aerobic and anaerobic wastewater treatment technology," *Water Research*, vol. 39, no. 17, pp. 4197–4203, 2005.

[45] J. Keller and K. Hartley, "Greenhouse gas production in wastewater treatment: process selection is the major factor," *Water Science and Technology*, vol. 47, no. 12, pp. 43–48, 2003.

[46] P. F. Greenfield and D. J. Batstone, "Anaerobic digestion: Impact of future greenhouse gases mitigation policies on methane generation and usage," *Water Science and Technology*, vol. 52, no. 1-2, pp. 39–47, 2005.

[47] H. D. Monteith, H. R. Sahely, H. L. MacLean, and D. M. Bagley, "A life-cycle approach for estimation of greenhouse gas emissions from canadian wastewater treatment," *Proceedings of the Water Environment Federation*, vol. 2003, no. 11, pp. 514–527, 2003.

[48] D. J. I. Gustavsson and S. Tumlin, "Carbon footprints of Scandinavian wastewater treatment plants," *Water Science and Technology*, vol. 68, no. 4, pp. 887–893, 2013.

[49] J. B. Van Lier, "High-rate anaerobic wastewater treatment: diversifying from end-of-the-pipe treatment to resource-oriented conversion techniques," *Water Science and Technology*, vol. 57, no. 8, pp. 1137–1148, 2008.

[50] D. Dionisi and I. M. O. Silva, "Production of ethanol, organic acids and hydrogen: an opportunity for mixed culture biotechnology?" *Reviews in Environmental Science and Bio/Technology*, vol. 15, no. 2, pp. 213–242, 2016.

[51] A. Mostafa, M. Elsamadony, A. El-Dissouky, A. Elhusseiny, and A. Tawfik, "Biological H2 potential harvested from complex gelatinaceous wastewater via attached versus suspended growth culture anaerobes," *Bioresource Technology*, vol. 231, pp. 9–18, 2017.

[52] S. Paudel, Y. Kang, Y.-S. Yoo, and G. T. Seo, "Hydrogen production in the anaerobic treatment of domestic-grade synthetic wastewater," *Sustainability*, vol. 7, no. 12, pp. 16260–16272, 2015.

[53] B. S. Fernandes, G. Peixoto, F. R. Albrecht, N. K. Saavedra del Aguila, and M. Zaiat, "Potential to produce biohydrogen from various wastewaters," *Energy for Sustainable Development*, vol. 14, no. 2, pp. 143–148, 2010.

[54] Y. Kawagoshi, N. Hino, A. Fujimoto et al., "Effect of inoculum conditioning on hydrogen fermentation and pH effect on bacterial community relevant to hydrogen production," *Journal of Bioscience and Bioengineering*, vol. 100, no. 5, pp. 524–530, 2005.

[55] P. Mohammadi, S. Ibrahim, M. S. M. Annuar, S. Ghafari, S. Vikineswary, and A. A. Zinatizadeh, "Influences of environmental and operational factors on dark fermentative hydrogen production: a review," *CLEAN—Soil, Air, Water*, vol. 40, no. 11, pp. 1297–1305, 2012.

[56] Y. Zhang and J. Shen, "Effect of temperature and iron concentration on the growth and hydrogen production of mixed bacteria," *International Journal of Hydrogen Energy*, vol. 31, no. 4, pp. 441–446, 2006.

[57] Y. Mu and H.-Q. Yu, "Biological hydrogen production in a UASB reactor with granules. I: Physicochemical characteristics of hydrogen-producing granules," *Biotechnology and Bioengineering*, vol. 94, no. 5, pp. 980–987, 2006.

[58] Y. Sharma and B. Li, "Optimizing energy harvest in wastewater treatment by combining anaerobic hydrogen producing biofermentor (HPB) and microbial fuel cell (MFC)," *International Journal of Hydrogen Energy*, vol. 35, no. 8, pp. 3789–3797, 2010.

[59] L. Singh and Z. A. Wahid, "Methods for enhancing biohydrogen production from biological process: A review," *Journal of Industrial and Engineering Chemistry*, vol. 21, pp. 70–80, 2015.

[60] A. Tawfik, F. El-Gohary, and H. Temmink, "Treatment of domestic wastewater in an up-flow anaerobic sludge blanket reactor followed by moving bed biofilm reactor," *Bioprocess and Biosystems Engineering*, vol. 33, no. 2, pp. 267–276, 2010.

[61] M. Tandukar, A. Ohashi, and H. Harada, "Performance comparison of a pilot-scale UASB and DHS system and activated sludge process for the treatment of municipal wastewater," *Water Research*, vol. 41, no. 12, pp. 2697–2705, 2007.

[62] N. Khalil, R. Sinha, A. Raghav, and A. Mittal, "UASB technology for sewage treatment in India: experience, economic evaluation and its potential in other developing countries," in *Proceedings of the in Twelfth International Water Technology Conference*, 2008.

[63] E. Alonso, A. Santos, and P. Riesco, "Micro-organism re-growth wastewater disinfected by UV radiation and ozone: a microbiological study," *Environmental Technology*, vol. 25, no. 4, pp. 433–441, 2004.

[64] H. Rizvi, N. Ahmad, A. Yasar, K. Bukhari, and H. Khan, "Disinfection of UASB-treated municipal wastewater by H2O2, UV, ozone, PAA, H2O2/sunlight, and advanced oxidation processes: regrowth potential of pathogens," *Polish Journal of Environmental Studies*, vol. 22, no. 4, pp. 1153–1161, 2013.

[65] T. E. Possmoser-Nascimento, V. A. J. Rodrigues, M. Sperling, and J.-L. Vasel, "Sludge accumulation in shallow maturation ponds treating UASB reactor effluent: results after 11 years of operation," *Water Science and Technology*, vol. 70, no. 2, pp. 321–328, 2014.

[66] M. Von Sperling, V. H. Freire, and C. A. De Lemos Chernicharo, "Performance evaluation of a UASB-activated sludge system treating municipal wastewater," *Water Science & Technology*, vol. 43, no. 11, pp. 323–328, 2001.

[67] L. Foresti, "Anaerobic treatment of domestic sewage: established technologies and perspectives," *Water Science and Technology*, vol. 45, no. 10, pp. 181–186, 2002.

[68] S. J. Zhang, N. R. Liu, and C. X. Zhang, "Study on the performance of modified UASB process treating sewage," *Advanced Materials Research*, vol. 610–613, pp. 2174–2178, 2013.

[69] F. I. Turkdogan-Aydinol and K. Yetilmezsoy, "A fuzzy-logic-based model to predict biogas and methane production rates in a pilot-scale mesophilic UASB reactor treating molasses wastewater," *Journal of Hazardous Materials*, vol. 182, no. 1–3, pp. 460–471, 2010.

[70] I. Bodík, B. Herdová, and M. Drtil, "Anaerobic treatment of the municipal wastewater under psychrophilic conditions," *Bioprocess Engineering*, vol. 22, no. 5, pp. 385–390, 2000.

[71] N. Azbar, F. T. Dokgöz, T. Keskin et al., "Comparative evaluation of bio-hydrogen production from cheese whey wastewater under thermophilic and mesophilic anaerobic conditions," *International Journal of Green Energy*, vol. 6, no. 2, pp. 192–200, 2009.

[72] K. S. Singh, H. Harada, and T. Viraraghavan, "Low-strength wastewater treatment by a UASB reactor," *Bioresource Technology*, vol. 55, no. 3, pp. 187–194, 1996.

[73] Y. Kalogo and W. Verstraete, "Development of anaerobic sludge bed (ASB) reactor technologies for domestic wastewater treatment: motives and perspectives," *World Journal of Microbiology and Biotechnology*, vol. 15, no. 5, pp. 523–534, 1999.

[74] J. B. Van Lier and G. Lettinga, "Appropriate technologies for effective management of industrial and domestic waste waters: the decentralised approach," *Water Science and Technology*, vol. 40, no. 7, pp. 171–183, 1999.

[75] M. Halalsheh, Z. Sawajneh, M. Zu'bi et al., "Treatment of strong domestic sewage in a 96 m 3 UASB reactor operated at ambient temperatures: two-stage versus single-stage reactor," *Bioresource Technology*, vol. 96, no. 5, pp. 577–585, 2005.

[76] R. F. Hickey, W.-M. Wu, M. C. Veiga, and R. Jones, "Start-up, operation, monitoring and control of high-rate anaerobic treatment systems," *Water Science and Technology*, vol. 24, no. 8, pp. 207–255, 1991.

[77] S. M. M. Vieira and A. D. Garcia Jr., "Sewage treatment by UASB-reactor. Operation results and recommendations for design and utilization," *Water Science and Technology*, vol. 25, no. 7, pp. 143–157, 1992.

[78] R. Rajakumar, T. Meenambal, J. R. Banu, and I. T. Yeom, "Treatment of poultry slaughterhouse wastewater in upflow anaerobic filter under low upflow velocity," *International Journal of Environmental Science and Technology*, vol. 8, no. 1, pp. 149–158, 2011.

[79] R. R. Liu, Q. Tian, B. Yang, and J. H. Chen, "Hybrid anaerobic baffled reactor for treatment of desizing wastewater," *International Journal of Environmental Science and Technology*, vol. 7, no. 1, pp. 111–118, 2010.

[80] V. N. Nkemka and M. Murto, "Evaluation of biogas production from seaweed in batch tests and in UASB reactors combined with the removal of heavy metals," *Journal of Environmental Management*, vol. 91, no. 7, pp. 1573–1579, 2010.

[81] M. R. Peña, D. D. Mara, and G. P. Avella, "Dispersion and treatment performance analysis of an UASB reactor under different hydraulic loading rates," *Water Research*, vol. 40, no. 3, pp. 445–452, 2006.

[82] H. Rizvi, *Sewage treatment by an upflow anaerobic sludge blanket reactor (UASB) under subtropical conditions [PhD Thesis]*, University of the Punjab, Lahore, India, 2011.

[83] N. Sato, T. Okubo, T. Onodera, A. Ohashi, and H. Harada, "Prospects for a self-sustainable sewage treatment system: a case study on full-scale UASB system in India's Yamuna River Basin," *Journal of Environmental Management*, vol. 80, no. 3, pp. 198–207, 2006.

[84] Y. J. Chan, M. F. Chong, C. L. Law, and D. G. Hassell, "A review on anaerobic-aerobic treatment of industrial and municipal wastewater," *Chemical Engineering Journal*, vol. 155, no. 1-2, pp. 1–18, 2009.

[85] S. C. Oliveira and M. Von Sperling, "Performance evaluation of UASB reactor systems with and without post-treatment," *Water Science and Technology*, vol. 59, no. 7, pp. 1299–1306, 2009.

[86] S. K. Khanal, *Anaerobic Biotechnology for Bioenergy Production: Principles and Applications*, John Wiley & Sons, 2011.

[87] A. A. Khan, R. Z. Gaur, V. K. Tyagi et al., "Sustainable options of post treatment of UASB effluent treating sewage: a review," *Resources, Conservation & Recycling*, vol. 55, no. 12, pp. 1232–1251, 2011.

[88] A. T. Nair and M. M. Ahammed, "The reuse of water treatment sludge as a coagulant for post-treatment of UASB reactor treating urban wastewater," *Journal of Cleaner Production*, vol. 96, article 3884, pp. 272–281, 2015.

[89] B. D. Shoener, I. M. Bradley, R. D. Cusick, and J. S. Guest, "Energy positive domestic wastewater treatment: the roles of anaerobic and phototrophic technologies," *Environmental Sciences: Processes & Impacts*, vol. 16, no. 6, pp. 1204–1222, 2014.

[90] S. Shastry, T. Nandy, S. R. Wate, and S. N. Kaul, "Hydrogenated vegetable oil industry wastewater treatment using UASB reactor system with recourse to energy recovery," *Water, Air, & Soil Pollution*, vol. 208, no. 1–4, pp. 323–333, 2010.

[91] G. Lettinga, J. B. Van Lier, J. C. L. Van Buuren, and G. Zeeman, "Sustainable development in pollution control and the role of anaerobic treatment," *Water Science and Technology*, vol. 44, no. 6, pp. 181–188, 2001.

[92] M. Von Sperling and C. A. De Lemos Chernicharo, *Biological Wastewater Treatment in Warm Climate Regions*, vol. 1, IWA publishing, 2005.

[93] J. B. Van Lier, A. Vashi, J. Van Der Lubbe, B. Heffernan, and H. Fang, "Anaerobic sewage treatment using UASB reactors: engineering and operational aspects," in *Environmental anaerobic technology; applications and new developments*, vol. 59, World Scientific, Imperial College Press, London, UK, 2010.

[94] B. C. Crone, J. L. Garland, G. A. Sorial, and L. M. Vane, "Significance of dissolved methane in effluents of anaerobically treated low strength wastewater and potential for recovery as an energy product: a review," *Water Research*, vol. 104, pp. 520–531, 2016.

[95] N. Sato, T. Okubo, T. Onodera, L. K. Agrawal, A. Ohashi, and H. Harada, "Economic evaluation of sewage treatment processes in India," *Journal of Environmental Management*, vol. 84, no. 4, pp. 447–460, 2007.

[96] A. Gnanadipathy and C. Polprasert, "Treatment of a domestic wastewater with UASB reactors," *Water Science and Technology*, vol. 27, no. 1, pp. 195–203, 1993.

[97] C. A. L. Chernicharo and M. Dos Reis Cardoso, "Development and evaluation of a partitioned upflow anaerobic sludge blanket (UASB) reactor for the treatment of domestic sewage from small villages," *Water Science and Technology*, vol. 40, no. 8, pp. 107–113, 1999.

[98] Z. Sawajneh, A. Al-Omari, and M. Halalsheh, "Anaerobic treatment of strong sewage by a two stage system of AF and UASB reactors," *Water Science and Technology*, vol. 61, no. 9, pp. 2399–2406, 2010.

[99] C. A. L. Chernicharo, P. G. S. Almeida, L. C. S. Lobato, T. C. Couto, J. M. Borges, and Y. S. Lacerda, "Experience with the design and start up of two full-scale UASB plants in Brazil: enhancements and drawbacks," *Water Science and Technology*, vol. 60, no. 2, pp. 507–515, 2009.

[100] N. Sundaresan and L. Philip, "Performance evaluation of various aerobic biological systems for the treatment of domestic wastewater at low temperatures," *Water Science and Technology*, vol. 58, no. 4, pp. 819–830, 2008.

[101] X.-G. Chen, P. Zheng, J. Cai, and M. Qaisar, "Bed expansion behavior and sensitivity analysis for super-high-rate anaerobic bioreactor," *Journal of Zhejiang University SCIENCE B*, vol. 11, no. 2, pp. 79–86, 2010.

[102] O. R. Zimmo, N. P. van der Steen, and H. J. Gijzen, "Effect of organic surface load on process performance of pilot-scale algae and duckweed-based waste stabilization ponds," *Journal of Environmental Engineering*, vol. 131, no. 4, pp. 587–594, 2005.

[103] H. Salvadó and M. P. Gracia, "Determination of organic loading rate of activated sludge plants based on protozoan analysis," *Water Research*, vol. 27, no. 5, pp. 891–895, 1993.

[104] D. H. Liu and B. G. Liptak, *Environmental Engineers' Handbook on CD-ROM*, CRC Press, 1999.

[105] B. Picot, J. Paing, J. P. Sambuco, R. H. R. Costa, and A. Rambaud, "Biogas production, sludge accumulation and mass balance of carbon in anaerobic ponds," *Water Science and Technology*, vol. 48, no. 2, pp. 243–250, 2003.

[106] H. Toprak, "Temperature and organic loading dependency of methane and carbon dioxide emission rates of a full-scale anaerobic waste stabilization pond," *Water Research*, vol. 29, no. 4, pp. 1111–1119, 1995.

[107] T.-T. Ren, Y. Mu, B.-J. Ni, and H.-Q. Yu, "Hydrodynamics of upflow anaerobic sludge blanket reactors," *AIChE Journal*, vol. 55, no. 2, pp. 516–528, 2009.

[108] S. Farajzadehha, S. Mirbagheri, S. Farajzadehha, and J. Shayegan, "Lab scale study of HRT and OLR optimization in UASB reactor for pretreating fortified wastewater in various operational temperatures," *APCBEE Procedia*, vol. 1, pp. 90–95, 2012.

[109] Y. Miron, *Anaerobic Treatment of Domestic Sewage with a Two-step Uasr-Uasb System*, Hebrew University of Jerusalem, 1997.

[110] C. Casserly and L. Erijman, "Molecular monitoring of microbial diversity in an UASB reactor," *International Biodeterioration & Biodegradation*, vol. 52, no. 1, pp. 7–12, 2003.

[111] B. Zhang, L.-L. Zhang, S.-C. Zhang, H.-Z. Shi, and W.-M. Cai, "The influence of pH on hydrolysis and acidogenesis of kitchen wastes in two-phase anaerobic digestion," *Environmental Technology*, vol. 26, no. 3, pp. 329–339, 2005.

[112] D. J. Batstone, J. L. A. Hernandez, and J. E. Schmidt, "Hydraulics of laboratory and full-scale upflow anaerobic sludge blanket (UASB) reactors," *Biotechnology and Bioengineering*, vol. 91, no. 3, pp. 387–391, 2005.

[113] J.-H. Tay, H.-L. Xu, and K.-C. Teo, "Molecular mechanism of granulation. I: H+ translocation-dehydration theory," *Journal of Environmental Engineering*, vol. 126, no. 5, pp. 403–410, 2000.

[114] S. Aiyuk and W. Verstraete, "Sedimentological evolution in an UASB treating SYNTHES, a new representative synthetic sewage, at low loading rates," *Bioresource Technology*, vol. 93, no. 3, pp. 269–278, 2004.

[115] J.-H. Tay and Y.-G. Yan, "Influence of substrate concentration on microbial selection and granulation during start-up of upflow anaerobic sludge blanket reactors," *Water Environment Research*, vol. 68, no. 7, pp. 1140–1150, 1996.

[116] H.-H. Chou and J.-S. Huang, "Role of mass transfer resistance in overall substrate removal rate in upflow anaerobic sludge bed reactors," *Journal of Environmental Engineering*, vol. 131, no. 4, pp. 548–556, 2005.

[117] R. G. Veronez, A. A. Orra, R. Ribeiro, M. Zaiat, S. M. Ratusznei, and J. A. D. Rodrigues, "A simplified analysis of granule behavior in ASBR and UASB reactors treating low-strength synthetic wastewater," *Brazilian Journal of Chemical Engineering*, vol. 22, no. 3, pp. 361–369, 2005.

[118] A. Vlyssides, E. M. Barampouti, and S. Mai, "Determination of granule size distribution in a UASB reactor," *Journal of Environmental Management*, vol. 86, no. 4, pp. 660–664, 2008.

[119] P. Kaparaju, I. Buendia, L. Ellegaard, and I. Angelidakia, "Effects of mixing on methane production during thermophilic anaerobic digestion of manure: lab-scale and pilot-scale studies," *Bioresource Technology*, vol. 99, no. 11, pp. 4919–4928, 2008.

[120] K. Karim, R. Hoffmann, K. T. Klasson, and M. H. Al-Dahhan, "Anaerobic digestion of animal waste: Effect of mode of mixing," *Water Research*, vol. 39, no. 15, pp. 3597–3606, 2005.

[121] S. R. Guiot, A. Pauss, and J. W. Costerton, "A structured model of the anaerobic granule consortium," *Water Science and Technology*, vol. 25, no. 7, pp. 1–10, 1992.

9

Lignitic Humic Acids as Environmentally-Friendly Adsorbent for Heavy Metals

Martina Klučáková[1] and Marcela Pavlíková[2]

[1]*Materials Research Centre, Faculty of Chemistry, Brno University of Technology, Purkyňova 118, 612 00 Brno, Czech Republic*
[2]*Institute of Chemistry, Faculty of Civil Engineering, Brno University of Technology, Žižkova 17, 602 00 Brno, Czech Republic*

Correspondence should be addressed to Martina Klučáková; klucakova@fch.vutbr.cz

Academic Editor: Ziya A. Khan

Humic acids are a part of humus material, are abundant in nature, and form a substantial pool of natural organic matter. They participate in the transport of both beneficial and harmful species. Due to their structure and properties, they can interact with metal ions and, with them, form relatively stable complexes. These substances are thus responsible for the so-called self-cleaning ability of soils. Lignite as a young coal type contains a relatively high amount of humic acids which can be used as an environmentally-friendly adsorbent for heavy metals. In this work, we compared the adsorption of single Cu^{2+} ions with the simultaneous adsorption of several different metal ions (Cd^{2+}, Cu^{2+}, Pb^{2+}, and Zn^{2+}). The adsorption efficiency of humic acids was very high, almost 100% in the case of the single adsorption of Cu^{2+} ions and more than 90% for the adsorption from the mixture of metal ions. The stability of formed complexes, considered on the basis of the leaching in different extraction agents, was higher than 80%; only 9–18% was in the mobile phase. After adsorption, metal ions are contained in humic acids after adsorption in mainly strongly bonded form (60–73%) and can be liberated from their structure only in low amounts.

1. Introduction

One of the most characteristic properties of humic substances is their adsorption ability [1–4]. They can bind to heavy metal ions, which are characterized by high toxicity and the ability to accumulate in the environment, and in this way, they can influence the effectiveness of regeneration and purification processes in soil and aqueous systems. Hence, the immobilization or removal of toxic metals is a research goal as well as an industrial task [5–7]. A cost-effective and environmentally-friendly approach is the use of sorption technology based on the use of sorbents prepared from some biomaterial [8, 9]. A potential low-cost sorbent of toxic metals and radionuclides for use in water treatment and groundwater remediation is lignite [2, 9–11]. When humic acids, as the most active constituents, were isolated from lignite and used as an adsorbent for metal ions, their effectivity and sorption capacity substantially increased [2]. In comparison with humic acids isolated from various soil types, lignitic humic acids exhibit a very high sorption capacity and a

low degree of desorption [6, 12]. On the other hand, humic acids isolated from oxyhumolite exhibited a lower sorption capacity in comparison with humic soil-derived humic acids [13]. The removal of heavy metals from aqueous systems such as waste streams employs various technologies which are often either expensive or inefficient, especially when very low residual concentrations compliant with health-based limits are required [7, 8]. On the basis of previous results, lignitic humic acids were used in this work for the environmentally-friendly adsorption of metal ions.

Interactions between humic acids and metal ions have a complex nature given by their heterogeneous, polyelectrolyte, and polydispersive character. Metal ions can bond with humic acids in several different ways from the purely electrostatic, nonspecific interaction of metal cation with the net negative charge on the surface of a humic particle to specific interactions in the formation of complexes and chelates with functional groups [1, 2, 4, 7, 14–16]. Due to the heterogeneous character of humic acids and their partial solubility in water, they can form water-soluble and water

TABLE 1: Elemental analysis of lignitic humic acids (normalized on dry ash-free sample).

C	H	N	S	O
42.7% at.	41.8% at.	0.9% at.	0.3% at.	14.3% at.

TABLE 2: Average adsorption efficiencies for single Cu^{2+} ions and Cd^{2+}, Cu^{2+}, Pb^{2+}, and Zn^{2+} ions adsorbed from their mixture.

Cu^{2+} (single)	Cd^{2+}	Cu^{2+}	Pb^{2+}	Zn^{2+}
99.9%	90.3%	94.4%	92.8%	92.0%

TABLE 3: Average participation in total adsorption amount for Cd^{2+}, Cu^{2+}, Pb^{2+}, and Zn^{2+} ions adsorbed from their mixture.

Cd^{2+}	Cu^{2+}	Pb^{2+}	Zn^{2+}
21.0%	23.7%	31.7%	23.6%

insoluble complexes with metal ions and other pollutants [4, 17–21]. Humic acids of low molecular weight had a much higher sorption capacity than those of high molecular weight or unfractionated humic material [13]. It was found that the binding of metals to humic acids depends on their type. The sorption capacity of humic acids for some metals is much higher than their possibilities given by the content of functional groups [4, 16]. Metal ions are bound to humic acids by different strengths and particular metals can exist in humic material in several different fractions [4, 22–26].

This work follows our previous adsorption studies [2, 4, 6] and is focused mainly on potential industrial applications such as the removal of heavy metals from waters, with an emphasis on the (often problematic) issue of low concentrations and the stability of the formed complexes.

2. Materials and Methods

Humic acids (HA) were extracted from South-Moravia lignite using a mixture of NaOH and $Na_4P_2O_7$ by means of a procedure described in our previous studies [4, 6] and characterized previously [4, 6, 27, 28]. Elemental analysis was carried out using a CHNSO Microanalyser Flash 1112 (Carlo Erba). The obtained values are listed in Table 1.

Humic acids and metal ion solutions ($CdCl_2$, $CuCl_2$, $PbCl_2$, and $ZnCl_2$, Sigma-Aldrich) were mixed in the ratio 1 g/50 mL and stirred to equilibrium (24 h). Then, solid humic particles were sequestered by centrifugation and the supernatant was analyzed. The quantity of the adsorbed amount was determined on the basis of the decrease in the metal ion concentration in the solution during sorption.

The stability of the formed complexes was studied using the extraction of metal ions by different agents. Water was used for the extraction of the mobile fraction (free unbound metal ions). A 1 M $MgCl_2$ solution was used as the extraction agent for the ion-exchangeable fraction (weakly bound metal ions). The strongly bound fraction of metal ions was extracted by a 1 M HCl solution. The extraction agents were chosen on the basis of results published in previous works [12, 29, 30].

The concentrations of metal ions in different solutions were measured by electrode coulometric titration (EcaFlow 150 GLP), based on the direct electrochemical conversion of the analyte species in the pores of the electrode or in a thin layer [17]. Metal ions were electrochemically deposited from the flowing sample solution onto the porous working electrode. The deposition was performed by applying a suitable potential. In the next step, the deposit was stripped galvanostatically and the stripping chronopotentiogram was recorded and evaluated.

All experiments were made in triplicate at 25°C (in an air-conditioned laboratory). The data are presented as average values with standard deviation.

3. Results and Discussion

In this work, the adsorption from low concentration metal ion solutions by solid humic acids was studied. We compared the adsorption of single Cu^{2+} ions with simultaneous adsorption of several different metal ions (Cd^{2+}, Cu^{2+}, Pb^{2+}, and Zn^{2+}). This concentration region was chosen on the basis of the potential utilization of humic substances as a low-cost sorbent for environmentally-friendly adsorption of toxic metal ions often occurring in low but still problematic and harmful concentrations. As we can see, the efficiency of adsorption in the case of single Cu^{2+} ions was almost 100% in the used concentration range. Its value slightly decreased with respect to adsorption from the mixture of metal ions, but the average efficiency was higher than 90% for all metal ions used (see Table 2).

The real efficiency fluctuated by 80 and 100% with the initial concentrations of metal ions (see Figure 1). Their values were thus very high over the whole concentration range and no specific trend was observed.

Comparing the percentages of metal ions that were adsorbed we can see that the most readily adsorbed metal was lead. Copper and zinc were adsorbed to a similar extent. Cadmium exhibited the lowest percentage, but its adsorption was only little worse than in the case of copper and zinc (see Table 3).

Since, the efficiency of adsorption was in many cases 100% and the used concentrations were low, it was not possible to fit experimental data by any of the usually used adsorption isotherms. We determined the "distribution coefficient" as the slope obtained from the dependence of adsorbed amount on equilibrium concentration, where data with efficiency lower than 100% were used (zero equilibrium concentrations for 100% adsorption efficiency were not included in the calculation). The highest values were obtained for the single adsorption of Cu^{2+} ions; in the case of adsorption from the mixture, this metal also had the highest distribution coefficient. Strong adsorption was also found for Pb^{2+}. On the other hand, the adsorption of Zn^{2+} was, according to the "distribution coefficient," relatively weak. The obtained values of the "distribution coefficient" are not fully in agreement with the results listed in Tables 2 and 3. The reason is the above-mentioned partial elimination of some adsorption data because of the zero equilibrium concentration. Therefore, these values can be considered only as complementary

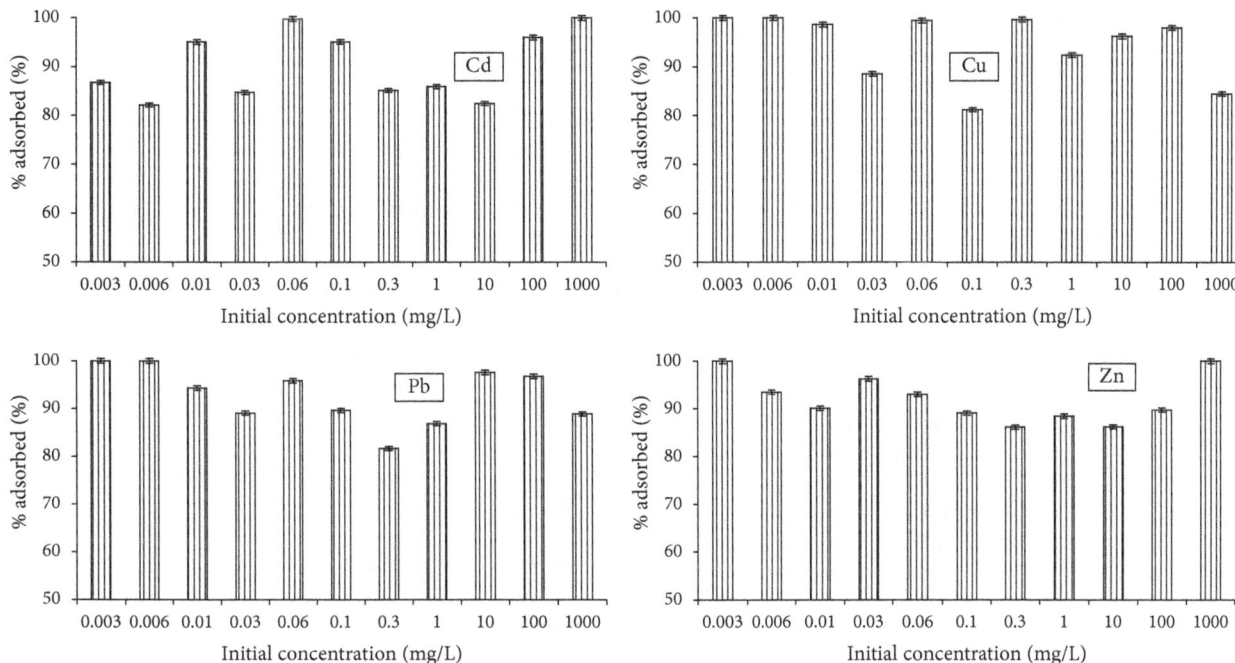

FIGURE 1: Amounts of metal ions adsorbed in dependence on their initial concentration in the mixture.

TABLE 4: Distribution coefficient for single Cu^{2+} ions and Cd^{2+}, Cu^{2+}, Pb^{2+}, and Zn^{2+} ions adsorbed from their mixture.

Cu^{2+} (single)	Cd^{2+}	Cu^{2+}	Pb^{2+}	Zn^{2+}
0.902 L/g	0.236 L/g	0.627 L/g	0.423 L/g	0.135 L/g

TABLE 5: Average contents of mobile phase, ion-exchangeable phase, strongly bound phase, and residual phase of single Cu^{2+} ions and Cd^{2+}, Cu^{2+}, Pb^{2+}, and Zn^{2+} ions adsorbed from their mixture.

	Cu^{2+} (single)	Cd^{2+}	Cu^{2+}	Pb^{2+}	Zn^{2+}
Mobile phase	9.8%	14.4%	11.2%	12.4%	17.8%
Ion-exchangeable phase	17.0%	25.7%	18.0%	19.3%	21.8%
Strongly bound phase	22.2%	29.2%	23.6%	23.1%	26.7%
Residual phase	51.0%	30.7%	47.2%	45.2%	33.7%

ones for the complex illustration of the studied adsorption (see Table 4).

Although the average proportions of copper and zinc in the total adsorption amount were practically the same, their distributions according to bond strength were different. As we can see in Table 5, zinc was bound less strongly and larger part was present in mobile or ion-exchangeable form. Cadmium exhibited a similar content in the weakly bound mobile or ion-exchangeable form. On the other hand, copper and lead were bound strongly and only a small amount of these metals could be leached, for example, to water in nature. As can be seen, the individual extraction agents were able to leach different amounts of metal ions, which correspond with

different affinities of humic acids to the metal. Water is a weak leaching agent, which can extract only mobile fractions of metal ions. In order to obtain the ion-exchangeable fraction of metal ions a 1 M $MgCl_2$ solution was used. Metal ions extractable by a 1 M HCl solution constitute the fraction of strongly bound metal ions [12, 22, 29, 30]. Garcia-Mina [31] showed that the solubility of metal-humic complexes decreased with an increase in the metal : humic ratio, which could be related to the gradual reduction of the free ionized functional group in line with this increase. Our results confirmed this trend only partially, probably because of the use of low concentrations of metal ions. The strongly bound and residual phases were highest for copper and lead. These two metals were in large part strongly complexed in the humic structure and only a very small amount could be washed under normal conditions, as also confirmed by Dević [32]. Dudare and Klavins [33] showed that the complex stability constants significantly changed and were well correlated with the age and humification degree of humic acids. The lignitic humic acids used in our work belonged to the group of well-humified, more aromatic ones with a relatively high content of acidic functional groups, which predestines them for having a high complexation capacity for forming stable metal-humic complexes [7, 12, 34].

Our results correspond with some other works. Janoš et al. [23, 24] stated that Cu^{2+} and Pb^{2+} ions were bound very strongly to the sorbent and that only small portions of these metals could be extracted by weak agents. Substantial proportions of Cu^{2+} and Pb^{2+} ions remained in the residual fraction or were liberated only after the destruction of the organic matrix. On the other hand, Cd^{2+} and Zn^{2+} ions were retained less strongly in the sorbent, probably by an ion-exchange mechanism, and could be leached from the sorbent in acid media. Town and Van Leeuwen [26] obtained similar results for the adsorption of Cd^{2+}, Cu^{2+}, and Pb^{2+} ions. Bosire et al. [35] showed that Zn^{2+} and Cu^{2+} ions were statistically different in their adsorption capacities. Their surface complexation models showed stronger lead and copper interactions with monodentate sites compared to bidentate sites. According to their results, the ratio of Pb^{2+} ions adsorbed on monodentate carboxylic acid sites to Pb^{2+} ions adsorbed on bidentate sites was approximately 1:1. A similar trend was observed for Cu^{2+} ions. The participation of monodentate sites increased with temperature and phenolic sites were involved in bonding only weakly. Yang and Van Den Berg [36] and Kostić et al. [37] confirmed that humic complexes formed with copper exhibited greater stability than those formed with zinc. The stability constant for the complex between lead and humic acids was greater than those for other metal-humic complexes. Terbouche et al. [38] showed that the degree of association between metal ions and humic acids was greater for zinc than for cadmium, which corresponds with our results. Ren et al. [39] divided Cd^{2+}, Cu^{2+}, Pb^{2+}, and Zn^{2+} ions occurring in contaminated soils into two groups: a fast exchangeable pool and a much slower exchangeable one. The distribution of the two exchangeable pools varied significantly among metals. The amount of cadmium related to the fast pool was dominant which could correspond with the highest amount of weakly bound Cd^{2+} ions found in this work. The authors stated that the exchange of cadmium was a relatively fast process, whereas the exchange of copper, lead, and zinc was more sluggish, which is of significant importance for the metal mobility, risk assessment, and management of contaminated soils.

Our results showed that the partitioning of metal ions according to bond strength depends on the presence of other metal ions. The strongly bound and residual fractions decreased in the case of adsorption from the mixture of metal ions, probably due to occupation of strong biding sites by other ions. This was connected with the decrease in adsorption efficiency when compared with the single adsorption of Cu^{2+} ions. On the other hand, the efficiency was, in all cases, higher than 80% and in some cases it achieved 100%. This is a very good result with respect to the potential application of lignitic humic acids as an environmentally-friendly adsorbent for heavy metals. It was found that the leaching of metal ions from humic acids into water was very low; in most of cases it was about 10% and did not exceed 20%. The majority of metal ions ($\geq$60%) were bound very strongly and only partially leachable under strongly acidic conditions.

4. Conclusions

Humic acids isolated from lignite were investigated in order to evaluate their use as an environmentally-friendly adsorbent of heavy metals. Lignite, as a young coal type, contains a relatively high amount of humic acids and can be used for the low-cost production of these valuable substances. It was found that their adsorption efficiency is very high, usually between 80 and 100%. The stability of the formed complexes was evaluated on the basis of leaching using different extraction agents. Copper and lead were bound very

strongly in metal ion-humic complexes and their leaching was low. On the other hand, the amount of cadmium and zinc in the mobile and ion-exchangeable phases was higher (in comparison with copper) but this amount did not exceed 40% of the total content of cadmium and zinc in humic acids. It was confirmed that lignitic humic acids are suitable adsorbent for heavy metals and can be used also for cleaning of waters, soils, and other systems with low or residual concentrations of metal ions. Their partial binding in the mobile and ion-exchangeable phases can be compensated for or eliminated by the use of fresh humic acids without the possibility of the backward leaching of metal ions into the cleaned system.

Competing Interests

The authors declare no conflict of interests regarding the publication of this paper.

Authors' Contributions

Marcela Pavlíková and Martina Klučáková performed the experiments; Marcela Pavlíková and Martina Klučáková analyzed the data; Martina Klučáková wrote the paper.

Acknowledgments

This research was supported by Materials Research Centre at FCH BUT-Sustainability and Development, REG LO1211, with financial support from National Programme for Sustainability I (Ministry of Education, Youth and Sports).

References

[1] M. A. Ferro-García, J. Rivera-Utrilla, I. Bautista-Toledo, and C. Moreno-Castilla, "Adsorption of humic substances on activated carbon from aqueous solutions and their effect on the removal of Cr(III) ions," *Langmuir*, vol. 14, no. 7, pp. 1880–1886, 1998.

[2] M. Klučáková and L. Omelka, "Study of sorption of metal ions on lignite and humic acids," *Chemical Papers*, vol. 58, no. 3, pp. 170–175, 2004.

[3] E. A. Ghabbour and G. Davies, "Environmental insights from Langmuir adsorption site capacities," *Colloids and Surfaces A: Physicochemical and Engineering Aspects*, vol. 381, no. 1–3, pp. 37–40, 2011.

[4] M. Klučáková and M. Pekař, "New model for equilibrium sorption of metal ions on solid humic acids," *Colloids and Surfaces A: Physicochemical and Engineering Aspects*, vol. 286, no. 1-3, pp. 126–133, 2006.

[5] E. Pehlivan and G. Arslan, "Removal of metal ions using lignite in aqueous solution—low cost biosorbents," *Fuel Processing Technology*, vol. 88, no. 1, pp. 99–106, 2007.

[6] M. Pekař and M. Klučáková, "Comparison of copper sorption on lignite and on soils of different types and their humic acids," *Environmental Engineering Science*, vol. 25, no. 8, pp. 1123–1128, 2008.

[7] M. Havelcová, J. Mizera, I. Sýkorová, and M. Pekař, "Sorption of metal ions on lignite and the derived humic substances," *Journal of Hazardous Materials*, vol. 161, no. 1, pp. 559–564, 2009.

[8] S. Babel and T. A. Kurniawan, "Low-cost adsorbents for heavy metals uptake from contaminated water: a review," *Journal of Hazardous Materials*, vol. 97, no. 1-3, pp. 219–243, 2003.

[9] L. Doskočil and M. Pekař, "Removal of metal ions from multi-component mixture using natural lignite," *Fuel Processing Technology*, vol. 101, pp. 29–34, 2012.

[10] J. Mizera, G. Mizerová, V. Machovič, and L. Borecká, "Sorption of cesium, cobalt and europium on low-rank coal and chitosan," *Water Research*, vol. 41, no. 3, pp. 620–626, 2007.

[11] D. Mohan and S. Chander, "Removal and recovery of metal ions from acid mine drainage using lignite—a low cost sorbent," *Journal of Hazardous Materials*, vol. 137, no. 3, pp. 1545–1553, 2006.

[12] M. Pekař, M. Klucáková, G. Barancíková, M. Madaras, and J. Makovníková, "Affinity of soil and lignitic humic acids for Cu(II) and Cd(II) ions," in *Humic Substances. Molecular Details and Applications in Land and Water Conservation*, E. Ghabbour and G. Davies, Eds., pp. 211–223, Taylor and Francis, New York, NY, USA, 2005.

[13] J. Čežíková, J. Kozler, L. Madronová, J. Novák, and P. Janoš, "Humic acids from coals of the North-Bohemian coal field—II. Metal-binding capacity under static conditions," *Reactive & Functional Polymers*, vol. 47, no. 2, pp. 111–118, 2001.

[14] A. G. S. Prado and C. Airoldi, "Humic acid-divalent cation interactions," *Thermochimica Acta*, vol. 405, no. 2, pp. 287–292, 2003.

[15] H. Martyniuk and J. Wieckowska, "Adsorption of metal ions on humic acids extracted from brown coals," *Fuel Processing Technology*, vol. 84, no. 1–3, pp. 23–36, 2003.

[16] V. N. Kislenko and L. P. Oliinyk, "Binding of copper(II), cobalt(II), and nickel(II) cations with humic acids and their sodium salts in aqueous media," *Russian Journal of Applied Chemistry*, vol. 76, no. 12, pp. 1962–1964, 2003.

[17] M. Klučáková, "Adsorption of nitrate on humic acids studied by flow-through coulometry," *Environmental Chemistry Letters*, vol. 8, no. 2, pp. 145–148, 2010.

[18] M. Klučáková, M. Kaláb, M. Pekař, and L. Lapčík, "Study of structure and properties of humic and fulvic acids. II. Complexation of Cu^{2+} ions with humic acid extracted from lignite," *Journal of Polymer Materials*, vol. 19, no. 3, pp. 287–294, 2002.

[19] M. Klučáková and M. Pekař, "Study of structure and properties of humic and fulvic acids. III. Study of complexation of Cu^{2+} ions with humic acids in sols," *Journal of Polymer Materials*, vol. 20, no. 2, pp. 145–153, 2003.

[20] A. Liu and R. D. Gonzalez, "Modeling adsorption of copper(II), cadmium(II) and lead(II) on purified humic acid," *Langmuir*, vol. 16, no. 8, pp. 3902–3909, 2000.

[21] B. Shi, H. E. Allen, M. T. Grassi, and H. Ma, "Modeling copper partitioning in surface waters," *Water Research*, vol. 32, no. 12, pp. 3756–3764, 1998.

[22] M. Kalina, M. Klučáková, and P. Sedláček, "Utilization of fractional extraction for characterization of the interactions between humic acids and metals," *Geoderma*, vol. 207-208, no. 1, pp. 92–98, 2013.

[23] P. Janoš, J. Sypecká, P. Mlčkovská, P. Kuráň, and V. Pilařová, "Removal of metal ions from aqueous solutions by sorption onto untreated low-rank coal (oxihumolite)," *Separation and Purification Technology*, vol. 53, no. 3, pp. 322–329, 2007.

[24] P. Janoš, L. Herzogová, J. Rejnek, and J. Hodslavská, "Assessment of heavy metals leachability from metallo-organic sorbent—iron humate—with the aid of sequential extraction test," *Talanta*, vol. 62, no. 3, pp. 497–501, 2004.

[25] M. A. Shaker and H. M. albishri, "Dynamics and thermodynamics of toxic metals adsorption onto soil-extracted humic acid," *Chemosphere*, vol. 111, pp. 587–595, 2014.

[26] R. M. Town and H. P. Van Leeuwen, "Intraparticulate speciation analysis of soft nanoparticulate metal complexes. the impact of electric condensation on the binding of $Cd^{2+}/Pb^{2+}/Cu^{2+}$ by humic acids," *Physical Chemistry Chemical Physics*, vol. 18, no. 15, pp. 10049–10058, 2016.

[27] V. Enev, L. Pospíšilová, M. Klučáková, T. Liptaj, and L. Doskočil, "Spectral characterization of selected humic substances," *Soil and Water Research*, vol. 9, no. 1, pp. 9–17, 2014.

[28] J. Peuravuori, P. Žbánková, and K. Pihlaja, "Aspects of structural features in lignite and lignite humic acids," *Fuel Processing Technology*, vol. 87, no. 9, pp. 829–839, 2006.

[29] A. Tessier, P. G. C. Campbell, and M. Blsson, "Sequential extraction procedure for the speciation of particulate trace metals," *Analytical Chemistry*, vol. 51, no. 7, pp. 844–851, 1979.

[30] G. Rauret, "Extraction procedures for the determination of heavy metals in contaminated soil and sediment," *Talanta*, vol. 46, no. 3, pp. 449–455, 1998.

[31] J. M. Garcia-Mina, "Stability, solubility and maximum metal binding capacity in metal-humic complexes involving humic substances extracted from peat and organic compost," *Organic Geochemistry*, vol. 37, no. 12, pp. 1960–1972, 2006.

[32] G. J. Dević, "Characterization of eluted metal ions by sequential extraction from Krepoljin coal basin, Serbia: mechanisms of metal interaction," *Energy Sources, Part A: Recovery, Utilization and Environmental Effects*, vol. 38, no. 13, pp. 1912–1917, 2016.

[33] D. Dudare and M. Klavins, "Changes in the humic acid-metal complex-forming characteristics depending on the humification degree," *Fresenius Environmental Bulletin*, vol. 22, no. 2, pp. 604–613, 2013.

[34] L. Pospíšilová, M. Komínková, O. Zítka et al., "Fate of humic acids isolated from natural humic substances," *Acta Agriculturae Scandinavica Section B: Soil and Plant Science*, vol. 65, no. 6, pp. 517–528, 2015.

[35] G. O. Bosire, B. V. Kgarebe, and J. C. Ngila, "Experimental and theoretical characterization of metal complexation with humic acid," *Analytical Letters*, vol. 49, no. 14, pp. 2365–2376, 2016.

[36] R. Yang and C. M. G. Van Den Berg, "Metal complexation by humic substances in seawater," *Environmental Science and Technology*, vol. 43, no. 19, pp. 7192–7197, 2009.

[37] I. S. Kostić, T. D. Andelković, R. S. Nikolić, T. P. Cvetković, D. D. Pavlović, and A. L. Bojić, "Comparative study of binding strengths of heavy metals with humic acid," *Hemijska Industrija*, vol. 67, no. 5, pp. 773–779, 2013.

[38] A. Terbouche, S. Djebbar, O. Benali-Baitich, and D. Hauchard, "Complexation study of humic acids extracted from forest and Sahara soils with zinc (II) and cadmium (II) by differential pulse anodic stripping voltammetry (DPASV) and conductimetric methods," *Water, Air, and Soil Pollution*, vol. 216, no. 1–4, pp. 679–691, 2011.

[39] Z.-L. Ren, Y. Sivry, J. Dai et al., "Exploring Cd, Cu, Pb, and Zn dynamic speciation in mining and smelting-contaminated soils with stable isotopic exchange kinetics," *Applied Geochemistry*, vol. 64, pp. 157–163, 2015.

10

Regional Groundwater Quality Management through Hydrogeological Modeling in LCC, West Faisalabad, Pakistan

Aamir Shakoor,[1] Zahid Mahmood Khan,[1] Muhammad Arshad,[2] Hafiz Umar Farid,[1] Muhammad Sultan,[1] Muhammad Azmat,[3] Muhammad Adnan Shahid,[4] and Zafar Hussain[5]

[1]Department of Agricultural Engineering, Bahauddin Zakariya University, Multan, Pakistan
[2]Department of Irrigation and Drainage, University of Agriculture, Faisalabad, Pakistan
[3]Geo-Informatics Engineering, National University of Sciences and Technology (NUST), Islamabad, Pakistan
[4]Water Management Research Center, University of Agriculture, Faisalabad, Pakistan
[5]Department of Forestry and Range Management, Bahauddin Zakariya University, Multan, Pakistan

Correspondence should be addressed to Hafiz Umar Farid; farid_vjr@yahoo.com

Academic Editor: Pedro Ávila-Pérez

The intensive abstraction of groundwater is causing a number of problems such as groundwater depletion and quality deterioration. To manage such problems, the data of 256 piezometers regarding groundwater levels and quality were acquired for the period of 2003 to 2012 in command area of Lower Chenab Canal (LCC), West Faisalabad, Pakistan. MODFLOW and MT3D models were calibrated for the period of 2003–2007 and validated for years 2008–2012 with respect to observed groundwater levels and quality data, respectively. After the successful calibration and validation, two pumping scenarios were developed up to year 2030: Scenario I (increase in pumping rate according to the historical trend) and Scenario II (adjusted canal water supplies and groundwater patterns). The predicted results of Scenario I revealed that, up to year 2030, the area under good quality groundwater reduced significantly from 50.35 to 28.95%, while marginal and hazardous groundwater quality area increased from 49.65 to 71.06%. Under Scenario II, the good quality groundwater area increased to 6.32% and 12.48% area possesses less hazardous quality of groundwater. It was concluded that the canal water supply should shift from good quality aquifer zone to poor quality aquifer zone for proficient management of groundwater at the study area.

1. Introduction

In Pakistan, groundwater is the second largest source of irrigated agriculture because of the arid climatic conditions [1, 2]. The reliance on groundwater has significantly increased during the past two decades to meet the food and fiber requirements of growing population [3]. Furthermore, the surface irrigation system has high degree of conveyance and application water losses. Thus, the system operates at an efficiency of less than 40% that meets only less than 40% of the crop water requirements [4]. The provision of irrigation water from other sources such as groundwater is indispensable for potential productivity [5]. The growth rate of private tube wells has increased significantly at the rate of 60% in Punjab province from 1991 to 2000. About 1.20 million tube wells of capacity ranging from 0.015 to 0.056 m^3/sec have been installed in Pakistan for irrigation at depth of 30–85 m having diameter of 15–30 cm, out of which 86% tube wells are installed in Punjab [6]. The farmers of tail ends of the distributaries and watercourses are forced to rely heavily on groundwater, particularly where canal water supplies are constrained. Without groundwater availability, not only Punjab but the whole country would face a severe water shortage that leads to food shortages as Punjab produces more than 90% of total grains [7]. However, the unchecked abstraction of groundwater has created serious negative concerns in terms of lowering water table and saltwater intrusion that may lead to the issues of sustainability of usable groundwater resources. As a result of saline groundwater intrusion, about 200 public tube wells initially installed in the fresh groundwater zone of

Punjab and Sindh provinces had been abandoned [8]. Shah [9] reported similar problems existed in most of the irrigated regions of the world; those were further increasing rapidly and negatively affecting agricultural productivity.

There are wide spatial and temporal variations in groundwater quality in the Indus basin, which is due to the pattern of groundwater movement in the aquifer [10]. The belt of fresh groundwater is generally available near the main rivers and canal on account of high recharge of fresh seepage water. But, then, the quality of groundwater changes to unfit as laterally away from the rivers [5, 11]. The continuous and intensive use of groundwater for irrigation is adding plenty of salts causing secondary salinization because groundwater generally has more salts than canal water. Bakhsh and Awan [12] reported that application of groundwater having total dissolved solids (TDS) of 1000 mgL^{-1} up to a soil depth 370 mm changed the top 300 mm depth of nonsaline into a saline soil that impaired crop productivity. In many agricultural areas of Pakistan, the usage of poor quality tube well water for irrigation is considered as one of the major causes of salinity and, consequently, lower food productivity [13]. According to an estimate in Pakistan, the secondary salinization degraded the crop land which reduced the production potential of major crops by 25%, valued at an estimated loss of US $250 million/year [14]. Low quality water and soil salinity can affect plant growth and soil structure in several ways, directly and indirectly [15].

Thus, the assessment of groundwater quality is important to ensure the safe use of these resources on a sustainable basis. Groundwater models are most widely used tools for efficient management of precious groundwater resources and to predict different future scenarios [16]. Different groundwater modeling codes are available, each with their own capabilities, operational characteristics, and limitations such as PMWIN, FEFLOW, SVFlux, and GWVistas. But the most extensively used three-dimensional groundwater flow model among the available models is PMWIN (Processing MODFLOW for Window) [17, 18]. Its popularity has continued, in part due to the modularity of the program, resulting ability, and user friendly interface [19]. It uses a block-centered finite-difference scheme for saturated zone. The advantages of PMWIN include numerous facilities for data preparation, easy exchange of data in standard form, extended worldwide experience, continuous development, availability of source code, and relatively low price or being freely available [20]. A number of research studies have been conducted regarding application of modeling approach for groundwater management in different part of the world. Moeck et al. [21] developed a 3D groundwater model for simulating existing and proposed water management strategies as a tool to ensure the utmost security for drinking water in Basel, Switzerland. Gebreyohannes et al. [22] developed regional groundwater flow model for Geba Basin, northern Ethiopia, and reported that none of the hydrogeological formations can be exploited for large-scale groundwater exploitation. Rahmawati et al. [23] conducted research to study saltwater intrusion from 1995 to 2108 in Semarang city based on well log data and MODFLOW numerical model was used. For salt intrusion projection in the future, the sea level rise projection also was

conceived. Kori et al. [17] linked groundwater flow model (MODFLOW) with solute transport model (MT3D), and several simulation runs were carried out after successful calibration of model for two sampled sites located at JRS-57 and JRS-60 tube wells at Nawabshah-Pakistan. Carretero et al. [24] applied groundwater modeling technique to study the impact of a possible rise of 1 m in sea level against the low-lying coast of Partido de La Costa, Argentina. Similarly, Abu-el-Shar and Hatamleh [25] simulated the groundwater model (PMWIM) for the Azraq Basin, Jordan, to manage groundwater. The large number of latest research papers—Asoka et al. [26], Galitskaya et al. [27], Kambale et al. [28], Abdullah and Morteza [29], and Durand et al. [30]—dealing with groundwater management using modeling techniques were published is a testimony to the important role played by the models.

Therefore, the assessment of groundwater quality through modeling is important to ensure the safe use of these resources on a sustainable basis. If unchecked pumping of groundwater continues to remain in the study area, the irrigation tube wells would not be able to lift water at their present quality. Hence, there is a dire need to investigate the impact of groundwater flow conditions and overexploitation on groundwater quality. Thus, the followings were the two main objectives of the current study: (1) to develop a regional hydrogeological groundwater flow model and observe its future trend and (2) to formulate groundwater management strategy for its proficient utilization.

2. Material and Methods

2.1. Study Area. The research study was carried out in the command area of Lower Chenab Canal (LCC), West Faisalabad, Pakistan, having longitude of 73.85° to 72.18° (E) and latitude of 32.32° to 30.85° (N) (Figure 1). The site has gross command area of 1.16 Mha and culturable command area of 0.98 Mha, respectively. It falls in rice-wheat agroecological zone of province Punjab, Pakistan. It is comprised of vast canal network such as main canals, branch canals, and minor distributaries. The River Chenab and Gugera Branch Canals are located at the northwest and southeast side, respectively. Similarly, Qadirabad-Baluki Link Canal is located in the northeast side and Trimmu-Sidhnai Link Canal is present in the southwest side of the study area (Figure 2).

2.2. Climatic. The summer season starts from April and continues till October. The temperature varies in the range of 21 to 51°C during the summer season. Similarly, the winter season starts from October and lasts till April with temperature ranges of 1 to 27°C. The average annual precipitation was estimated to be 439 mm. The ten-year average value of ET$_o$ was 1413 mm/year.

2.3. Hydrogeological Conditions. Lithology of aquifer system in the study area has different classification according to different textural characteristics. The surface soil textures are largely fine and moderately medium with good permeability properties. The aquifers of study area were formed as a result of sediment deposition due to flat topography. These

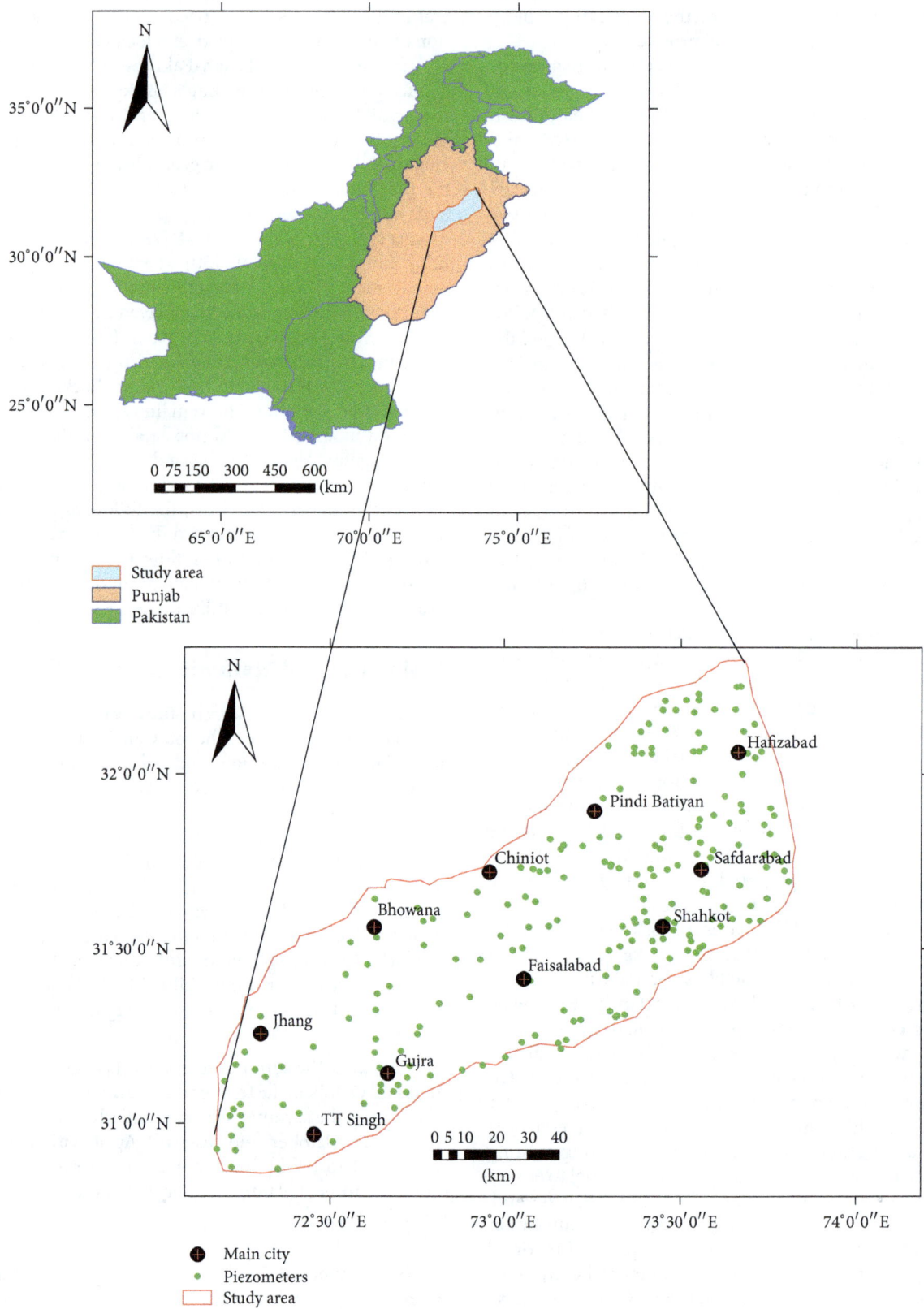

FIGURE 1: Geographical location of study area.

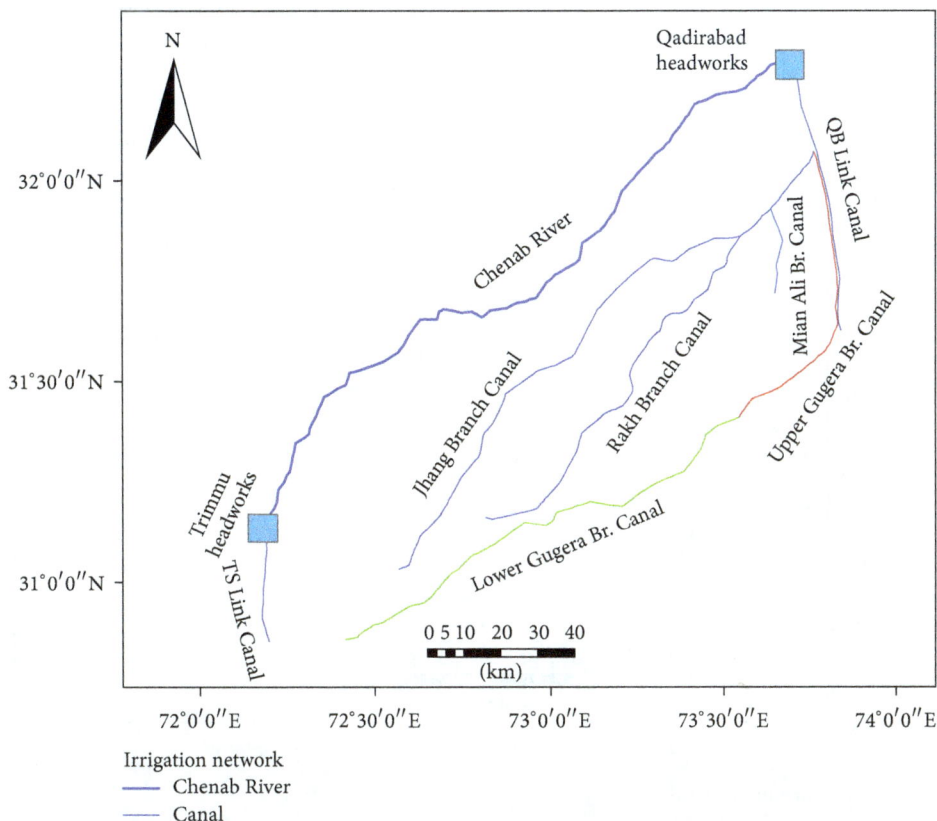

FIGURE 2: Canal network of the study area.

sediments were carried by the river waters from the vast alluvial basin of rivers as a result of materials washed down from Himalayan Mountain. In general, like other aquifers in Punjab, the aquifers of study area are unconfined. The groundwater levels in the year 2012 were present in the range of 145 to 205 m (Figure 3). The groundwater level was higher in the upper part and gradually decreased towards the lower side, which means that the groundwater flow was from upper to lower side.

The good quality groundwater is present in the northern part whereas poor quality groundwater is in the southern part of the study area. The average spatial variation of measured groundwater TDS in the entire study area for the year 2012 is given in Figure 4. A calibrated TDS meter (TDS-98302) was used to measure the TDS of groundwater with an accuracy of ±2%. The calibration of TDS meter was accomplished in the standard solution in the Water Quality and Environment Laboratory, Department of Irrigation and Drainage, University of Agriculture, Faisalabad. The measured TDS values indicated that almost half of the aquifer of the study area has marginal to hazardous quality of groundwater. A 30 km wide strip along the Chenab River had groundwater of good quality, whereas hazardous quality groundwater existed in 20 km wide strip at a distance of 40 km away from the Chenab River. The groundwater quality zones of fresh and saline water were formulated according to criteria developed by WAPDA [31] for irrigation water quality (good: TDS < 1000 mg/L; marginal: TDS of 1000–1700 mg/L; and hazardous: TDS > 1700 mg/L),

Shakoor et al. [6]. The northern and northwestern part of the study area (along Chenab River) contained good quality groundwater. The marginal and hazardous quality was laid in the southern part of the study area.

The annual water balance of the whole study area from the year 2003 to the year 2012 is shown in Figure 5. The water balance describes the volume of water entering, subtraction, and net storage in the aquifer system. The water entering parameters are recharge and river leakage, while subtraction parameters are wells and ET. The positive storage term showed that volume of water extracted from the aquifer was more than the recharge which means crop water demand was fulfilled from aquifer storage and vice versa. The storage in the aquifer in 2003 and 2012 was −444.5 and 326.6 MCM, respectively, clearly indicating the replenishment of aquifer in 2012. Similarly, the volume of water withdrawn through tube wells increased from 3298 to 3759 MCM from years 2003 to 2012, respectively, also showing increased demand of groundwater.

2.4. Model Selection. PMWIN 5.3 (Processing MODFLOW for Window) was used in this research. It is a simulation system based on the modular three-dimensional finite-difference technique for modeling groundwater flow and contamination in groundwater with a wide range of natural systems. PMWIN is used widely throughout the world and it was applied in many groundwater modeling applications [17, 23, 25].

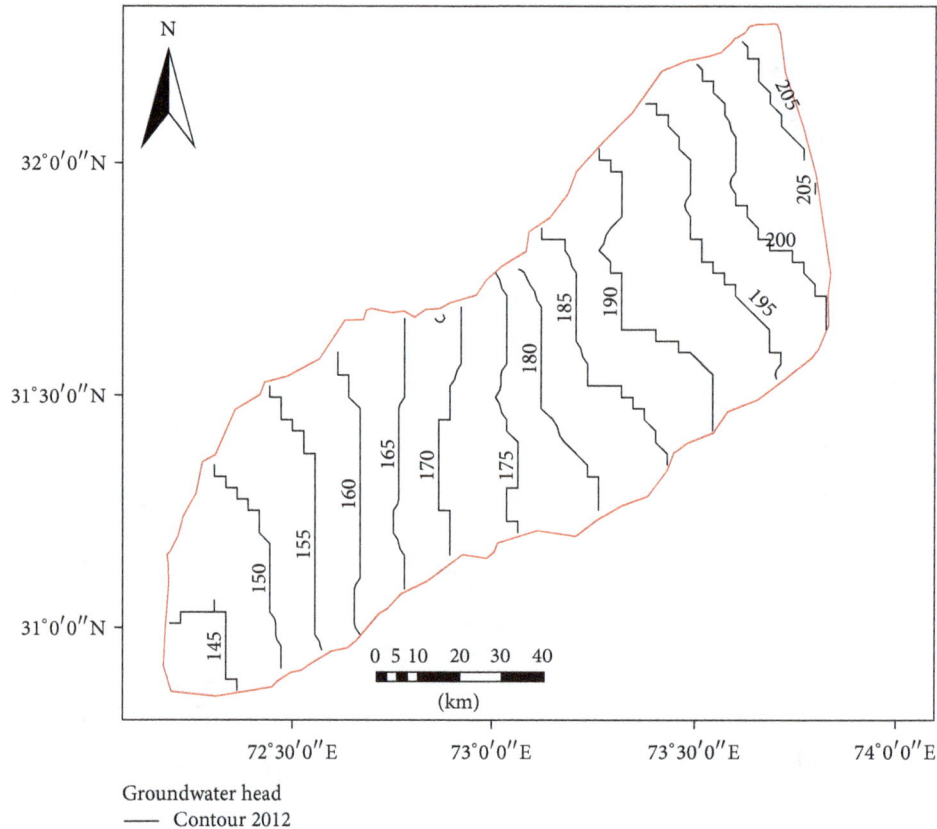

Groundwater head
—— Contour 2012

FIGURE 3: Spatial variation in measured groundwater levels for the year 2012.

2.5. Model Input and Calibration. The MODFLOW and MT3D models are packages of PMWIN 5.3. They were calibrated with respect to observed groundwater level and quality data, respectively, using inverse modeling method. The data of 256 piezometers regarding groundwater level and quality were acquired from the Department of Land and Reclamation, Faisalabad, for the period from 2003 to 2012 to achieve the model calibration and validation (Figure 1). The piezometer wells have maximum depths of 55 m. The depth of piezometers varied at some location depending upon the water table conditions. The piezometer wells have diameters of 5 cm and strainer lengths of 3 m.

The model area had a square geometry and the whole area was divided into 76 columns and 76 rows. The total number of cells was of 5776 cells. Each square cell has a dimension of 2.5 km × 2.5 km. The model area had 3653 inactive cells, which were outside the boundary of the study area, while the model area had 2123 active cells located within the boundary of study area. The cell size of 2.5 km × 2.5 km (6.25 km^2) was also used for groundwater modeling studies by Al-Fatlawi [32] in Umm Er Radhuma, the Western Desert, Iraq, and by Khan et al. [33] in Rechna Doab. Abu-el-Shar and Hatamleh [25] developed groundwater model for the Azraq Basin and the biggest cell size of 8.69 km^2 was selected. Similarly, Schoups et al. [34] used cell size of 2 km × 2 km to calibrate groundwater model of the Yaqui Valley, having

6800 km^2 irrigated agricultural region located along the Sea of Cortez in Sonora, Mexico.

Lithology of aquifer system in the study area was obtained from the Water and Power Development Authority (WAPDA) [31]. The soils have different classification according to different textural characteristics. The surface soil textures are largely fine and moderately medium, with good permeability properties. The areas of 4451.3 (38.48%), 4987.3 (43.08%), 1621 (14%), 464.1 (4%), and 52.3 km^2 (0.45%) have soil texture of fine, moderately medium, medium, moderately course, and course, respectively. The aquifer of the study area was defined with four different layers depending upon their lithological data [31]. The spatial domain represented in the model consisted of four layers (0–7, 7–30, 30–90, and 90 m to bedrock). The horizontal (K_h) and vertical (K_v) hydraulic conductivities have large variation from one to the other side of the study area (Table 1). The minimum and maximum values of 1–265 m/day and 1–15 m/day, respectively, for horizontal and vertical hydraulic conductivities were used in the model, as majority of the tube wells were installed in second layer because 80% part of study area has K_h in the range of 70–100 m/day. The similar range of values for hydraulic conductivity within Punjab province domain was used by Jehangir et al. [35]; Ahmad [36]; Arshad [37]; and Khan et al. [33]. The specific storage values for layer 1 ranged as 0.0001–0.001 m^{-1} and for the remaining layers the values

TABLE 1: Lithological data of all layers.

Subsurface layers	K_h (m/day)	% area	K_v (m/day)	% area	Specific storage (m^{-1})	% area	Specific yield	% area
(Layer 1)	1–15	20	>0-1	80	$1E-4$–$5E-4$	40	0.05–0.013	15
	15–70	60	1–5	10	$5E-4$–$6E-4$	25	0.013–0.17	60
	70–160	20	5–15	10	$6E-4$–$1E-3$	35	0.17–0.25	25
(Layer 2)	1–70	10	>0-2	50	$1E-5$–$2E-4$	5	0.05–0.15	10
	70–100	80	4–10	25	$3E-4$–$1E-3$	15	0.2–0.25	80
	100–165	10	10–15	25	$2E-4$–$3E-4$	80	0.15–0.2	10
(Layer 3)	20–80	20	>0-2	50	$1E-5$–$2E-4$	5	—	—
	80–120	30	4–10	25	$3E-4$–$1E-3$	15	—	—
	120–265	20	10–15	25	$2E-4$–$3E-4$	80	—	—
(Layer 4)	20–80	20	>0-2	50	$1E-5$–$2E-4$	5	—	—
	80–120	30	4–10	25	$3E-4$–$1E-3$	15	—	—
	120–265	20	10–15	25	$2E-4$–$3E-4$	80	—	—

K_h: horizontal hydraulic conductivity. K_v: vertical hydraulic conductivity.

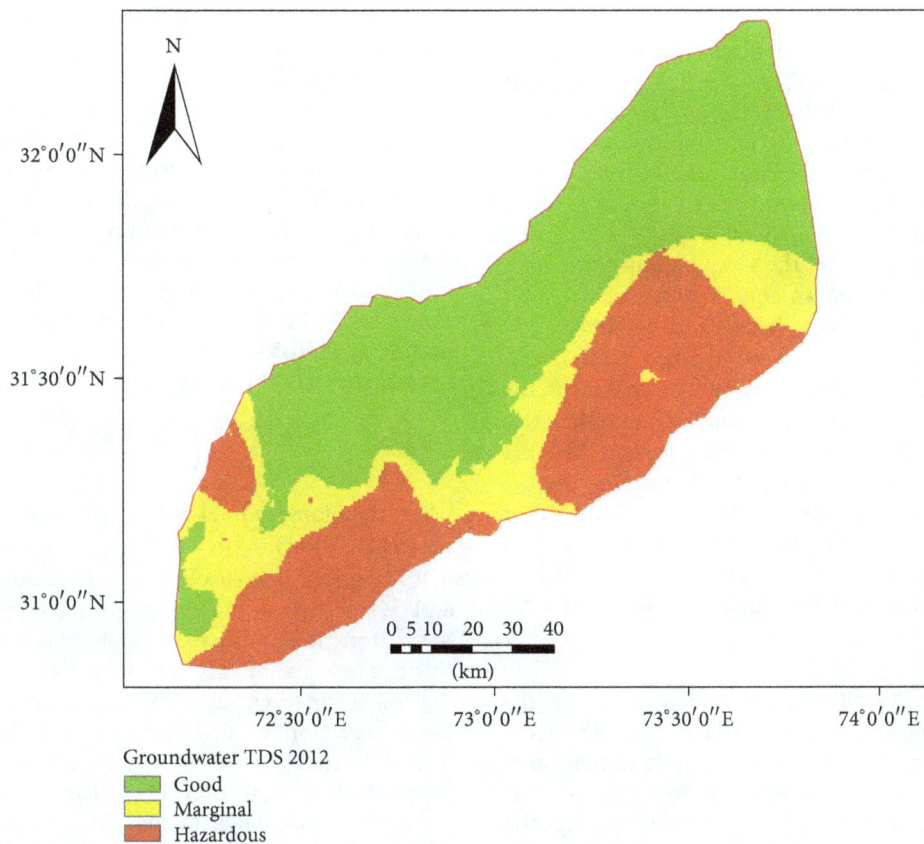

FIGURE 4: Spatial variations of measured groundwater TDS in the year 2012.

of 0.00001–$0.0003\,m^{-1}$ were used. The values of specific yield for layer 1 ranged as 00.05–0.25 and for the remaining layers the values of 0.05–0.20 were used. The effective porosity of 0.25 was given to all layers [31]. The simulation time unit "days" and simulation flow type "transient" was selected. Two stress periods in each year were considered to represent the Kharif and Rabi seasons having 183 and 182 days, respectively, with six time steps in each stress period, as there are two main cropping seasons based on agroclimatic conditions in

Pakistan, Kharif and Rabi. Kharif starts from June and July and goes to October and November, while the Rabi season starts from September and October and continues to April and May.

The cumulative evapotranspiration during a period (Kharif or Rabi) was divided by its duration in days and thus the evapotranspiration rate per day was calculated using CROPWAT. The evapotranspiration rates of 0.006 and 0.003 m/day were used for odd (Kharif) and even (Rabi)

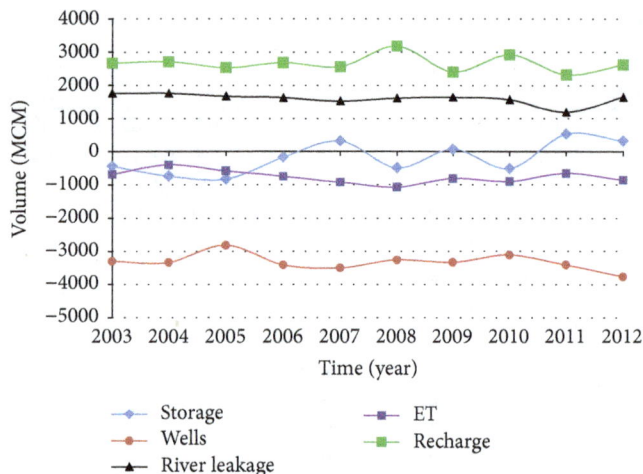

FIGURE 5: Water balance for the years 2003 to 2012.

stress periods, respectively, throughout the model stress periods for model calibration. The recharge package was used to simulate spatial distribution of recharge water at field from rainfall and irrigation to the groundwater system. The minimum and maximum recharge values from 0.00065 to 0.0013 m/day were used for odd (Kharif) stress periods, whereas values from 0.00026 to 0.0005 m/day were used for even (Rabi) stress periods. The river package was used to simulate the flow between an aquifer and surface-water features such as rivers, canals, lakes, and reservoir. The hydraulic features, such as canal length, bed width, full-supply level, hydraulic conductance, and discharge of the canals of the study area, were used. The recharge flux was computed by the model through the canal system. The net groundwater demand was calculated by subtracting the net canal water supplies from net crop water requirement. After providing all the input data, the calibration process was started. In this calibration process, input parameters such as recharge, pumping rate, and hydraulic conductivities were adjusted in MODFLOW.

2.6. Solute Transport Model. The MT3D numerical model with PMWIN 5.3 user interface was used to simulate advection and dispersion contaminants in the three-dimensional groundwater flow system of the study area. The MT3D simulated groundwater flow by numerically solving the groundwater flow and solute transport equations. The partial differential equation describing the three-dimensional transport of dissolved solutes in the groundwater [38] can be written in

$$\frac{\partial}{\partial x_i}\left(D_i\frac{\partial C}{\partial X_i}\right) - \frac{\partial}{\partial x_i}\left(v_iC\right) \pm \frac{q_s}{\theta}c_s + \sum_{k=1}^{N}R_k = \frac{\partial C}{\partial t}, \quad (1)$$

where C is the concentration of contaminants dissolved in groundwater [ML^{-3}], x_i is the distance along the respective Cartesian coordinate axis [L], D_i is the hydrodynamic dispersion coefficient [L^2T^{-1}], v_i is the seepage or linear pore water velocity [LT^{-1}], q_s is the volumetric flux of water per unit

volume of aquifer representing sources (positive) and sinks (negative) [T^{-1}], C_s is the concentration of sources or sinks [ML^{-3}], $\sum_{K=1}^{N}R_k$ is the chemical reaction term [$ML^{-3}T^{-1}$], and t is time [T].

The calibration of MT3D was achieved by adjusting the values of advection and dispersion values. Advection depicts mass transport basically due to the bulk flow of water in which the mass is dissolved or movement of solute as a result of groundwater flow. Advection is the dominant process of solute transport due to groundwater flow. For a given time step, the movement of solutes due to advection was simulated using the corresponding field velocity which was computed by the calibrated MODFLOW model and the effective porosity of the aquifer. MXPART of maximum 1300000 particles was allowed in a simulation while the value PERCEL (Courant number or number of cells in any particle) was selected as 0.79. It was allowed to move in any direction within one transport step and generally ranged from 0.5 to 1 [39]. The default values of the remaining parameters were used for MT3D model.

The longitudinal dispersivity (α_L) and transverse dispersivity (α_T) are required to estimate dispersive transport of the solute. The dispersivity is a characteristic property of the porous medium that describes the spreading of the solute in a medium. Longitudinal dispersivity is used when the spreading of solute is in the direction of bulk flow, whereas the transverse dispersivity is the perpendicular (vertical and horizontal) spreading of solute to the direction of bulk flow. The longitudinal dispersivity (α_L) was computed from molecular diffusion and local dispersion coefficient based on heterogeneity of the medium [40, 41] as given in

$$\alpha_L = \left(\frac{D_d}{v} + \frac{D_L}{v}\right) + A_L, \quad (2)$$

where α_L is longitudinal dispersivity (cm), D_d is coefficient of molecular diffusion (16×10^{-6} cm^2/s), V is velocity of flow (6.8×10^{-6} cm/s), D_L is dispersion coefficient (30×10^{-6} cm^2/s), and A_L is asymptotic longitudinal dispersivity (500 cm). Asymptotic longitudinal dispersivity is due to heterogeneity of the medium, mainly dependent on the variance of the log transformed conductivity and correlation length in the mean direction of flow. As no field data were available for solute transport in aquifer media, the value of A_L was adopted as 500 cm after [42]. The Peclet number, $P_e = \Delta x/\alpha_L$, was calculated to measure the relative significance of advection and dispersion in the study area, where Δx is the size of a cell equal to 2500 m. The unitless Peclet number is usually used to decide the dominant factor in solute transport among advection and dispersion [43]. After the successful calibration of both models (MODFLOW and MD3D) for years 2003–2007, the models validated for the years 2008–2012 and two future scenarios were simulated.

2.7. Future Scenarios. The model prediction was accomplished in order to investigate the response of the model for two future scenarios regarding the pumping rate up to year 2030. It was assumed that there will be no uncertain change or tragedy in climate and irrigation system. In Scenario I,

TABLE 2: Statistical analysis of the field and modeled data.

Parameters	Formula	Results	
		MODFLOW (m)	MT3D (mg/L)
Mean error	$ME = \dfrac{1}{n}\sum_{i=1}^{n}(h_o - h_s)$	−1.10	19.5
Mean absolute error	$MAE = \dfrac{1}{n}\sum_{i=1}^{n}|h_o - h_s|$	1.72	139
Root mean square error	$RMSE = \sqrt{\dfrac{1}{n}\sum_{i=1}^{n}(h_o - h_s)_i^2}$	2.24	184
Model efficiency	$MEF = \dfrac{\sum_{i=1}^{n}(h_o - h_s)^2}{\sum_{i=1}^{n}(h_o - \overline{h})^2}$	0.98	0.91
(R^2)		0.89	0.87

the pumping will increase according to the historical trend. As in Punjab province of Pakistan, about 1.20 million tube wells have been installed and increasing at 5.5% annually [44]. This indicated that the amount of withdrawal from aquifer may increase from 3738 MCM in year 2012 to 6008 MCM in year 2030 whereas recharge would be about 2664 MCM according to historic trend. In *Scenario II*, one of the water management options for the study area was proposed. In the upper part of the study area (Pindi Bhatiyan-Safdrabad) where groundwater has good quality, the rate of groundwater abstraction was increased and recharge through an irrigation system was decreased by 35%, while in the lower part, where groundwater has poor quality, the groundwater abstraction was decreased and recharge through an irrigation system was increased by the same rate.

3. Results and Discussion

3.1. Calibration and Validation of Models. The degree of fit between model simulations and field measurements was quantified by statistical means (Table 2) and all parameters were found in acceptable range. The minus sign of mean error (ME) represented that the model simulated values were higher than the measured head. The calibration criterion for hydraulic head is root mean squire error (RMSE) which is less than or equal to 10% of head variation within the aquifer being modeled [45] and the same criterion is also followed in groundwater TDS. The head in the aquifer within the study area varies from approximately 145 to 205 m, resulting in an acceptable RMSE of 6 m or less. Similarly, the TDS in the aquifer at selected points are varied from 384 to 3768 mg/L, resulting in an acceptable RMSE of 339 mg/L or less. Anderson and Woesner [46] and Moriasi et al. [47] reported that the RMSE is generally considered the best calibration indicator. Asghar et al. [48] said that the negative value of Model Efficiency (MEF) indicates the high variability between the observed and simulated values. A zero value of MEF indicates a poor simulation. If the model simulated values exactly match the observed value then the MEF = 1. So, all the values showed a good fit between measured

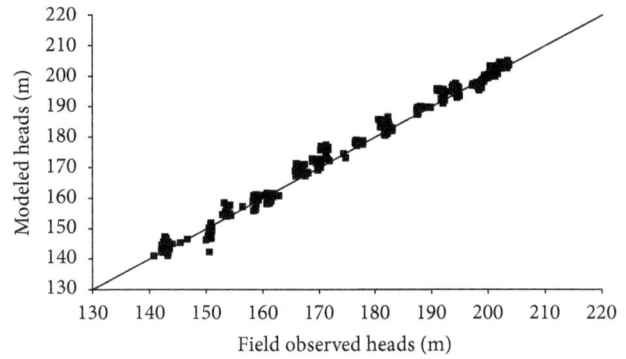

FIGURE 6: Scattergram of measured versus modeled heads in MODLOW.

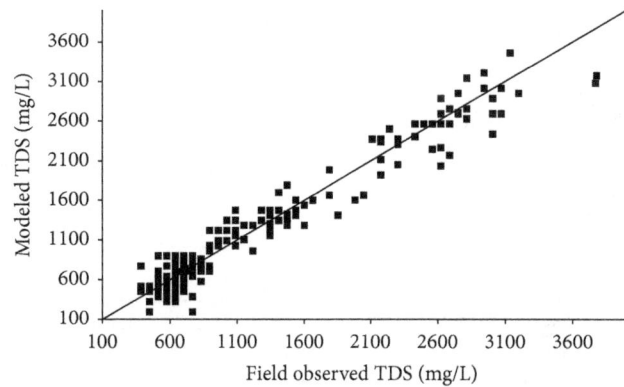

FIGURE 7: Scattergram of measured versus modeled TDS in MT3D.

and simulated values. The average coefficient of determination (R^2) for the selected piezometers for calibration and validation period was calculated as 0.89 and 0.87 (Figures 6 and 7) for MODFLOW and MT3D models, respectively, which indicated a close agreement between the calculated and measured groundwater head. Hagos [49] calibrated PMWIN model for Raya valley, Ethiopia, with respect to groundwater level. The values of ME, MAE, and RMSE of calibrated results were −1.4 m, 7.8 m, and 10.7 m, respectively, with coefficient of determination of 0.97 and reported to be satisfactory.

3.2. Sensitivity Analysis. The sensitivity analysis showed that recharge and transmissivity of the aquifer were most sensitive parameters. The factors of 0.5, 0.8, 0.9, 1.1, 1.2, 1.3, and 1.5 were multiplied with the calibrated values of recharge and transmissivity. The resulting hydraulic heads were then compared with the observed heads and RMSE was calculated for each parameter. It was observed that the minor variation in transmissivity or recharge rate values affected the hydraulic head impressively. The resulting plots of sensitivity showed the nonlinear response to recharge and transmissivity (Figures 8 and 9).

3.3. Predicted Groundwater Level. The contour lines of predicted groundwater level for the year 2030 under Scenario I

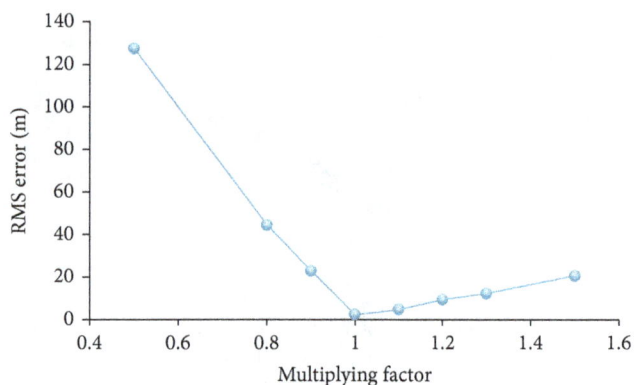

FIGURE 8: Sensitivity analysis with respect to recharge.

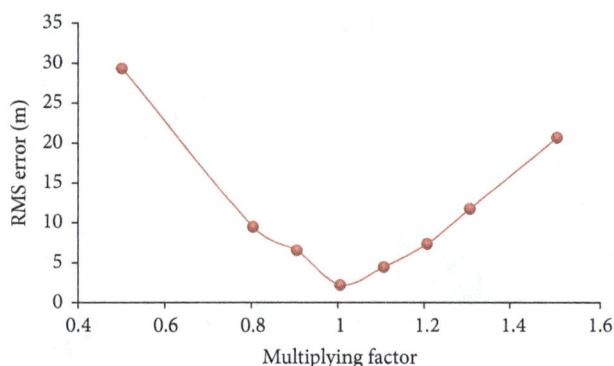

FIGURE 9: Sensitivity analysis with respect to transmissivity.

TABLE 3: Dispersivity values used in MT3D.

Parameter	Used value	Possible range and source
Longitudinal dispersivity (m)	5.06	3–15.24 [42] 1–10 [54] 1.89–5 [53]
Horizontal transverse dispersivity (m)	0.45	0.01–10 [42, 55]
Vertical transverse dispersivity (m)	0.015	0.01–10 [42, 55]

TABLE 4: Value of parameters for dispersive transport.

Layer number	TRPT (—)	TRPV	DMCOEF
1	0.09	0.003	0.1
2	0.09	0.003	0.1
3	0.09	0.003	0.1

TRPT is the ratio of the horizontal transverse dispersivity to the longitudinal dispersivity; TRPV is the ratio of the vertical transverse dispersivity to the longitudinal dispersivity; DMCOEF is the effective molecular diffusion coefficient [L^2T^{-1}].

and Scenario II are presented in Figures 10 and 11, respectively. Under Scenario I, the maximum depletion in groundwater level would be up to 18 m near Bhawana, where irrigation is mainly dependent upon groundwater. Qureshi et al. [7] reported that the depletion of groundwater was more pronounced in uncommand areas of the Punjab, where surface-water supplies were constrained and agriculture was heavily dependent on groundwater. In the upper part of the study area, comparatively less depletion (2 m on the average) in the groundwater level was observed. Under Scenario II, decline of groundwater level by 3-4 m was observed. In comparison with Scenario I the difference is of about 1-2 m, which is not too high. In the lower part near Jhang and TT Singh, there would be a rise of groundwater level by up to 2 m (Figure 11). Shifting of canal water supplies would have good effect for replenishing groundwater.

The depletion of groundwater level has direct and indirect impact on pumping cost. Kori et al. [17] discussed that the decline in groundwater level would increase the abstraction cost. The construction cost of a deep electric tube well (>20 m) was reported as US $5000 as compared to US $1000 for a shallow (<6 m) tube well. Obrien et al. [50] reported that cost of pumping 10^3 m^3 water was US $8.61 and US $18.78 for 31 m and 91 m lift, respectively. Basharat [51] reported that cost of pumping per cubic meter of groundwater increased about 3.5 times because the depth of water table dropped from 6 to 21 m. Qureshi et al. [52] calculated that the installation

cost of private tube well in Pakistan was US $530 for the areas where the water table depth was less than 6 m and it was US $3206 for the areas with more than 24 m depth.

3.4. Transport of Salts. The value of longitudinal dispersivity was calculated as 5.06 m (Table 3). Ahmad [53] reported that the values of longitudinal dispersivity were between 1.89 and 5 m for the aquifer of the Indus Basin of Pakistan. Gelhar et al. [42] reviewed various researches and reported that the value of longitudinal dispersivity was between 3 and 15.24 m, whereas Engesgaard et al. [54] reported the range of longitudinal dispersivity from 1 to 10 m. The value of horizontal and vertical transverse dispersivity was found to be 0.45 and 0.015 m, respectively, based on soil type and hydraulic conductivity [42]. Shieh et al. [55] concluded that the transverse (horizontal and vertical) dispersivity was between 0.01 and 10 m. Narayan et al. [56] developed SUTRA, a solute transport model for the Lower Burdekin Delta, North Queensland, and found longitudinal dispersivity of 2.5 m and transverse dispersivity of 0.5 m.

The dispersivity values used in the MT3D model are given in Table 4. The value of TRPT (ratio of the horizontal transverse dispersivity to the longitudinal dispersivity) and TRPV (ratio of the vertical transverse dispersivity to the longitudinal dispersivity) was 0.09 and 0.003, respectively. The effective molecular diffusion coefficient (DMCOEF) describes the diffusive flux of a solute in water from an area of greater concentration towards an area where it is less concentrated. The molecular diffusion coefficient is generally very small and negligible compared to the mechanical dispersion, so it was 0.1 [17, 39]. The value of Peclet number was found to be 494. For such higher value of Peclet number, the solute transport is dominated by the advection, that is, the transport of solute due to bulk flow of water [43, 57].

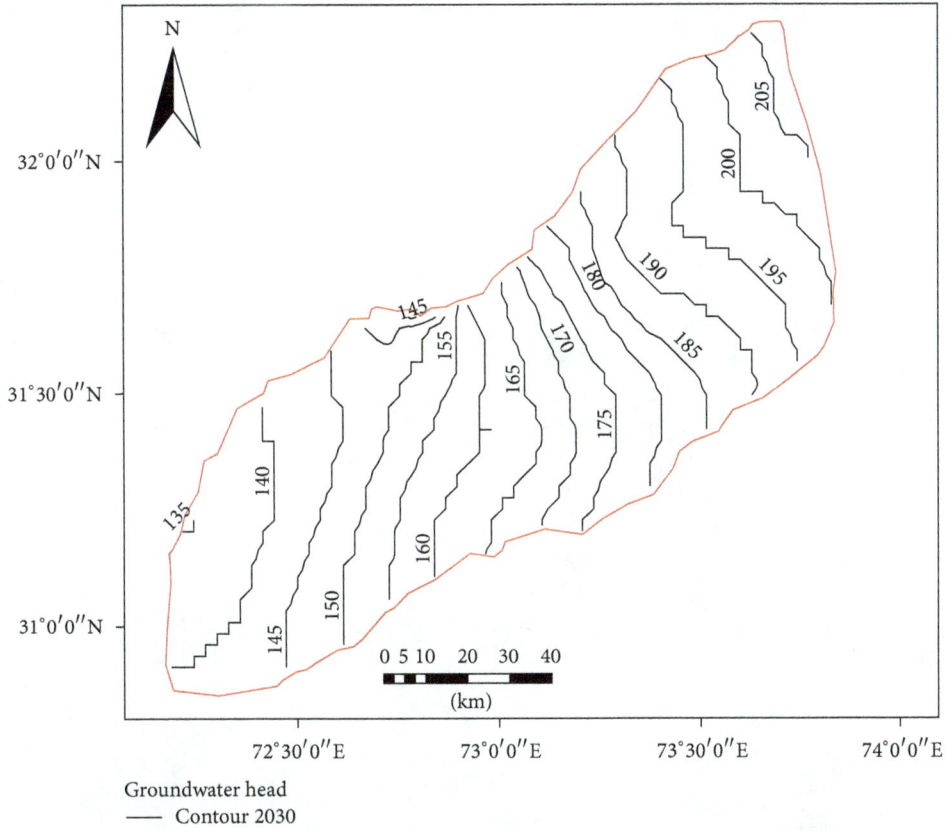

Groundwater head
— Contour 2030

FIGURE 10: Predicted groundwater level in 2030 (Scenario I).

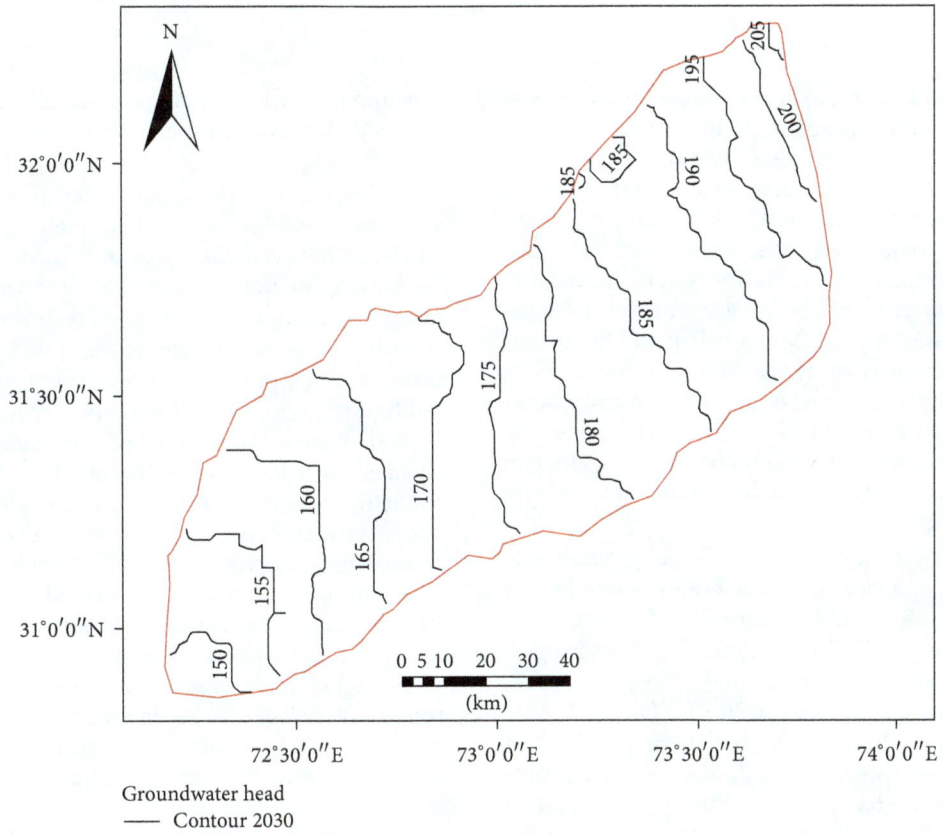

Groundwater head
— Contour 2030

FIGURE 11: Predicted groundwater level in 2030 (Scenario II).

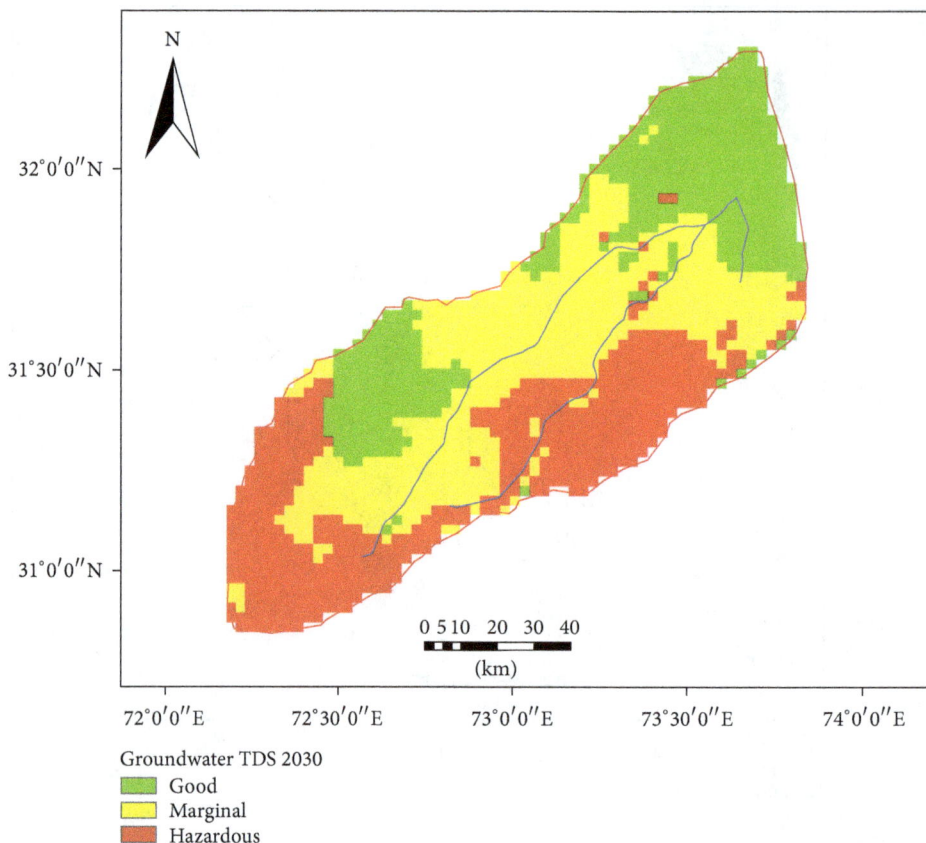

FIGURE 12: Predicted groundwater quality in year 2030 (Scenario I).

3.5. Predicted Groundwater Quality. Predicted variations in groundwater TDS under Scenario I up to year 2030 are presented in Figure 12. For the year 2030, the 28.95% area will have good quality groundwater, 38.35% area will have marginal quality, and the remaining 32.71% area will have hazardous quality groundwater of the aquifer with TDS concentration of greater than 1000 mg/L. The results showed that 21.40% good quality area will be converted from marginal into poor quality after 18 years. So, marginal and hazardous quality groundwater areas are expected to increase by 19.28 and 2.13%, respectively. The area of hazardous quality water almost remained the same but its severance increased due to less recharge of fresh water than discharge. The change in groundwater TDS concentration under Scenario I up to year 2030 is shown in Figure 13.

The negative sign indicated the increase in groundwater TDS concentration and vice versa. The groundwater TDS of about 85% of the aquifer would increase up to 1500 mg/L and the remaining area up to 2500 mg/L located at southern part of the study area near Jhang and TT Singh. The groundwater quality of <1% area would increase up to 140 mg/L near Hafizabad because of having relatively high potential of recharge irrigation network and rainfall. The predicted results revealed that groundwater quality of the northern part of the study area will not be affected with the increase in the pumping due to balance between discharge and recharge. The irrigation networks in that upstream area are well established and generally received more canal water than the allocated share [50, 51].

Simultaneously, the groundwater quality in middle and lower part will be affected severely by salinity due to imbalance between discharge and recharge of fresh water. The farmers of downstream areas received about 21% less canal water than designed discharge imposed farmers to use groundwater as a primary source [51, 52]. This abundant abstraction of groundwater is deteriorating the groundwater quality due to both horizontal and vertical saline intrusion [58, 59]. Prinos [60] described the multiple pathways for saltwater intrusion into freshwater zone in Florida near the well field. Khan et al. [33] said that overpumping of groundwater from aquifer induced the lateral saltwater intrusion into the fresh groundwater area and the vertical upconing of the saline interface is resulting in degradation of aquifers. The farmers in the study area abstracted groundwater without any quality check. The continuous use of such poor quality water is causing secondary salinization and ultimately reduces crop productivity [6, 51]. The groundwater quality results of Scenario I clearly indicated that an increase in the number of tube wells in the future could cause the problem of salinization; therefore, the groundwater regulation aimed at protecting the quality and quantity of groundwater resource must be implemented [61–63].

TABLE 5: Summary of groundwater TDS under Scenarios I and II.

Scenario	Year	Groundwater quality					
		Good		Marginal		Hazardous	
		Area (km^2)	Area (%)	Area (km^2)	Area (%)	Area (km^2)	Area (%)
Current status	2012	5829.27	50.35	2207.98	19.07	3538.75	30.58
I	2030	3350.95	28.95	4438.92	38.35	3786.14	32.71
II	2030	4083.51	35.27	5149.72	44.48	2342.76	20.23

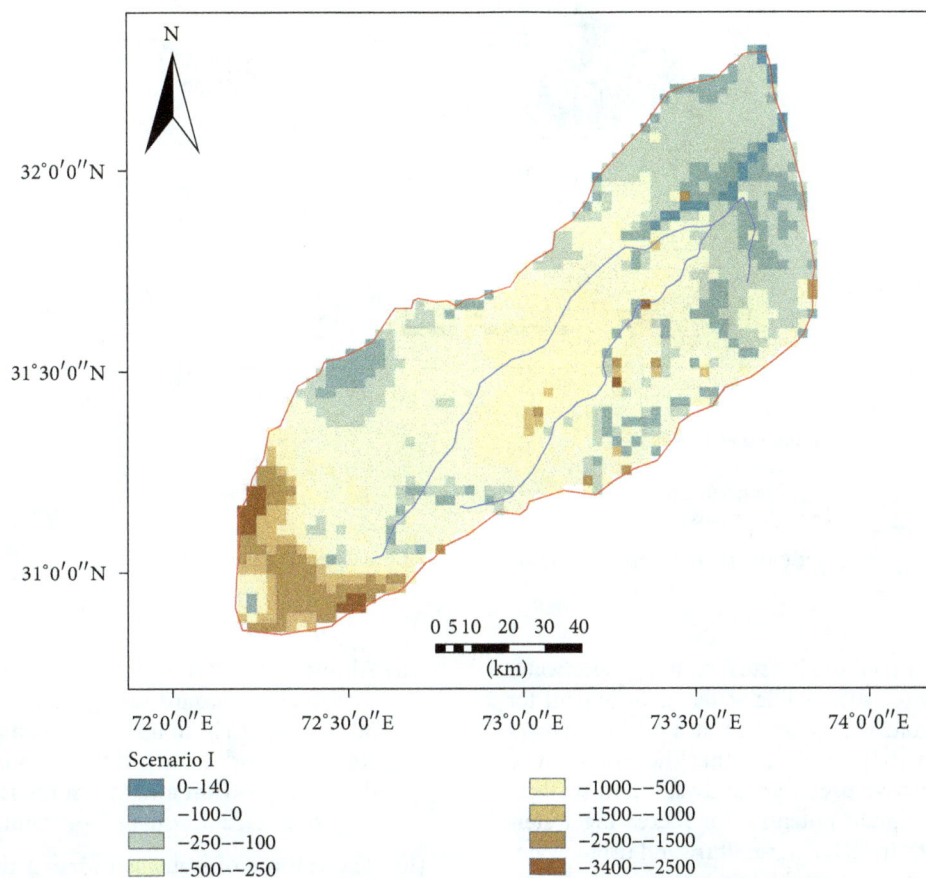

FIGURE 13: Change in groundwater TDS in the year 2030 (Scenario I).

The predicted results of the spatial variation of groundwater salinity up to year 2030, under Scenario II, are shown in Figure 14. It shows that 35.27% area of the aquifer will have good quality, 44.48% area will have marginal quality, and the remaining 20.23% area will have hazardous quality groundwater. The analysis of the groundwater quality results indicated that if the groundwater abstraction is increased by 35% and irrigation recharge is decreased by 35% in the upper part of the study, the overall increase in groundwater salinity will remain under the safe limit, while, in the lower part of the study area, if abstraction is decreased by 35% and irrigation recharge is increased by 35%, there will be an improvement in groundwater quality up to 500 mg/L (Figure 15). Under Scenario II, the area under good and hazardous quality

increased by 15.08 and 10.35%, respectively, compared with current status.

The results of groundwater quality for Scenarios I and II are summarized in Table 5. The analysis of the model results of both scenarios up to year 2030 showed that 6.32% area has good quality water and 12.48% has less hazardous quality in Scenario II compared with Scenario I. In Scenario II, 6.13% more area is under marginal quality, possibly due to shifting of water quality from hazardous quality due to higher recharge from increased canal water supply in the lower part of study area. Hence the results of Scenario II revealed that the need of the hour is a need for a shift in surface irrigation water to the lower part of the study area to reduce groundwater salinity problems. This will also provide formers

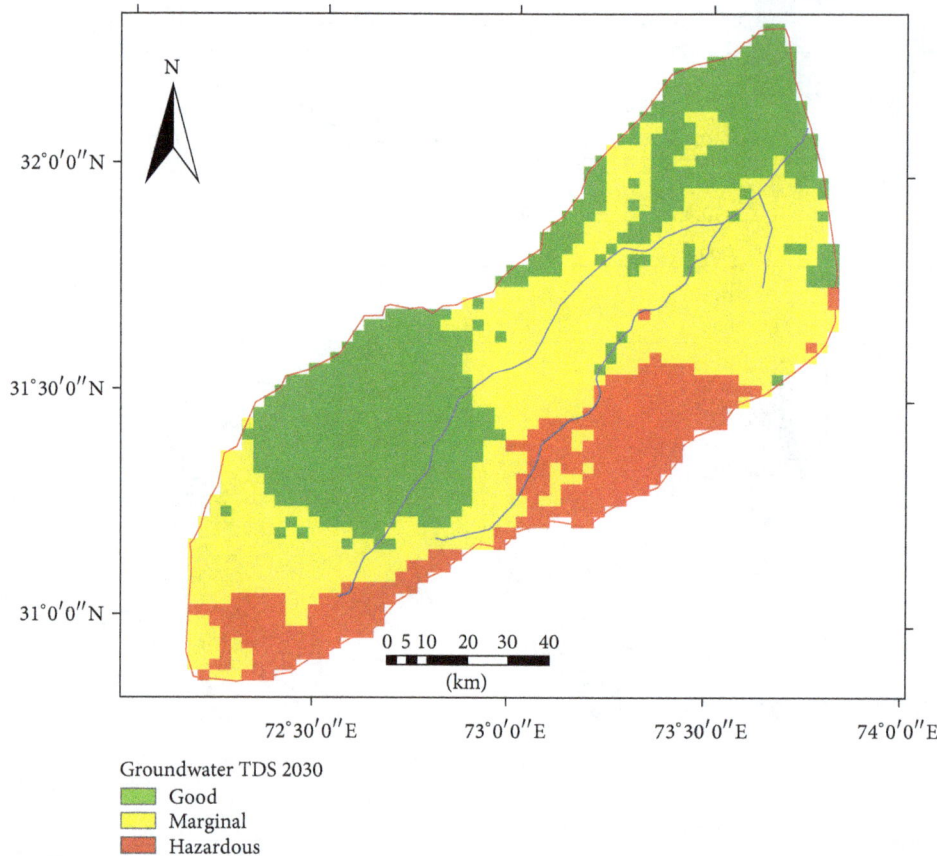

Groundwater TDS 2030
- 🟩 Good
- 🟨 Marginal
- 🟥 Hazardous

FIGURE 14: Predicted groundwater quality in the year 2030 (Scenario II).

with a way forward for conjunctive surface and groundwater irrigation technology for sustainable water management for this region and to avoid secondary salinization [59]. Foster and van Steenbergen [64] concluded that the evolution to more planned conjunctive use of groundwater and surface-water resources offers great potential for increasing water-supply security in both irrigated agriculture and urban water supply across the developing world, especially on large alluvial plains which are often major centers of population and economic development. It was also reported that there are spontaneous conjunctive groundwater and surface-water use in Indian, Pakistani, Moroccan, and Argentinean irrigation-canal commands which have largely arisen due to inadequate surface-water supply to meet irrigation demand.

4. Conclusions

The MODFLOW and MT3D models were calibrated for the period of 2003–2007 and validated for the years 2008–2012 using the measured groundwater level and quality data, respectively, for regional groundwater management in Punjab province of Pakistan. The main conclusions drawn from this study were as follows:

(i) The surface soil textures are largely fine and moderately medium with good permeability properties. In general, the aquifers of the study area are unconfined.

(ii) Almost half of the aquifer of the study area has marginal to hazardous quality of groundwater. A 30 km wide strip along the Chenab River had groundwater of good quality. It was also observed that, at a distance of 40 km away from the river, a 20 km wide strip of hazardous quality was found.

(iii) The volume of water withdrawn through tube wells increased by about 14% from the years 2003 to 2012 and caused the rapid lowering of water table.

(iv) The predicted results of Scenario I revealed that the area good quality groundwater will reduce to 21.4%, while marginal and areas hazardous quality water will increase as 19.28 and 2%, respectively.

(v) The most affected areas will be found in the middle and lower parts of the study area such as Faisalabad, Jhang, Gujra, and TT Singh.

(vi) Under Scenario II, area of good quality groundwater will increase to 6.32% area and 12.48% area will possess less hazardous quality of groundwater.

These results indicated that the canal water supply should shift from good to poor quality groundwater region for proficient management of aquifer water. This can be achieved by regulating the surface-water supplies from irrigation headworks through irrigation department. It is also recommended

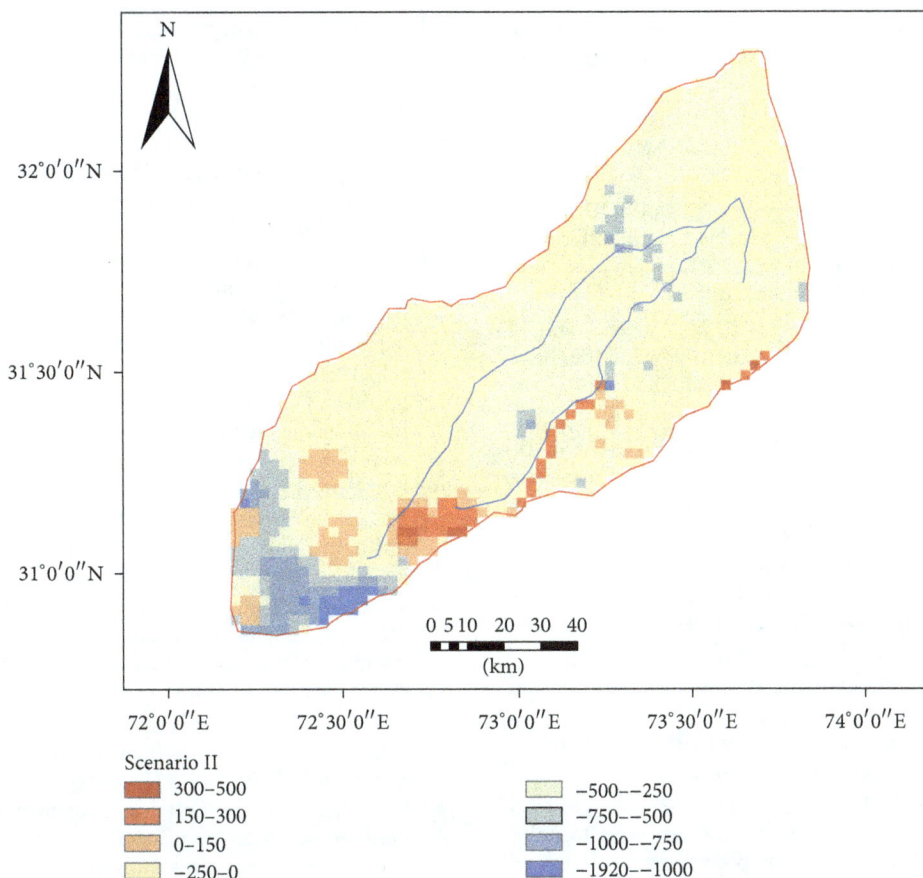

FIGURE 15: Change in groundwater TDS in the year 2030 (Scenario II).

that the detailed field survey is required to determine the exact advective and dispersive solute transport parameters for more accurate results of MT3D model.

Conflicts of Interest

The authors declared that there are no conflicts of interest regarding the publication of this paper.

Acknowledgments

The authors acknowledge the financial support from Higher Education Commission (HEC), Pakistan, under Ph.D. Indigenous Scholarship Program. They are also grateful to WAPDA, Pakistan Meteorology Department, and Department of Land and Reclamation, Faisalabad, for providing data to accomplish this research.

References

[1] Q. Javed, M. Arshad, A. Bakhsh, A. Shakoor, Z. A. Chatha, and I. Ahmad, "Redesigning of drip irrigation system using locally manufactured material to control pipe losses for orchard," *Pakistan Journal of Life and Social Sciences*, vol. 13, no. 1, pp. 16–19, 2015.

[2] A. S. Mongat, M. Arshad, A. Bakhsh et al., "Design, installation and evaluation of solar drip irrigation system at mini dam command area," *Pakistan Journal of Agricultural Sciences*, vol. 52, no. 2, pp. 483–490, 2015.

[3] M. S. Shafique, F. D. Johnson, and E. W. Jackson, "Sulfurous acid generator: A technology to mitigate drought conditions," in *National Symposium on Drought and Water Resources in Pakistan*, Lahore, Pakistan, 2002.

[4] M. Arshad, A. Shakoor, I. Ahmad, and M. Ahmad, "Hydraulic transmissivity determination for the groundwater exploration using vertical electric sounding method in comparison to the traditional methods," *Pakistan Journal of Agricultural Sciences*, vol. 50, no. 3, pp. 487–492, 2013.

[5] M. N. Bhutta and L. K. Smedema, "One hundred years of waterlogging and salinity control in the indus valley, Pakistan: A historical review," *Irrigation and Drainage*, vol. 56, no. 1, pp. S81–S90, 2007.

[6] A. Shakoor, M. Arshad, A. Bakhsh, and R. Ahmed, "Gis based assessment and delineation of groundwater quality zones and its impact on agricultural productivity," *Pakistan Journal of Agricultural Sciences*, vol. 52, no. 3, pp. 837–843, 2015.

[7] A. S. Qureshi, P. G. McCornick, M. Qadir, and Z. Aslam, "Managing salinity and waterlogging in the Indus Basin of Pakistan," *Agricultural Water Management*, vol. 95, no. 1, pp. 1–10, 2008.

[8] A. L. Qureshi, B. K. Lashari, S. M. Kori, and G. A. Lashari, "Hydro-salinity behavior of shallow groundwater aquifer

underlain by salty groundwater in Sindh Pakistan," in *Fifteenth international water technology conference, IWTC-15*, Alexandria, Egypt, 2011.

[9] T. Shah, "Groundwater governance and irrigated agriculture," Tec Background Papers No. 19. Global Water Partnership Technical Committee (TEC), 2014.

[10] A. S. Qureshi, P. G. McCornick, A. Sarwar, and B. R. Sharma, "Challenges and Prospects of Sustainable Groundwater Management in the Indus Basin, Pakistan," *Water Resources Management*, vol. 24, no. 8, pp. 1551–1569, 2010.

[11] H. Farid, Z. Mahmood-Khan, A. Ali, M. Mubeen, and M. Anjum, "Site-Specific Aquifer Characterization and Identification of Potential Groundwater Areas in Pakistan," *Polish Journal of Environmental Studies*, vol. 26, no. 1, pp. 17–27, 2017.

[12] A. Bakhsh and Q. A. Awan, *Water Issues in Pakistan And Their Remedies. In, National Symposium on Drought And Water Resources in Pakistan*, Cewre, university of engineering and technology Lahore, Pakistan, 2002.

[13] M. Irfan, M. Arshad, A. Shakoor, and L. Anjum, "Impact of irrigation management practices and water quality on maize production and water use efficiency," *Journal of Animal and Plant Sciences*, vol. 24, no. 5, pp. 1518–1524, 2014.

[14] G. Haider and G. Hussain, "Problems associated with ground water utilization and their management in scarp areas. In proceedings of the workshop on membrane biophysics and salt tolerance in plants, Faisalabad," Pakistan, 2001.

[15] V. Ezin, R. Delapena, and A. Ahanchede, "Physiological and agronomical criteria for screening tomato genotypes for tolerance to salinity," *Electronic Journal of Environmental, Agricultural and Food Chemistry*, vol. 9, no. 10, p. 1641, 2010.

[16] A. Ashraf and Z. Ahmad, "Regional groundwater flow modelling of Upper Chaj Doab of Indus Basin, Pakistan using finite element model (Feflow) and geoinformatics," *Geophysical Journal International*, vol. 173, no. 1, pp. 17–24, 2008.

[17] S. M. Kori, A. L. Qureshi, B. K. Lashari, and N. A. Memon, "Optimum strategies of groundwater pumping regime under scavanger tubewells in lower indus basin, sindh," *Pakistan. Inter. Water tech. J. (IWTJ*, vol. 3, no. 3, p. 138, 2013.

[18] R. Lalehzari, S. H. Tabatabaei, and M. Kholghi, "Hydrodynamic coefficients estimation and aquifer simulation using PMWIN model," in *Fourteenth International Water Technology Conference (IWTC-14)*, Cairo, Egypt, 2010.

[19] C. P. Kumar, "Groundwater Modelling Software – Capabilities and Limitations," *IOSR Journal of Environmental Science, Toxicology and Food Technology*, vol. 1, no. 2, pp. 46–57, 2012.

[20] M. Arshad, N. Ahmad, and M. Usman, "Simulating seepage from branch canal under crop, land and water relationships," *International Journal of Agriculture and Biology*, vol. 11, article 529, 2009.

[21] C. Moeck, A. Affolter, D. Radny, A. Auckenthaler, P. Huggenberger, and M. Schirmer, "Improved water resource management using three dimensional groundwater modelling for a highly complex environmental," in *Proceedings of the InEGU General Assembly Conference Abstracts*, vol. 19, 19056, 2017.

[22] T. Gebreyohannes, F. de Smedt, K. Walraevens et al., "Regional groundwater flow modeling of the Geba basin, northern Ethiopia," *Hydrogeology Journal*, pp. 1–17, 2017.

[23] N. Rahmawati, J.-F. Vuillaume, and I. L. S. Purnama, "Salt intrusion in coastal and lowland areas of semarang city," *Journal of Hydrology*, vol. 494, pp. 146–159, 2013.

[24] S. Carretero, J. Rapaglia, H. Bokuniewicz, and E. Kruse, "Impact of sea-level rise on saltwater intrusion length into the coastal aquifer, Partido de La Costa, Argentina," *Continental Shelf Research*, vol. 61-62, pp. 62–70, 2013.

[25] W. A. Abu-el-Shar and R. I. Hatamleh, "Using Modflow and mt3d groundwater flow and transport models as a management tool for the Azraq groundwater system," *Jordan Journal of Civil Engineering*, vol. 1, no. 2, p. 153, 2007.

[26] A. Asoka, T. Gleeson, Y. Wada, and V. Mishra, "Relative contribution of monsoon precipitation and pumping to changes in groundwater storage in India," *Nature Geoscience*, vol. 10, no. 2, pp. 109–117, 2017.

[27] I. Galitskaya, K. R. Mohan, A. K. Krishna et al., "Assessment of soil and Groundwater Contamination by Heavy Metals and Metalloids in Russian and Indian Megacities," *Procedia Earth and Planetary Science*, vol. 17, pp. 674–677, 2017.

[28] J. B. Kambale, D. K. Singh, and A. Sarangi, "Impact of climate change on groundwater recharge in a semi-arid region of northern India," *Applied Ecology and Environmental Research*, vol. 15, no. 1, pp. 335–362, 2017.

[29] T. Abdullah and K. Morteza, "Groundwater modeling by MODFLOW model In Toyserkan aquifer and evaluation of hydrogeological state under present and future conditions," *Water Engineering*, vol. 9, no. 31, pp. 45–60, 2017.

[30] V. Durand, V. Léonardi, G. de Marsily, and P. Lachassagne, "Quantification of the specific yield in a two-layer hard-rock aquifer model," *Journal of Hydrology*, vol. 551, pp. 328–339, 2017.

[31] WAPDA, Hydrogeological data of Rechna Doab. Volume-I, publication No. 25. Project planning organization (N.Z) Water and Power Development Authority (WAPDA), 27-E/I Gulberg III, Lahore, 1978.

[32] A. N. Al-Fatlawi, The application of the mathematical model (MODFLOW) to simulate the behavior of groundwater flow in Umm er Radhuma unconfined aquifer. Euphrates Journal of Agriculture Science, vol 3, article 1, no. 1, 2011.

[33] S. Khan, T. Rana, H. F. Gabriel, and M. K. Ullah, "Hydrogeologic assessment of escalating groundwater exploitation in the Indus Basin, Pakistan," *Hydrogeology Journal*, vol. 16, no. 8, pp. 1635–1654, 2008.

[34] G. Schoups, C. L. Addams, and S. M. Gorelick, "Multi-objective calibration of a surface water-groundwater flow model in an irrigated agricultural region: Yaqui Valley, Sonora, Mexico," *Hydrology and Earth System Sciences*, vol. 9, no. 5, pp. 549–568, 2005.

[35] W. A. Jehangir, A. S. Qureshi, and N. Ali, Conjunctive water management in the Rechna doab: an overview of resources and issues. Working paper 48. International water management institute Lahore, Pakistan, 13, 1, 2002.

[36] M. D. Ahmad, Estimation of net groundwater use in irrigated river basins using geo-information techniques: a case study in Rechna doab, Pakistan (Ph.D. Thesis), Wageningen University, Wageningen, 2002.

[37] M. Arshad, Contribution of irrigation conveyance system components to the recharge potential in Rechna Doab under lined and unlined options. PhD Thesis, Faculty of Agricultural Engineering and Technology, University of Agriculture, Faisalabad-Pakistan, 2004.

[38] C. Zheng, MT3D-A modular three-dimensional transport model for simulation of advection, dispersion and chemical reactions of contaminants in groundwater systems. Report to the u.s. Environmental protection agency, ADA, Oklahoma, 1990.

[39] W. H. Chiang and W. Kinzelbach, "Processing Modflow: a simulation system for modeling groundwater flow and pollution. User's manual," User's manual. U.S. Department of the interior, U.S. Geological survey, 1998.

[40] P. A. Domenico and F. W. Schwartz, *Physical and chemical hydrogeology*, John Wiley, New York, USA, 2nd edition, 1990.

[41] CSIRO, "Investigation conjunctive water management options using a dynamic surface-groundwater modeling approach: a case study of Rechna doab. Commonwealth scientific and industrial research organization (CSIRO), land and water," Technical report 35/03, IWMI, 2003.

[42] L. W. Gelhar, C. Welty, and K. R. Rehfeldt, "A critical review of data on field-scale dispersion in aquifers," *Water Resources Research*, vol. 28, no. 7, pp. 1955–1974, 1992.

[43] M. Huysmans and A. Dassargues, "Review of the use of Péclet numbers to determine the relative importance of advection and diffusion in low permeability environments," *Hydrogeology Journal*, vol. 13, no. 5-6, pp. 895–904, 2005.

[44] PES. Pakistan economic survey 2008-09, Agriculture, Table 107, PP 171-172. Federal bureau of statistics, statistics division, ministry of economic affairs and statistics, govt. Of Pakistan, Islamabad, Pakistan, 2009.

[45] M.-S. Tsou, S. P. Perkins, X. Zhan, D. O. Whittemore, and L. Zheng, "Inverse approaches with lithologic information for a regional groundwater system in southwest Kansas," *Journal of Hydrology*, vol. 318, no. 1-4, pp. 292–300, 2006.

[46] M. P. Anderson and W. W. Woesner, *Applied Groundwater Modeling: Simulation of Flow And Advective Transport*, Academic press, San Diego, Calif, USA, 1992.

[47] D. N. Moriasi, J. G. Arnold, M. W. Van Liew, R. L. Bingner, R. D. Harmel, and T. L. Veith, "Model evaluation guidelines for systematic quantification of accuracy in watershed simulations," *Transactions of the ASABE*, vol. 50, no. 3, pp. 885–900, 2007.

[48] M. N. Asghar, S. A. Prathapar, and M. S. Shafique, "Extracting relatively-fresh groundwater from aquifers underlain by salty groundwater," *Agricultural Water Management*, vol. 52, no. 2, pp. 119–137, 2002.

[49] M. A. Hagos, "Groundwater flow modeling assisted by gis and rs techniques (raya valley- ethiopia)," *M.Sc. Thesis in international institutes for geo-information science and earth observation Enschede*, 2010.

[50] D. M. Obrien, T. J. Dumler, and D. H. Rogers, "Irrigation capital requirements and energy costs," in *Farm Management Guide, MF-836*, pp. 1–4, Kansas State University., 2011.

[51] M. Basharat, "Spatial and temporal appraisal of groundwater depth and quality in LBDC command-issues and options," *Pakistan Journal of Engineering and Applied Sciences*, vol. 11, article 14, 2012.

[52] A. S. Qureshi, T. Shah, and M. Akhtar, *The groundwater economy of Pakistan*, Working Paper 64. Lahore, Pakistan: International Water Management Institute (IWMI), vol. 64, pp. 1-31, 2003.

[53] N. Ahmad, *Groundwater resources of Pakistan*, 61-b/2, Gulberg III, Lahore, Pakistan, 1974.

[54] P. Engesgaard, K. H. Jensen, J. Molson, E. O. Frind, and H. Olsen, "Large-scale dispersion in a sandy aquifer: Simulation of subsurface transport of environmental tritium," *Water Resources Research*, vol. 32, no. 11, pp. 3253–3266, 1996.

[55] H.-Y. Shieh, J.-S. Chen, C.-N. Lin, W.-K. Wang, and C.-W. Liu, "Development of an artificial neural network model for determination of longitudinal and transverse dispersivities in a convergent flow tracer test," *Journal of Hydrology*, vol. 391, no. 3-4, pp. 367–376, 2010.

[56] K. A. Narayan, C. Schleeberger, P. B. Charlesworth, and K. L. Bristow, *Effects of groundwater pumping on saltwater intrusion in the lower burdekin delta, north queensland*, vol. 201, article 224, Modelling and simulation society of Australia and Newzealand Inc., 2002.

[57] A. Fiori, "Finite Peclet extensions of Dagan's solutions to transport in anisotropic heterogeneous formations," *Water Resources Research*, vol. 32, no. 1, pp. 193–198, 1996.

[58] S. Asghar, "Farming system analysis of irrigated farms in Faisalabad, Pakistan," Thesis in natural resources management, school of environment, resources and development, Asian Institute of Technology, Thailand, 2014.

[59] D. Mekonnen, A. Siddiqi, and C. Ringler, "Drivers of groundwater use and technical efficiency of groundwater, canal water, and conjunctive use in Pakistan's Indus Basin Irrigation System," *International Journal of Water Resources Development*, vol. 32, no. 3, pp. 459–476, 2016.

[60] S. T. Prinos, "Saltwater intrusion monitoring in Florida. Special Issue: Status of Florida's Groundwater Resources," *Florida Scientist*, vol. 79, no. 4, 2016.

[61] A. S. Qureshi, "Improving food security and livelihood resilience through groundwater management in Pakistan," *Global Advanced Research Journal of Agricultural Science*, vol. 4, article 687, no. 10, 2015.

[62] A. S. Qureshi, "Water management in the Indus basin in Pakistan: Challenges and opportunities," *Mountain Research and Development*, vol. 31, no. 3, pp. 252–260, 2011.

[63] E. C. Dogrul, C. F. Brush, and T. N. Kadir, "Groundwater modeling in support ofwater resources management and planning under complex climate, regulatory, and economic stresses," *Water (Switzerland)*, vol. 8, no. 12, article no. 592, 2016.

[64] S. Foster and F. V. van Steenbergen, "Conjunctive groundwater use: A 'lost opportunity' for water management in the developing world?" *Hydrogeology Journal*, vol. 19, no. 5, pp. 959–962, 2011.

11

Study of Carbonaceous and Nitrogenous Pollutant Removal Efficiencies in a Hybrid Membrane Bioreactor

Victor S. Ruys, Kamel Zerari, Isabelle Seyssiecq, and Nicolas Roche

Aix-Marseille University, CNRS, Centrale Marseille, M2P2 UMR 7340, 13541 Marseille Cedex 13, France

Correspondence should be addressed to Nicolas Roche; nicolas.roche@univ-amu.fr

Academic Editor: Carlos Alberto Lberto Martínez-Huitle

A hybrid membrane bioreactor (HMBR) comprises activated sludge (free biomass), a biofilm (supported biomass), and a membrane separation. A laboratory pilot-scale HMBR was operated for seven months with high organic loads of both carbonic and nitrogen pollutants. Several experiments were conducted to investigate the influence of the height of the packing bed (27 cm, 50 cm, and 0 cm) and the effect of the concentration of dissolved oxygen (DO) on the organic removal rate, total nitrogen removal rate (TN), and ammonium removal. The organic removal rate was always >95% and mostly >98%. The NH_4^+-N and TN removal rates were directly related to DO. NH_4^+-N removal rate reached 100% and was mostly >99% with a concentration of DO > 0.1 mg/L, whereas the NO_3^--N removal rate was differentially affected depending on the level of DO. The removal rate increased when the concentration of DO was optimal for simultaneous nitrification and denitrification, which was between 0.1 and 0.5 mg/l, and the TN removal rate was consequently high. The removal rate decreased when DO was high and denitrification was consequently low thereby reducing the TN removal rate. This implies that high levels of DO (>1 mg/L) limit the denitrification process and low levels of DO (<0.1 mg/L) limit the nitrification process and hence total nitrogen removal in the bioreactor.

1. Introduction

Today, an increasing number of surface water resources suffer from eutrophication. The increased nutriment loads in the sewage and the discharge of inadequately treated wastewater and industrial water into water resources are the main causes of this phenomenon.

It is therefore important to find more efficient processes to remove the increased pollution and to protect water resources. Processing using a hybrid membrane bioreactor (HMBR) first depends on the combined effect of a suspended activated sludge (free bacteria) and a biofilm (supported bacteria) which make it possible to enhance removal efficiency compared to a conventional activated sludge (AS) process [1]. The use of membrane technology for the separation of free bacteria from the liquid phase also improves the efficiency of treatment. Carriers are used to minimize the negative effect of suspended solids, reduce the formation of a membrane cake layer thanks to the scouring effects of suspended carriers, minimize fouling, and improve nutrient removal and filterability [2–8]. In a previous work [8] the benefits of gas-liquid mass transfer and of adding a packing bed in the bioreactors for different solutions (water, suspension, and non-Newtonian fluid) were demonstrated.

HMBR combines classical organic carbon depollution with simultaneous nitrification and denitrification. Biological nitrification-denitrification is a development process for nitrogen removal from wastewater. On the one hand and during the nitrification step, ammonia-oxidizing bacteria (AOB) aerobically oxidize the ammonium to nitrite after which the nitrite is oxidized to nitrate by nitrite-oxidizing bacteria (NOB). On the other hand, during the denitrification step, nitrate is reduced to gaseous nitrogen by denitrifying microorganisms [9]. Shortcut biological nitrogen removal (SBNR) [10] is based on partial nitrification to nitrite followed by nitrite denitrification (Figure 1).

However, the stability of nitrite accumulation, which is a key prerequisite for successful SBNR, is a major challenge to the implementation of this process. One major problem is that nitrifiers with low growth rate can be washed out in the conventional activated sludge process, resulting in a reduction in the nitrification performance. To overcome this

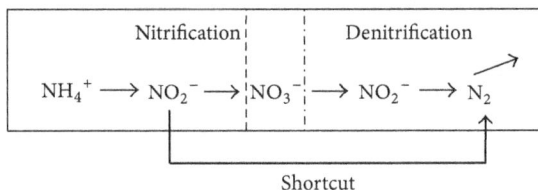

FIGURE 1: Biological nitrogen removal by nitrification-denitrification.

problem, a submerged membrane bioreactor (MBR) is introduced to improve nitrification efficiency because membrane filtration enables complete separation of solids and liquids and maintains a high level of autotrophic biomass in the reactor [11].

Degradation is influenced by the filling fraction of carriers, the extent of biofilm growth, the concentration of MLSS in suspended growth, and biomass activity. The 1/3 lower concentration of biomass in the attached form can reach the same removal rates as biomass in the suspended form [12].

The aBF-MBR has higher specific oxygen uptake rates (SOUR) compared to conventional AS-MBR [13, 14]. But, with similar hydraulic retention times (HRTs) and sludge retention times (SRTs), both systems can achieve 95–99% of chemical oxygen demand (COD) removals and there is no significant difference in the nitrification rate, which is >96%. However, some authors found that AS-MBR has 2–4% lower nitrification rate compared to aBF-MBR [6, 14].

aBF-MBR enables higher total nitrogen removal because of simultaneous nitrification/denitrification (SND) in deeper biofilm layers [2, 12, 14].

According to Gupta and Sharma [15], biological oxidation of high strength nitrogenous wastewater (TKN = 1.0 to 1.1 kg/m^3) was successful after the feed was supplemented with an external source of organic carbon to raise the COD : TKN ratio to nearly 1. Over 96% nitrification and up to 99% TKN removal were achieved with a SRT of 30 d and a HRT of 2 d.

Mines and Sherrard [16] observed incomplete nitrification (40 to 65% with SRTs of, resp., 17 and 21 d) with an influent of 0.5 kg TKN·m^{-3}. In a reciprocating jet bioreactor, 85 to 95% of the influent ammonium (0.65–0.85 kg NH$_4^+$-N·m^{-3}) was converted at a low HRT (4.2 h) [17]. In this case, an antifoaming agent should be introduced to avoid excessive formation of foam.

According to Charmot-Charbonnel et al. [18], aerated submerged fixed beds have a high nitrification capacity which enables distribution of the wastewater over the whole biofilm. In the case of the treatment of fertilizer waste containing 0.5 kg NH$_4^+$-N·m^{-3} in an upflow biofilm reactor, removal rates can reach up to 0.6 kg NH$_4^+$- N·m^{-3}·d^{-1} (or 13.2 kg NH$_4^+$- N·m^{-2}·d^{-1}) and 90.6 to 99% of ammonium removal was achieved with an ammonium volumetric loading of 0.4 to 0.6 kg·m^{-3}·d^{-1}. Nitrification stopped when the concentration of ammonium-N reached 5 kg·m^{-3} (or B_v = 2.6 kg NH$_4^+$-N·m^{-3}·d^{-1}), as indicated by the low ammonium removal rate (7%).

Simultaneous nitrification-denitrification has already been observed in other biofilm reactors, for example, rotating biological contactors (RBC) [19].

Tang et al. [20] used a multihabitat membrane bioreactor to investigate the DO distribution and the factors influencing the mass transfer of DO. Different proportions of COD : TN : TP of synthetic water and low aeration rate were used to obtain an aerobic or an anoxic status. The accumulation of biomass was the main factor which influenced the distribution of DO to the different zones. Organic and n-containing substances could be removed simultaneously, but the effective removal of TN was only possible after obvious anoxic and aerobic zones were formed within the bioreactor.

The objective of the present study was to evaluate the impact of the concentration of dissolved oxygen (DO) on simultaneous nitrification and denitrification (SND) in a HMBR for the treatment of high load wastewater. The effect of enriched attached culture seeded on the packing in HMBR was also studied to understand their role in SND.

2. Material and Methods

Experiments were carried out in a bioreactor (a cylindrical clear PVC column) with and without a packing bed. The column used was 15.3 cm in diameter (ID), 126.5 cm in total height (HT), filled with 91 cm of liquid (HL), and packed with either 27 cm or 54 cm (HS) of filling material (Figure 2).

Cylindrical solid packing rings (AnoxKaldnes® rings) made of polyethylene (about 10 mm in diameter and 7 mm in height) were used in the experiments. Their main geometrical characteristics are summarized in Table 1.

In order to prevent the loss of packing material due to recirculating liquid, the packing rings were held in place by two grids with 0.5 mm apertures located just above and under the bed. Compressed air was supplied by a compressor and was injected at the bottom of the column using a porous disc diffuser.

Air flow rates were measured using a flowmeter (SHO-RATE). The liquid phase was recirculated from the bottom to the top of the bioreactor by a centrifuge pump and the flow rate was measured using a ROSE-MOUNT 4 X flowmeter. The concentration of dissolved oxygen in the bioreactor was measured using an oxygen electrode oximeter HQ30D (Hach Lange), equipped with a temperature sensor. Gas input ranged between 60 L/h and 180 L/h. The flow rate of the recirculated liquid was set between 200 and 300 L/h to prevent sludge accumulating in the membrane channels and to ensure complete mixing. It was previously shown that an increase in the recirculated flow rate had no significant effects on the measured $K_L a$ [8]. A ceramic microfiltration membrane Carbosep® (pore size: 0.1 μm, area: 0.0226 m^2) with six channels was used in the system to separate the activated sludge from the purified water during normal operation (i.e., in presence of activated sludge in the bioreactor and continuous feeding with substrate).

The substrate, which was prepared every two or three days in a 50 L tank, consisted of sugar, meat extract (Viandox®), ammonium chloride (NH$_4$Cl), sodium bicarbonate (NaHCO$_3$), and phosphoric acid (H$_3$PO$_4$). For a COD equal

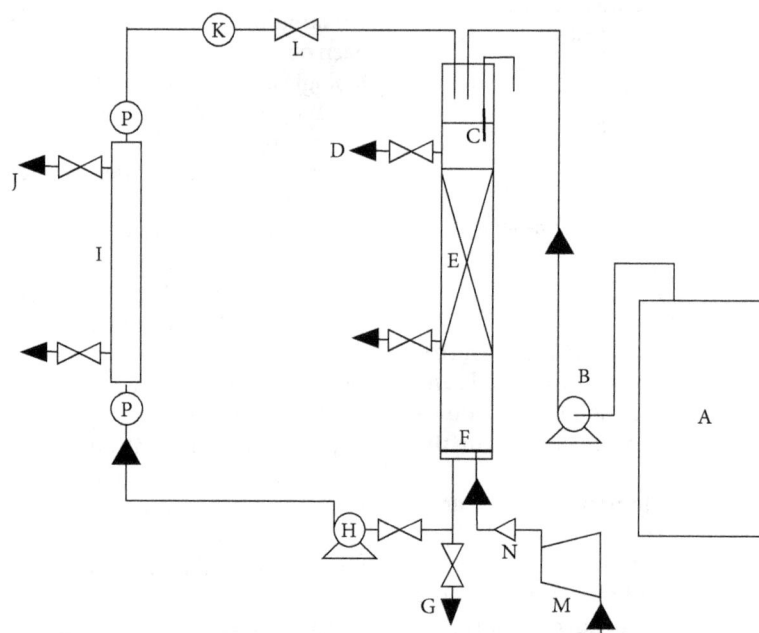

FIGURE 2: Experimental setup of laboratory pilot scale; A: feeding tank; B: dosing pump; C: oximeter; D: sampling; E: fixed bed; F: porous diffuser; G: discharge valve; H: recirculation pump; I: membrane; J: outlet valve; K: liquid flowmeter; L: valve water; M: compressor; N: gas flowmeter.

TABLE 1: Physical packing characteristics (AnoxKaldnes rings).

Parameter	Density ρ_s (kg/m^3)	Porosity ε (—)	Diameter d_p (mm)	Specific area a (m^2/m^3)
Value	968	0,79	10	1000

to 1,000 mg/L, the following proportions were used: 39 g, 42.5 g, 11.5 g, 33.33 g, and 0.75 mL of sugar, meat extract, NH$_4$Cl, NaHCO$_3$, and H$_3$PO$_4$, respectively. A constant and continuous feeding inlet flowrate of about 1 L/h was applied. The same proportions but with bigger quantities were used for higher pollutant loads.

The ammonium N-NH$_4^+$ and the nitrates N-NO$_3^-$ were measured by an ammonium meter SC200 (Hach Lange). pH was measured by a pH-meter HANNA (pH210). Mixed liquor suspended solids (MLSS) for suspended bacteria: 30 mL of mixed liquor was centrifuged for 15 minutes at 13500 rpm with a SIGMA 2-16 centrifuge; the solid part was then dried at 105°C for 24 hours. The MLSS was the percentage difference in weight before and after drying. The mass of the fixed bacteria on the packing material was calculated monthly by measuring the difference in weight in some pieces of packing before and after drying. The chemical oxygen demand (COD) of the substrate, of the supernatant, and of the permeate was measured using one of the two following methods: first, using CHE Metrics COD tubes and an Aquamate ThermoSpectronic spectrophotometer; second, using measured total organic carbon (TOC) to deduce the COD using an experimental calibration curve between COD values and TOC. Total organic carbon (TOC) and total nitrogen were measured by TOC-VCPH (Total Organic Carbon Analyzer) SHIMADZU.

At the beginning of the pilot operation, activated sludge from the recirculation line of the Aix-en-Provence (France) WWTP was seeded into the bioreactor to provide a concentration of mixed liquor suspended solids (MLSS) of 2.7 g/L which increased after several days to reach about 3.5 g/L MLSS concentration.

A mean MLSS concentration of 3.0–4.5 g/L was maintained by removing a precise volume of sludge from the bioreactor every day. MLVSS was calculated several times over the whole operation period. The MLVSS/MLSS ratio was always around 0.92.

The membrane was cleaned when the permeate flow rate dropped to about 0.5 L/h.

Operating Conditions. The pilot operation was run for 166 consecutive days divided into three phases according to the packing height, phase I (0–79 d) with a bed height of 27 cm, phase II (80–153 d) with a bed height of 54 cm, and phase III (153–170) with no packing bed. Each phase comprised several periods of one or two weeks corresponding to specific operating conditions in terms of the concentration of inlet pollutant and the air flowrate (i.e., one operating parameter was changed from one period to the other). The different operating conditions are detailed in Table 2. The applied loading rates (from 2.9 to 8.2 kgCOD/(m3·d) for carbonaceous pollution and from 0.19 to 0.55 kgN/(m3·d) for nitrogenous pollution) can be considered as high.

TABLE 2: Operational conditions of the HMBR.

Height of packing bed (cm)	Number of period	Number of days in the period	Accumulated time from day to day	Concentration of COD at the inlet Kg/m³	Concentration of N_{total} Kg N/m³	Air flowrate L/h	Average COD loading rate kgCOD/(m³·d)	Average N_{total} loading rate kgN/(m³·d)
	1	16	1–16	250–1000	16–65	120		
	2	10	17–26	2000	135		2,87	0,19
	3	7	27–33			60		
Phase I	4	8	34–41	2500	165		3,59	0,24
27 cm	5	8	42–49			120		
	6	10	50–59	3000	202		4,30	0,29
	7	7	60–66			60		
	8	7	67–73	5000	335		7,17	0,48
	9	6	74–79			120		
	10	10	80–89	3000	202	180	4,30	0,29
	11	14	90–103			60		
	12	7	104–110					
Phase II	13	8	111–118	5700	380	120	8,18	0,55
54 cm	14	11	119–129			180		
	15	5	130–134					
	16	5	135–139	4000	270	60	5,74	0,39
	17	9	140–148			120		
	18	5	149–153			60		
Phase III	19	6	154–159	4000	270		5,74	0,39
0 cm	20	7	160–166			120		

The applied loading rates, the removal efficiencies, and the removal rates have been, respectively, calculated by the classical formulas with for

(i) the applied loading rate (L_v in kg/m³·d): $L_v = (([S]_{in} \cdot Q)/V) \cdot 24$,

(ii) the removal efficiencies (E in %): $E = (1-[S]_{out}/[S]_{in}) \cdot 100$,

(iii) the removal rate (R_v in kg/m³·d): $R_v = (L_V \cdot E)/100$,

where $[S]_{in}$ is the concentration (kg/m³) of COD, or TN, or NH_4^+-N in the influent, $[S]_{out}$ is the concentration (kg/m³) of COD, or TN, or NH_4^+-N at the outlet, Q is the influent flowrate (m³/h), and V is the volume of the bioreactor (m³).

3. Results and Discussion

The results of the whole experiment are summarized for each period (mean values) in Table 3 including DO concentration, COD removal rates, and nitrogen removal rates and in Figure 3 as a function of time.

On the one hand, the difference in treatment efficiency between phase I and phase II revealed a notable effect of bed height. Phase III allowed us to compare the difference between CMBR and HMBR in terms of pollutant removal efficiency. On the other hand, the concentration of the feed and the dissolved oxygen concentration appeared to be the key to simultaneous nitrification and denitrification efficiency.

The first period is not discussed because it corresponds to acclimation of the activated sludge to the synthetic substrate. Acclimation was considered to be complete when the COD removal efficiency and activated growth yield were constant, in this case, after 16 days.

Because an easily biodegradable synthetic substrate was used, whatever the operating conditions, the COD removal efficiency was very high in all the periods. Increasing the height of the packing bed made it possible to enhance the aerobic conditions in the bioreactor because of the increase of oxygen mass transfer in the bioreactor as it has been shown in the previous work of Zerari et al. [8] and hence to increase the volumetric efficiency of the bioreactor, for the aerobic biodegradation of the pollutants, with a COD removal rate of about 8 kgCOD/m³·d for periods 12, 13, and 14 compared to the higher level of COD removal rate obtained in phase I (periods 8 and 9) of about 7 kgCOD/m³·d.

For the nitrogen compounds, removal efficiencies were linked to the aerobic or anoxic conditions in the bioreactor. In periods 8 and 9, nitrification efficiency was reduced to about 68%. In this configuration, there was clearly a limit to nitrification with a maximum TN removal rate of about 0.33 kgN/m³·d. It was possible to enhance the removal rate, that is, the volumetric efficiency of the bioreactor, by increasing of the height of the packing bed. In periods 13, 14, and 18, with

TABLE 3: Experimental removal rates and removal efficiencies.

Packing height Cm	Number of period	Air flow rate L/h	Average DO mg/L	COD removal rate kg COD/m^3·d	TN removal rate kg N/m^3·d	COD removal efficiency%	NH$_4^+$-N removal efficiency%	N$_{total}$ removal efficiency%
Phase I 27 Cm	1	120	5.5	—	—	61,1	62.2	2.2
	2	120	4.3	2,71	0,015	94,6	95.1	7.7
	3	60	0.5	2,68	0,18	93,5	97.7	92.8
	4	60	0.59	3,46	0,228	96,6	99.1	96.3
	5	120	1.62	3,51	0,13	97,9	99.6	55.02
	6	120	1.66	4,22	0,18	98	99.76	62.1
	7	60	0.36	4,24	0,28	98,6	98.4	96.6
	8	60	0.28	7,04	0,315	98,2	66.8	65.5
	9	120	1.7	7,05	0,322	98,3	68.5	67
Phase II 54 Cm	10	180	4.17	4,23	0,235	98,2	84.2	81.1
	11	60	0.86	4,2	0,279	97,5	99.5	96.3
	12	60	0.11	8,04	0,41	98,3	76.4	75.2
	13	120	0.18	8,06	0,401	98,6	75.4	73.6
	14	180	1.23	8,08	0,374	98,8	71.4	68.6
	15	180	2.14	5,66	0,218	98,7	59.2	56.4
	16	60	0.66–0.1	5,64	0,192	98,3	56	49.6
	17	120	0.16	5,65	0,304	98,5	82.9	78.6
	18	60	0.39	5,65	0,365	98,4	96.8	94.2
Phase III 0 Cm	19	60	0.73	5,63	0	98,2	97	Continuous decrease
	20	120	4.08	5,6	0	97,6	99.6	Continuous decrease

a packing bed height of 54 cm, nitrogen removal rates were 0.4 kgN/m^3·d. With this HMBR, it was impossible to obtain higher nitrogen removal rates. The packing bed in the bioreactor had two combined effects; it enabled an increase in the concentration of the biomass in the bioreactor with the presence of a biofilm and also in the rate of oxygen mass transfer in the bioreactor.

Nitrification efficiency, that is, NH$_4^+$-N removal efficiency, depended to a great extent on the aerobic conditions in the bioreactor. The highest efficiency rate, >95%, was obtained in periods 3, 4, 5, 6, 7, 18, 19, and 20 with a DO concertation > 0.4 mg/l of dissolved oxygen in the bioreactor. When the DO level was too low, nitrification in the bioreactor was limited due to lack of oxygen, as observed in periods 8, 12, and 16.

Denitrification was only achieved in anoxic conditions. With a DO level < 1.0 mg/L, the biofilm could be considered to be in anoxic conditions and denitrification occurred (periods 3, 4, 7, 8, 11, 12, 13, 16, 17, and 18).

It was also possible to remove all the nitrogen at the same time (simultaneous nitrification and denitrification) with aerobic conditions in the activated sludge and anoxic conditions in the biofilm obtained by controlling the level of DO in the bioreactor. DO levels of between 0.5 and 1 mg/L were observed in periods 4, 11, and 18, with simultaneous and

complete nitrification and denitrification in the HMBR with nitrogen removal efficiency observed of about 95% for those periods. These periods have to be compared, respectively, to the periods 5, 10, and 17 with the same loading rates and higher air flow rates and DO levels. The assumption of simultaneous nitrification and denitrification could be made because of a constant amount of carbonaceous consumption observed (same COD removal efficiencies) and because it was not necessary to add carbonaceous compounds to achieve complete nitrification-denitrification in those periods, possibly due to a shortcut in the biological nitrogen removal process (NO$_2^-$ to N$_2$, as it has been described in the literature: Figure 1).

In periods 7 and 11, the height of the packing bed was seen to affect the efficiency of the oxygenation transfer. In the same operating conditions, that is, with the same loading rates and air flow rate, a higher concentration of DO was observed with a 54 cm packing bed (0.86 mg/L in period 11) than with a 27 cm packing bed (0.36 mg/L in period 7). It confirms the fact that the height of the packing bed has a positive effect on the oxic conditions in the bioreactor.

In phase III, the bioreactor was operated with no packing bed, that is, like a classical AS process. A continuous decrease in nitrification efficiency was observed because with no biofilm in the reactor, it is impossible to obtain anoxic conditions

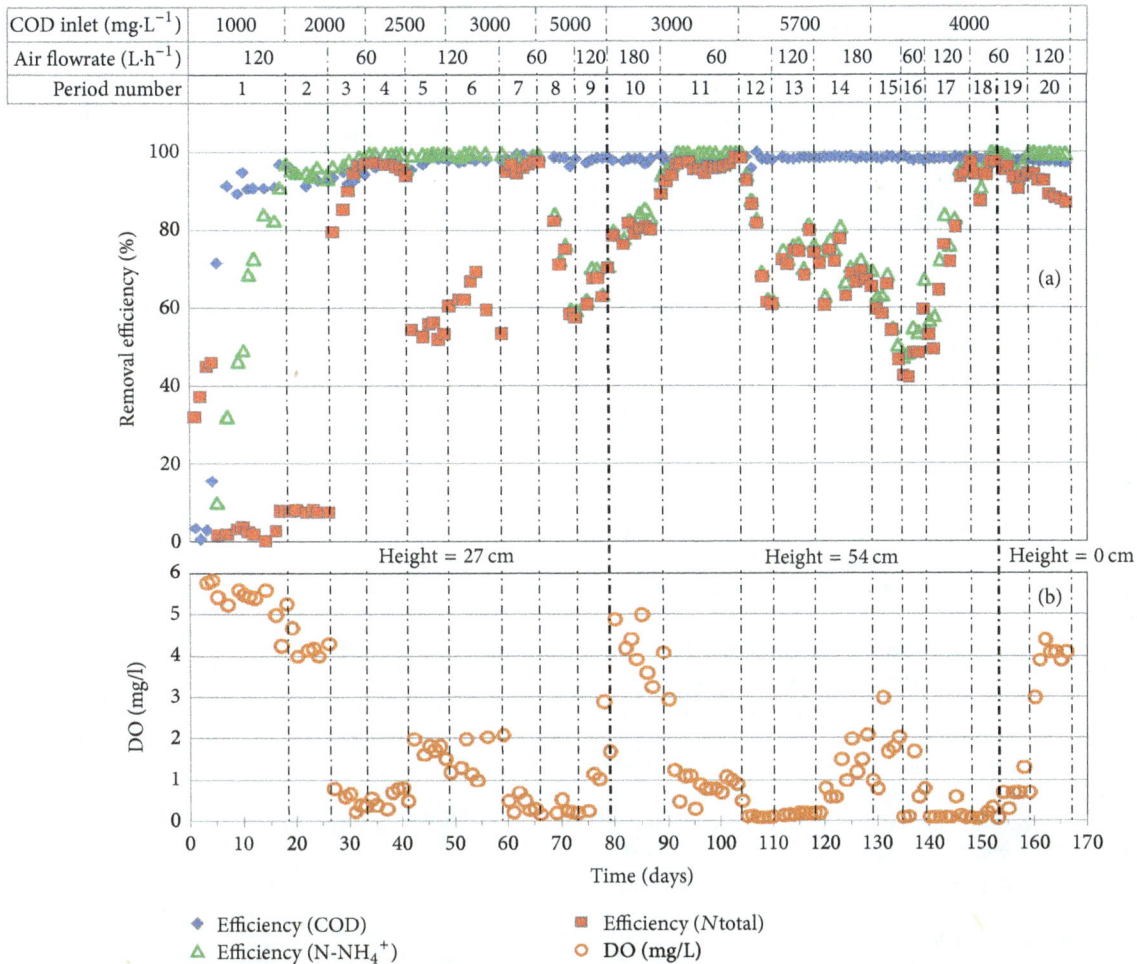

COD inlet (mg·L^{-1})	1000	2000	2500	3000	5000	3000		5700		4000			
Air flowrate (L·h^{-1})	120	60	120	60	120	180	60	120	180	60	120	60	120
Period number	1	2	3	4	5	6	7	8	9	10	11	12	13	14	15	16	17	18	19	20

FIGURE 3: Removal efficiencies for a COD; N-NH$_4^+$; Ntotal; and (b) DO concentrations as a function of time.

and simultaneous nitrification and denitrification, due to continuous aeration.

4. Conclusions

An innovative HMBR configuration enabled improved depollution efficiency without increasing the amount of suspended bacteria in the mixed liquor. Compared to classical membrane bioreactors (MBR), HMBR thus makes it possible to treat highly loaded wastewaters while reducing problem of fouling caused the use of low concentration of MLSS.

HMBR also enabled highly efficient nitrification and denitrification thanks to the development of a biofilm on the inert surface of the filling material and the presence of both aerobic and anoxic conditions in the reactor. This configuration efficiently removes nitrates and nitrites by controlling the concentrations of dissolved oxygen in the reactor. The most important factors controlling the correct functioning of the reactor are thus the concentration of dissolved oxygen and the height of the packing bed. Simultaneous nitrification and denitrification enabled high removal efficiencies with concentrations of dissolved oxygen between 0.4 and 0.9 mL/L.

The HMBR configuration made it possible to increase the efficiency of oxygen transfer with no additional energy consumption and with an increase in the volumetric efficiency of pollutant removal in the bioreactor.

Disclosure

An earlier version of this work was presented as an abstract at 10th European Congress of Chemical Engineering, France, 2014.

Conflicts of Interest

The authors declare that there are no conflicts of interest regarding the publication of this paper.

References

[1] M. Kraume and A. Drews, "Membrane bioreactors in waste water treatment—status and trends," *Chemical Engineering & Technology*, vol. 33, no. 8, pp. 1251–1259, 2010.

[2] J. Lee, W.-Y. Ahn, and C.-H. Lee, "Comparison of the filtration characteristics between attached and suspended growth

microorganisms in submerged membrane bioreactor," *Water Research*, vol. 35, no. 10, pp. 2435–2445, 2001.

[3] W.-N. Lee, I.-J. Kang, and C.-H. Lee, "Factors affecting filtration characteristics in membrane-coupled moving bed biofilm reactor," *Water Research*, vol. 40, no. 9, pp. 1827–1835, 2006.

[4] X. C. Wang, Q. Liu, and Y. J. Liu, "Membrane fouling control of hybrid membrane bioreactor: Effect of extracellular polymeric substances," *Separation Science and Technology*, vol. 45, no. 7, pp. 928–934, 2010.

[5] K. Sombatsompop, C. Visvanathan, and R. Ben Aim, "Evaluation of biofouling phenomenon in suspended and attached growth membrane bioreactor systems," *Desalination*, vol. 201, no. 1-3, pp. 138–149, 2006.

[6] Z. Liang, A. Das, D. Beerman, and Z. Hu, "Biomass characteristics of two types of submerged membrane bioreactors for nitrogen removal from wastewater," *Water Research*, vol. 44, no. 11, pp. 3313–3320, 2010.

[7] X. Huang, C.-H. Wei, and K.-C. Yu, "Mechanism of membrane fouling control by suspended carriers in a submerged membrane bioreactor," *Journal of Membrane Science*, vol. 309, no. 1-2, pp. 7–16, 2008.

[8] K. Zerari, I. Seyssieq, D.-E. Akretche, and N. Roche, "Enhancement of oxygen mass transfer coefficients in a hybrid membrane bioreactor," *Journal of Chemical Technology and Biotechnology*, vol. 88, no. 6, pp. 1007–1013, 2013.

[9] G. Dotro, B. Jefferson, M. Jones, P. Vale, E. Cartmell, and T. Stephenson, "A review of the impact and potential of intermittent aeration on continuous flow nitrifying activated sludge," *Environmental Technology*, vol. 32, no. 15, pp. 1685–1697, 2011.

[10] L. Qiu and J. Ma, "A novel approach to remove nitrogen in a biological aerobic filter (BAF)," *Environmental Technology*, vol. 26, no. 8, pp. 923–930, 2005.

[11] S.-H. Yoon and S. Lee, "Critical operational parameters for zero sludge production in biological wastewater treatment processes combined with sludge disintegration," *Water Research*, vol. 39, no. 15, pp. 3738–3754, 2005.

[12] Q. Liu, X. C. Wang, Y. Liu, H. Yuan, and Y. Du, "Performance of a hybrid membrane bioreactor in municipal wastewater treatment," *Desalination*, vol. 258, no. 1-3, pp. 143–147, 2010.

[13] S. Jamal Khan, S. Ilyas, S. Javid, C. Visvanathan, and V. Jegatheesan, "Performance of suspended and attached growth MBR systems in treating high strength synthetic wastewater," *Bioresource Technology*, vol. 102, no. 9, pp. 5331–5336, 2011.

[14] S. Yang, F. Yang, Z. Fu, and R. Lei, "Comparison between a moving bed membrane bioreactor and a conventional membrane bioreactor on organic carbon and nitrogen removal," *Bioresource Technology*, vol. 100, no. 8, pp. 2369–2374, 2009.

[15] S. K. Gupta and R. Sharma, "Biological oxidation of high strength nitrogenous wastewater," *Water Research*, vol. 30, no. 3, pp. 593–600, 1996.

[16] R. O. Mines and J. H. Sherrard, "Biological treatment of a high strength nitrogenous wastewater," *Journal of Environmental Science and Health*, vol. 32, no. 5, pp. 1353–1375, 1997.

[17] K. H. Radwan and T. K. Ramanujam, "Influence of COD/NH$_3$-N ratio on organic removal and nitrification using a modified RBC," *Bioprocess Engineering*, vol. 16, no. 2, pp. 77–81, 1997.

[18] M.-L. Charmot-Charbonnel, S. Herment, N. Roche, and C. Prost, "Nitrification of high strength ammonium wastewater in an aerated submerged fixed bed," *Environmental Progress*, vol. 18, no. 2, pp. 123–129, 1999.

[19] S. K. Gupta, S. M. Raja, and A. B. Gupta, "Simultaneous nitrificationdenitrification in a rotating biological contactor," *Environmental Technology*, vol. 15, no. 2, pp. 145–153, 1994.

[20] B. Tang, B. Qiu, S. Huang et al., "Distribution and mass transfer of dissolved oxygen in a multi-habitat membrane bioreactor," *Bioresource Technology*, vol. 182, pp. 323–328, 2015.

12

Valorization of Toxic Weed *Lantana camara* L. Biomass for Adsorptive Removal of Lead

Vipin Kumar Saini,[1] Surindra Suthar,[1] Chaudhari Karmveer,[1] and Kapil Kumar[2]

[1]*School of Environment & Natural Resources, Doon University, Dehradun, Uttarakhand 248001, India*
[2]*Department of Environmental Engineering, National Institute of Technology, New Delhi 110040, India*

Correspondence should be addressed to Surindra Suthar; suthariitd@gmail.com

Academic Editor: Daryoush Afzali

Valorization of *Lantana camara* L., which is a recognized invasive plant, as a potential source of activated carbon is proposed in this study. Its stem and leaf have been utilized for the preparation of activated carbon (ACL and ACS) by following acid-impregnation technique, followed by thermal treatment. The developed activated carbon samples were characterized for their structural and surface related properties by low-temperature nitrogen adsorption isotherm, SEM techniques, and pH_{PZC} method. The samples show reasonable high surface area and pore volume; nonetheless, these properties are higher in case of ACL as compraed to ACS. Both of these samples developed negative charge on their surface due to acid treatment that resulted in an increase in adsorption at pH > 5. The batch adsorption studies on these samples shows the Pb(II) ion adsorption capacities of ACL and ACS were 36.01 and 32.24 $mg \cdot g^{-1}$, respectively, at 25°C. The kinetics of adsorption with both the sample systems follow the pseudo-second-order model, whereas the experimental equilibrium isotherm data of ACL and ACS were explained by Freundlich and Langmuir models, respectively. For these samples, the HCl shows maximum desorption with which the recycling test on these samples shows that ACS has better recycling potential over ACL samples.

1. Introduction

Lantana camara is a flowering plant that belongs to Verbenaceae family and is often planted in gardens. It is native to Central and South America and spread to around more than 40 different countries [1], where now it has become an invasive species [2]. It was brought by Dutch explorers to Europe from America, where it was cultivated widely and soon it spread into Asia and other countries, where it became the world's most notorious weed. It grows impassable thickets which suppress the growth of native species. Its growth often competes with, suppresses, and crowds out more desirable species, which leads to a loss of plant diversity, in affecting the area. Other problems with this species include its toxicity to livestock and if it invades agricultural land it may cause a decrease in productivity [3]. In India, Australia, and South Africa, it is widespread occupying millions of hectares of land [1]. There is a quest if we can fight the spread of invasive species or we need to develop strategies for their adaptive management.

According to literature, its spread is being controlled by several methods on a large scale, including mechanical (stickraking, bulldozing, plowing, and grubbing), chemical (using fluroxypyr and glyphosate), and biological methods (using different bug, beets, and seed-feeding fly) [4]. Use of this plant for a pollution abatement technique would be a most cost-effective management tool, in its control. One of the possible uses of converting this biomass into applicable materials is shown in [2]. One way of its valorization could be its use in environmental applications like wastewater treatment.

Among various technologies to treat industrial wastewater, the adsorption is an established technique for its efficiency and cost of operation. In adsorption technique, an extremely porous material (adsorbent) is used, which selectively adsorb the impurities from the aqueous phase by noncovalent bonding (H-bond, electrostatic bonding, and coordination bonding) [5]. The key challenge in the use of adsorption technique for water treatment applications is the cost of adsorbent. Activated carbon is a well-known adsorbent particularly due to its microporous structure and high surface area. It is prepared from a variety of carbonaceous sources like nutshells, coconut husk, peat, coir, lignite, coal, and wood. Physical or chemical activation processes are generally used for its development. The cost of activated carbon depends upon the choice of the process as well as the price of starting carbonaceous material (raw material) [5, 6]. So far, numerous plant materials and agricultural, industrial wastes have been explored for the development of activated carbon and their application in water treatment. But, surprisingly, there were no such studies on invasive weeds in general and *Lantana camara* in particular. Only recently have few researchers started to investigate its use as a source of activated carbon and study its potential for removal of water pollutants. But these studies show limited adsorption capacity and are limited to dyes [7] and phenol [8]. In this context, it is apparent that if processed and activated suitably, this *Lantana camara* L. can become a potential source of activated carbon, and this can valorize its biomass for adsorptive removal of water pollutants.

Like many other heavy metals, lead is also toxic and known for bioaccumulation that affects various body systems and is predominantly unsafe to young children [9]. It comes into water through the combustion of fossil fuels and the smelting of sulfide ore and into lakes and streams by acid mine drainage. Process industries, such as battery manufacturing and metal plating and finishing, are also a prime source of lead pollution [10]. Lead accumulates mainly in bones, brain, kidney, and muscles and may cause many serious disorders like anemia, kidney disease, nervous disorders, and sickness and even death. There is no known level of lead exposure that is considered safe. Current EPA drinking water standard for lead is 0.05 ppm, but a level of 0.02 ppm has been proposed and is under review [11].

In this study, leaves and stem of *Lantana camara* L. were used to prepare activated carbon by chemical activation method. The developed carbon was characterized by its structural and surface properties. The adoptions potentials of both the prepared materials were studied compared with selective adsorption of lead from aqueous solutions. The isotherm and kinetic modeling were used to explore adsorption behavior of these materials. Sustainability of optimized material in adsorption application was investigated in the light of their reusability in different adsorption cycle. To the best of our knowledge, this is the first study where an invasive weed like *Lantana camara* L. has been used as a precursor for activated carbon and used for adsorptive removal of pollutants.

2. Experimental

2.1. Materials.
All chemicals used in this work were of analytical grade and were purchased from Merck (India). The solutions were prepared by using deionized water. Analytical grade lead nitrate ($Pb(NO_3)_2$), 99% pure, was used to prepare a lead solution. All the experiments were conducted at a constant temperature of $25 \pm 0.1°C$ unless otherwise specified.

2.2. Adsorbent Preparation.
The leaves and stems samples of *Lantana camara* were collected locally, from the premises of Doon University, Dehradun. To remove dirt particles from leaves and stems, it was washed several times with tap water and then in the end with double distilled water. The washed samples were air dried, followed by drying in a hot air oven for 24 hours at 80°C. The dried samples were shredded and then crushed to powdered form. For chemical activation, the method was adopted from literature [12]. In this method, the dried powders of sample were mixed with concentrated H_2SO_4 (*Lantana* leaves or stem: H_2SO_4 ratio; 1 : 1.5 w/v) and were kept at 200°C for 24 h. The chemically burned (carbonized) samples were then washed several times with distilled water to remove any free acid and were then soaked overnight in 1% sodium bicarbonate solution to remove any residual acid from pores. The material then was again washed twice with distilled water and dried in hot air oven at 105°C. The dried samples were then finely ground and sieved to get a particle size of 150 μm. The as-prepared samples of activated carbon from leaves and stem were labeled as ACL and ACS, respectively, and stored in desiccators until required.

2.3. Adsorbent Characterization.
The morphological structure of the prepared carbon samples was observed by means of scanning electron microscopy (SEM) using a ZEISS EVO Series Microscope EVO 50. BET surface areas of samples were measured by liquid nitrogen adsorption method at 77 K using Micromeritics ASAP 2020. To characterize the surface charge of adsorbent, the experiment of the point of zero-charge pH_{zpc} was carried out. The methodology for determination of pH_{zpc} was adopted from an earlier study [13]. For this purpose, 0.025 g of adsorbent sample was kept in contact with 25 mL of a solution under 11 different pH conditions (2, 3, 4, 5, 6, 7, 8, 9, 10, 11, and 12) adjusted with solutions of 0.1 M NaOH or HNO_3. The samples were kept stirring on a water bath at 25°C. After this step, the pH of each solution was measured and a plot of the initial pH versus the final pH was obtained. From this plot, the pH at which the values of initial and final pH are found identical is testified as pH_{PZC}, the point of zero charge.

2.4. Adsorption Studies.
Several parameters individually influence the adsorption properties of a given adsorbent. It is, therefore, essential to study the variation in the amount of adsorption as a function of the different parameter, like time, pH, temperature, and adsorbate (lead ion in the present case) concentration. The effect of pH on Pb(II) ion concentrations in solutions without adsorbent was also examined so as to evaluate the attribution of precipitation to the sorption process. Thereafter the % precipitation was deducted from total removal present to calculate the correct % adsorption amount. The adsorption studies were performed in the form of batch experiments. The samples obtained from batch adsorption experiments were filtered through Millipore SLHN033NB Millex HN Syringe Filter

with Nylon Membrane (0.45 μm) and analyzed by iCE™ 3300 AAS Atomic Absorption Spectrometer. The adsorbed amount was calculated with the following mass balance equation, where q_e is the amount of adsorption (mg·g^{-1}), C_0 and C_e are the initial and equilibrium concentrations (mg·L^{-1}), m denotes the mass (g) of adsorbent, and V is volume (mL) of adsorbate solution.

$$q_e = \frac{(C_0 - C_e) V}{m \times 1000}. \tag{1}$$

The equilibrium time of adsorption was observed with both the adsorbents. For this, 2.5 g of sample was added to 500 mL lead solution 200 mg·L^{-1} in a conical flask. The solution was stirred with a magnetic stirrer in a water bath at 25°C. Periodically, the aliquots of 3 mL were sampled from this flask, up to 4 h. The effect of pH on the adsorption was studied by adjusting the pH of the lead solution from 2 to 12, using 0.1 M NaOH or HCl solutions. To study the effect of temperature on adsorption the batch adsorption experiments on two temperatures 25 and 45°C were carried out.

2.5. Adsorption Isotherms.

The mechanism of adsorption in terms of adsorbate-adsorbent interaction and maximum adsorption capacity can be explored with the help of adsorption isotherms. These isotherms are characterized by certain constants and describe the mathematical relationship between the amount of adsorbate adsorbed per unit mass of adsorbent and the equilibrium concentration of adsorbate in the solution. For this study, Pb(II) solutions (each 100 mL) of different concentration ranging from 50 to 500 mg·L^{-1} were stirred with 0.5 g of adsorbents until equilibrium adsorption time, after which the samples were collected, filtered, and analyzed by AAS to determine C_e and q_e. To observe an adsorption isotherm for a given adsorbent-adsorbate system, q_e are plotted as a function of C_e. Then different mathematical models of adsorption were applied to these isotherms and compared for better fitting [14]. In this study, nonlinear models of Langmuir (see (2)) and Freundlich (see (3)) were applied.

$$q_e = \frac{q_{max} \cdot b \cdot C_e}{1 + b \cdot C_e}, \tag{2}$$

$$q_e = K_f \cdot C_e^{1/n}. \tag{3}$$

In these equations, q_{max} and "b" are the Langmuir constants that denote maximum adsorption potential and equilibrium constant. Similarly, K_f and "n" are the Freundlich constant pertaining to the adsorption capacity and adsorption intensity, respectively.

2.6. Adsorption Kinetics.

The adsorption kinetics were investigated with the help of contact time data. This data shows the progress of adsorption (mg·g^{-1}) with respect to contact time, t (min). This kinetic data was then fitted to linear models of the pseudo-first-order (see (4)), pseudo-second-order (see

(5)), and intraparticle diffusion (see (6)) models [15, 16]. The linear forms of these equations are expressed as follows:

$$\log (q_e - q) = \log q_e - \frac{k_1}{2.303} t, \tag{4}$$

$$\frac{t}{q} = \frac{1}{k_2 q_e^2} + \frac{1}{q_e} t, \tag{5}$$

$$q = k_{id} t^{0.5} + C, \tag{6}$$

where q is the amount of adsorption at any time (t, min), k_1 (min^{-1}) is the pseudo-first-order adsorption rate constant, k_2 (g·mg^{-1}·min^{-1}) is the pseudo-second-order adsorption rate constant, k_{id} (mg·g^{-1}·min$^{-1/2}$) is the intraparticle diffusion rate constant, and C (mg·g^{-1}) is a constant in the intraparticle diffusion model that reflects the significance of the boundary layer on mass transfer effect.

2.7. Evaluation of Desorption Using Different Desorbing Solvents.

Desorption of adsorbed metal from the surface of adsorbent is necessary for the sake of recovery of metal as well as reuse of adsorbent. For this purpose, different desorbent solutions were tested. In this experiment, the assay was prepared in two steps. In the first step, 0.5 g of activated carbon was stirred with 100 mL of 200 mg·L^{-1} Pb(II) solution at 25°C up to equilibrium time. In the end, the AC samples were collected by filtration on a glass filter, washed with distilled water, and placed in an oven for 6 h at 60°C. The filtrate was analyzed to observe C_e. In the next step, the dry AC samples were placed in contact with 100 mL of different solution independently, like 0.1 M HNO$_3$, HCl, H$_3$PO$_4$, NaCl, and distilled water for 6 h on a magnetic stirrer in a water bath at 25°C. The liquid phase was filtered through membrane filters and analyzed for desorbed Pb(II) ion concentration. The desorbed percentage was determined according to (7). The solution which achieves the maximum percentage of Pb(II) desorption from the loaded adsorbent was identified and used in successive adsorbent reuse test.

$$\% \text{ desorbed} = \frac{\text{amount desorbed}}{\text{amount adsorbed}} \times 100. \tag{7}$$

2.8. Adsorbent Reuse Test.

The reusability is critical for the cost-effectiveness of a new adsorbent [27]. The reuse assay was performed in cycles, starting with the adsorption and ending with the desorption of the Pb(II) ions using identified desorbing solvent. After each cycle, the mass of AC samples was washed with distilled water and dried in an oven at 80°C for 2 h. Four cycles were conducted with the same mass of AC samples.

3. Results and Discussions

3.1. Textural and Structural Characterization.

Both the samples of prepared activated carbon were characterized for their surface and other adsorption related physicochemical properties. Porosity and surface area of prepared activated carbon samples were assessed by nitrogen adsorption at

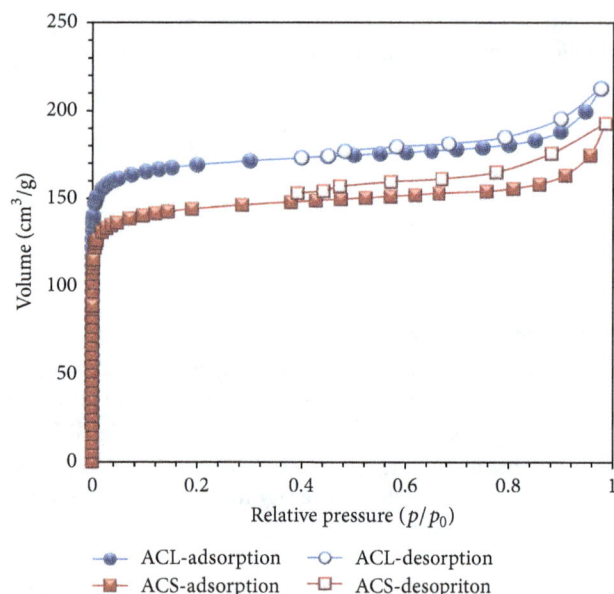

FIGURE 1: Low-temperature nitrogen adsorption isotherms, at −196°C, obtained from ACL and ACS samples.

TABLE 1: The surface properties of prepared activated carbon samples obtained from low-temperature nitrogen adsorption isotherm data.

Sample	Specific Surface area A_{BET}	Total pore volume V_{total}	Micropore volume $V\mu$
	m^2/g	cm^3/g	cm^3/g
ACL	634	0.31	0.22
ACS	523	0.26	0.19

the surface. Likewise, the stem sample surface also shows variation before and after activation; however, this variation is not as noticeable as in the case of leaves sample. It is likely due to the variation in the biochemical composition of leaves and stem [31]. In brief, it is evident from the SEM images that the surface of raw materials gets changed after acid treatment, though the extent of change is different for each sample.

In the chemical as well as physical activation method, the variation in original chemical composition resulted in the difference in surface functionality, because different compounds present in the surface and matrix of the sample give rise to different charge group (cationic or anionic) on the surface after carbonization and activation [32]. To observe this phenomenon in the context of the present samples pH_{PZC} analysis has been carried out and the results are presented in Figure 3. It shows that pH_{PZC} is approximately 5.0 for both the samples, which means at pH > 5.0 the activated carbons would have predominantly a negative charge at the surface while below this value the surface is positively charged. Typically, the pH at the point of zero charges on activated carbon is approximately 7 [33, 34]. However, this was not observed from the AC samples prepared in the present study. The value of 5.0 can be explained by acid treatment to raw material which was used to prepare activated carbon. The H^+ ions present on the surface after treatment will be released into the solution which resulted in lowering of pH. The adsorption of cations, such as metal ions, is favored at pH > pH_{PZC} while the adsorption of anions is favored at pH < pH_{PZC} [35, 36]. Thus, the lead adsorption process should be carried out at pH equal to or greater than 5.0, to identify the optimal pH range.

−196°C (Figure 1). The isotherm of activated carbon prepared from leaf (ACL) as well as stem (ACS) shows similar shape, which is of type I and slightly type IV with H1 loop, which are the characteristics of micro- and mesoporous materials [28]. Both the isotherms have slow condensation step at a relative pressure ($p/p_0 = 0.8-0.99$) which corresponds to the capillary condensation of N_2 within uniform mesopores. These results show the essential micro- and mesoporous nature of the prepared activated carbon samples. However, a decrease in adsorption amount in ACS sample is noted, as compared to ACL sample. From N_2 adsorption data the specific surface area, A_{BET}, was determined through BET model (within p/p_0 0.05-0.15) [29]; likewise the micropore volume V_{micro} and total pore volume V_{total} of the materials are listed in Table 1. The anatomical differences in leaves and stem give rise to a change in their structural composition. The same difference has been translated into their surface area and pore volume properties after activation. The specific surface areas of ACL and ACS are 634 and 523 m^2/g, respectively. It is evident that chemical activation has produced large numbers of new pores in the samples by continuous devolatilization of the char and carbon burn-off. The dried leaves contain spongy mesophyll as compared to cortex and pith in the stem [30]. The total pore volume of ACL is higher compared to ACS; both the activated carbon samples are microporous in nature which is confirmed by fraction of their micropore volume with respect to their total pore volume.

SEM technique has been used to observe the change in samples morphology, before and after the treatment. The changes can be observed in Figure 2, where SEM images of raw materials (*Lantana* leaf and stem dust) are compared with, respectively, activated carbons (ACL and ACS). The surface of leaf sample seems uniform before activation and after activation it shows emergence of porous structure on

3.2. Effect of Contact Time. The effect of contact time on the amount of adsorption by ACL and ACS is shown in Figure 4. It was observed that after more than 3 h at 25°C ACL and ACS adsorb approximately 90% and 81%, respectively, from 200 mg·L^{-1} Pb(II) solution. It also shows the maximum adsorption capacity of ACL and ACS as 36.3 and 32.2 mg·g^{-1}, respectively. Beyond 3 h up to 24 h, there was a small variation <2% in the remaining concentration of Pb(II), which indicate adsorption equilibrium point. Hence, a minimum contact time of 3 h was adopted in all the subsequent tests. The data shows that >50% of adsorption takes place in the initial 15 mins. During initial 10 mins of adsorption both ACS and ACL show the same rate of adsorption, but, subsequently, the rate of ACS system gets slightly slower compared to ASL system. This is possibly due to the higher surface area in the

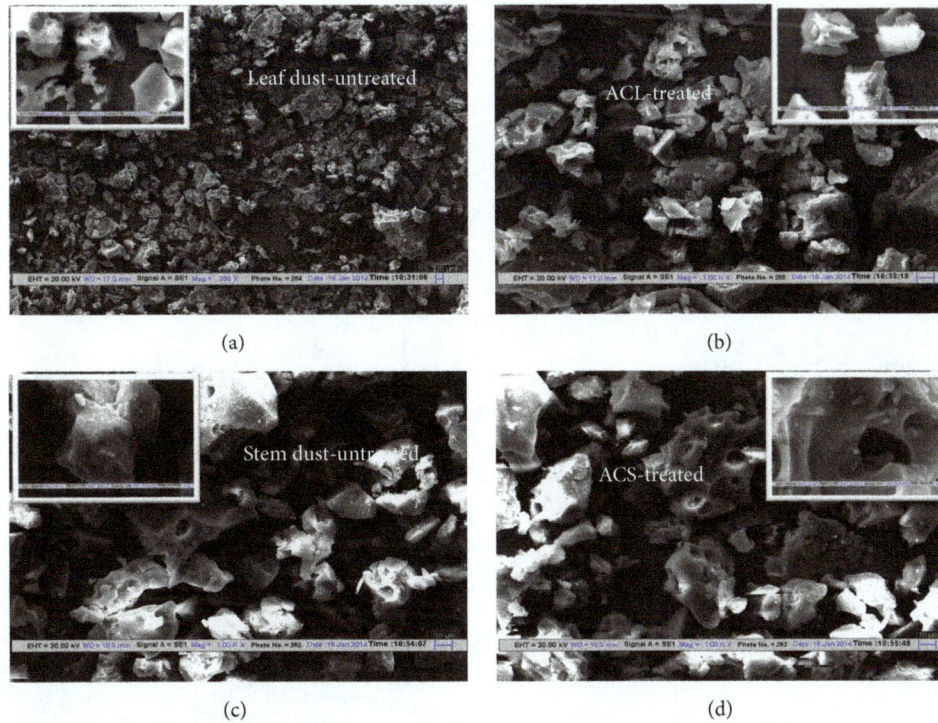

(a)

(b)

(c)

(d)

FIGURE 2: SEM images of different magnification of *L. camara* (stem and leaves) samples before and after acidic treatment.

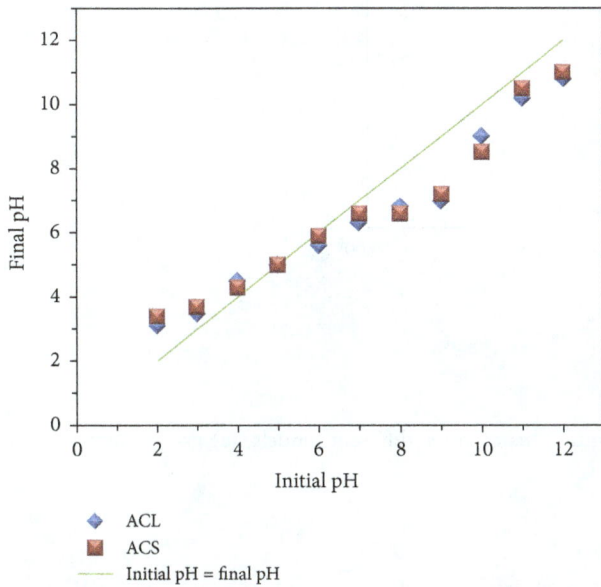

ACL
ACS
Initial pH = final pH

FIGURE 3: Determination of pH_{ZPC} of ACL and ACS samples ($T = 25°C$).

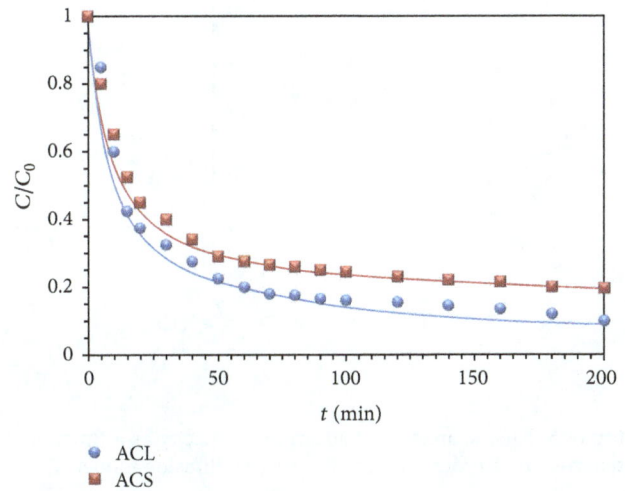

ACL
ACS

FIGURE 4: Effect of contact time on the adsorption of lead from aqueous solution, using ACS and ACL ($T = 25°C$).

latter sample (ACL), as the rate of adsorption is influenced by the fraction of surface area available for adsorption.

3.3. Adsorption Kinetics Study. The adsorption data obtained from the experiments on the effect of contact time were applied to three different kinetic models to study the kinetics of Pb(II) adsorption by ACL and ACS.

Figures 5(a), 5(b), and 5(c) show the fitting in three models, pseudo-first-order, pseudo-second-order, and intraparticle diffusion model, respectively. The experimental and calculated parameters of the pseudo-first-order and pseudo-second-order models have also been summarized in Table 2.

According to the data presented in this table, the calculated linear regression correlation coefficients for the pseudo-first-order model are relatively small (Figure 5(a)), and the experimental q_e values differ significantly from the values obtained from the linear plots. Therefore, it could be said

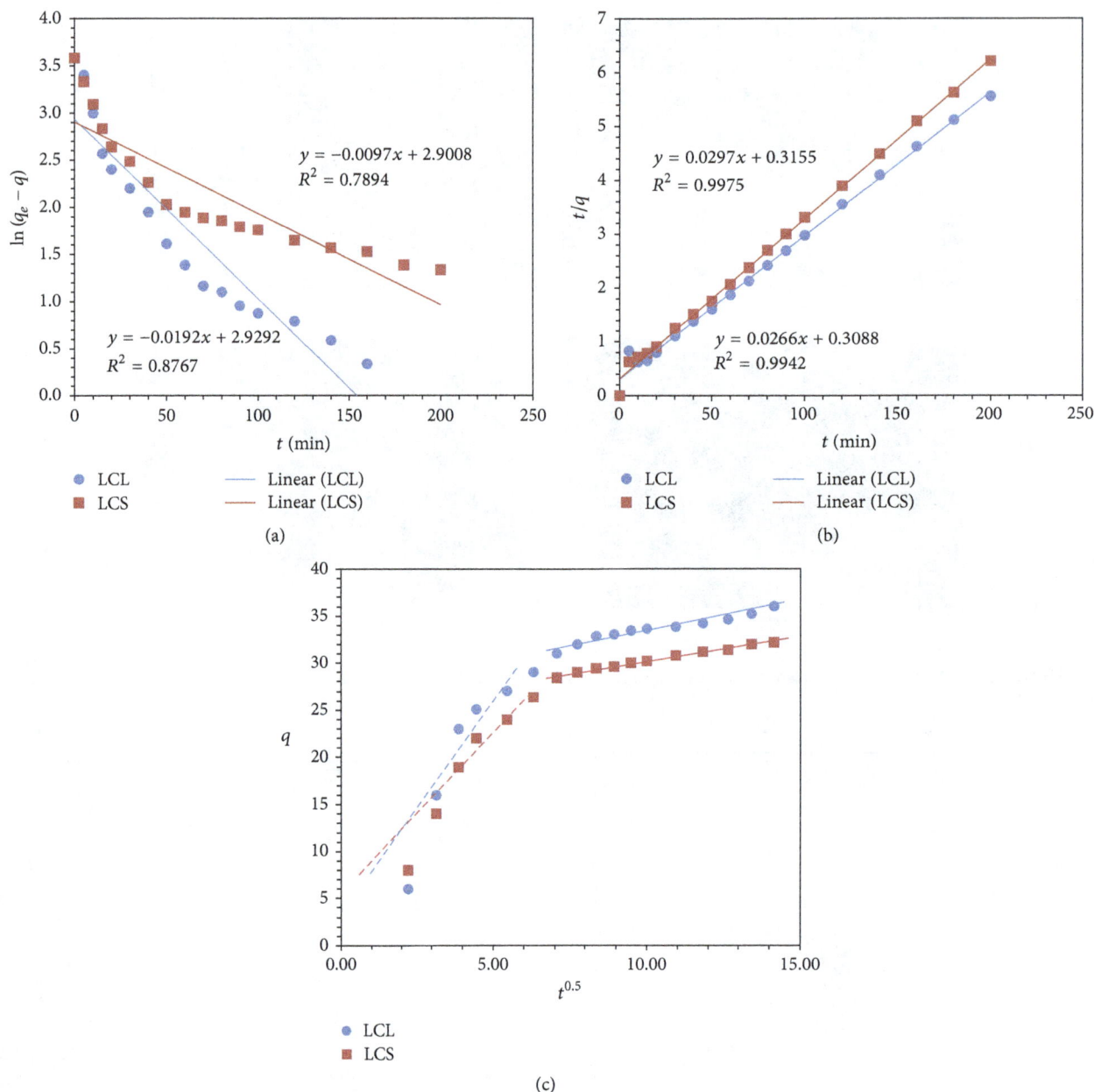

FIGURE 5: Kinetic analysis of adsorption capacity as a function of time (min^{-1}), using three different models: (a) pseudo-first-order, (b) pseudo-second-order, and (c) intraparticle diffusion model ($T = 25°C$).

that the pseudo-first-order equation is not suitable for lead adsorption on ACL and ACS system. As is evident from Figure 5(b) and Table 2, the fitting of experimental data to the pseudo-second-order model is reasonably good, $R^2 > 0.99$, and the calculated q_e values agree well with the experimental values. This shows that Pb(II) ions uptake via both of the ACL and ACS can be estimated with the pseudo-second-order kinetics model.

The intraparticle diffusion model is among the most commonly used techniques for the prediction of the rate-controlling step of the adsorption process. Figure 5(c) shows q versus $t^{0.5}$ plot for Pb(II) ions adsorption onto the ACL

and ACS. The plot shows two distinct kinetic steps which suggest the existence of at least two stages in the Pb(II) ions sorption process. It is likely that, in the sharper first portion, the adsorbate ions get bound or anchored with the active sites on the adsorbent surface while the second step was governed by the rate of intraparticle diffusion. In the first stage, due to the larger adsorbent surface area of the ACL sample, the adsorption rate onto this adsorbent was considerably higher than that for ACS sample. Considering the second stage, again the computed intraparticle diffusion parameters show a higher adsorption rate constant for ACL as compared to ACS. The remarkably slow diffusion into the intrapores of the ACS

TABLE 2: Calculated kinetic parameters for pseudo-first-order, second-order, and intraparticle diffusion models for the adsorption of Pb(II).

Samples	$q_{e,exp}$ (mg·g⁻¹)	Pseudo-first-order			Pseudo-second-order			Intraparticle diffusion model		
		$q_{e,calc}$ (mg·g⁻¹)	k_1 (min⁻¹)	R^2	$q_{e,calc}$ (mg·g⁻¹)	k_2 (g·mg⁻¹·min⁻¹)	R^2	k_i (mg·g⁻¹·min^{1/2})	C (mg·g⁻¹)	R^2
ACL	36.01	18.71	1.9×10^{-2}	0.88	38.46	2.2×10^{-3}	0.99	0.55	28.32	0.98
ACS	32.24	18.17	0.9×10^{-2}	0.79	34.48	2.7×10^{-3}	0.99	0.52	25.36	0.99

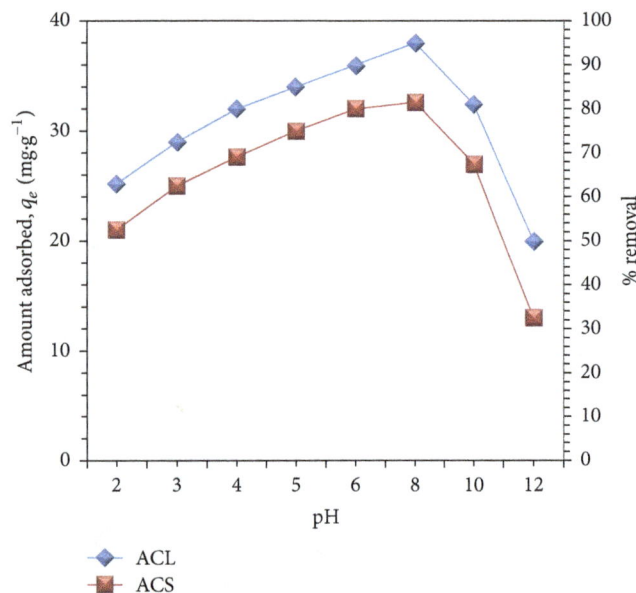

FIGURE 6: Effect of pH on the adsorption capacity and % removal of ACL and ACS sample ($T = 25°C$).

samples may have been due to lack of microporosity in the sample that could have facilitated diffusion of adsorbate ions in the rate determination step.

3.4. Effect of pH.

The pH of adsorbate solution determines the surface charge of adsorbents, the speciation of adsorbate species, and the degree of ionization and that is why it is one of the most important variables in adsorption investigations [5]. It is also directly related to precipitation conditions, particularly in the case of heavy metals, where it plays a critical role in the removal of heavy metal ions from aqueous solutions. The effect of solution pH on the % precipitation of the lead is shown in Figure 6. It was found that there is precipitation of the metal throughout the pH range examined. However, at pH > 5.0 over 30% precipitation was observed, which would modify the adsorption result if such pH values were applied in the tests. This precipitation occurs due to Pb(II) ion association with OH-ions of the basic solution to form lead hydroxides. At pH < 3.0, it is possible that the lead is present in the form of $PbOH^+$, even in small amounts, which may explain the low precipitation in this pH range.

The difference between the total removal percentage of Pb(II) ions and the percentage removed by precipitation yields the adsorption percentage of lead, which is also given in Figure 6. It shows an increase in the removal of lead ions with increases in solution pH up to 5, after which a gradual and then fast decrease were observed up to pH 10. Maximum lead removal values were obtained at pH values of 5-6. The reduced metal uptake witnessed at low pH values may have been due to excessive protonation (H^+ ions) of the carbon surface, which render the adsorption of Pb(II) ions [37]. At higher pH values, the concentration of H^+ becomes low, and the surface becomes negatively charged carbon surface (see Section 3.1 and Figure 3) that causes

further attraction of metal cations. Nonetheless, at alkaline conditions, a decrease in adsorption was observed which is because of the formation of several hydroxide species by Pb(II) ions which are precipitated and lead to the decrease in free metal ions available in solution and [38] reported an optimum pH of 5 for lead adsorption. Bhattacharyya and Sharma obtained a good retention of lead in the pH 7 and noted that at pH < 4.5 there was excessive protonation of the carbon surface, resulting in a decrease in the adsorption of Pb(II) [39]. Considering the interference due to precipitation during the adsorption experiments with AC, the pH_{PZC} value, and the lead speciation diagram, the pH selected for subsequent tests in this study was 5.0.

3.5. Adsorption Isotherms and Modeling of Isotherm Data.

Figure 7 displays the experimental data of Pb(II) adsorption isotherms on both ACL and ACS at three different temperatures (25, 35, and 45°C) and nonlinear fitting of Langmuir and Freundlich models on experimental values. Table 3 shows the model specific constants at different temperatures, which were calculated from the fitting parameters. The relative difference in adsorption amounts at different temperature shows that, with both the activated carbon samples, the adsorption increases with a decrease in temperature. It is likely due to increasing in surface energy of the activated carbon samples, which would render the interaction of adsorbing Pb(II) ions with the surface [40]. The Langmuir and Freundlich adsorption models were used for fitting the experimental data (Figure 7). A quick comparison in the fitting pattern shows that the Freundlich model shows better fitting on adsorption data by ACL. However, in contrast, ACS-adsorption data shows better fitting with Langmuir model and q_{max} calculated from this model (37.93 mg·g^{-1}) is close to q_{expt} (37.6 mg·g^{-1}) values. The application of the Langmuir model is based on the assumption that the binding of the adsorbate onto the surface of the adsorbent occurs primarily through a chemical reaction and that all sites have equal affinity for the metal. The Freundlich isotherm model is an empirical equation based on adsorption onto heterogeneous surfaces (Li et al. 2010). Most assumptions related to the Langmuir model (such as adsorption onto a homogenous surface, with an identical fixed number of adsorption sites) are invalid for the heterogeneous surface of activated carbon. Thus, it is appropriate to use the Langmuir equation for qualitative and descriptive, rather than quantitative, purposes.

3.6. Regeneration and Reuse Study.

Figure 8 shows a comparison in desorption% of the lead with six different desorbing agents in both the samples. None of the desorbing solutions caused apparent physical alterations or impairment to the samples, enabling its reuse. For both the samples HCl showed maximum effectiveness, with approximately 66.1% and 55.6% removal from ACL and ACS, respectively. The efficiency of the rest of the desorbents follows a common order of $HNO_3 > H_3PO_4 \gg NaOH > NaCl > H_2O$. The H^+ from an acidic medium may actively replace the metal ions (Pb ion) adsorbed on the carbon. Therefore, considering these results the HCl was used as a desorbing agent because it is of low

TABLE 3: Parameters of Langmuir and Freundlich adsorption models obtained from fitting of adsorption data at a different temperature.

Sample	Temp (°C)	Langmuir			Freundlich		
		q_{max} mg·g^{-1}	b L·mg^{-1}	R^2	K_f (mg·g^{-1}) (L·mg^{-1})$^{1/n}$	$1/n$	R^2
ACL	25	45.79	0.12	0.9639	13.92	0.23	0.995
	35	42.08	0.07	0.9716	10.81	0.25	0.9948
	45	38.08	0.06	0.9653	9.36	0.25	0.9896
ACS	25	37.93	0.19	0.999	14.74	0.18	0.9106
	35	32.75	0.13	0.999	12.12	0.18	0.899
	45	28.14	0.12	0.998	11.13	0.17	0.855

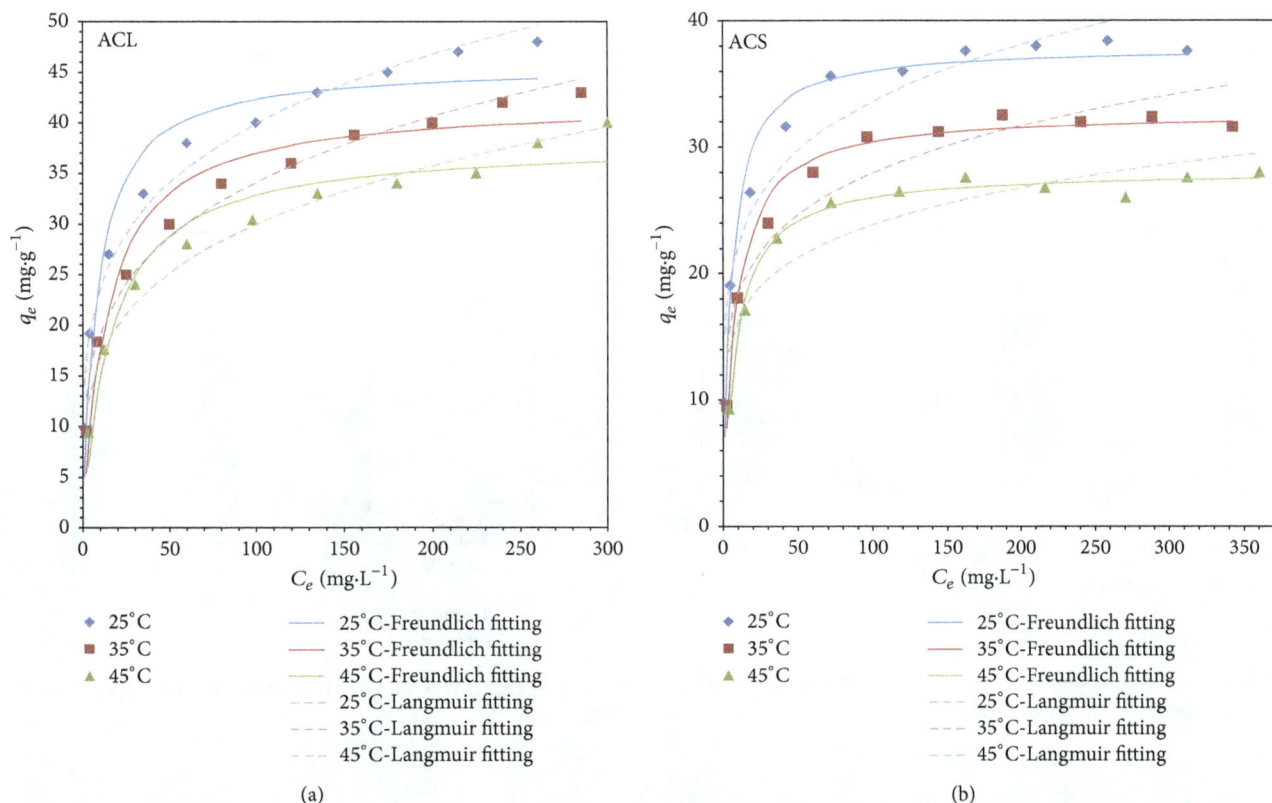

FIGURE 7: Adsorption isotherm of Pb(II) ions on ACL and ACS at different temperatures and fitting of adsorption data on Langmuir and Freundlich models.

cost compared to other materials and offers better desorbing efficiency.

The results of reuse experiments obtained from both activated carbon samples were of contrast to each other. The % of Pb(II) adsorption and desorption falls steeply in the case of ACL samples, which makes it almost unsuitable for reuse after two cycles (figure not given). On the other hand, the ACS sample shows reasonably good results of % adsorption and desorption up to four successive cycles; see Figure 9. It shows that during first cycle 90% adsorption and 66% desorption were achieved with respect to the value of $C_0 = 200$ mg·L^{-1}. For the second, third, and fourth cycles 80%, 67.5%, and 50% for adsorption and 56.2%, 40.7%, and 20% desorption, respectively, were observed. The 50% loss

in adsorption capacity during four cycles could be due to cumulative effect of the inadequate desorption. It may be crucial to use an acid solution with a higher strength or larger amounts of desorbent solutions.

In the view of sustainability, the desorbed lead from ACS samples could be reutilized and reverted to factories as a raw material, in spite of being deposited in sanitary landfills as waste. Electrolysis is one of the key processes for its recovery and can be useful to transform lead solution into the elemental lead (using stainless steel electrode).

3.7. Comparison of Adsorption Capacity with Other Adsorbents. The adsorption capacities of ACL and ACS were compared (Table 4) with other reported studies on activated

TABLE 4: Comparison between adsorption capacities of several adsorbents, obtained from natural organic materials, for Pb(II) ions.

Sorbent*	Adsorption capacity q_{max} $(mg \cdot g^{-1})$	Optimum pH	Equilibrium time T (min)	Reference
Cashew nut shell AC	28.9	6.0–6.5	30	[17]
Hazelnut husk AC	13.05	5.7	60	[18]
Coconut shell AC	26.5	4.5	35	[19]
Apricot Stone AC	22.85	4	300	[20]
Van apple pulp AC	15.96	5	50	[21]
Polygonum orientale AC	98.39	5	30	[22]
Pine cone AC	27.53	5	60	[23]
Cow bone AC	32.1	4	360	[24]
Dehydrated hazelnut husks carbon	133.3	4.5	60	[25]
Hazelnut husks AC	109.9	5	1200	[26]
ACL	48.1	5.0–6.0	180	This study
ACS	37.6	5.0–6.0	180	This study

*AC = activated carbon.

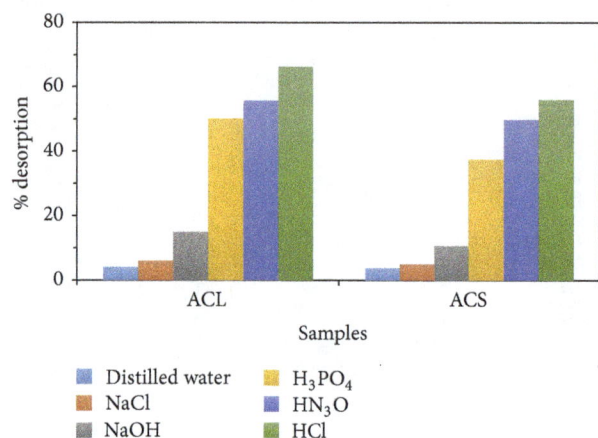

FIGURE 8: Lead desorption from activated carbon samples using several desorbent chemical agents ($T = 25°C$).

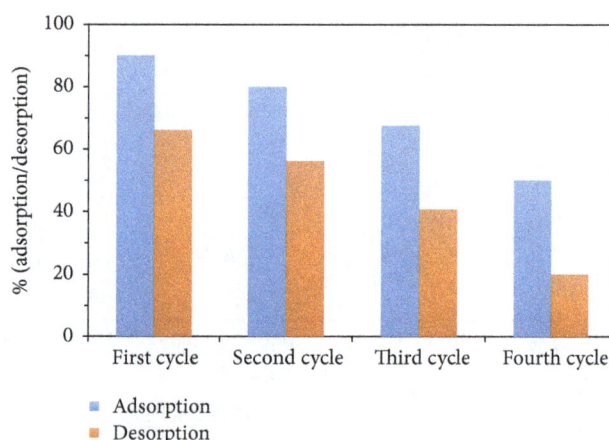

FIGURE 9: Percentage of Pb(II) adsorbed-desorbed per cycle ($T = 25°C$), with ACS sample.

carbon for lead removal. These studies have used a variety of starting materials like nutshells, pulp, seeds, and animal bones for activated carbon preparation. The comparison shows that adsorption capacities of the present materials are better than some of the reported studies. It shows that *Lantana camara* L. can be used as a lucrative source of activated carbon.

4. Conclusions

The results of this study show that valorization of *Lantana Camara* L. can be achieved by converting it into an activated carbon with the moderate surface area, by using H_2SO_4 impregnation method. The compositional differences in leaf and stem of plant give rise to a difference in textural and surface properties of resulting activated carbon. Therefore, the surface area and adsorption capacity of activated carbon obtained from its leaf samples are higher than those from stem samples. The acid treatment increases the negative charge on the surface in both the samples. Consequently, the pH becomes a critical factor in the lead adsorption, at pH > 5.0. For both the activated carbon samples, the pseudo-second-order kinetic model best represented the adsorption kinetics data. ACL-adsorption equilibrium system was best described by Langmuir model, whereas ACS system was better explained by Freundlich model. The adsorbed lead on the samples can be efficiently desorbed with the help of HCl. The reusability of ACL samples is not satisfactory as compared to ACS samples. Concluding this study, it may be mentioned that although ACS has <12% less adsorption capacity compared to ACL, but it shows better reusability and thereby shows the better potential of a new low-cost adsorbent. This study opens the route to further studies where the selectivity of this *Lantana camara* L. based activated carbon can be investigated for other emerging pollutants like endocrine disruptors and pharmaceuticals.

Competing Interests

The authors declare that they have no competing interests.

Acknowledgments

The authors acknowledge the financial support by School of Environment and Natural Resources, Doon University, Dehradun, and Department of Environmental Engineering, National Institute of Technology, New Delhi India.

References

[1] O. P. Sharma, H. P. S. Makkar, and R. K. Dawra, "A review of the noxious plant *Lantana camara*," *Toxicon*, vol. 26, no. 11, pp. 975–987, 1988.

[2] S. A. Bhagwat, E. Breman, T. Thekaekara, T. F. Thornton, and K. J. Willis, "A battle lost? Report on two centuries of invasion and management of *Lantana camara* L. in Australia, India and South Africa," *PLoS ONE*, vol. 7, no. 3, Article ID e32407, 2012.

[3] S. Raghu, O. O. Osunkoya, C. Perrett, and J.-B. Pichancourt, "Historical demography of *Lantana camara* L. reveals clues about the influence of land use and weather in the management of this widespread invasive species," *Basic and Applied Ecology*, vol. 15, no. 7, pp. 565–572, 2014.

[4] R. J. Fensham, "Land clearance and conservation of inland dry rainforest in north Queensland, Australia," *Biological Conservation*, vol. 75, no. 3, pp. 289–298, 1996.

[5] I. Ali and V. K. Gupta, "Advances in water treatment by adsorption technology," *Nature Protocols*, vol. 1, no. 6, pp. 2661–2667, 2007.

[6] S. Babel and T. A. Kurniawan, "Low-cost adsorbents for heavy metals uptake from contaminated water: a review," *Journal of Hazardous Materials*, vol. 97, no. 1–3, pp. 219–243, 2003.

[7] R. K. Gautam, P. K. Gautam, S. Banerjee et al., "Removal of tartrazine by activated carbon biosorbents of Lantana camara: kinetics, equilibrium modeling and spectroscopic analysis," *Journal of Environmental Chemical Engineering*, vol. 3, no. 1, pp. 79–88, 2015.

[8] C. R. Girish and V. Ramachandra Murty, "Mass transfer studies on adsorption of phenol from wastewater using *Lantana camara*, forest waste," *International Journal of Chemical Engineering*, vol. 2016, Article ID 5809505, 11 pages, 2016.

[9] A. D. Beattie, M. R. Moore, A. Goldberg, and W. T. Devenay, "Environmental lead pollution in an urban soft-water area," *British Medical Journal*, vol. 2, no. 5812, pp. 491–493, 1972.

[10] P. M. Félix, S. M. Almeida, T. Pinheiro, J. Sousa, C. Franco, and H. T. Wolterbeek, "Assessment of exposure to metals in lead processing industries," *International Journal of Hygiene and Environmental Health*, vol. 216, no. 1, pp. 17–24, 2013.

[11] Ö. Aktaş and F. Çeçen, "Adsorption and cometabolic bioregeneration in activated carbon treatment of 2-nitrophenol," *Journal of Hazardous Materials*, vol. 177, no. 1–3, pp. 956–961, 2010.

[12] C. Duran, D. Ozdes, A. Gundogdu, M. Imamoglu, and H. B. Senturk, "Tea-industry waste activated carbon, as a novel adsorbent, for separation, preconcentration and speciation of chromium," *Analytica Chimica Acta*, vol. 688, no. 1, pp. 75–83, 2011.

[13] J. E. Samad, S. Hashim, S. Ma, and J. R. Regalbuto, "Determining surface composition of mixed oxides with pH," *Journal of Colloid and Interface Science*, vol. 436, pp. 204–210, 2014.

[14] B. H. Hameed, J. M. Salman, and A. L. Ahmad, "Adsorption isotherm and kinetic modeling of 2,4-D pesticide on activated carbon derived from date stones," *Journal of Hazardous Materials*, vol. 163, no. 1, pp. 121–126, 2009.

[15] N. M. Kovalchuk, O. K. Matar, R. V. Craster, R. Miller, and V. M. Starov, "The effect of adsorption kinetics on the rate of surfactant-enhanced spreading," *Soft Matter*, vol. 12, no. 4, pp. 1009–1013, 2016.

[16] N. Woitovich Valetti and G. Picó, "Adsorption isotherms, kinetics and thermodynamic studies towards understanding the interaction between cross-linked alginate-guar gum matrix and chymotrypsin," *Journal of Chromatography B: Analytical Technologies in the Biomedical and Life Sciences*, vol. 1012-1013, pp. 204–210, 2016.

[17] N. I. S. Tangjuank, J. Tontrakoon, and V. Udeye, "Adsorption of Lead(II) and Cadmium(II) ions from aqueous solutions by adsorption on activated carbon prepared from cashew nut shells," *International Journal of Chemical, Molecular, Nuclear, Materials and Metallurgical Engineering*, vol. 3, pp. 221–227, 2009.

[18] M. Imamoglu and O. Tekir, "Removal of copper (II) and lead (II) ions from aqueous solutions by adsorption on activated carbon from a new precursor hazelnut husks," *Desalination*, vol. 228, no. 1–3, pp. 108–113, 2008.

[19] M. Sekar, V. Sakthi, and S. Rengaraj, "Kinetics and equilibrium adsorption study of lead(II) onto activated carbon prepared from coconut shell," *Journal of Colloid and Interface Science*, vol. 279, no. 2, pp. 307–313, 2004.

[20] L. Mouni, D. Merabet, A. Bouzaza, and L. Belkhiri, "Adsorption of Pb(II) from aqueous solutions using activated carbon developed from Apricot stone," *Desalination*, vol. 276, no. 1–3, pp. 148–153, 2011.

[21] T. Depci, A. R. Kul, and Y. Önal, "Competitive adsorption of lead and zinc from aqueous solution on activated carbon prepared from Van apple pulp: study in single- and multi-solute systems," *Chemical Engineering Journal*, vol. 200–202, pp. 224–236, 2012.

[22] L. Wang, J. Zhang, R. Zhao, Y. Li, C. Li, and C. Zhang, "Adsorption of Pb(II) on activated carbon prepared from Polygonum orientale Linn.: kinetics, isotherms, pH, and ionic strength studies," *Bioresource Technology*, vol. 101, no. 15, pp. 5808–5814, 2010.

[23] M. Momčilović, M. Purenović, A. Bojić, A. Zarubica, and M. Randelovid, "Removal of lead(II) ions from aqueous solutions by adsorption onto pine cone activated carbon," *Desalination*, vol. 276, no. 1-3, pp. 53–59, 2011.

[24] M. A. P. Cechinel, S. M. A. G. Ulson De Souza, and A. A. Ulson De Souza, "Study of lead (II) adsorption onto activated carbon originating from cow bone," *Journal of Cleaner Production*, vol. 65, pp. 342–349, 2014.

[25] M. Imamoglu, A. Vural, and H. Altundag, "Evaluation of adsorptive performance of dehydrated hazelnut husks carbon for Pb(II) and Mn(II) ions," *Desalination and Water Treatment*, vol. 52, no. 37-39, pp. 7241–7247, 2014.

[26] M. Imamoglu, H. Şahin, Ş. Aydın, F. Tosunoğlu, H. Yılmaz, and S. Z. Yıldız, "Investigation of Pb(II) adsorption on a novel activated carbon prepared from hazelnut husk by K_2CO_3 activation," *Desalination and Water Treatment*, vol. 57, no. 10, pp. 4587–4596, 2016.

[27] T. Yokoi, T. Tatsumi, and H. Yoshitake, "Fe^{3+} coordinated to amino-functionalized MCM-41: an adsorbent for the toxic oxyanions with high capacity, resistibility to inhibiting anions,

and reusability after a simple treatment," *Journal of Colloid and Interface Science*, vol. 274, no. 2, pp. 451–457, 2004.

[28] R.-L. Tseng, "Physical and chemical properties and adsorption type of activated carbon prepared from plum kernels by NaOH activation," *Journal of Hazardous Materials*, vol. 147, no. 3, pp. 1020–1027, 2007.

[29] B. Streppel and M. Hirscher, "BET specific surface area and pore structure of MOFs determined by hydrogen adsorption at 20 K," *Physical Chemistry Chemical Physics*, vol. 13, no. 8, pp. 3220–3222, 2011.

[30] R. K. Verma and S. K. Verma, "Phytochemical and termiticidal study of Lantana camara var. aculeata leaves," *Fitoterapia*, vol. 77, no. 6, pp. 466–468, 2006.

[31] N. R. Tesch, F. Mora, L. Rojas et al., "Chemical composition and antibacterial activity of the essential oil of Lantana camara var. moritziana," *Natural Product Communications*, vol. 6, no. 7, pp. 1031–1034, 2011.

[32] D. Angin, E. Altintig, and T. E. Köse, "Influence of process parameters on the surface and chemical properties of activated carbon obtained from biochar by chemical activation," *Bioresource Technology*, vol. 148, pp. 542–549, 2013.

[33] K. G. Sreejalekshmi, K. A. Krishnan, and T. S. Anirudhan, "Adsorption of Pb(II) and Pb(II)-citric acid on sawdust activated carbon: kinetic and equilibrium isotherm studies," *Journal of Hazardous Materials*, vol. 161, no. 2-3, pp. 1506–1513, 2009.

[34] A. Geethakarthi and B. R. Phanikumar, "Characterization of tannery sludge activated carbon and its utilization in the removal of azo reactive dye," *Environmental Science and Pollution Research*, vol. 19, no. 3, pp. 656–665, 2012.

[35] D. N. Shah, J. R. Feldkamp, J. L. White, and S. L. Hem, "Effect of the pH–zero point of charge relationship on the interaction of ionic compounds and polyols with aluminum hydroxide gel," *Journal of Pharmaceutical Sciences*, vol. 71, no. 2, pp. 266–268, 1982.

[36] E. C. Scholtz, J. R. Feldkamp, J. L. White, and S. L. Hem, "Point of zero charge of amorphous aluminum hydroxide as a function of adsorbed carbonate," *Journal of Pharmaceutical Sciences*, vol. 74, no. 4, pp. 478–481, 1985.

[37] W. Liu, Y. Liu, Y. Tao, Y. Yu, H. Jiang, and H. Lian, "Comparative study of adsorption of Pb(II) on native garlic peel and mercerized garlic peel," *Environmental Science and Pollution Research*, vol. 21, no. 3, pp. 2054–2063, 2014.

[38] J. Anwar, U. Shafique, Waheed-uz-Zaman, M. Salman, A. Dar, and S. Anwar, "Removal of Pb(II) and Cd(II) from water by adsorption on peels of banana," *Bioresource Technology*, vol. 101, no. 6, pp. 1752–1755, 2010.

[39] K. G. Bhattacharyya and A. Sharma, "Adsorption of Pb(II) from aqueous solution by Azadirachta indica (Neem) leaf powder," *Journal of Hazardous Materials*, vol. 113, no. 1-3, pp. 97–109, 2004.

[40] D. Mohan, P. Singh, A. Sarswat, P. H. Steele, and C. U. Pittman, "Lead sorptive removal using magnetic and nonmagnetic fast pyrolysis energy cane biochars," *Journal of Colloid and Interface Science*, vol. 448, pp. 238–250, 2015.

13

Co-composting of Olive Mill Waste and Wine-Processing Waste: An Application of Compost as Soil Amendment

Z. Majbar ⓘ,[1] K. Lahlou,[1] M. Ben Abbou,[2] E. Ammar,[3] A. Triki,[4] W. Abid,[3] M. Nawdali,[5] H. Bouka,[2] M. Taleb,[1] M. El Haji,[6] and Z. Rais[1]

[1]Laboratory of Engineering, Electrochemistry and Modeling Environment (LEEME), Faculty of Sciences, Fez, Morocco
[2]Laboratory of Natural Resources and Environment, University Sidi Mohammed Ben Abdallah, Faculty Polydisciplinary, Taza, Morocco
[3]Research Unity: Urban and Costal Environments, University of Sfax, National Engineering School of Sfax, Sfax, Tunisia
[4]Research Laboratory: Improvement and Protection of Olive Tree Genetic Resources, Institute of the Olive Tree, Sfax BP1087, Tunisia
[5]Laboratory of Chemistry of Condensed Matter (LCCM), University Sidi Mohammed Ben Abdallah, Faculty of Sciences and Technology, Fez, Morocco
[6]Laboratory Engineering Research-OSIL Team Optimization of Industrial and Logistics Systems, University Hassan II, Superior National School of Electricity and Mechanic (ENSEM), Casablanca, Morocco

Correspondence should be addressed to Z. Majbar; zinebmajbar@gmail.com

Academic Editor: Gassan Hodaifa

In order to decrease the environmental harm produced by the agro industries' wastes', an investigation of the co-composting of olive mill waste (olive mill wastewater (OMW), olive mill sludge (OMS)) and wine by-products (grape marc and winery wastewater) was done. Three aerated windrows of variable compositions were performed; these windrows differ in terms of their initial composition and the liquid used for their humidification; OMW and wastewater winery were used for humidification to replace water for windrow moistening. Moreover, the main physicochemical parameters (temperature, pH, electrical conductivity, and C/N) were monitored to evaluate the co-composting process. The latter lasted around three months. The elaborated composts were characterized by low C/N ratio, and they were rich in fertilizing and nutriment elements and of low heavy metal contents. The humidification of the windrows with OMW showed effectiveness in improving the windrows temperature, reflected by the high temperatures monitored during the composting process in comparison with the windrow humidified with winery wastewater. Furthermore, a longer thermophilic phase was held in windrows carrying OMS. The valorization of the produced composts for soil amendment significantly improved the soil fertility. Indeed, field experiments showed an increase in radish yield by 10%, the composts were harmless and did not have any phytotoxic effect on radish growth.

1. Introduction

The activity of the agro-food industries causes a lot of harm to environment due to the production of huge amounts of waste. In Morocco, the industrial sector generates yearly more than 1.2 million tons of wastes, among which the agro-food industry represents 67% [1]. Consequently, the management and the valorization of these harmful residues are necessary for the environment"s preservation.

The olive oil extraction is an important activity that generates a huge amount of effluent, namely, olive mill wastewater, estimated to 400 000 m^3 per year [2]. The OMW is an acidic and dark liquid effluent, with a high organic matter load and high conductivity [3, 4]; its composition varies both qualitatively and quantitatively according to the olive variety, climate condition, cultivation and harvesting practices, the olive storage time, and the olive oil extraction system [3].

In Morocco, the OMW management practice regulated and adopted by the majority of olive oil industries is the storage in evaporation ponds. However, this technique does not reduce the OMW toxicity since large quantities of olive mill sludge (OMS) are produced. These huge quantities represent a significant problem in olive oil industries; therefore, more effective solutions must be developed to remedy this problem.

The wine industry is another activity that generates significant quantity of waste; this activity produces large amount of by-products that are represented by solid organic residues (grape marc) and wastewater. The production of one hectoliter of wine produces 18 kg of grape marc [5], which gives an annual production of 7200 tons [6]. Still, the quantity of wine wastewater depends largely on the process used in the industry. The different residues from the wine industry are characterized by low pH and electrical conductivity and high organic matter content [7]. The richness of these by-products in organic matter allows their use for soil fertility improvement [8, 9]. The direct incorporation of these residues could cause serious environmental problems if they are added excessively to the soil in an uncontrolled manner. Consequently, their treatment is crucial before their discharge or use for agricultural purposes. The suitable management of these agro-food wastes is an important strategy for the environment protection. Many investigations have focused on the study of treatment and valorization techniques of these residues at an experimental scale [10–16]. At the industrial level, the treatment of these residues is rare, and the problem of all the available wastes has not yet been solved.

Recently, researchers have shown that composting is one of the effective alternatives for the recycling and the valorization of organic wastes [8, 17–22]. It is a degradation process of the organic matter, allowing the achievement of a stable product, rich in humic substances and in fertilizing elements, serving as soil organic amendment [23–25]. The application of compost can have strong ecological environmental values, allowing not only the removal of very expensive chemical fertilizers but also the improvement of the quality of agricultural soils and carbon sequestration [8, 9, 26].

Diverse studies had developed composting of organic waste, such as OMW [19, 24, 27–30], sludge from wastewater treatment [31], and agro-industrial waste [32]. While other studies on the co-composting of olive mill wastewater with poultry manure [25, 27, 33], olive mill wastewater with solid organic waste [20, 34], and date palm with activated sludge [35] have been carried out. The evaluation of the composting process and the study of its environmental impact due to gas emissions have also been the subject of numerous studies [35–38].

Considering the fact that co-composting olive oil waste (OMW and OMS) and wine industry have not gained much attention in previous studies; the aims of this work are as follows: firstly, to valorize the olive mill wastes (OMW and OMS) and by-products wine industry by co-composting; secondly, to study the evolution of the parameters describing the co-composting of mixtures of OM, OMS, and green waste; and lastly, to test the effect of the different composts produced on the performance and yield of radish in the field.

2. Materials and Methods

2.1. Raw Materials. The organic wastes used as substrates for the composting procedure included OMW, OMS, grape marc, winery wastewater, and household waste. The OMW and OMS were collected from a natural evaporation basin of continuous three-phase olive oil extraction unit located in Meknassa Ben Ali (8 km form Taza, Morocco). The basin was made of concrete with a storage capacity of 240 m^3. The collected OMW had a maximum storage time of 30 days; the OMS were collected at the end of August, after a total evaporation of the quantities of water contained in the OMW. The grape marc and the winery wastewater were collected from an industrial company "Celliers de Meknès" located in Meknes (Morocco), and the green waste was gathered from markets located in Taza (Morocco). They are composed mainly of fruit and vegetable residues. The waste physicochemical characterization is presented in Table 1.

2.2. Composting Procedure. The raw materials were mixed in order to be co-composted in open area, using 3 windrows of 0.6 m height, 2 m length, and 1.2 m diameter base [30]. These windrows differed in terms of their initial composition and the liquid used for their humidification (Table 2). The proportions of the raw materials were calculated according to the methods described by Proietti et al. and Soudi [26, 39] to obtain a physicochemical characterization of the initial mixture for starting the composting process.

The moisture content was adjusted to around 50–60% (optimal moisture content for composting); this operation was performed during the turning of the windrows. The windrows were turned once every 3 days at the beginning of the process, then once every 7 days, and once every 15 days for the remaining composting period.

Representative and homogenous samples were collected during the composting procedure from the windrows for analysis. Each sample was obtained by mixing 6 subsamples taken from six different points of windrows, according to ISO 8633 [40].

2.3. Physicochemical Analysis. The raw materials and compost samples were analyzed for pH and electrical conductivity (EC) in a 1:5 (w/v) water soluble extract in [41, 42]. The dry matter content was assessed by drying at 105°C for 24 h using a drying oven type WTB Binder ED 115, and the OM was determined by measuring the loss of ignition at 550°C for 4 h using muffle furnace type Lenton EF11/8B. The organic carbon (OC) was calculated using the following equation:

$$CO(\%) = \frac{\%MO}{1,724}. \qquad (1)$$

The total Kjeldahl nitrogen (TKN) was analyzed using the Kjeldahl method according to AFNOR standard (1981) [43]. The determination of heavy metals and oligoelements were first extracted by heating 2 g of compost with HNO$_3$, and then the filtrate was analyzed using ICP-AES method (inductively coupled plasma atomic emission spectroscopy).

TABLE 1: Physicochemical characteristics of the raw materials used for composting process.

Parameters	OMS	OMW	Grape marc	Winery wastewater	Green waste
pH	5.42 ± 0.03	5.43 ± 0.05	7.42 ± 0.20	6.66 ± 0.02	5.88 ± 0.02
EC (mS·cm^{-1})	8.72 ± 0.02	8.02 ± 0.02	4.81 ± 0.03	0.76 ± 0.02	1.63 ± 0.02
Moisture (%)	22.66 ± 3.69	92.47 ± .0.56	72.42 ± 0.58	99.86 ± 0.01	36.98 ± 4.59
Dry matter (%)	77.34 ± 3.69	7.53 ± 0.56	27.58 ± 0.58	0.14 ± 0.01	63.02 ± 4.59
Mineral matter (%)	15.05 ± 5.91	25.25 ± 5.46	30.19 ± 1.65	45.20 ± 6.63	35.15 ± 1.95
Organic matter (%)	84.95 ± 5.91	74.75 ± 5.46	69.81 ± 1.65	54.80 ± 6.63	64.85 ± 1.95
Organic carbon (%)	49.28 ± 3.43	43.36 ± 3.16	40.49 ± 0.96	31.78 ± 3.84	37.61 ± 1.13
TKN (%)	0.74 ± 0.12	0.86 ± 0.15	0.97 ± 0.10	3.75 ± 0.30	1.53 ± 0.06
C/N	67.77 ± 9.73	ND	42.06 ± 3.47	ND	24.69 ± 1.19
COD (g·O$_2$/l)	ND	96.07 ± 0.5	ND	14.07 ± 0.1	ND
BOD (g·O$_2$/l)	ND	25.67 ± 0.51	ND	14.40 ± 0.4	ND

ND: not determined.

TABLE 2: Composition of the different windrows.

Windrow	A (600 kg)	B (600 kg)	C (1 ton)
Proportion of solid raw materials*			
(i) OMS	1/3	1/3	0
(ii) Grape marc	1/3	1/3	1/2
(iii) Green waste	1/3	1/3	1/2
Humidification effluent	OMW	Winery wastewater	OMW
Total volume used (m^3)	0.8	1.0	1.3

*Proportion expressed in weight/weight.

The characterization of the effluents was performed according to the experimental protocols described by Rodier et al. [44]. pH and electrical conductivity were directly measured using multiparameter consort C335. The chemical oxygen demand (COD) was determined according to NF T90-101 standard [45]. The biological oxygen demand (BOD) was determined using BOD meter-type OxiTop.

The windrow temperature was measured using multi-parameter consort C535 daily during the mesophilic and the thermophilic composting phases and weekly until its maturity. The presented temperature value consists of a mean of 6 measurements at different points and depths of the windrow.

2.4. Phytotoxicity Test.

The compost phytotoxicity was determined by evaluating its aqueous extract on seed germination. The germination test was carried out on 10 seeds of cress (*Lepidium sativum*) experimented at different dilutions in Petri dishes, including filter paper soaked in the compost extract. The test was conducted in dark at 25°C for 72 hours [32, 33]. Three repetitions were performed.

The germination index (GI) was determined considering the number of sprouts and root growth, using the following equation:

$$GI(\%) = \frac{GB}{GT} * \frac{LB}{LT} * 100, \qquad (2)$$

where GB is the number of germinated seeds in the case of aqueous extract, GT is the number of germinated seeds in the case of the control where the distillated water was used, LB is the root length on compost extract, and LT is the root length control.

2.5. Radish Production.

The effects of the different produced composts were assessed on radish (*Raphanus sativus*), of a *National* variety. Nine parcels of 1.5 m^2 arranged in a random design were tested using compost and different combinations of compost mixed with farm manure, according to Table 3. These various combinations of composts and manure were used for the soil amendment at a rate of 6 t/ha; they were incorporated in the soil not only spread on the surface. The parcel, which was only amended with farm manure, was considered as the control. No mineral fertilization has been done in the parcels.

The radish seeds were sowed on 29 November 2015, and the radish was harvested on 24 January 2016. The composts effects were assessed by measuring radish yield at the end of the crop cycle.

2.6. Statistical Analysis.

Statistical analyses of data were made by IBM SPSS Statistics Version 20. The analysis of the means comparison was carried out by Student's *t*-test to compare the evolution profiles of the composting parameters for the different windrows. The Duncan test was used to calculate the variance and to compare the average values of germination indices for the different composts produced.

All analyses were performed in triplicate. However, for agronomic yields, the results were obtained in the form of one replicate due to the limitation of the plots to a single field. Also, minerals and fertilizing elements and heavy metals were in only one replicate because of the limited cost of investigation of the analysis.

3. Results and Discussion

3.1. Composting Process

3.1.1. Temperature Evolution.

During the composting process, the evolution of the temperature in the different windrows is presented in Figure 1. The temperature profiles allow the observation of four conventional stages: a mesophilic, a thermophilic, a cooling, and a maturation stage.

TABLE 3: Treatments used for parcel amendment.

Parcel	P_0	P_1	P_2	P_3	P_4	P_5	P_6	P_7	P_8	P_9
Compost (%)	0	100	100	100	75	75	75	50	50	50
Compost type		C_A	C_B	C_C	C_A	C_B	C_C	C_A	C_B	C_C
Farm manure (%)	100	0	0	0	25	25	25	50	50	50

C_A, C_B, and C_C are the different mature composts prepared, respectively, from the windrows A, B, and C.

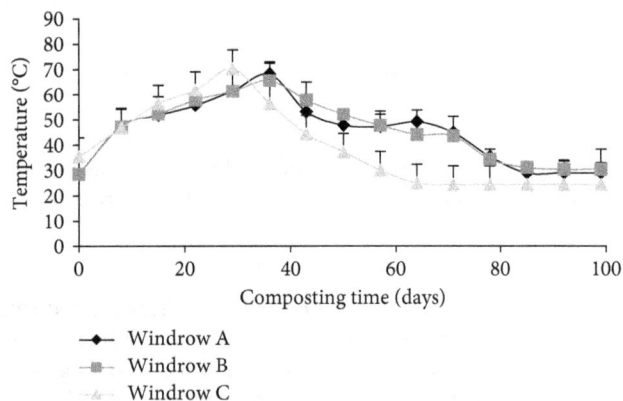

FIGURE 1: The windrows' temperature profiles during the composting process.

In the first stage, an increase of the temperature to 47°C was recorded during the first week. This observation resulted from the rapid colonization and activity of the mesophilic microorganisms, which degraded organic matter and released heat, thus increasing the windrows' temperature. Then, the temperature rose progressively to reach values above 60°C, corresponding to the second stage, the thermophilic phase, and remained approximately 6 weeks. This result was due to the intense microbial activity reflecting high degradation rates occurring during the first stage [20]. The high temperatures may result in the reduction of pathogens and enhanced windrow sanitation [18]. At the 6th week, the third stage began, and the temperature decreased progressively to reach 30°C. These low temperatures were the result of the less intense microbial activity due to the depletion of easily degradable organic matter [21]. Finally, at the maturity stage, the temperature stabilized at 30°C, the process of composting reached its end, and the produced compost was mature. Moreover, the windrow temperature was highly related to the ambient temperature at this stage.

The long phases observed in windrows A and B can be attributed to the initial composition of the mixtures, in particular the presence of OMS. Indeed, the OMS presents a source of organic matter and therefore an availability of additional degradable substances in these mixtures (windrows A and B) contrary to mixture C. Hachicha et al. have found the similar results suggesting that the self-insulating capacity of OMS during the degradation process can lead to a long thermophilic phase [27]. Such a phase reflects an abnormal degradation process and a delayed transition to the stabilization phase [46].

Additionally, these results also show a significant difference between the temperature profile of windrows A and C and windrows B and C, confirmed by the means comparison test by Student's t-test ($p < 0.05$. This difference can be attributed to the activity of microorganisms during the degradation process depending on the initial composition of the substrates.

3.1.2. pH Evolution. The pH evolution of the composted materials followed the same trend in the different windrows (Figure 2). At the beginning of the process, a slight decrease in pH was noted. This result could be explained by the production of organic acids, dissolved CO_2 in the medium and by-products from the degradation of easily biodegradable compounds [46]. Then, an increase in pH from 6.5 to 6.7, from 6.44 to 6.59, and from 6.48 to 7.66 was observed for windrows A, B, and C, respectively. This could be the result of ammonia production from the degradation of amines [35, 47]. Finally, the pH decreased progressively and stabilized at a neutral pH for windrows A and B and alkali pH for C. These results suggest the formation of humic substances which act as buffers, as confirmed by Zenjari et al. and Amir et al. [48, 49].

The comparison between the three pH profiles shows a significant difference ($p < 0.05$) between windrows A and C and windrows B and C. This can be explained by the nature of the initial substrates put to compost. However, the addition of OMW during the composting process influences the pH of the mixtures.

3.1.3. EC Evolution. The progress of the electrical conductivity (EC) during the composting process is presented in Figure 3. During the first weeks of composting process, the EC values increased from 2.11 to 3.42 mS·cm^{-1}, from 2.04 to 3.95 mS·cm^{-1}, and from 2.09 to 3.87 mS·cm^{-1}, respectively, in the windrows A, B, and C. This increase revealed the extent of mineralization of the organic substrate and the release of ions [35]. Then, a progressive decrease in the EC was observed, even in windrows moistened by OMW exhibiting salts loss by the leaching phenomenon as well as by decreasing their extractability because of their fixation on the stabilized organic matter [49].

Moreover, the EC was measured on the extract of organic matter which is sensitive to the solvent extraction ratio and the temperature at which the method was carried out. It could be noticed that each windrow composition presented a particular progress, with significant differences attributed to phenomena of mineralization, leaching, and fixation depending on the composition of the substrates and the activity of microorganisms.

3.1.4. Organic Matter Degradation. The evolution of the organic matter content of the mixture during the composting process is considered as an essential parameter of biodegradation and transformation of organic matter during composting. Figure 4 presents the organic matter contents of the three mixtures at the beginning and end of the

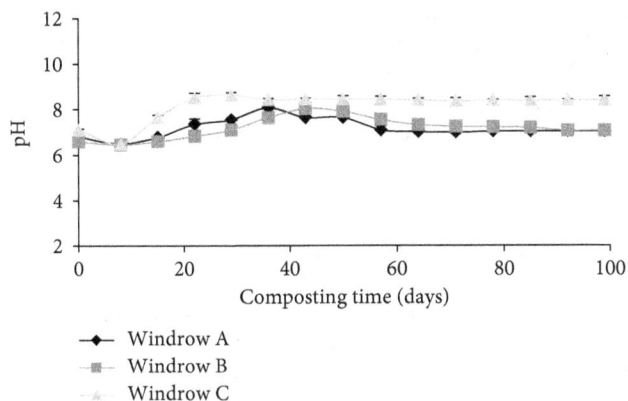

FIGURE 2: pH evolution during the composting process.

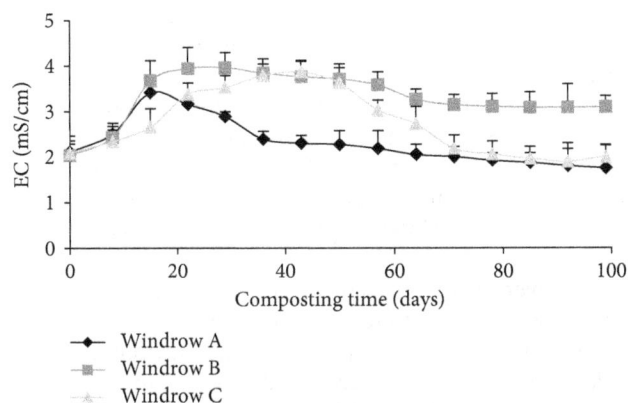

FIGURE 3: EC evolution during the composting process.

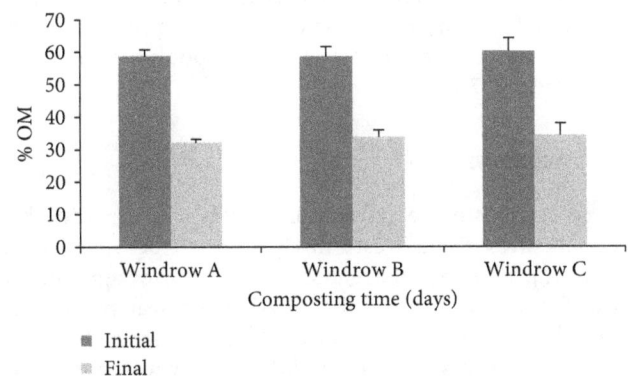

FIGURE 4: Organic matter degradation during the composting process.

composting process. These results show that all the experimented windrows were characterized by a high rate of organic matter (>55%) at the beginning of the composting process. It could be noticed that the windrow C has the highest organic content. During the composting process, the organic matter content gradually decreased, resulting in a degradation rate exceeding 40% for the three mixtures.

This degradation is highlighted during composting by a mass loss of the initial mixtures or even a remarkable reduction in the volumes of co-composted waste. The organic matter decomposition was the result of the microbial activity, allowing the transformation of the organic matter into stable humic substances [50]. The similar results were observed by El Fels et al., indicating a good degree of compostability of the mixtures [35].

Furthermore, no significant difference was shown by Student's t-test between the three windrows. So, one can conclude that the degradation of organic matter does not depend on the composition or nature of the co-composted waste.

3.1.5. C/N Ratio Evolution. The evolution of the C/N ratio is directly related to the biodegradation of organic matter, resulting in the lowering of the total carbon rate associated with the increase in nitrogen concentration. The initial C/N for the three windrows represented values between 25 and 35 (Figure 5). These C/N ratios were ideal for co-composting process [26, 51]. The C/N ratio evolution showed decreases of the ratios from 27.7 to 12.0, from 34.5 to 13.45, and from 34.3 to 12.2 for the windrows A, B, and C, respectively (Figure 5). This is closely related to the loss of organic carbon due to mineralization of the organic material, with a CO_2 production increasing the nitrogen concentration in the windrows during the biodegradation.

At the end of the co-composting, the C/N ratio reached values exceeding 8 for all windrows, indicating the maturity of the final compost [52]. Moreover, according to many authors, the C/N ratio is considered to be a maturity index; generally, the compost is mature if it presents a C/N ratio less than 20 [53–55].

The three profiles of the C/N ratio evolution are of the same speed and represent no significant difference between the three windrows confirmed by Student's t-test ($p > 0.05$).

3.2. Maturity and Quality of the Produced Composts

3.2.1. Characterization of the Produced Composts. The quality of the composts was assessed by comparing the final characteristics of the conceived composts to those of the French standard NF U44-051 (Table 4). The composts produced were neutral, except compost C was characterized by alkaline pH. This result is attributed to the nature of the co-composted waste. Also, they were characterized by C/N ratio around 12, exhibiting their stability.

The comparison of the three produced composts to the French standard showed significant amendment properties. Indeed, the produced composts were rich in fertilizing and mineral elements and low contents of heavy metals. These results have the same meaning with other previous studies valorizing OMW by co-composting, and the produced composts showed significant properties of organic fertilizers compared with the manure [18, 53, 56].

3.2.2. Phytotoxicity. The phytotoxicity of the produced composts was evaluated by the germination test; the result of this test is shown in Figure 6. This figure revealed the effects

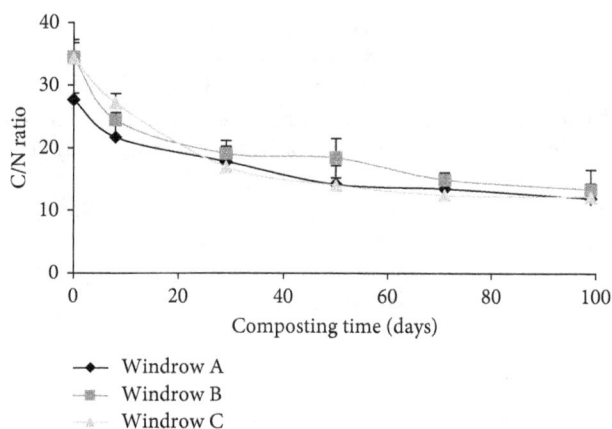

FIGURE 5: The C/N ratio evolution during the composting process.

TABLE 4: Physicochemical characteristics of produced composts.

Parameters	Compost A	Compost B	Compost C	NF U44-051
pH	7.02 ± 0.02	7.09 ± 0.03	8.30 ± 0.02	ND
CE mS·cm^{-1}	1.75 ± 0.02	3.04 ± 0.03	1.92 ± 0.03	ND
C/N	12.00 ± 1.23	13.45 ± 2.08	12.24 ± 1.15	>8
Minerals and fertilizing elements (mg/kg)				
P	855.99	158.37	346.15	ND
K	2763.56	2368.35	2264.55	ND
Mg	897.68	901.90	810.35	ND
Ca	10153.95	9237.85	9760.55	ND
Fe	5414.84	5101.55	6121.70	ND
Na	225.67	259.65	309.45	ND
Mn	49.65	31.65	33.35	ND
Heavy metals (mg/kg)				
Zn	57.83	11.6	12.2	600
Cu	22.26	5.95	6.55	300
Cr	9.35	7.60	5.95	12
Ni	<0.01	0.06	0.06	2
Cd	1.5	1.0	1.2	3
As	<0.01	<0.01	<0.01	18
Se	<0.01	<0.01	<0.01	12

ND: not determined.

of the extracts from produced composts on the germination of cress seeds at different dilutions.

The comparison between the different dilutions shows that a dilution rate of 25% shows the best germination for the different composts produced, marked by a higher rate for the compost and a significant difference between the different compost products C. However, a dilution rate of 50% shows a nonsignificant effect between the different compost products, whereas a dilution of 75% and the crude extract shows significant effects whose compost C represents the best germination index.

These differences can be explained by the nature of the raw materials, by the germination conditions of the watercress seeds and by the high conductivity of the compost extract. Indeed, high values in electrical conductivity inhibit the germination of watercress seeds, which results in low germination indices in raw compost extracts. According to

Lasaridi et al., compost that is too saline can be harmful to plants; acceptable EC values should not exceed 2 to 3 mS/cm [57]. However, the three composts do not represent any phytotoxic effect because all these results showed a germination rate higher than 50%, indicating a maturity of the compost produced [38, 58].

3.2.3. Effect of the Produced Composts on Radish Production.
The radish yields were assessed at the end of the crop cycle, while harvesting. The yields are presented in Figure 7. On farm manure, the yield was equal to 52.5 t/ha, but on the other parcels amended with the composts, the yield ranged from 52.8 t/ha to 57.7 t/ha. The produced composts significantly improved the productivity of the radish crop. However, the highest yields remained in parcels amended with the composts obtained from windrows C and A of 10%. It could be concluded that the OMW used for humidification played an important agronomic role. Indeed, the positive effect of composts based on olive mill waste on radish yields depends particularly on their richness in nutrients and fertilizers element particulates N, P, K, Ca, Mg, and Fe [19, 25]. These elements play a very important role in the growth of plants, thus promoting their vegetative activity, which results in an improvement in radish production yield.

However, Regni et al. suggest that long-term application of composted OMW to the soil improves vegetative activity and olive yield. They confirm that the contribution of nutrients to composted olive mill waste through their amendments associated with other indirect factors such as organic and dynamic matter content, water retention, and microbial biomass activity stimulates olive tree growth and improves olive yield and quality. These results are consistent with those found by Hachicha et al. showing the significant effect of composts based on OMS and poultry manure on potato yield [27].

4. Conclusion

The co-composting of olive mill wastes and the wine byproducts with green waste has proved to be an effective means of producing an organic amendment for agricultural soils. The monitoring of the physicochemical parameters during this process has revealed a good progress of the cocomposting process, a biodegradation of organic matter and a bioconversion of unstable matter into a stable product rich in humic substances. This biotransformation was also confirmed by the phytotoxicity test of the compost extracts produced, which showed that the various composts produced are mature and show no phytotoxic effect.

Moreover, the physicochemical characterization has proved that these composts are of good quality, rich in nutrients and particularly N, P, K, Ca, Mg, and Fe, and conforms to the standards of an organic amendment NF U44-051. The application of compost produced in the fields as organic soil improvers for radish cultivation has had a positive effect by increasing radish production yield by 10% compared to manure. Given the slow and complex dynamics

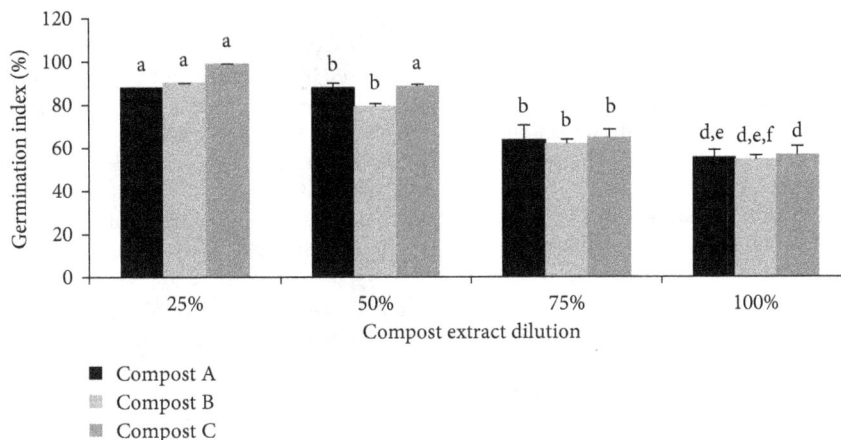

FIGURE 6: Effect of the compost extract on germination of cress seeds at different dilutions ($n = 5$, $p = 0.05$). Different letters (a–f) indicate significant differences ($p < 0.05$) between compost and dilution rates.

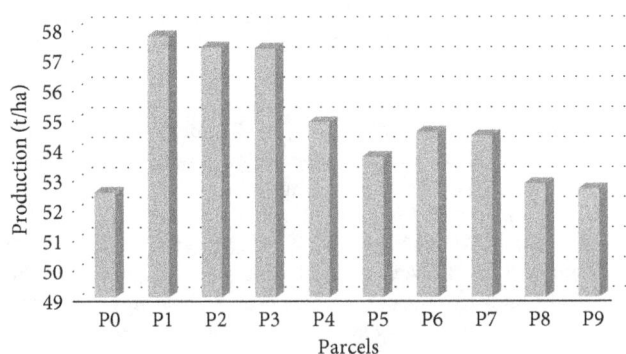

FIGURE 7: Production yield of harvested radish.

of organic substance transformation, long-term amendment experiments would be necessary and will be the subject of further research.

To conclude, one could say that the valorization of olive mill waste and the wine by-products is a promise strategy for the sustainable management of this type of waste allowing the transformation of environmental threats into a valuable product assuring fertility to agricultural soils.

Conflicts of Interest

The authors declare that they have no conflicts of interest.

Acknowledgments

This work was supported by the funds of the PPR project type B (2015–2017) funded by the National Center for Scientific and Technical Research (CNRST). The authors would like to express their deep gratitude to all the professors and members of the doctoral research center who contributed to the improvement of this article. They extend their thanks to all of them for their advice, knowledge, insightful criticisms, and suggestions. Many thanks are also expressed to the persons who provided the material and made their laboratories available whenever needed.

References

[1] N. Perkins, A. Ajir, and L. El Ouazzani, *Rapport sur la Gestion des Déchets Solides au Maroc*, Scribd, San Francisco, CA, USA, 2014.

[2] M. Rahmani, *Examen National de l'Export vert du Maroc: Produits Oléicoles, Romarin et Thym*, United Nations, Geneva, 2017.

[3] S. Dermeche, M. Nadour, C. Larroche, F. Moulti-Mati, and P. Michaud, "Olive mill wastes: biochemical characterizations and valorization strategies," *Process Biochemistry*, vol. 48, no. 10, pp. 1532–1552, 2013.

[4] M. Ben Abbou, R. Rheribi, M. El Haji, Z. Rais, and M. Zemzami, "Effect of using olive mill wastewater by electrocoagulation process on the development and germination of tomato seeds," *Physical and Chemical News*, vol. 74, no. 10, pp. 37–43, 2014.

[5] Institut Français de la Vigne et de Vin, *Marcs de Raisins, Lies de Vin et Bourbes : Quelle Gestion des Sous-Produits Vinicoles?*, Institut Français de la Vigne et de Vin, Le Grau-du-Roi, France, 2013.

[6] Direction du Développement Des Filières de Production, *Ministère de l'Agriculture, de la Pêche Maritime, du Développement Rural et des Eaux et des Forêts Rabat*, 2016.

[7] M. A. Bustamante, R. Moral, C. Paredes, A. Pérez-Espinosa, J. Moreno-Caselles, and M. D. Pérez-Murcia, "Agrochemical characterisation of the solid by-products and residues from the winery and distillery industry," *Waste Management*, vol. 28, no. 2, pp. 372–380, 2008.

[8] L. Regni, L. Nasini, L. Ilarioni et al., "Long term amendment with fresh and composted solid olive mill waste on olive grove affects carbon sequestration by prunings, fruits, and soil," *Frontiers in Plant Science*, vol. 7, 2017.

[9] P. Proietti, E. Federici, L. Fidati et al., "Effects of amendment with oil mill waste and its derived-compost on soil chemical and microbiological characteristics and olive (*Olea europaea* L.) productivity," *Agriculture, Ecosystems and Environment*, vol. 207, pp. 51–60, 2015.

[10] E. Bertran, X. Sort, M. Soliva, and I. Trillas, "Composting winery waste: sludges and grape stalks," *Bioresource Technology*, vol. 95, no. 2, pp. 203–208, 2004.

[11] F. Hanafi, O. Assobhei, and M. Mountadar, "Detoxification and discoloration of Moroccan olive mill wastewater by electrocoagulation," *Journal of Hazardous Materials*, vol. 174, no. 1–3, pp. 807–812, 2010.

[12] N. Hytiris, I. E. Kapellakis, R. L. R. de, and K. P. Tsagarakis, "The potential use of olive mill sludge in solidification process," *Resources, Conservation and Recycling*, vol. 40, no. 2, pp. 129–139, 2004.

[13] M. Madani, M. Aliabadi, B. Nasernejad, R. K. Abdulrahman, M. Y. Kilic, and K. Kestioglu, "Treatment of olive mill wastewater using physico-chemical and Fenton processes," *Desalination and Water Treatment*, vol. 53, no. 8, pp. 2031–2040, 2015.

[14] M. Mosca, F. Cuomo, F. Lopez, G. Palumbo, G. Bufalo, and L. Ambrosone, "Adsorbent properties of olive mill wastes for chromate removal," *Desalination and Water Treatment*, vol. 54, no. 1, pp. 275–283, 2015.

[15] S. P. Tsonis, V. P. Tsola, and S. G. Grigoropoulos, "Systematic characterization and chemical treatment of olive oil mill wastewater," *Toxicological and EnvironmentalChemistry*, vol. 20-21, no. 1, pp. 437–457, 1989.

[16] A. Zirehpour, A. Rahimpour, and M. Jahanshahi, "The filtration performance and efficiency of olive mill wastewater treatment by integrated membrane process," *Desalination and Water Treatment*, vol. 53, no. 5, pp. 1254–1262, 2015.

[17] L. E. Fels, M. Hafidi, and Y. Ouhdouch, "Date palm and the activated sludge co-composting actinobacteria sanitization potential," *Environmental Technology*, vol. 37, no. 1, pp. 129–135, 2016.

[18] S. Hachicha, F. Sallemi, K. Medhioub, R. Hachicha, and E. Ammar, "Quality assessment of composts prepared with olive mill wastewater and agricultural wastes," *Waste Management*, vol. 28, no. 12, pp. 2593–2603, 2008.

[19] S. Masmoudi, R. Jarboui, H. E. Feki, T. Gea, K. Medhioub, and E. Ammar, "Characterization of olive mill wastes composts and their humic acids: stability assessment within different particle size fractions," *Environmental Technology*, vol. 34, no. 6, pp. 787–797, 2013.

[20] C. Paredes, A. Roig, M. P. Bernal, M. A. Sánchez-Monedero, and J. Cegarra, "Evolution of organic matter and nitrogen during co-composting of olive mill wastewater with solid organic wastes," *Biology and Fertility of Soils*, vol. 32, no. 3, pp. 222–227, 2000.

[21] L. Zhang and X. Sun, "Influence of bulking agents on physical, chemical, and microbiological properties during the two-stage composting of green waste," *Waste Management*, vol. 48, pp. 115–126, 2016.

[22] L. Zhang and X. Sun, "Changes in physical, chemical, and microbiological properties during the two-stage co-composting of green waste with spent mushroom compost and biochar," *Bioresource Technology*, vol. 171, pp. 274–284, 2014.

[23] C. S. Akratos, A. G. Tekerlekopoulou, I. A. Vasiliadou, and D. V. Vayenas, "Chapter 8-Cocomposting of olive mill waste for the production of soil amendments," in *Olive Mill Waste*, C. M. Galanakis, Ed., pp. 161–182, Academic Press, Cambridge, MA, USA, 2017.

[24] S. Hachicha, M. Chtourou, K. Medhioub, and E. Ammar, "Compost of poultry manure and olive mill wastes as an alternative fertilizer," *Agronomy for Sustainable Development*, vol. 26, no. 2, pp. 135–142, 2006.

[25] F. Sellami, R. Jarboui, S. Hachicha, K. Medhioub, and E. Ammar, "Co-composting of oil exhausted olive-cake, poultry manure and industrial residues of agro-food activity for soil amendment," *Bioresource Technology*, vol. 99, no. 5, pp. 1177–1188, 2008.

[26] P. Proietti, R. Calisti, G. Gigliotti, L. Nasini, L. Regni, and A. Marchini, "Composting optimization: integrating cost analysis with the physical-chemical properties of materials to be composted," *Journal of Cleaner Production*, vol. 137, pp. 1086–1099, 2016.

[27] S. Hachicha, F. Sellami, J. Cegarra et al., "Biological activity during co-composting of sludge issued from the OMW evaporation ponds with poultry manure—physico-chemical characterization of the processed organic matter," *Journal of Hazardous Materials*, vol. 162, no. 1, pp. 402–409, 2009.

[28] N. Abid and S. Sayadi, "Detrimental effects of olive mill wastewater on the composting process of agricultural wastes," *Waste Management*, vol. 26, no. 10, pp. 1099–1107, 2006.

[29] I. Aviani, Y. Laor, S. Medina, A. Krassnovsky, and M. Raviv, "Co-composting of solid and liquid olive mill wastes: management aspects and the horticultural value of the resulting composts," *Bioresource Technology*, vol. 101, no. 17, pp. 6699–6706, 2010.

[30] Z. Majbar, Z. Rais, M. El Haji, M. B. Abbou, H. Bouka, and M. Nawdali, "Olive mill wastewater and wine by-products valorization by co-composting," *Journal of Materials and Environmental Science*, vol. 8, no. 9, pp. 3162–3167, 2017.

[31] K. Lahlou, M. Ben Abbou, Z. Majbar et al., "Recovery of sludge from the sewage treatment plant in the city of Fez (STEP) through the composting process," *Journal of Materials and Environmental Sciences*, vol. 8, no. 12, pp. 4582–4590, 2017.

[32] D. Raj and R. S. Antil, "Evaluation of maturity and stability parameters of composts prepared from agro-industrial wastes," *Bioresource Technology*, vol. 102, no. 3, pp. 2868–2873, 2011.

[33] W. Abid, E. Ammar, M. A. Triki, M. Ben Abbou, and M. El Haji, "Gestion et valorisation des margines par co compostage avec les déchets verts et amendements des sols agricoles pour l'amélioration des rendements," *Brevet déposé à l'Office Marocain de la Propriété Industrielle et Commerciale*, MA 20150445 A1, 2015.

[34] C. Paredes, M. P. Bernal, J. Cegarra, and A. Roig, "Bio-degradation of olive mill wastewater sludge by its co-composting with agricultural wastes," *Bioresource Technology*, vol. 85, no. 1, pp. 1–8, 2002.

[35] L. El Fels, M. Zamama, A. El Asli, and M. Hafidi, "Assessment of biotransformation of organic matter during co-composting of sewage sludge-lignocelullosic waste by chemical, FTIR analyses, and phytotoxicity tests," *International Biodeterioration and Biodegradation*, vol. 87, pp. 128–137, 2014.

[36] L. E. Fels, M. Zamama, A. Aguelmous et al., "Assessment of organo-mineral fraction during co-composting of sewage sludge-lignocellulosic waste by XRD and FTIR analysis," *Moroccan Journal of Chemistry*, vol. 5, no. 4, pp. 730–739, 2017.

[37] L. El Fels, F. Z. El Ouaqoudi, L. Lemee et al., "Identification and assay of microbial fatty acids during co-composting of active sewage sludge with palm waste by TMAH-thermochemolysis coupled with GC-MS," *Chemistry and Ecology*, vol. 31, no. 1, pp. 64–76, 2015.

[38] L. Nasini, G. De Luca, A. Ricci et al., "Gas emissions during olive mill waste composting under static pile conditions,"

International Biodeterioration and Biodegradation, vol. 107, pp. 70–76, 2016.

[39] B. Soudi, "Le compostage des déchets des cultures sous serre et du fumier," *Bullletin Mensuel d'Information et de Liaison du PNTTA*, vol. 129, pp. 1-6, 2005.

[40] ISO 8633, *Solid Fertilizers-Simple Samplingmethod for Small Lots*, ISO, Geneva, Switzerland, 1992, https://www.iso.org/standard/15994.html.

[41] International Organization for Standardization, *ISO 10390-Soil Quality-Determination of pH*, ISO, Geneva, Switzerland, 2005, https://www.iso.org/standard/40879.html.

[42] International Organization for Standardization, *ISO 11265-Soil Quality-Determination of the Specific Electrical Conductivity*, ISO, Geneva, Switzerland, 1994, https://www.iso.org/standard/19243.html.

[43] French Association for Standardization, *Determination of Nitrogen Kjeldahl-Titrimetric Determination Method after Mineralization and Distillation*, NF T90–110, French Association for Standardization, France, 1981.

[44] J. Rodier, B. Legube, N. Merlet, and R. Brunet, L'Analyse de l'Eau, Dunod, Paris, France, 9th edition, 2009, https://www.fichier-pdf.fr/2012/12/08/l-analyse-de-l-eau-9e-edition-entierement-mise-a-jour-jean-rodie/l-analyse-de-l-eau-9e-edition-entierement-mise-a-jour-jean-rodie.pdf.

[45] NF T90-101, "Water quality-determination of chemical oxygen demand (COD)-qualité de l'eau," 2001, https://www.boutique.afnor.org/norme/nf-t90-101/qualite-de-l-eau-determination-de-la-demande-chimique-en-oxygene-dco/article/778836/fa111555.

[46] C. Tognetti, M. J. Mazzarino, and F. Laos, "Improving the quality of municipal organic waste compost," *Bioresource Technology*, vol. 98, no. 5, pp. 1067–1076, 2007.

[47] L. Baeta-Hall, M. CéuSàágua, M. Lourdes Bartolomeu, A. M. Anselmo, and M. Fernanda Rosa, "Bio-degradation of olive oil husks in composting aerated piles," *Bioresource Technology*, vol. 96, no. 1, pp. 69–78, 2005.

[48] B. Zenjari, H. El Hajjouji, G. AitBaddi et al., "Eliminating toxic compounds by composting olive mill wastewater–straw mixtures," *Journal of Hazardous Materials*, vol. 138, no. 3, pp. 433–437, 2006.

[49] S. Amir, M. Hafidi, G. Merlina, and J.-C. Revel, "Sequential extraction of heavy metals during composting of sewage sludge," *Chemosphere*, vol. 59, no. 6, pp. 801–810, 2005.

[50] C. Francou, *Stabilisation de la Matière Organique au Cours du Compostage de Déchets Urbains: Influence de la Nature des Déchets et du Procédé de Compostage-Recherche d'indicateurs Pertinents*, INAPG (AgroParisTech), Paris, France, 2003, https://pastel.archives-ouvertes.fr/pastel-00000788/.

[51] J. Singh and A. S. Kalamdhad, "Assessment of compost quality in agitated pile composting of water hyacinth collected from different sources," *International Journal of Recycling of Organic Waste in Agriculture*, vol. 4, no. 3, pp. 175–183, 2015.

[52] NF U44-051, "Organic soil improvers-designations, specifications and marking," 2006, https://www.boutique.afnor.org/standard/nf-u44-051/organic-soil-improvers-designations-specifications-and-marking/article/686933/fa125064.

[53] J. A. Alburquerque, J. Gonzálvez, G. Tortosa, G. A. Baddi, and J. Cegarra, "Evaluation of "alperujo composting based on organic matter degradation, humification and compost quality," *Biodegradation*, vol. 20, no. 2, pp. 257–270, 2009.

[54] J. A. Alburquerque, J. Gonzálvez, D. García, and J. Cegarra, "Measuring detoxification and maturity in compost made from "alperujo," the solid by-product of extracting olive oil by the two-phase centrifugation system," *Chemosphere*, vol. 64, no. 3, pp. 470–477, 2006.

[55] A. K. M. Muktadirul Bari Chowdhury, C. S. Akratos, D. V. Vayenas, and S. Pavlou, "Olive mill waste composting: a review," *International Biodeterioration and Biodegradation*, vol. 85, pp. 108–119, 2013.

[56] F. Sellami, S. Hachicha, M. Chtourou, K. Medhioub, and E. Ammar, "Bioconversion of wastes from the olive oil and confectionary industries: spectroscopic study of humic acids," *EnvironmentalTechnology*, vol. 28, no. 11, pp. 1285–1298, 2007.

[57] K. Lasaridi, I. Protopapa, M. Kotsou, G. Pilidis, T. Manios, and A. Kyriacou, "Quality assessment of composts in the Greek market: the need for standards and quality assurance," *Journal of Environmental Management*, vol. 80, no. 1, pp. 58–65, 2006.

[58] M. Chikae, R. Ikeda, K. Kerman, Y. Morita, and E. Tamiya, "Estimation of maturity of compost from food wastes and agro-residues by multiple regression analysis," *Bioresource Technology*, vol. 97, no. 16, pp. 1979–1985, 2006.

14

Aluminum in Effluents and Sludges from the Preliminary Coagulation in Dairy Factory Aerated Balancing Tanks: An Analytical and Sorption Study

Michał Sadowski ⓘ, Piotr Anielak ⓘ, and Wojciech M. Wolf

Institute of General and Ecological Chemistry, Faculty of Chemistry, Lodz University of Technology, 116 Zeromskiego Street, 90-924 Lodz, Poland

Correspondence should be addressed to Michał Sadowski; michal.sadowski@edu.p.lodz.pl

Academic Editor: Hassan Arida

The coagulation process is used as the first step for the chemical treatment of liquid waste. Resulting wet sludge is treated with coagulants or polyelectrolytes to improve dewatering characteristics of its so-called conditioning. The coagulants such as aluminum sulphate or polyaluminum chloride $(Al_2Cl_n(OH)_{6-n})$ are widely used for dairy wastewater treatment systems. The pretreatment dairy sludge can be applied in agriculture as fertilizer containing valuable nutrients. Hence, a simple, cost effective, and rapid method for the determination of aluminum content in the sludge is essential for calculation of the appropriate coagulant dose at the sludge pretreatment process. In this paper either colorimetric or atomic absorption spectrometric methods as applied for the determination of aluminum concentration in the dairy wastewater are compared. For colorimetric method, the optimum experimental conditions such as pH, reaction time, and concentration of other ions in the sample were determined. The sorption isotherms of aluminum on the activated sludge were determined for diverse aluminum species.

1. Introduction

Dairy manufacturing has become one of the fastest growing and most profitable sectors of food industry in Poland. Domestic milk production approached 12.9 mln liters in 2015 [1]. Unfortunately, dairy farms and factories are also one of the largest sewage producers in Poland. Assuming that each $1 m^3$ of processed milk provides almost $3.2 m^3$ of wastewater, it can be extrapolated that total amount of dairy sewage in 2015 approached 41.3 mln liters [2].

The coagulation process is widely used as the first step for chemical treatment of liquid waste. Resulting wet sludge is treated with coagulants or polyelectrolytes in a process known as conditioning to improve its dewatering characteristics. The coagulants such as aluminum sulphate or polyaluminum chloride are commonly used for dairy wastewater treatment systems to enhance sedimentation rate of organic matter. The European Union promotes usage of the sewage sludge in agriculture [3]. The utilization by the

thermal treatment i.e., pyrolysis, gasification, or combustion is also supported [4]. This also concerns aluminum ions because in relatively low concentrations, they may become toxic either to plants or to the aquatic environment. Aluminum and its compounds can interact with metals and nonmetals leading to changes in the biological availability of essential elements which are necessary for the proper functioning of living organisms. Dewatered dairy sludge is an important source of inorganic and organic nutrients and may be used in agriculture for the soil fertilization and improvement. Therefore, a simple, cost effective, and rapid method for the determination of aluminum content in the sludge is essential for calculation of the appropriate coagulant dose at the sludge pretreatment process. It may also be useful for controlling the final aluminum level just before the soil fertilization.

For many years, the basic analytical method being used to determine the concentration of aluminum in industrial wastewater was the colorimetric analysis. It is based on the well-recognized ability of aluminum to form colourful complexes with organic compounds such as eriochrome cyanine R (3″-sulfo-3,3′-dimethyl-4-hydroxyfuchson-5,5′-dicarboxylic (ECR)) (Figure 1), catechol violet [5], or aluminon [6].

Nowadays, modern analytical techniques such as atomic absorption spectrometry (AAS) and inductively coupled plasma spectrometry (ICP) have challenged classic colorimetric analysis. Both methods are quite expensive and require considerable expertise. In this study, aluminum concentration in wastewater has been determined by modified spectrophotometric method which is characterized by good precision, high reproducibility, low cost, and simplicity. Flame and graphite furnace atomic absorption spectrometries [7] were used as reference methodologies.

Aluminum is an amphoteric element with the lowest solubility at neutral pH, when it precipitates as aluminum hydroxide Al(OH)$_3$. Its solubility increases in either acidic or alkaline solutions. Aluminum can form complexes with ECR giving stable compounds such as Al-ECR$_3$ or Al-ECR$_2$OH in the range of pH 5 to 6 [8] (Figure 2). Therefore, during the analysis, it is necessary to take into account the impact of solution pH on the structure of aluminum-ECR complexes.

An important issue affecting the metals' determination in complex samples is the interfering effect of matrix components, especially in highly saline samples, including seawaters, dialysis solutions, or mineralisates. This matrix effect can be corrected by the careful sample mineralization and the standard addition calibration method. The pH optimization for the signal to noise enhancement is mandatory [9].

2. Experimental

2.1. The Dairy Sewage Materials. The examined sewage was generated in the production of cottage cheese in the dairy factory located in Lodz region. Postproduction sewage is drained to the factory sewage pretreatment plant, which consists of three aerated balancing tanks followed by the sedimentation tank. Aluminum coagulants were added to the former to keep the pH at the 6.5 level. Wastewater was discharged into the municipal sewage system after the four-day treatment time. Samples were collected from balancing tanks in a two-week period over the whole 2017 year.

2.1.1. The pH Optimization. The experimental studies were carried out to get the information about impact of acetate buffer concentration and pH value on spectrophotometric ECR method for determination of aluminum content in dairy sewage samples. The correct solution pH was maintained by using acetate buffer. The various concentrations of acetate buffer as well as different ratios of buffer components were investigated. The buffer was prepared by mixing CH$_3$COONa·3H$_2$O with CH$_3$COOH (Chempur, Piekary Sl., Poland).

2.1.2. Measurements of the UV-Vis Spectra. Maximum absorbance wavelength for both eriochrome cyanine R and

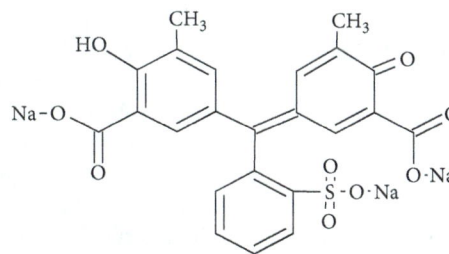

FIGURE 1: The chemical structure of eriochrome cyanine R (ECR).

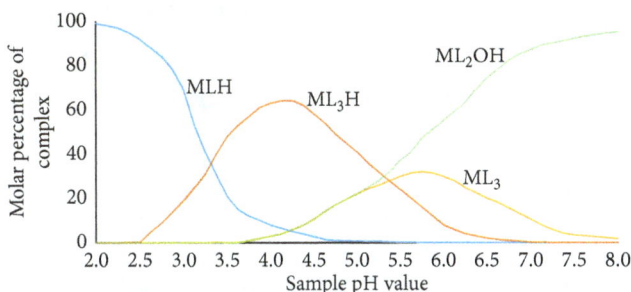

FIGURE 2: The types of complexes Al^{3+}-ECR depending on solution pH, $\mu = 0.1$ M KNO$_3$ [8].

complex of Al-ECR was determined. The ECR was supplied by POCH (Gliwice, Poland). The spectra were measured on a UV-Vis spectrophotometer (2041-PC, Shimadzu, Japan). The reference sample was deionized water, and the analyzed samples contained aluminum in the form of either KAl(SO$_4$)$_2$·12H$_2$O (Chempur) or polyaluminum chloride (Kemira PAX) at concentration of 7.4·10^{-2} mM. The samples were prepared in accordance with the procedure for determining the aluminum content in environmental samples as described in the Polish Standard [10].

2.1.3. Mineralization of the Samples. Due to the high content of organic compounds in the samples, the dried samples of the dairy sludge were mineralized before the aluminum content was determined. The each sludge mineralization was made by closed microwave reaction system Multiwave 300 (Anton Paar, USA). The average mass of the tested samples was about 0.1 g. The mixture of 6 : 1 mL (86.6 mM HNO$_3$ (Chempur) and 11.7 mM HCl (Chempur)) was applied. In the next step, the mineralisates were filled up to 100 ml with deionized water. The same acid mixture was also added in the course of preparing the calibration curves for mineralisates. The oxidative effect of HNO$_3$/HCl mixture on absorbance of emergent complex Al-ECR$_3$ was also investigated. The acid mixture concentration used for calibration curves was gradually increased from 0.1 mM to 10 mM. The constant aluminum and the ECR doses (7.4·10^{-2} and 3.73·10^{-2} mM, respectively) were applied.

2.1.4. Measurements of the Absorption Colorimetric Complex. The absorption intensity of the tested aluminum complex was measured using Spekol 11 colorimeter (Carl-Zeiss, Jena,

Germany). Samples for analysis were prepared in accordance with the procedure described in Polish Standards [10], taking into account the differences in concentrations of acetate buffer dependent on type of sample physical state and organic matter content. When determining the Al concentration in samples of nonmineralized supernatant liquid, sulfuric acid was used (0.01 M), while in other samples, it was used a mixture of nitric acid and hydrochloric acid (0.05 M). All pH value measurements were made by a pH meter (Model N-517, Mera-Elwro, Poland).

In colorimetric method, it is necessary to prepare calibration curves before every analysis. It is the simplest way to reduce influence of random errors on the analysis. For aluminum determination in concentration range of $0.01-0.50\,mg\cdot L^{-1}$ was prepared a calibration curve following the procedure as described in Hermanowicz et al. [11]. The calibration solutions were obtained by dissolving $KAl(SO_4)_2\cdot 12H_2O$ (Chempur) in double distilled water. The absorbance was measured at 535 nm against a blank solution containing only the reagents used in analysis.

The formation of the $Al(ECR)_3$ complex takes place in pH range 5.0–6.0. The pH value of supernatant liquor was adjusted to the optimum range by adding sulfuric acid (0.01 M), according to the Polish Standard [10] and Hermanowicz et al. [11]. The ECR concentration in all studies was kept constant at $3.73\cdot 10^{-2}\,mM$ [12].

Based on earlier studies on the Al-ECR complex stability [6] as well as our own analyses, it was concluded that the most intensive signal appeared in 10 min after adding the eriochrome reagent. This time was applied to all investigated samples.

2.1.5. The Filtered Supernatant Liquid. In the first step, dairy slurry was homogenized by magnetic stirrer and then was filtered through the medium filter paper. The aluminum residual content in filtrate was determined by graphite furnace atomic adsorption spectrometry (GFAAS). The aluminum concentration never exceeded the value of $0.015\,mg\cdot L^{-1}$. The difference between the blank sample containing the supernatant and the blank sample for deionized water was subtracted from analysis data.

2.2. Atomic Absorption Spectrometry. The basic methods to determining aluminum content in sewage sludge are instrumental techniques mainly based on atomic absorption spectrometry. The Polish Standard [7] recommends using the flame atomic absorption spectrometry (FAAS) for determining the concentration of aluminum in the range from $5\,mg\cdot L^{-1}$ to $100\,mg\cdot L^{-1}$. The content of aluminum was analyzed by Thermo Elemental SOLAAR S2 Flame AA Spectrometer (Thermo Scientific, Waltham, USA). Studies have shown that at concentrations above $5\,mg\cdot L^{-1}$, this method is completely satisfactory. At low concentrations, less than $4\,mg\cdot L^{-1}$, the accuracy of the assay is unsuitable, and SD value reaches exceeding 5%. The graphite furnace atomic absorption spectrometry (GFAAS) is recommended for small or trace aluminum contents, from 0.01 to

$0.1\,mg\cdot L^{-1}$ [7]. GFAAS spectrometer Sensa AA (GBC Scientific Equipment, Braeside, VIC, Australia) was applied. The introduced division is applied to analyze both sewage sludge samples and supernatant liquid.

Graphite furnace atomic absorption spectrometry is a method based on atomization of aluminum as a result of heating the graphite tube. The nitrous oxide-acetylene flame is a oxidant/fuel combination, which has a maximum temperature of about 2900°C [13]. The liquid aluminum sample was heated stepwise (up to 2500°C for 5.0 seconds) to evaporate [14]. Aluminum determination was made at an excitation wavelength of 309.35 nm, at a slit width of 0.5 nm [7]. All atomic absorption measurements were made, using a single-element aluminum hollow-cathode lamp (Photron, Narre Warren, Australia). The GFAAS calibration curves were prepared by adequate dilutions of a $1000\,mg\cdot L^{-1}$ aluminum standard solution for ASS (Merck, Germany) in range 0.01; 0.05; $0.10\,mg\cdot L^{-1}$.

The main disadvantage in GFAAS aluminum analysis is matrix interference and the formation of refractory carbides [15]. From among three chemical modifiers, i.e., phosphoric acid (Chempur) [16], $Mg(NO_3)_2$ (Merck) [17], and $Pd(NO_3)_2$ (Merck), the last was chosen. The mixture of $Pd(NO_3)_2/Mg(NO_3)_2$ as a chemical modifier is recommended too [18, 19]. $Pd(NO_3)_2$ can be used to stabilise Al ions at higher pyrolysis temperature [20]. Palladium modifier gives correct results of aluminum content, when the samples have low silica content [21]. The interference of easily decomposing chlorides was eliminated by a nitric acid acidification of the sample [22].

3. Results and Discussion

3.1. The Colorimetric Method. Maximum absorbance wavelength was determined for both eriochrome cyanine R and Al-ECR complex. As shown in Figure 3, the absorption spectrum of ECR has maximum absorbance at wavelength of 440 nm, and for Al-ECR complex, the absorption peak is shifted to longer wavelengths (535 nm).

3.2. The Buffer Capacity of the System. Following Hermanowicz et al. [11], the acetate buffer at concentration 0.20 M was used to obtain pH = 6.15, but the standard method for spectrophotometric determination of aluminum recommends the usage of stronger acetate buffer of 0.46 M, which gives pH = 6.35 [10]. This issue is of particular importance when applied strongly acidic mixtures in the microwave mineralization when the capacity of the weaker acetate buffer was practically exhausted (Figure 4).

The acid concentration above 6 mM leads to decomposition of the chromogenic reagent if a weaker acetate buffer was used. The stronger buffer solution of 0.46 M was sufficient to neutralize the damaging acid impact.

The solutions' pH values in this analysis were within range from 6.7 to 2.5, and the maximum solution absorbance occurred at pH of 6.1 when the most stable complexes between aluminum and triple molecules ECR-like ligands are formed [8]. The absorbance decreases with increase of

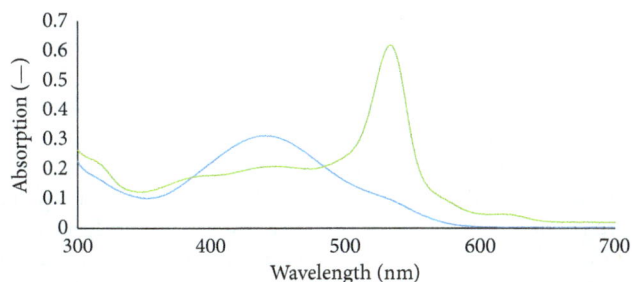

FIGURE 3: Spectra of eriochrome cyanine R (blue) and Al-ECR complexes (green) against distilled water with the following conditions: pH, 5.20; acetate buffer, 0.2 M; Al^{3+}, $7.4 \cdot 10^{-2}$ mM.

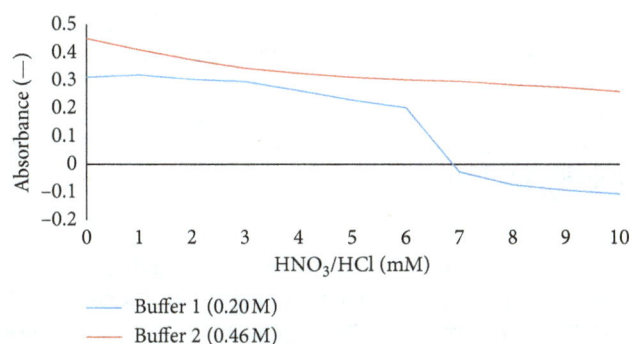

FIGURE 5: The pH effect on Al-ECR$_3$ signal intensity: acetate buffer (0.2 M Al^{3+} ($7.4 \cdot 10^{-2}$ mM)).

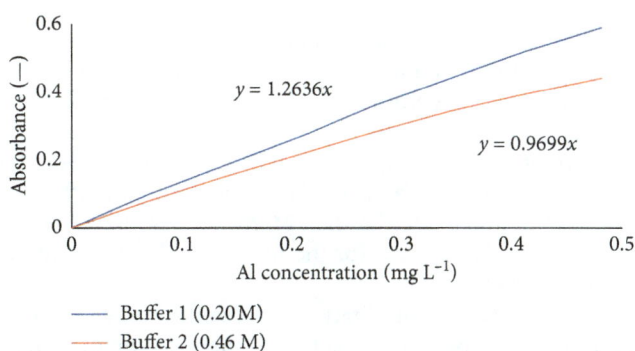

FIGURE 4: The effect of acid concentration on absorbance.

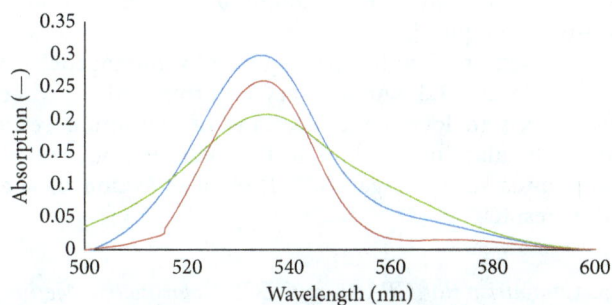

FIGURE 6: Influence of acetate buffer concentrations on absorbance.

the hydrogen ion concentration, and finally the oxidation of ECR occurs at the pH value 3.5–3.7 (Figure 5).

Figure 6 shows the influence of acetate buffer concentration on the absorbance of the Al(ECR)$_3$ complex. The concentration of a HCl and HNO$_3$ mixture was fixed at 1.74 mM for all analysis in this experiment.

The buffer at concentration of 0.20 M gives a higher absorbance values, despite lower pH of the sample, i.e., 5.17 ± 0.03. The stronger buffer (0.46 M) allows to obtain pH value of 5.55 ± 0.05. However, its greater ionic strength disturbs the ECR complex formation and, in effect, reduces the sensitivity of absorbance measurements.

As shown in Figure 7, the buffer concentration has an impact on the absorbance value, but has no influence on the optimum wavelength for particular measurement. The red line corresponds to the high pH and high concentration of the acetate buffer (0.46 M), while the blue and green lines represent absorbance obtained with the 0.20 M buffer.

3.3. The Organic Matter in ECR Colorimetric Method. Three calibration curves were prepared by diluting the standard aluminum solution in the form KAl(SO$_4$)$_2$·H$_2$O. The first curve (series 1) was made according to the Polish Standard [10], in a range of aluminum concentrations from 0.07 mg·L^{-1} to 0.49 mg·L^{-1}. The second curve (series 2) was prepared in a similar way but instead deionized water, 20 ml of supernatant liquid was added. The third curve (series 3) was based on the supernatant liquid and treated by mineralization according to the Polish Standard [10], where

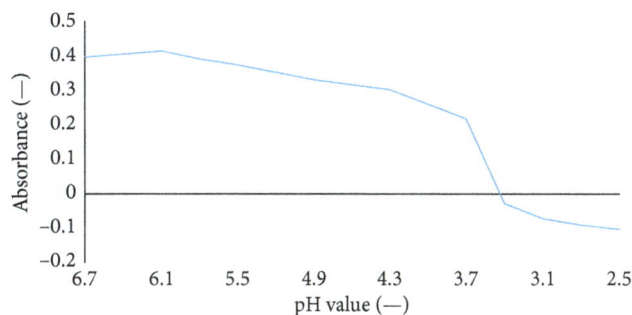

FIGURE 7: Spectra of Al-ECR complexes at various concentrations of acetate buffer and pH, in relation to blank probe: blue, 5.2; orange, 5.5; green, 4.9; Al^{3+} ($7.4 \cdot 10^{-2}$ mM).

oxidation acids were evaporated at the end. The pH of samples was adjusted to value 6.3 by addition of acetate buffer.

Figure 8 shows differences between the slopes of calibration curves. In real samples, the organic matrix components interact with aluminum. The compounds of aluminum and organic matter have often greater stability than the complex of aluminum and the eriochrome cyanine R [14]. Additionally, there was the intensification of interactions of different ions, which appeared in samples as a consequence of mineralization process. As a result, there is a decrease in the signal of Al(ECR)$_3$ complex. The absorbance signals of curve 2 and 3 were lower against the standard calibration curve 1.

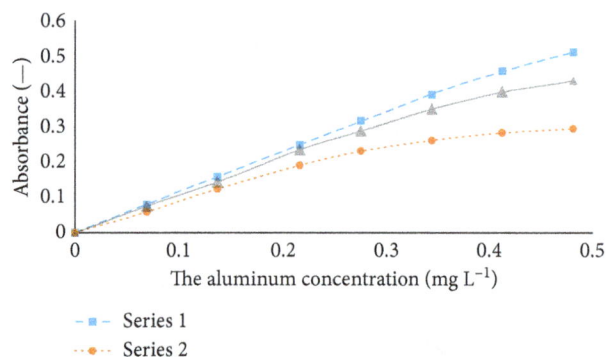

FIGURE 8: Dependence between absorbance and aluminum concentration.

3.4. The Graphite Furnace Atomic Adsorption Spectroscopy. Determination of aluminum content in samples by GFAAS brings a lot of difficulties in implementing the analysis. The first experimental problem is the purity of reagents used to measurements. Only high analytical grade reagents could be applied to avoid interferences of different ions and influence of aluminum contained in that. Many analyses have shown that this problem occurs for the first step when calibration curve was prepared.

The quality and the lifetime of graphite tube are particularly important because many analyses on the same tube lead to irremediable adsorption of trace content aluminum on the tube surface. The optimum number of measurement with a single tube is difficult to estimate because it depends on the aluminum concentration in measured samples [23].

The next problem is a prerequisite for homogeneity of the sample [24]. GFAAS is a very sensitive method, where fluctuations in local concentrations of aluminum result from the aluminum binding by hydrophobic organic compounds such as ligands. It leads to rejection of analytical results.

3.5. Comparison of GFAAS and ECR Colorimetric Method. The analyzed samples, both supernatant and solid sludge, have a complex chemical composition, and substantial matrix effects are very likely. The interfering effect of matrix components can be minimalized by the samples' mineralization and using standard addition calibration method. For this reason, the aluminum determination by atomic absorption spectrometry and by colorimetric method with ECR ligands was evaluated.

Results of aluminum concentration analyses in samples of mineralized supernatant for both measurement methods were compared. A calibration curve was prepared with the addition of 20 mL of mineralized supernatant, analogously to the previously described procedure (p. 3.1.3). A calibration curve for GFAAS was also prepared, according to the Polish Standard [7], for the aluminum standard and for mineralisate with the addition of standard at 5-fold dilution of the samples, so that the aluminum concentrations ranged from $0.05\,mg\cdot L^{-1}$ to $0.5\,mg\cdot L^{-1}$. For prepared samples, a 3-fold measurement was made. The obtained

results indicate that deviations from the actual signal value increase with the increase of the concentration of matrix components. The main reason of the increase in the deviation from the true absorbance value is the increase of the ionic strength of solutions.

Figure 9 shows the difference between the reading value (rhombus red dashed line for GFAAS method and square blue dashed line for colorimetric complex method) and the real the aluminum concentration in the studied supernatant liquid (green line). The red and yellow columns (from GFAAS and Al(ECR)$_3$, respectively) in the bottom shows deviation from the real aluminum content in percent standard deviation scale. The colorimetric method read-out results are lower in the whole measuring range compared to real contents.

It should be noted that similar results can be obtained for samples of mineralized supernatant liquid, when the concentration of Al^{+3} ions is up to $0.25\,mg\cdot L^{-1}$. Based on the analyses, it can be concluded that the GFAAS method better determines the actual content in the samples than the Al(ECR)$_3$ method.

3.6. Adsorption Study. Experiments were conducted to determine the adsorptive capacity of activated sludge towards aluminum. The sewage sludge was prepared by washing several times with deionized water to remove the impurities and then filtration through standard filter papers. The sludge was mixed with water by the magnetic stirrer. The aluminum concentration in filtrate samples for each batch never exceeded $15\,\mu g\cdot L^{-1}$, and they could be treated as a blank sample. Such low aluminum concentrations could be determined only by GFAAS.

About 0.1 g of activated sludge (dry weight) was shaken with 100 mL of aluminum solutions of various concentrations in the range 10–$170\,mg\cdot L^{-1}$, for a contact time of 5 h. Aluminum was used in the form of various chemical compounds listed below. The solutions were filtered and the supernatants were analyzed for Al residues using FAAS method. The adsorption efficiency was calculated by dividing the amount of adsorbed metal by the mass of activated sludge.

Figure 10 indicates high adsorption of aluminum in the form of sulphate at a pH of about 4.5 (yellow line). A neutralization of the solution by sodium hydroxide to pH about 7.0 leads to the formation of insoluble aluminum hydroxide Al(OH)$_3$, and it causes almost total metal adsorption (blue line). The analyzed samples measured by GFAAS method gives aluminum concentration read-outs below $0.05\,mg\cdot L^{-1}$. The same samples analyzed by FAAS give unreal aluminum content in level exceeding $0.4\,mg\cdot L^{-1}$. The flame atomic absorption spectrometry gives reproducible results for the aluminum concentration above $5\,mg\cdot L^{-1}$.

The use of acidic coagulants type PAX 25 (green line) and PAX 19H (purple line) shows lower adsorption because pH value of tested samples is low in the range of 3.5 ± 0.3. The aluminum adsorption for PAX 19H is particularly low and does not exceed $10\,mg\cdot L^{-1}$. It showed also worst

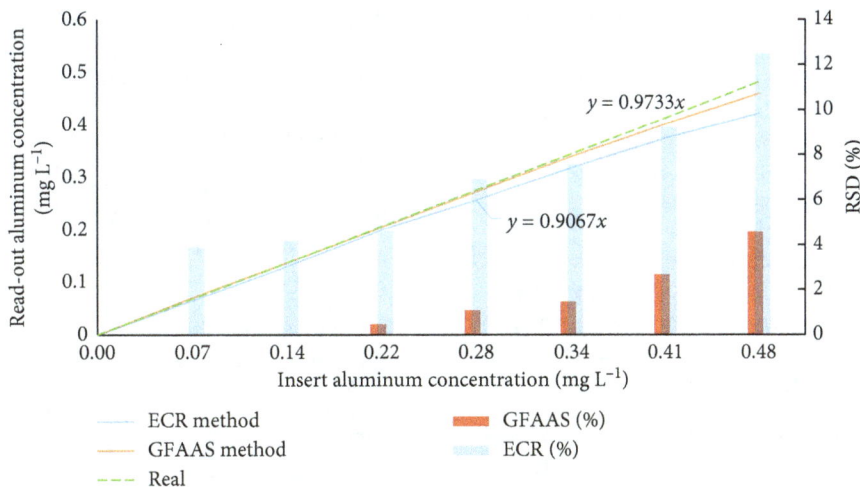

FIGURE 9: Comparison of GFAAS aluminum concentration determinations and colorimetric method in relation to the actual value.

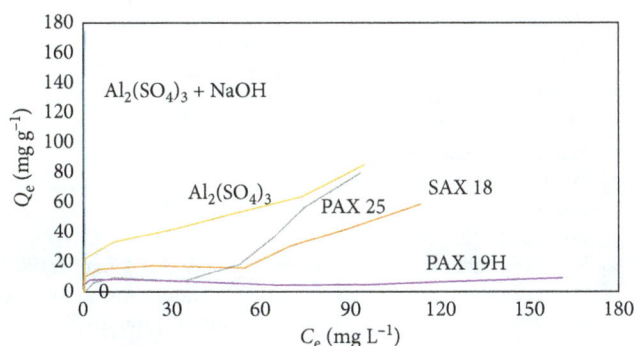

FIGURE 10: Adsorption of aluminum by sludge for different types of coagulants.

sludge sedimentation, probably due to higher solution acidity [25]. The samples with the addition of the alkaline coagulant SAX 18 (red line) showed pH value from 6.5 to 9.5. For SAX 18, aluminum adsorption efficiency incipiently was better, but in final coagulants, doses were much worse in comparison with PAX 25 (reaction sample close to 3.5).

It should be noted that both aluminum sulphate and PAX 25, at a dose exceeding $10 \, mg \cdot L^{-1}$ aluminum, show an acidic pH of the solution of similar value. Then, acidic forms of aluminum $Al(OH_2)_6^{+3}$ and $AlOH(OH_2)_5^{+2}$ are formed [26]. So, such large differences in the adsorption of aluminum by sludge must therefore be caused by differences in the structure of aluminum compounds used, i.e., polyaluminum chloride. In the case of basic SAX 18, the pH value of the solution increases, leading to the formation of soluble aluminum forms such as $Al(OH)_2^+$. Hence, the conclusion is that ionic reactions have a significant role in the sorption of aluminum by the activated sludge during the coagulation process.

The sorption of aluminum dependent on dosed coagulant type is an important but not the only factor affecting the efficiency of sewage sludge treatment. It should also take into account such measures of treated wastewater as TOC, TN, NTU, size, and viscosity of the resulting aggregates of sludge.

4. Conclusions

Wastewater produced in dairy plants is commonly treated with aluminum sulphate or polyaluminum chloride coagulants. They enhance sedimentation rate of organic matter component. This paper evaluates common methods for the determination of aluminum concentration in wastewater. The main results are summarized as follows.

Considering the measuring range of used analytical methods, the GFAAS allows detection of the aluminum content in samples ranging from 0.01 to $0.1 \, mg \cdot L^{-1}$, but analysis of mineralized samples can be stretched to range up to $0.5 \, mg \cdot L^{-1}$ by diluting 5 times in very good effect. Increasing the dilution multiplicity leads to a reduction in the accuracy of the results to an unacceptable level. Colorimetric analysis can give similar results to GFAAS within range up to $0.25 \, mg \cdot L^{-1}$. The detection range of aluminum content in sludge samples can be extended up to 2.0 mg in the case of mineralized and diluted samples, but read-out results were indicative. Whereas the FAAS method is recommended for samples with a aluminum concentration of 5 to $100 \, mg \cdot L^{-1}$. At low aluminum concentrations, less than $4 \, mg \cdot L^{-1}$, the accuracy of the assay is unsuitable for FAAS. The detection limits determined in this work were, respectively, $0.02 \, mg \cdot L^{-1}$ for GFAAS method, $0.07 \, mg \cdot L^{-1}$ for colorimetric method, and $4.0 \, mg \cdot L^{-1}$ for FAAS.

The aluminum absorbance depends on the samples' pH and the form of aluminum which is introduced into the system. Aluminum hydroxide is not fully atomized at temperatures generated in the standard acetylene flame. It has to be raised by the addition of N_2O. The flame atomic absorption spectrometry gives reproducible results for the aluminum concentration above $5 \, mg \cdot L^{-1}$, and it was too high to compare the aluminum content with other methods for the same analyzed samples. Therefore, it is necessary to analyze supernatant samples using both types of atomic absorption spectrometry, when an analyte content fluctuates between one of the detection limit scales.

In the colorimetric method, the absorbance value depends not only on the aluminum concentration but also on the reaction time and pH. A simple modification of the colorimetric method, consisting in adjustment of suitable buffer capacity of the sample, allows to determine the aluminum content in sewage sludge after its acidic mineralization. To neutralize acidity impact, it is necessary to use the stronger acetate buffer (0.46 mM), which protected eriochrome cyanine R before the decomposition.

The low value of pH decreased the absorbance. High acetate buffer concentrations led to absorption decrease, despite the relatively high pH value of the sample. We speculate that high ionic strength of the acetate buffer hampered the aluminum complex formation.

Determination of aluminum content in samples by GFAAS is a more challenging method. Particular attention should be paid to the inhomogeneity of analyzed samples which may lead to substantial absorption intensity variations. Reagents of high analytical grade, ultra pure water, and high quality graphite tube should be applied.

Determination of aluminum content by spectrophotometric method may be a useful and cheap alternative for analyses carried out by atomic adsorption spectroscopies, especially when concentrations of an analyte at the upper limit of the GFAAS or the lower limit of the FAAS are concerned. However, it should be noticed that atomic absorption spectrometry determines the total content of aluminum either in a free ionic or complexed forms. The latter may be characterized by lower stability than the $Al(ECR)_3$.

Conflicts of Interest

The authors declare that they have no conflicts of interest.

Acknowledgments

The authors wish to thank Mr. Jakub Kubicki (Institute of General and Ecological Chemistry, Lodz University of Technology) and Mrs. Jadwiga Albińska (Institute of General and Ecological Chemistry, Lodz University of Technology) for technical assistance.

References

[1] M. Kuliś and A. Tylkowska-Siek, *Fizyczne Rozmiary Produkcji Zwierzęcej w 2015 r*, Główny Urząd Statystyczny, Departament Rolnictwa, Warsaw, Poland, 2016.

[2] D. Boruszko and W. Dąbrowski, "Badania efektywności procesu flotacji i właściwości osadów poflotacyjnych z podczyszczania ścieków mleczarskich," *Inżynieria i Ochrona Środowiska*, vol. 17, no. 2, pp. 269–280, 2014.

[3] K. Szwedziak, "Charakterystyka osadów ściekowych i rolnicze wykorzystanie," *Inzynieria Rolnicza*, vol. 4, pp. 297–302, 2006.

[4] J. B. Bień and K. Wystalska, "Energetyczne wykorzystanie osadów ściekowych," *Przemysł Chemiczny*, vol. 94, no. 9, pp. 1496–1499, 2008.

[5] D. Hawke, J. Powell, and S. Simpson, "Equilibrium modelling of interferences in the visible spectrophotometric determination of aluminium(III): comparison of the chromophores chrome azurolS, eriochrome cyanine R and pyrocatechol violet, and stability constants for eriochrome cyanine R-aluminium complexes," *Analytica Chimica Acta*, vol. 319, no. 3, pp. 305–314, 1996.

[6] W. Dougan and A. Wilson, "The absorptiometric determination of aluminium in water. A comparison of some chromogenic reagents and the development of an improved method," *Analyst*, vol. 99, no. 1180, pp. 413–430, 1974.

[7] Polish Standard, *Jakość Wody-Oznaczanie Glinu-Metody Atomowej Spektrometrii Absorpcyjnej*, PN-EN ISO 12020, Polish Standard, Warsaw, Poland, 2002.

[8] A. Shokrollahi, M. Ghaedi, M. S. Niband, and H. R. Rajabi, "Selective and sensitive spectrophotometric method for determination of sub-micro-molar amounts of aluminium ion," *Journal of Hazardous Materials*, vol. 151, no. 2-3, pp. 642–648, 2008.

[9] A. G. G. Dionisio, A. M. Dantas de Jesus, R. S. Amais et al., "Old and new flavors of flame (furnace) atomic absorption spectrometry," *International Journal of Spectroscopy*, vol. 2011, Article ID 262715, 30 pages, 2011.

[10] Polish Standard, *Oznaczanie Glinu Metodą z Eriochromocyjaniną R, PN-C-04605-02*, Polish Standard, Warsaw, Poland, 1992.

[11] W. Hermanowicz, J. Dojlido, and W. Dożańska, *Fizyczno-Chemiczne Badanie Wody i Ścieków*, Arkady, Warsaw, Poland, 1999.

[12] W. Siriangkhawut, S. Tontrong, and P. Chantiratikul, "Quantitation of aluminium content in waters and soft drinks by spectrophotometry using eriochrome cyanine R," *Research Journal of Pharmaceutical, Biological and Chemical Sciences*, vol. 4, no. 3, pp. 1154–1161, 2013.

[13] B. Perkin-Elmer, *Analytical Methods for Atomic Absorption Spectroscopy*, The Perkin-Elmer Corp., Norwalk, CT, USA, 1996.

[14] I. Narin, M. Tuzen, and M. Soylak, "Aluminium determination in environmental samples by graphite furnace atomic absorption spectrometry after solid phase extraction on Amberlite XAD-1180/pyrocatechol violet chelating resin," *Talanta*, vol. 63, no. 2, pp. 411–418, 2004.

[15] J. Tria, E. C. V. Butler, P. R. Haddad, and A. R. Bowie, "Determination of aluminium in natural water samples," *Analytica Chimica Acta*, vol. 58, no. 2, pp. 153–165, 2007.

[16] C. L. Craney, K. Swartout, F. W. Smith, and C. D. West, "Improvement of trace aluminum determination by electrothermal atomic absorption spectrophotometry using phosphoric acid," *Analytical Chemistry*, vol. 58, no. 3, pp. 656–658, 1986.

[17] C. Marin, A. Tudorache, and L. Vladescu, "Aluminium determination and speciation modelling in groundwater from the area of a future radioactive waste repository," *Revistade Chimie (Bucharest)*, vol. 5, pp. 431–438, 2010.

[18] Agilent, *Technical Overview, The Role of Chemical Modifiers in Graphite Furnace Atomic Absorption Spectrometry*, 5991-9286EN, Agilent, Santa Clara, CA, USA, 2018.

[19] P. Bermejo-Barrera, E. Beceiro-González, and A. Bermejo-Barrera, "Use of Pd-Mg(NO$_3$)$_2$ as matrix modifier for the determination of aluminum in water by electrothermal atomization atomic absorption spectrometry," *Microchemical Journal*, vol. 45, no. 1, pp. 90–96, 1992.

[20] L. M. Voth-Beach and D. E. Shrader, "Investigations of a reduced palladium chemical modifier for graphite furnace

atomic absorption spectrometry," *Journal of Analytical Atomic Spectrometry*, vol. 2, no. 1, pp. 45–50, 1987.

[21] S. Noremberg, M. Veiga, D. Bohrer et al., "Determination of aluminum and silicon in bovine liver by graphite furnace atomic absorption spectrometry after dissolution with tetramethylammonium hydroxide," *Analytical Methods*, vol. 7, no. 2, 2014.

[22] L. Pszonicki and A. M. Essed, "Palladium and magnesium nitrate as modifiers for the determination of lead by graphite furnace atomic absorption spectrometry," *Chemia Analityczna (Warsaw)*, vol. 38, pp. 771–778, 1993.

[23] R. Sturgeon, "Graphite furnace atomic absorption spectrometry: fact and fiction," *Fresenius' Zeitschrift für analytische Chemie*, vol. 324, no. 8, pp. 807–818, 1986.

[24] N. R. Bader and B. Zimmermann, "Sample preparation for atomic spectroscopic analysis: an overview," *Advances in Applied Science Research*, vol. 3, pp. 1733–1737, 2012.

[25] X. H. Guan, G. H. Chen, and C. Shang, "ATR-FTIR and XPS study on the structure of complexes formed upon the adsorption of simple organic acids and aluminum hydroxide," *Journal of Environmental Sciences*, vol. 19, no. 4, pp. 438–443, 2007.

[26] J. D. Hem and C. E. Roberson, "Form and Stability of aluminum hydroxide complexes in dilute solution," in *Chemistry of Aluminum in Natural Water, Geological Survey Water-Supply Paper 1827-A*, United States Government Printing Office, Washington, DC, USA, 1967.

15

Application of Multivariate Statistical Analysis in Evaluation of Surface River Water Quality of a Tropical River

Teck-Yee Ling,[1] Chen-Lin Soo,[1] Jing-Jing Liew,[1] Lee Nyanti,[2] Siong-Fong Sim,[1] and Jongkar Grinang[3]

[1]*Department of Chemistry, Faculty of Resource Science and Technology, Universiti Malaysia Sarawak, 94300 Kota Samarahan, Sarawak, Malaysia*
[2]*Department of Aquatic Science, Faculty of Resource Science and Technology, Universiti Malaysia Sarawak, 94300 Kota Samarahan, Sarawak, Malaysia*
[3]*Institute of Biodiversity and Environmental Conservation, Universiti Malaysia Sarawak, 94300 Kota Samarahan, Sarawak, Malaysia*

Correspondence should be addressed to Teck-Yee Ling; teckyee60@gmail.com

Academic Editor: Athanasios Katsoyiannis

The present study evaluated the spatial variations of surface water quality in a tropical river using multivariate statistical techniques, including cluster analysis (CA) and principal component analysis (PCA). Twenty physicochemical parameters were measured at 30 stations along the Batang Baram and its tributaries. The water quality of the Batang Baram was categorized as "slightly polluted" where the chemical oxygen demand and total suspended solids were the most deteriorated parameters. The CA grouped the 30 stations into four clusters which shared similar characteristics within the same cluster, representing the upstream, middle, and downstream regions of the main river and the tributaries from the middle to downstream regions of the river. The PCA has determined a reduced number of six principal components that explained 83.6% of the data set variance. The first PC indicated that the total suspended solids, turbidity, and hydrogen sulphide were the dominant polluting factors which is attributed to the logging activities, followed by the five-day biochemical oxygen demand, total phosphorus, organic nitrogen, and nitrate-nitrogen in the second PC which are related to the discharges from domestic wastewater. The components also imply that logging activities are the major anthropogenic activities responsible for water quality variations in the Batang Baram when compared to the domestic wastewater discharge.

1. Introduction

The Batang Baram ("batang" denotes big river) (coordinates: 4°35′5.28″N and 113°58′44.256″E) is located on the northern part of Sarawak where it flows 400 km westwards, mostly through primary and secondary forest to the South China Sea. The river is the second longest river in Sarawak and the third longest river in Malaysia. The Baram area was once a pristine area but it has undergone profound changes associated with the population growth and development. Increasing residential area, numerous longhouses, and swidden agriculture are found along the river. Commercial logging has also been carried out actively in the area for decades where the logged forest was then converted to commercial oil palm and acacia plantations [1].

Although development continues to grow in this area, the study on the water quality of the river is relatively scarce despite the river serving as an important source for drinking water for the rural community. The discharges of domestic sewage and agricultural runoff can lead to eutrophication [2–4] while deforestation can cause sedimentation and nutrient enrichment in the river [5–8]. In the year 1995, sampling was conducted in the uppermost catchment of the Baram River basin. The study revealed that the overall water quality of the river was relatively good at that time but was subjected to high suspended solids which came from soil erosions due to land

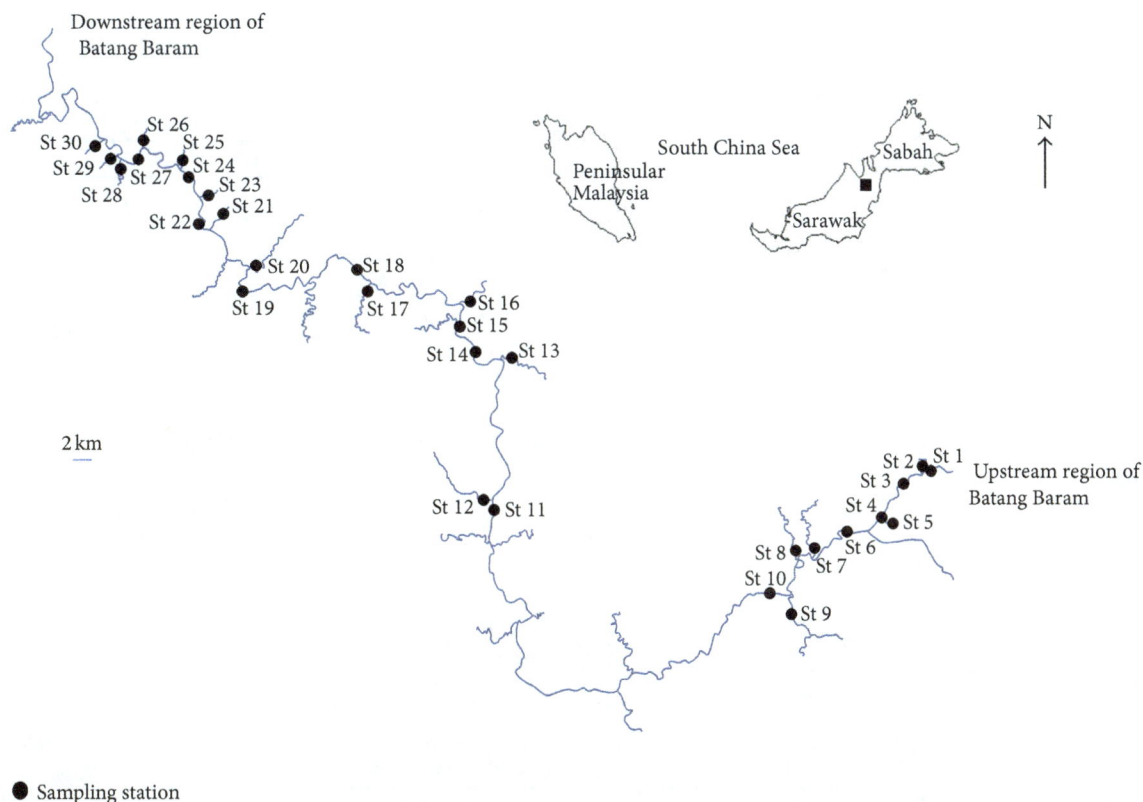

FIGURE 1: The study area in Sarawak state and location of the 30 sampling stations along the Batang Baram and its tributaries in the present study.

clearing and timber harvesting [9]. The author also pointed out that elevated ammonia was found near to the domestic and animal waste discharges.

Water quality assessment and monitoring on large river basin like the Batang Baram potentially generate a large data set. Numerous studies have shown that multivariate statistical analysis is useful for the assessment of the spatial water quality variations in a river [10–16]. Cluster analysis could reveal similarities among the large number of sampling stations in a river while principal component analysis assists in identifying important factors accounting for most of the variances in water quality of a river. Hence, the aim of the present study was to apply the multivariate statistical analysis in the interpretation of the physicochemical characteristics of the Batang Baram and its tributaries. The analysis output would provide valuable information for the decision making in the river basin management.

2. Materials and Methods

2.1. Field Collection. In situ and ex situ parameters were collected at 30 sampling stations located along the Batang Baram and its tributaries covering a distance of approximately of 172 km (Figure 1). Table 1 shows the details of the samplings from upstream to downstream directions that were carried out in the year 2015. The water level of river was high during samplings due to the rain before each sampling. The whole study area was subjected to logging activities. Numerous longhouses and plantation activities were included in Table 1. In situ parameters including temperature, pH, conductivity, oxygen saturation (DOsat), dissolved oxygen (DO), and turbidity were measured using a multiparameter water quality sonde (YSI6920 V2-2). Transparency, depth, and flow velocity were measured using a Secchi disc with a measuring tape, a depth sounder (PS-7, Hondex), and a stream flow meter (Geopacks), respectively. Total discharge, mean velocity, and mean depth were calculated according to [17]. The water samples were taken for the analyses of chlorophyll a (chl a), total suspended solids (TSS), five-day biochemical oxygen demand (BOD_5), chemical oxygen demand (COD), total phosphorus (TP), total ammonia nitrogen (TAN), nitrite-nitrogen (NO_2^--N), nitrate-nitrogen (NO_3^--N), organic nitrogen (Org-N), and total sulphide (TS). All sampling bottles were acid-washed, cleaned, and dried before use. Analyses of chl a, TSS, and BOD_5 began in the field immediately after sampling while NO_2^--N, NO_3^--N, and TS analyses were completed in the field after sampling. Water samples were acidified to pH < 2 for COD, TP, TAN, and Org-N analyses. The samples were placed in an ice box and transported to the laboratory for further analysis [18].

2.2. Laboratory Analysis. All the analyses were conducted according to the standard methods [16, 17]. Chl a was determined from samples filtered through a 0.7 μm glass

TABLE 1: The details of the sampling regime and sampling locations surveyed in the present study.

Sampling	Station	Location	Remark
29-30 July 2015	St 1	Batang Baram N 03°10′10.8″ E 115°11′47.6″	Lio Mato longhouse
	St 2	Sungai Selungo N 03°10′15.0″ E 115°11′46.7″	
	St 3	Batang Baram N 03°09′34.0″ E 115°10′51.0″	Long Tungan longhouse
	St 4	Sungai Temendan N 03°07′16.7″ E 115°09′15.2″	
	St 5	Batang Baram N 03°07′21.8″ E 115°09′10.5″	
	St 6	Batang Baram N 03°06′54.0″ E 115°07′21.8″	Long Semiyang longhouse
	St 7	Sungai Sebatu N 03°05′50.6″ E 115°05′46.3″	
	St 8	Sungai Sela'an N 03°05′50.5″ E 115°04′52.7″	
	St 9	Sungai Moh N 03°02′34.1″ E 115°04′35.9″	
	St 10	Batang Baram N 03°03′37.5″ E 115°03′23.3″	Long Moh longhouse Replantation
3-4 September 2015	St 11	Batang Baram N 03°08′23.6″ E 114°48′52.7″	Long Apu longhouse
	St 12	Sungai Lasa N 03°08′40.8″ E 114°48′49.2″	Residential area
	St 13	Sungai Menuang N 03°16′19.1″ E 114°49′22.4″	
	St 14	Batang Baram N 03°16′56.6″ E 114°47′53.1″	
	St 15	Batang Baram N 03°17′49.6″ E 114°47′08.1″	Long San longhouse
	St 16	Sungai Akah N 03°19′11.3″ E 114°47′32.7″	Residential area
	St 17	Sungai Keluang N 03°19′58.6″ E 114°42′21.8″	
	St 18	Batang Baram N 03°21′19.0″ E 114°41′31.5″	Long Sangah longhouse
	St 19	Batang Baram N 03°19′50.8″ E114°35′43.1″	Long Naha'a longhouse
	St 20	Sungai Patah N 03°21′02.9″ E114°36′22.3″	
21-22 January 2015	St 21	Sungai Kahah N 03°23′12.9″ E 114°34′05.7″	
	St 22	Batang Baram N 03°23′08.4″ E 114°33′35.8″	
	St 23	Sungai Piping N 03°24′47.4″ E 114°33′39.0″	
	St 24	Batang Baram N 03°25′45.9″ E 114°33′00.5″	
	St 25	Sungai Jertang N 03°26′19.3″ E114°32′28.3″	

Sampling	Station	Location	Remark
	St 26	Sungai Kesseh N 03°27′26.2″E 114°30′44.5″	Long Kesseh longhouse
	St 27	Batang Baram N 03°27′11.3″E 114°30′29.4″	Long Kesseh longhouse
	St 28	Sungai Nakan N 03°26′29.6″E114°29′20.6″	Long Nakan longhouse Oil palm plantation
	St 29	Sungai Liyans N 03°27′07.8″E 114°29′15.3″	
	St 30	Sungai Kemenyeh N 03°27′46.7″E 114°28′29.4″	

microfibre filter (Whatman GF/F) and extracted for 24 h using 90% (v/v) acetone. For TSS, filtration of an adequate sample through a 1.0 μm glass microfibre filter (Whatman GF/B) was carried out in the field and drying of the filter was conducted in the laboratory in an oven at 105°C until a constant weight was obtained. It was then determined by calculating the difference between the initial and final weight of the sample and expressed as milligram per liter of sample. For BOD$_5$, it was determined as the difference between the initial and final DO content, after a five-day period of incubation of the sample. The initial DO content was measured in the field. Whenever the in situ DO value was deemed too low, it was raised by vigorous aeration. COD was determined by the closed reflux method followed by the titrimetric method. For TP analysis, persulfate digestion of samples was conducted followed by the ascorbic acid method. TAN, NO_2^--N, and NO_3^--N were determined by Nessler's method, diazotization method (low range), and cadmium reduction method, respectively. Before the analyses of NO_2^--N and NO_3^--N, the water sample was filtered through a 0.7 μm glass microfibre filter (Whatman GF/F). Org-N was determined by the Macro-Kjeldahl Method where ammonia was removed from the water sample before digestion and distillation. Subsequently, ammonia was analyzed by using Nessler's method. TS was analyzed using the methylene blue method. H_2S was calculated according to [18] with the following equation:

$$H_2S = \frac{TS}{\left(1 + K'_1 / [H^+]\right)}, \quad (1)$$

where H_2S is the unionized hydrogen sulphide, TS is the total sulphide, K' is the conditional ionization constant, and $[H^+]$ is the hydrogen ion concentration.

A calibration curve was constructed for each chemical analysis. The blank and standard solutions were treated in the same way as the sample.

2.3. Water Quality Index (WQI).
Water quality index (WQI) which combines the six variables of DO, BOD, COD, TSS,

AN, and pH was calculated according to the following equation:

$$WQI = 0.22 * SI_{DO} + 0.19 * SI_{BOD} + 0.16 * SI_{COD} + 0.15 * SI_{AN} + 0.16 * SI_{SS} + 0.12 * SI_{pH}, \quad (2)$$

where SI_{DO} is the subindex for DO (% saturation), SI_{BOD} is the subindex for BOD (mg/L), SI_{COD} is the subindex for COD (mg/L), SI_{AN} is the subindex for AN (mg/L), SI_{SS} is the subindex for SS (mg/L), and SI_{pH} is the subindex for pH [19].

2.4. Statistical Analysis.
Comparison of physicochemical parameters between the stations in the Batang Baram was conducted using one-way ANOVA and Tukey's pairwise comparisons with 5% significance level. The independent samples t-test was used to compare the physicochemical parameters between the main river and tributary stations. Pearson's correlation analysis was performed to determine the relationship among all the parameters. Cluster analysis (CA) was used to investigate the grouping of the sampling stations by using the physicochemical parameters collected in the river. Z-score standardization of the variables and Ward's method using Euclidean distances as a measure of similarity were used. The cluster was considered statistically significant at a linkage distance of <60% and the number of clusters was decided by the practicality of the outputs [12]. Principal component analysis (PCA) was conducted to characterize the loadings of all physicochemical parameters for each of the PCs obtained having eigenvectors higher than one (Kaiser criterion). The component has significant loading on a variable when the loading is greater than 0.4 [20]. The data were square-rooted and standardized prior to the analysis. The quality of data for PCA was confirmed with Kaiser-Meyer-Olkin (KMO) measure of sampling adequacy test and Bartlett's test of sphericity. All the statistical analyses were carried out by using the Statistical Software for Social Sciences (SPSS Version 22, SPSS Inc., 1995).

3. Results and Discussion

3.1. The Physicochemical Characteristics of the Batang Baram and Its Tributaries.
Figures 2 and 3 show the mean values of the physicochemical parameters of the Batang Baram and its tributaries from upstream to downstream regions. During the

(a)

(b)

(c)

(d)

(e)

(f)

(g)

(h)

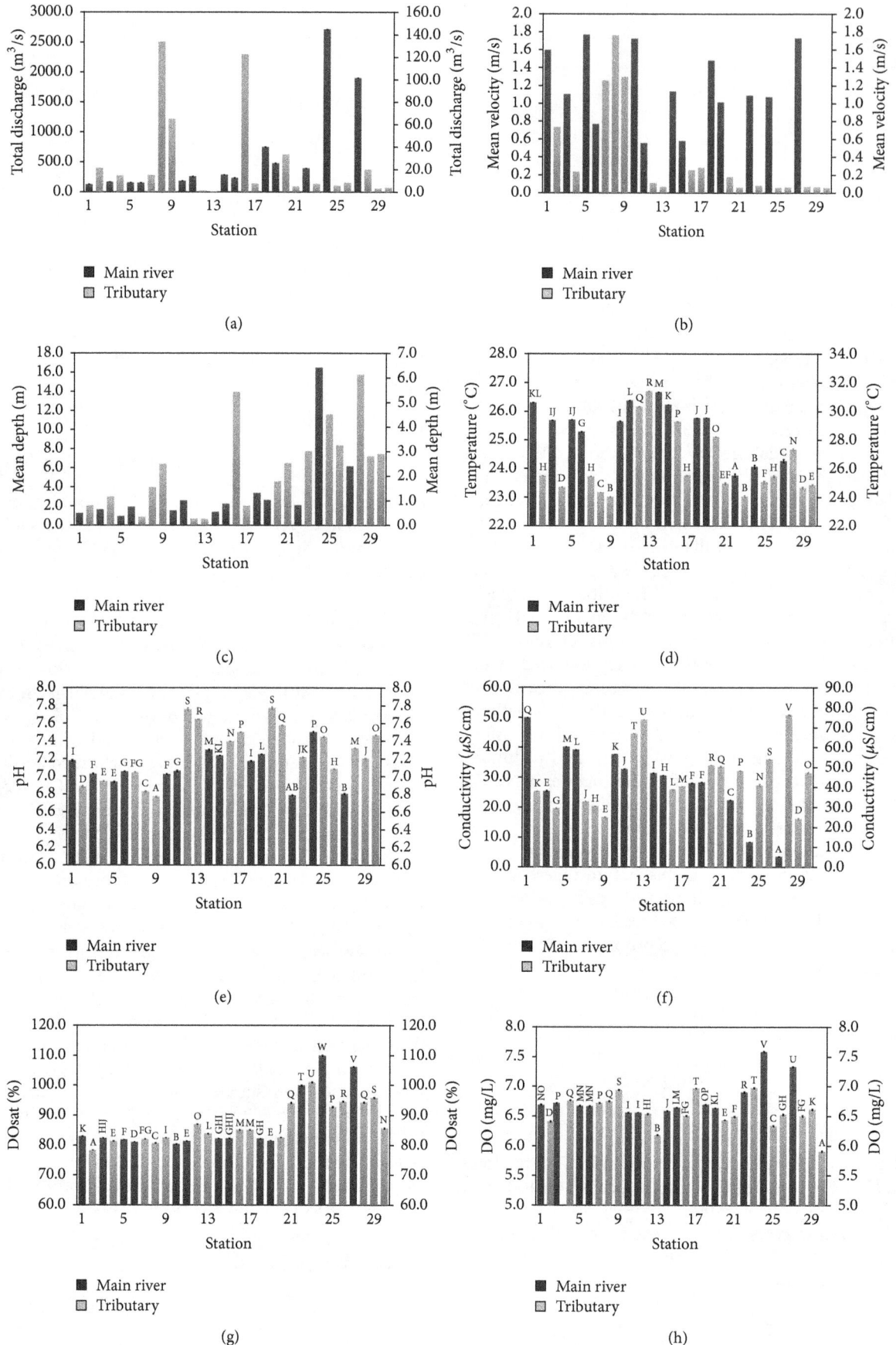

FIGURE 2: Continued.

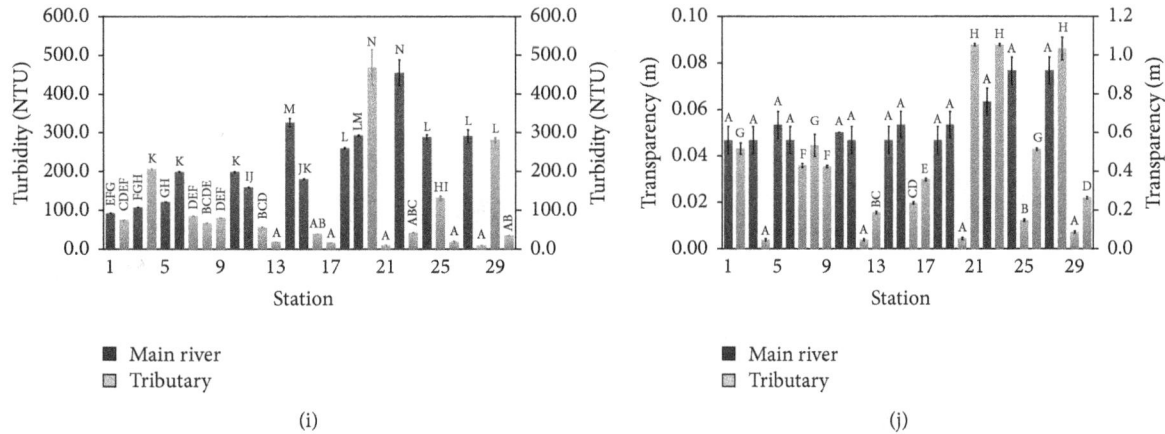

FIGURE 2: In situ parameters of (a) total discharge, (b) mean velocity, (c), mean depth, (d) temperature, (e) pH, (f) conductivity, (g) DOsat, (h) DO, (i) turbidity, and (j) transparency measured at the 30 sampling stations located along the Batang Baram (left axis) and its tributaries (right axis) (different letters indicate significant difference at p value ≤ 0.05).

sampling, total discharge of the Batang Baram ranged from $126.3 \, \mathrm{m^3/s}$ to $2711.8 \, \mathrm{m^3/s}$ and from $0.5 \, \mathrm{m^3/s}$ to $133.6 \, \mathrm{m^3/s}$ in main river and tributaries, respectively. Figure 2 illustrates that total discharge of main river showed an increasing trend towards downstream regions whereas the highest total discharge in tributaries was observed at station 8 followed by station 16. Mean velocity of the river was relatively consistent in main river with a mean value of $1.2 \, \mathrm{m/s}$. High mean velocity ($>1 \, \mathrm{m/s}$) was also observed in some of the tributaries located upstream but most of the tributaries were slow flowing ($\approx 0.2 \, \mathrm{m/s}$). Mean depth of the Batang Baram ranged from $0.9 \, \mathrm{m}$ to $16.5 \, \mathrm{m}$ and from $0.2 \, \mathrm{m}$ to $6.1 \, \mathrm{m}$ in main river and tributaries, respectively. Both main river and tributaries were relatively deeper downstream compared to upstream.

The results of ANOVA showed that all of the parameters demonstrated significant variations (p value ≤ 0.05) from one sampling station to another. The physicochemical parameters showed different distribution patterns along the main river. The turbidity, TSS, and H_2S values increased significantly (p value ≤ 0.05) towards downstream with the highest values of turbidity ($468.8 \pm 45.4 \, \mathrm{NTU}$) and TSS ($320.0 \pm 19.3 \, \mathrm{mg/L}$) which were both observed at station 20 while the highest value of H_2S ($0.83 \pm 0.01 \, \mathrm{mg/L}$) was observed at station 22. The high turbidity and TSS downstream indicate the accumulation of sediment in the river. Reference [21] reported that a spit was formed in the Baram River mouth and continued to expand due to the erosion associated with deforestation and land use changes in the upstream region. The similar distribution pattern and significant positive correlation (p value ≤ 0.05) between H_2S, turbidity, and TSS (Table 2) indicated that H_2S was associated with suspended solids in the river.

On the other hand, the conductivity, BOD_5, TP, NO_3^--N, and Org-N showed higher values at the upper part of the river and decreased significantly (p value ≤ 0.05) towards downstream region. In contrary, [22] demonstrated that TP, TN, and NH_3-N concentrations tend to increase from upstream to downstream regions in the Qiantang River, East China. In the present study, the highest conductivity value was observed at station 1 ($50.0 \, \mu S/cm$) and steadily decreased

to $3.5 \, \mu S/cm$ at station 27. The highest values of BOD_5 ($5.7 \pm 0.2 \, \mathrm{mg/L}$) and Org-N ($2.74 \pm 0.01 \, \mathrm{mg/L}$) were observed at stations 1 and 4, respectively, while the highest values of TP ($2.2 \pm 0.1 \, \mathrm{mg/L}$) and NO_3^--N ($0.07 \pm 0.01 \, \mathrm{mg/L}$) were observed at station 6. The conductivity value ($82 \, \mu S/cm$– $133 \, \mu S/cm$) in the uppermost part of the Baram River basin reported by [9] was relatively higher than the present study which agrees with the present result that conductivity value was higher in the upper part of the river. However, the author also reported the concentrations of the BOD_5 ($0.7 \, \mathrm{mg/L}$ to $2.0 \, \mathrm{mg/L}$) and NO_3^--N ($0.01 \, \mathrm{mg/L}$–$0.02 \, \mathrm{mg/L}$) which were lower than the present study.

Significantly higher COD value (p value ≤ 0.05) was observed in the middle section of the river ($110.1 \, \mathrm{mg/L}$– $181.8 \, \mathrm{mg/L}$) whereas NO_2^--N ($0.001 \, \mathrm{mg/L}$–$0.002 \, \mathrm{mg/L}$) and TAN ($0.12 \, \mathrm{mg/L}$–$0.33 \, \mathrm{mg/L}$) values were significantly lower (p value ≤ 0.05) there. Significantly higher (p value ≤ 0.05) TAN was observed at stations 21 ($1.57 \pm 0.07 \, \mathrm{mg/L}$) and 22 ($1.49 \pm 0.20 \, \mathrm{mg/L}$) while significantly higher (p value ≤ 0.05) NO_2^--N was observed at station 20 ($0.055 \pm 0.001 \, \mathrm{mg/L}$). Similar to BOD_5 and NO_3^--N, the NH_3-N concentration in the uppermost part of the Baram River basin which ranged from $0.7 \, \mathrm{mg/L}$ to $2.0 \, \mathrm{mg/L}$ [9] was lower than the present study. The author attributed the high ammonia concentration in his study to the sewage discharge from the longhouse and animal waste. The higher nutrients concentration in the present study indicated the deterioration of water quality over time due to the increase in population and land development in the area.

Table 3 shows that the river temperature, pH, conductivity, transparency, chl a, and NO_2^--N were significantly higher (p value ≤ 0.05) in tributaries than in the main river. The high water temperature in tributaries particularly at stations 12, 13, and 16 ($>29°C$) indicated that direct solar radiation due to the forest canopy exposure after logging had increased the river temperature in those tributaries [6]. The Baram River basin contained high dissolved ions which gave the high conductivity values in the river [9]. Besides, significant positive correlation (p value ≤ 0.05) between temperature and conductivity indicated that the high temperature in

TABLE 2: Correlation matrix (p value ≤ 0.05) of the in situ and ex situ parameters collected from the 30 sampling stations.

	Discharge	Velocity	Depth	Temp	pH	Cond	DOsat	DO	Turb	Trans	Chl a	TSS	BOD$_5$	COD	TP	TAN	NO$_2^-$-N	NO$_3^-$-N	Org-N	H$_2$S
Discharge																				
Velocity	.392																			
Depth	.834																			
Temp																				
pH	-.546	-.391		.658																
Cond		-.511		.610	.549															
DOsat	.601				-.539															
DO			.641																	
Turb	.706	.381	.513	-.480	-.650			.527												
Trans		-.389			.413															
Chl a										-.582										
TSS	.389				-.608		.915		.366	-.605										
BOD$_5$		.555				-.444		-.544												
COD						.478			-.493											
TP		.491				-.467		-.456												
TAN				-.528																
NO$_2^-$-N																-.528				
NO$_3^-$-N		.454		-.390		-.554							.545	-.392	.540					
Org-N								-.547					.523		.564			.515		
H$_2$S				-.403	-.566	-.454	.714	.362	.399	-.361		.596								

(a)

(b)

(c)

(d)

(e)

(f)

(g)

(h)

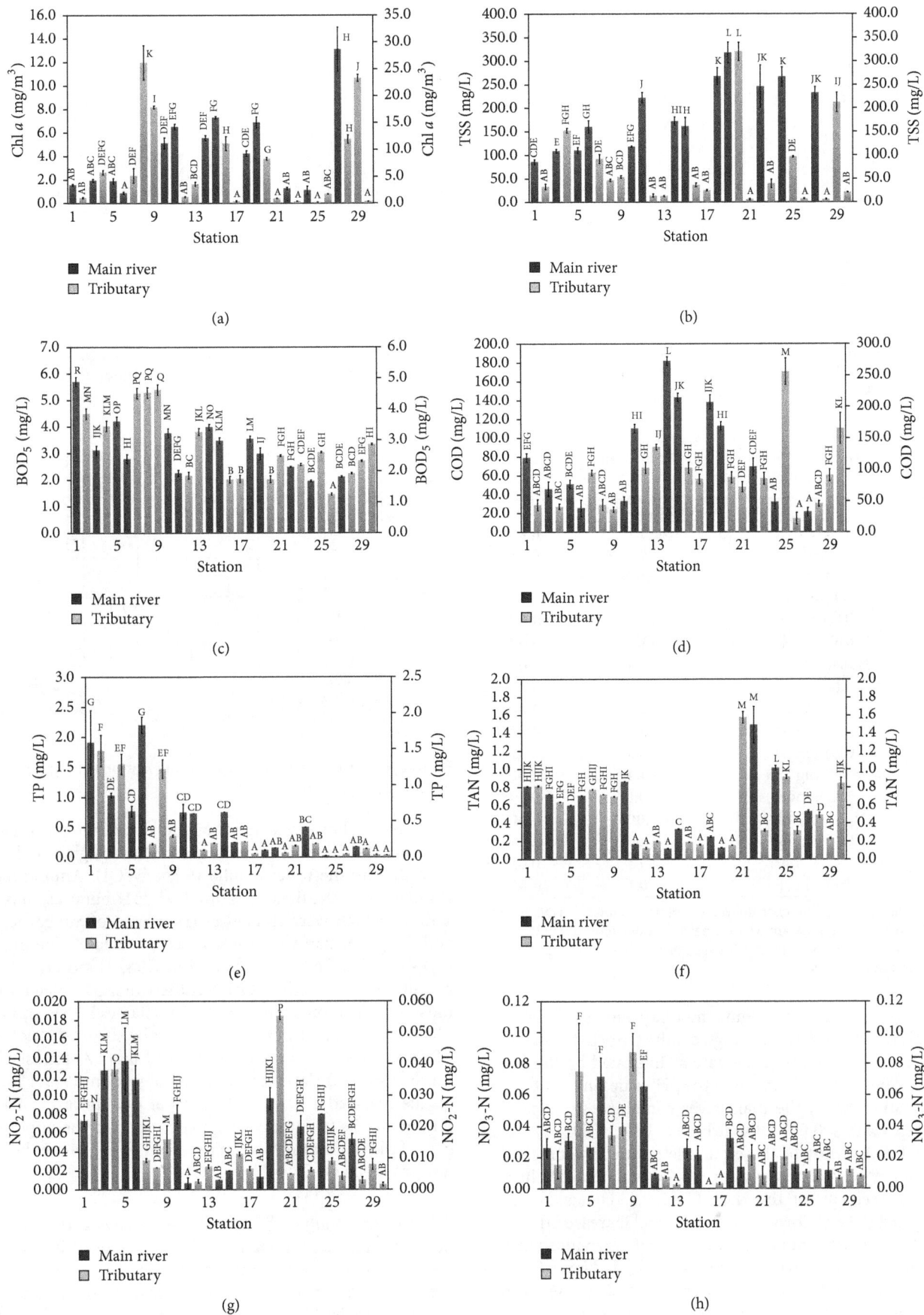

FIGURE 3: Continued.

(i)

(j)

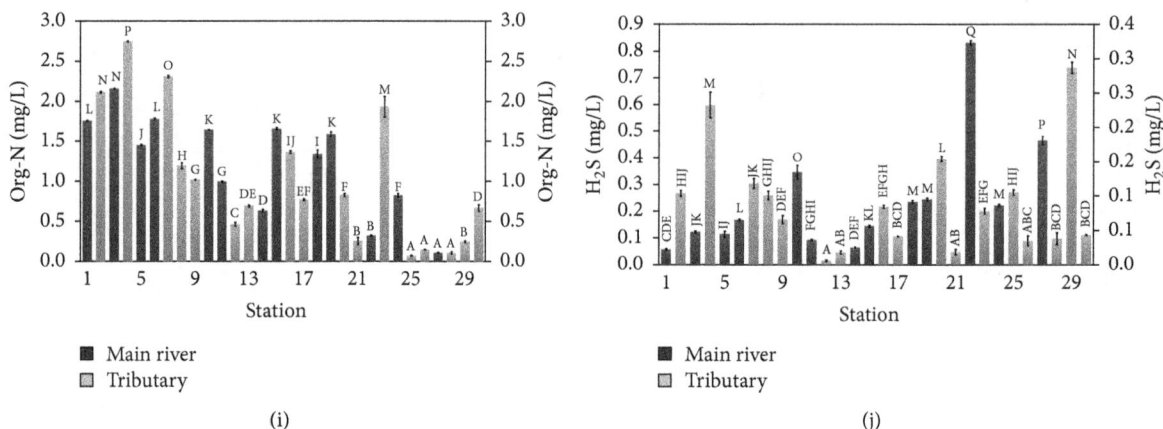

FIGURE 3: Ex situ water quality of (a) chl *a*, (b) TSS, (c), BOD$_5$, (d) COD, (e) TP, (f) TAN, (g) NO$_2^-$-N, (h) NO$_3^-$-N, (i) Org-N, and (j) H$_2$S measured at the 30 sampling stations located along the Batang Baram (left axis) and its tributaries (right axis) (different letters indicate significant difference at *p* value ≤ 0.05).

TABLE 3: Mean difference of in situ and ex situ water quality parameters between the main river of the Batang Baram and its tributaries.

	Parameter	Mean difference	*p* value
	Temperature, °C	**−0.7**	**0.050**
	pH	**−0.2**	**0.001**
	Conductivity, μS/cm	**−16.2**	**0.000**
In situ	DOsat, %	−0.2	0.915
	DO, mg/L	**+0.2**	**0.000**
	Turbidity, NTU	**+132.0**	**0.000**
	Transparency, m	**−0.4**	**0.000**
	Chl *a*, mg/m^3	**−2.6**	**0.048**
	TSS, mg/L	**+120.7**	**0.000**
	BOD$_5$, mg/L	**+0.5**	**0.026**
	COD, mg/L	−8.3	0.474
	TP, mg/L	**+0.4**	**0.002**
Ex situ	TAN, mg/L	+0.1	0.497
	NO$_2^-$-N, mg/L	**−0.006**	**0.006**
	NO$_3^-$-N, mg/L	+0.007	0.169
	Org-N, mg/L	+0.3	0.093
	H$_2$S, mg/L	**+0.1**	**0.000**

Positive value of mean difference indicates parameter studied was higher in the main river of Batang Baram whereas negative value indicates parameter studied was higher in the tributary. The significant difference at *p* value ≤ 0.05 was indicated in bold.

tributaries increased the ionic mobility and solubility of minerals which is reflected in high conductivity in tributaries. Also, the high photosynthesis rate as indicated by the high chl *a* in tributaries had increased the pH values in tributaries. On the other hand, the main river contained significantly higher (*p* value ≤ 0.05) DO, turbidity, TSS, BOD$_5$, TP, and H$_2$S (Table 3). Most of these parameters were significantly and positively correlated (*p* value ≤ 0.05) with total discharge and mean velocity of the river (Table 2). Hence, we can assume that the fast flowing main river had increased the DO content due to more rapid aeration and had introduced more pollutants into the river via surface runoff. Nevertheless, tributaries of the Batang Baram were also well aerated as all of the stations were recorded with DO content of more than 5 mg/L and DOsat more than 80%.

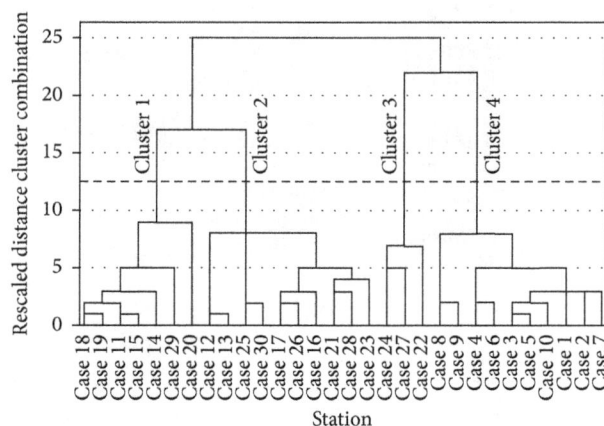

FIGURE 4: Clustering of the 30 sampling stations along the Batang Baram and its tributaries.

Table 4 shows that most of the sampling stations were classified as Class III and categorized as "slightly polluted" according to the water quality index (WQI). Among the 30 stations along the Batang Baram and its tributaries, only two stations which were located at tributaries of Sungai Kesseh and Sungai Nakan were categorized as "clean." The pH and DO were classified as Class I and/or Class II indicating good condition whereas the COD was the worst parameter where most of the stations were classified as Classes III, IV, and/or V. The river also possesses pollution risk by suspended solids as TSS was classified as Class III and/or Class IV at most of the stations and was classified as Class V at stations 19 and 20. The results revealed a deteriorating water quality of the Batang Baram and its tributaries when compared to the uppermost part of the Baram River basin reported by [9] where the river was grouped as a Class II river.

3.2. Cluster Analysis (CA).

Cluster analysis was used to detect similarities among the sampling stations in the study area. The dendogram shows that sampling stations in the present study can be grouped into four significant clusters as illustrated by Figure 4. The clustering pattern shows that physicochemical characteristics of the Batang Baram changed

TABLE 4: Classification of water quality of the Batang Baram from upstream to downstream regions according to WQI.

Station	AN	BOD$_5$	COD	DO	pH	TSS	WQI	Status
1	III	III	IV	II	I	III	III	Slightly polluted
2	III	III	III	II	II	II	III	Slightly polluted
3	III	III	III	II	I	III	III	Slightly polluted
4	III	III	III	II	II	IV	III	Slightly polluted
5	III	III	IV	II	II	III	III	Slightly polluted
6	III	II	III	II	I	IV	III	Slightly polluted
7	III	III	IV	II	I	III	III	Slightly polluted
8	III	III	III	II	II	II	III	Slightly polluted
9	III	III	III	II	II	III	III	Slightly polluted
10	III	III	III	II	I	III	III	Slightly polluted
11	II	II	V	II	I	IV	III	Slightly polluted
12	II	II	V	II	I	I	II	Slightly polluted
13	II	III	V	II	I	I	III	Slightly polluted
14	II	III	V	II	I	IV	III	Slightly polluted
15	III	III	V	II	I	IV	III	Slightly polluted
16	II	II	V	II	I	II	II	Slightly polluted
17	II	II	IV	II	I	II	II	Slightly polluted
18	II	III	V	II	I	IV	III	Slightly polluted
19	II	II	V	II	I	V	III	Slightly polluted
20	II	II	IV	II	I	V	III	Slightly polluted
21	IV	II	IV	II	I	I	III	Slightly polluted
22	IV	II	IV	II	II	IV	III	Slightly polluted
23	III	II	IV	II	I	II	II	Slightly polluted
24	IV	II	III	I	I	IV	III	Slightly polluted
25	IV	II	V	II	I	III	III	Slightly polluted
26	III	II	II	II	I	I	II	Clean
27	III	II	II	I	II	IV	III	Slightly polluted
28	III	II	III	II	I	I	II	Clean
29	II	II	IV	II	I	IV	III	Slightly polluted
30	III	II	V	II	I	I	III	Slightly polluted

from upstream to downstream regions as demonstrated by the grouping of upstream stations, middle stations, and downstream stations into different clusters. Cluster 1 consists of stations that were located mostly in the middle section of the river (stations 11, 14, 15, 18, 19, and 20) except station 29 which was located downstream. Tributaries located from middle to downstream regions of the river shared similar characteristics as demonstrated by the grouping of stations 12, 13, 16, 17, 21, 23, 25, 26, 28, and 30 in cluster 2. The main river located downstream (stations 22, 24, and 27) also showed similarity and grouped together as cluster 3. These two clusters show that main river and tributaries of the Batang Baram which were located downstream shared no similarity. Finally, cluster 4 consists of stations that were located upstream of the river including stations that were located at the main river and tributaries (station 1 to station 10). This analysis suggests that a reduced number of sampling stations in each cluster may serve as a rapid assessment of the water quality of the Batang Baram and leads to a more cost-effective monitoring study in the future.

3.3. Principal Component Analysis (PCA). The PCA was used to explore the most important factors determining the spatial variations in physicochemical parameters of the Batang Baram. A total of six principal components (PCs) were obtained with eigenvalues more than one which accounted for around 83.6% of the total variance in the 20 physicochemical parameters of the Batang Baram (Table 5). The first component (PC1), accounting for 30.0% of the total variance in the data sets of the river water, has significant positive loadings on turbidity, TSS, and H$_2$S and negative loadings on conductivity and transparency. These factors imply that soil erosion occurred in the present study area, and a high loading of turbidity and H$_2$S is associated with the presence of suspended solids [23]. Similarly, strong positive loadings on turbidity and suspended solids were also observed in a Mekong Delta area of Vietnam [11] which is a result of soil erosion from disturbed land. The Sarawak forest is subjected to high timber harvesting pressure rendering sedimentation problem in its forest streams [1, 6, 24–26]. The present study shows that the Batang Baram in Sarawak state is no

TABLE 5: Loadings of the physicochemical parameters on the first six varimax-rotated PCs (eigenvalue > 1) along the Batang Baram and its tributaries.

| Parameter | Rotated Component Matrix[a] | | | | | |
| | Component | | | | | |
	1	2	3	4	5	6
Total discharge			+0.900			
Mean velocity		+0.674	+0.640			
Mean depth			+0.712			
Temperature				−0.904		
pH		−0.451		−0.607		+0.444
Conductivity	−0.605		−0.476			
DOsat		−0.750	+0.488			
DO			+0.678			
Turbidity	+0.948					
Transparency	−0.791					
Chl a						−0.906
TSS	+0.956					
BOD$_5$		+0.854				
COD			−0.423		−0.692	
TP		+0.850				
TAN				+0.780		
NO2					+0.784	
NO3		+0.612		+0.563		
Org-N		+0.716				
H$_2$S	+0.759			+0.427		
Initial eigenvalue	6.0	3.7	2.7	1.7	1.4	1.2
% of variance	30.1	18.6	13.5	8.7	6.8	5.9
Cumulative %	30.1	48.7	62.2	70.9	77.7	83.6

[a]Rotation converged in 11 iterations.

exception. Logging activities in the surrounding area have caused sedimentation and increased suspended solids level in the river. The PC1 has the largest proportion of the total variance indicating that logging activities are the major source of river water contamination in the Batang Baram.

The PC2 accounting for 18.6% of the total variance has significant positive loadings on mean velocity, BOD$_5$, TP, NO$_3^-$-N, and Org-N and negative loadings on pH and DOsat. These factors indicate an inflow of effluent from longhouses and residential area largely consisting of organic pollutants; and a negative loading of pH and DOsat is attributed to the process of decomposition of the organic matter. Similarly, the analysis of PCA was applied in the Qiantang River which indicated that TN, NO$_3^-$, NH$_3$-N, and TP were the dominant pollution factors in the river [22]. The authors attributed the pollutions to the domestic sewage, discharge of poultry and animal feces, and fertilizer that were flushed into the river. Also, an "organic" factor that positively loaded with COD, BOD$_5$, TON, TP, and PO$_4^{3-}$ was reported in a main river system in northern Greece which represented the influence of municipal and industrial effluents [16]. The PC3 accounting for 13.6% of total variance has significant positive loadings on total discharge, mean velocity, mean depth, DOsat, and DO and negative loadings on conductivity and COD suggesting a dilution of chemically oxidizable material in the river

associated with high volume of river water and high dissolved oxygen level. The high COD/BOD$_5$ ratio in the Batang Baram indicates a large nonbiodegradable fraction of organic matter in the river.

The PC4 (8.7% of the total variance) has significant positive loadings on transparency, TAN, NO$_3^-$-N, and H$_2$S and negative loadings on temperature, pH, and conductivity. In an anaerobic condition, the high loading of organic matter in river can lead to the formation of ammonia and organic acids which coupled with the production of hydrogen sulphide and carbon dioxide during decomposition [27, 28] can cause acidification of water. By employing the PCA for the data interpretation, [10] also revealed that parameters related to organic pollutants and temperature were the most important parameters contributing to water quality variation in the Sava River, Croatia. The PC5 (6.8% of the total variance) is significantly and negatively loaded on COD but positively loaded on NO$_2^-$-N. Again, the high loading of organic matter in the river likely led to the build-up of NO$_2^-$-N in the water. Similar to PC2, both PC4 and PC5 can be explained as influences from domestic discharges which contained high nutrients and organic matter. As PC2 has a larger proportion of the total variance than PC4 and PC5, we can assume that organic pollution in the Batang Baram is more severe than the inorganic pollution. Reference [29] also reported that organic pollutants, followed by nutrients and salt concentration, were the most important parameters contributing to water quality variation in the Wen-Rui Tang River watershed, China. Finally, the PC6 (5.9% of the total variance) is significantly and positively loaded on pH but negatively loaded on chl a indicating the decomposition of dead phytoplankton and that most of the variability in the data is due to the pH changes as pH plays a significant role in mineralization process.

4. Conclusions

The present study revealed that the Batang Baram and its tributaries demonstrated sign of pollution as indicated by high suspended solids and nutrients in the river. Most of the stations along the river were classified as Class III and categorized as "slightly polluted." In particular, the COD and TSS were the most severe parameters where they were classified as Class V at some of the stations in the river. A large number of sampling stations in the present study were grouped into four clusters which divided the river into upstream, middle, and downstream main river and tributaries located middle and downstream of the river. The PCA revealed that parameters related to logging activities and domestic sewage discharge were the most important parameters contributing to water quality variation in the river.

Competing Interests

The authors declare that there is no conflict of interests regarding the publication of this paper.

Acknowledgments

The authors appreciate the financial support provided by the Sarawak Energy Berhad through Grant no. GL(F07)/SEB/3C/2013(22) and the facilities provided by Universiti Malaysia Sarawak.

References

[1] M. Ichikawa, "Degradation and loss of forest land and land-use changes in Sarawak, East Malaysia: a study of native land use by the Iban," *Ecological Research*, vol. 22, no. 3, pp. 403–413, 2007.

[2] H. P. Jarvie, C. Neal, and P. J. A. Withers, "Sewage-effluent phosphorus: a greater risk to river eutrophication than agricultural phosphorus?" *The Science of the Total Environment*, vol. 360, no. 1–3, pp. 246–253, 2006.

[3] M. B. Rothenberger, J. M. Burkholder, and C. Brownie, "Long-term effects of changing land use practices on surface water quality in a coastal river and lagoonal estuary," *Environmental Management*, vol. 44, no. 3, pp. 505–523, 2009.

[4] C. Soo, T. Ling, N. Lee, and K. Apun, "Assessment of the characteristic of nutrients, total metals, and fecal coliform in Sibu Laut River, Sarawak, Malaysia," *Applied Water Science*, vol. 6, no. 1, pp. 77–96, 2016.

[5] F. Gökbulak, Y. Serengil, S. Özhan, N. Özyuvaci, and N. Balci, "Effect of timber harvest on physical water quality characteristics," *Water Resources Management*, vol. 22, no. 5, pp. 635–649, 2008.

[6] T.-Y. Ling, C.-L. Soo, J.-R. Sivalingam, L. Nyanti, S.-F. Sim, and J. Grinang, "Assessment of the water and sediment quality of tropical forest streams in upper reaches of the Baleh River, Sarawak, Malaysia, subjected to logging activities," *Journal of Chemistry*, vol. 2016, Article ID 8503931, 13 pages, 2016.

[7] J. Schelker, K. Eklöf, K. Bishop, and H. Laudon, "Effects of forestry operations on dissolved organic carbon concentrations and export in boreal first-order streams," *Journal of Geophysical Research: Biogeosciences*, vol. 117, no. 1, pp. 1–12, 2012.

[8] A. Nor Zaiha, M. S. M. Ismid, Salmiati, and M. S. S. Azri, "Effects of logging activities on ecological water quality indicators in the Berasau River, Johor, Malaysia," *Environmental Monitoring and Assessment*, vol. 187, no. 8, article 493, 2015.

[9] L. Seng, *Water in the Environment: Tainted Life Source Hungers for Cures*, UNIMAS, 2011.

[10] A. Marinović Ruždjak and D. Ruždjak, "Evaluation of river water quality variations using multivariate statistical techniques: Sava River (Croatia): A Case Study," *Environmental Monitoring and Assessment*, vol. 187, no. 4, pp. 1–14, 2015.

[11] D. Phung, C. Huang, S. Rutherford et al., "Temporal and spatial assessment of river surface water quality using multivariate statistical techniques: a study in Can Tho City, a Mekong Delta area, Vietnam," *Environmental Monitoring and Assessment*, vol. 187, no. 5, article no. 229, 2015.

[12] S. Muangthong and S. Shrestha, "Assessment of surface water quality using multivariate statistical techniques: case study of the Nampong River and Songkhram River, Thailand," *Environmental Monitoring and Assessment*, vol. 187, no. 9, article 548, 2015.

[13] R. L. Olsen, R. W. Chappell, and J. C. Loftis, "Water quality sample collection, data treatment and results presentation for principal components analysis—literature review and Illinois River watershed case study," *Water Research*, vol. 46, no. 9, pp. 3110–3122, 2012.

[14] S. Shrestha and F. Kazama, "Assessment of surface water quality using multivariate statistical techniques: a case study of the Fuji river basin, Japan," *Environmental Modelling and Software*, vol. 22, no. 4, pp. 464–475, 2007.

[15] P. Zeilhofer, E. B. N. R. Lima, and G. A. R. Lima, "Spatial patterns of water quality in the cuiabá River Basin, Central Brazil," *Environmental Monitoring and Assessment*, vol. 123, no. 1–3, pp. 41–62, 2006.

[16] V. Simeonov, J. A. Stratis, C. Samara et al., "Assessment of the surface water quality in Northern Greece," *Water Research*, vol. 37, no. 17, pp. 4119–4124, 2003.

[17] S. C. Chapra, *Surface Water-Quality Modeling*, McGraw-Hill, New York, NY, USA, 1997.

[18] D. Jenkins, J. J. Connors, and A. E. Greenberg, *Standard Methods for the Examination of Water and Wastewater*, American Public Health Association, Washington, DC, USA, 21st edition, 2005.

[19] Department of Environment, *Malaysia Environmental Quality Report 2014*, DOE, Kuala Lumpur, Malaysia, 2015.

[20] R. C. Jones, D. P. Kelso, and E. Schaeffer, "Spatial and seasonal patterns in water quality in an embayment-mainstem reach of the tidal freshwater Potomac River, USA: A Multiyear Study," *Environmental Monitoring and Assessment*, vol. 147, no. 1–3, pp. 351–375, 2008.

[21] R. Nagarajan, M. P. Jonathan, P. D. Roy, G. Muthusankar, and C. Lakshumanan, "Decadal evolution of a spit in the Baram river mouth in eastern Malaysia," *Continental Shelf Research*, vol. 105, pp. 18–25, 2015.

[22] H. Yuan, E. Liu, W. Pan, and S. An, "Water pollution characteristics and assessment in different functional zones," *Polish Journal of Environmental Studies*, vol. 23, no. 2, pp. 541–549, 2014.

[23] G. S. Bilotta and R. E. Brazier, "Understanding the influence of suspended solids on water quality and aquatic biota," *Water Research*, vol. 42, no. 12, pp. 2849–2861, 2008.

[24] T. Jinggut, C. M. Yule, and L. Boyero, "Stream ecosystem integrity is impaired by logging and shifting agriculture in a global megadiversity center (Sarawak, Borneo)," *Science of the Total Environment*, vol. 437, pp. 83–90, 2012.

[25] J. Hon and S. Shibata, "A review on land use in the Malaysian state of Sarawak, Borneo and recommendations for wildlife conservation inside production forest environment," *Borneo Journal of Resource Science and Technology*, vol. 3, no. 2, pp. 22–35, 2013.

[26] J. E. Bryan, P. L. Shearman, G. P. Asner, D. E. Knapp, G. Aoro, and B. Lokes, "Extreme differences in forest degradation in borneo: comparing practices in Sarawak, Sabah, and Brunei," *PLoS ONE*, vol. 8, no. 7, Article ID e69679, 2013.

[27] M. Holmer and E. Kristensen, "Impact of marine fish cage farming on metabolism and sulfate reduction of underlying sediments," *Marine Ecology Progress Series*, vol. 80, no. 2-3, pp. 191–201, 1992.

[28] K. I. A. Kularatne, D. P. Dissanayake, and K. R. R. Mahanama, "Contribution of dissolved sulfates and sulfites in hydrogen sulfide emission from stagnant water bodies in Sri Lanka," *Chemosphere*, vol. 52, no. 5, pp. 901–907, 2003.

[29] K. Mei, L. Liao, Y. Zhu et al., "Evaluation of spatial-temporal variations and trends in surface water quality across a rural-suburban-urban interface," *Environmental Science and Pollution Research*, vol. 21, no. 13, pp. 8036–8051, 2014.

16

Preparation and Evaluation of a Profile Control Agent Base on Waste Drilling Fluid

Xiaoping Qin, Haiwei Lu, Yilin Li, Tong Peng, Lijie Xing, Haixi Xue, and Jing Xu

Drilling and Production Technology Research Institute, PetroChina Jidong Oilfield Company, Tangshan 063004, China

Correspondence should be addressed to Xiaoping Qin; 948801727@qq.com

Academic Editor: Nicolas Roche

The waste drilling fluid was treated by a flocculant and a pH regulator. And a novel profile control agent base on waste drilling fluid (PCAWDF) was prepared using polymer, formaldehyde, resorcinol, and thiourea as raw materials under mild conditions. PCAWDF was characterized by infrared (IR) spectroscopy and scanning electron microscope (SEM). Compared with the profile control agent prepared by the recirculated water (PCARW), PCAWDF exhibited comparable or better stability, salt resistance, and viscoelasticity. The results of parallel core plugging experiments showed that the profile improvement capability of PCAWDF was stronger than that of PCARW (for 3000 mg/L: 84.6% versus 83.1%; for 5000 mg/L: 91.8% versus 90.2%). The main performance indexes of PCAWDF could meet the need of profile control for the water injection wells. The method could solve the problem of waste drilling fluid treatment in an economic and environmental way.

1. Introduction

Waste drilling fluid contains a large number of additives, treatment agents, and other chemical agents, so random discharge of untreated waste drilling fluid will cause serious damage to the ecological environment [1–3]. However, achieving standard emissions requires a high cost of treatment [4–6]. Therefore, in order to reduce disposal cost of waste drilling fluid and protect the ecological environment, waste drilling fluid treatment technology needs to be further studied[7–10].

At present, the drilling fluid systems used in China's oil fields are usually polymer drilling fluid, polysulfonate drilling fluid, KCl film-forming drilling fluid, KCl anti-high temperature drilling fluid, and so on. Among them, polysulfonate drilling fluid is the most widely used one. The curing with drilling treatment technology is usually adopted in the waste polysulfonate drilling fluid, which is achieved by adding cement, lime, and fly ash into the waste drilling fluid [11–14]. Chemical substances in the treatment products will pollute the ground water and air [15–18]. In addition, this method greatly increases the processing cost of waste drilling fluid (around 10 dollars/square meter).

Profile modification in water injection wells can improve the water injection profile, thereby increasing the swept volume of injected water. The main profile control agents are cement profile control agent, resin profile control agent, inorganic salt precipitation type profile control agent, gel profile control agent, and so forth. Gel profile control agent has many advantages, such as easy injection, low cost, and good sealing ability, so it is widely used in water injection wells profile control.

Keeping in mind all above points, herein, a novel gel profile control agent base on waste drilling fluid was prepared. This method not only solves the problem of high treatment cost of waste drilling fluid but also provides a water source for profile control of water injection well. After all, surface water resources are limited.

2. Experimental

2.1. Chemicals and Reagents. Aluminum sulfate (AR), sodium hydroxide (AR), formaldehyde (mass concentration 37% water solution), resorcinol (AR), thiourea (AR), sodium chloride (NaCl, AR), calcium chloride anhydrous (CaCl$_2$, AR), magnesium chloride hexahydrate (MgCl$_2$·6H$_2$O, AR),

TABLE 1: The reagent dosage of polymer solutions and profile control agents.

Preparation water	Polymer mg/L	Formaldehyde mg/L	Resorcinol mg/L	Thiourea mg/L	Code
TWDF	3000	/	/	/	3000-TWDF
TWDF	5000	/	/	/	5000-TWDF
Recirculated water	3000	/	/	/	3000-RW
Recirculated water	5000	/	/	/	5000-RW
TWDF	3000	2000	600	100	3000-PCAWDF
TWDF	5000	3000	1000	120	5000-PCAWDF
Recirculated water	3000	2000	600	100	3000-PCARW
Recirculated water	5000	3000	1000	120	5000-PCARW

TABLE 2: TDS and chemical composition of the formation water and the recirculated water.

Items	$Na^+ + K^+$	Ca^{2+}	Mg^{2+}	SO_4^{2-}	Cl^-	HCO_3^-	TDS
Content of the recirculated water (mg/L)	2093.3	73.8	41.2	94.5	3224.5	417.6	5944.9
Content of the formation water (mg/L)	3071.6	131.1	53.6	124.9	4580.7	573.2	8535.1

potassium chloride (KCl, AR), sodium sulfate (Na_2SO_4, AR), and sodium bicarbonate ($NaHCO_3$, AR) were purchased from Chengdu Kelong Chemical Reagent Factory (Sichuan, China). All chemicals and reagents were used as received without any further purification. Partially hydrolyzed polyacrylamide (HPAM, degree of hydrolysis: 25%, viscosity molecular weight: 22×10^6) was purchased from Daqing Lianhua Technology Co., Ltd.

2.2. Treatment of Waste Drilling Fluid. The waste drilling fluid was taken from X88 well, and the polymer sulfonate drilling fluid was used in this well. This drilling fluid was composed of a base slurry, a partially hydrolyzed polyacrylamide, a partially hydrolyzed polyacrylamide ammonium salt, a liquid lubricant, an ultralow permeability treatment agent, and a pH regulator. Solid fraction in the waste drilling fluid was removed with high frequency sieve. The flocculant aluminum sulfate was added to remove the colloidal particles and suspended matter in the liquid phase. Then sodium hydroxide was added to regulate the pH of solution as 7-8, and the treated waste drilling fluid (TWDF) was obtained.

2.3. Preparation of Profile Control Agent. Profile control agents were prepared using preparation water (TWDF or recirculated water), HPAM, formaldehyde, resorcinol, and thiourea at room temperature. The dosage of each agent is shown in Table 1. The polymer was added to the preparation water, and the polymer was stirred until being completely dissolved. Then formaldehyde, resorcinol, and thiourea were added to the polymer solution, and the mixture was stirred for 20 min. Finally, these profile control agents were obtained and loaded into sealed bottles.

2.4. Stability. Different concentrations of profile control agents were prepared by the given methods and loaded into sealed bottles. These profile control agents were put into a constant temperature box (65°C). Every 10 days, the apparent

viscosity of these profile control agents was tested using Brookfiled DV3T viscometer at 65°C.

2.5. Salt Resistance. The salt resistance performance was studied by increasing NaCl concentration in the recirculated water or TWDF, and then profile control agents prepared using these waters were aged 72 h at 65°C. The apparent viscosity of these profile control agents was measured via Brookfiled DV3T viscometer at 65°C.

2.6. Viscoelasticity. Profile control agents were formulated with the preparation water and placed for 72 h at 65°C. Viscoelasticity measurements of these profile control agents were conducted on an Anton Paar MCR102 Rheometer. The test system was plate, and the rotor was PP50 at 65°C. The range of scanning frequency (f) was 0.5–10 Hz, and the stress was 0.5 Pa by using the same test system and rotor in viscoelasticity measurements.

2.7. Parallel Core Plugging Experiment. Sandstone cores were used to study the profile improvement ability of these profile control agents. Total dissolved solids (TDS) and chemical composition of the formation water and the recirculated water are listed in Table 2.

According to the flow chart (Figure 1), the experimental devices were connected. If not otherwise noted, all experiments were carried out at 65°C. The cores were placed into the Hassler core holders with 1.0 MPa backpressure and 3.0 MPa confining pressure. These cores were saturated with the formation water at 0.5 mL/min, and then 0.5 PV (pore volume) profile control agent was injected into the cores. After aging for 72 h, the recirculated water was injected into the cores at 0.5 mL/min until the flow rate was stable.

The profile improvement rate of profile control agent is calculated with the following equation:

$$f = \frac{Q_{hb}/Q_{lb} - Q_{ha}/Q_{la}}{Q_{hb}/Q_{lb}}, \tag{1}$$

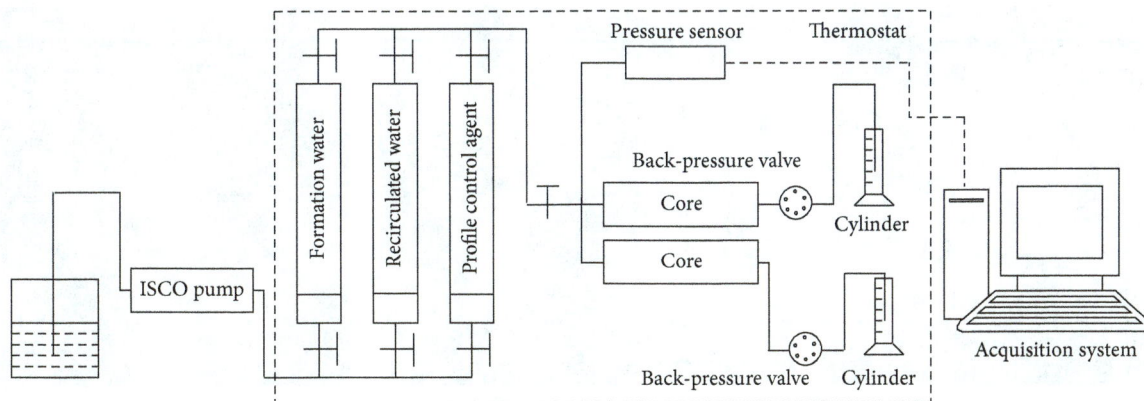

FIGURE 1: Flow chart of the parallel core plugging tests.

where f is the profile improvement rate of profile control agent, %; Q_{hb} is the percentage of water absorption before high permeability core profile control, %; Q_{ha} is the percentage of water absorption after high permeability core profile control, %; Q_{lb} is the percentage of water absorption before low permeability core profile control, %; and Q_{la} is the percentage of water absorption after low permeability core profile control, %.

3. Results and Discussion

3.1. Viscosity of Polymer Solution Prepared Using TWDF.
As shown in Figure 2, the viscosity difference of these polymer solutions prepared with TWDF or recirculated water was very small under the same conditions. Viscosity comparison results showed that the flocculant and pH modifier could effectively improve the water quality of the waste drilling fluid. In this way, the polymer solution could be prepared with TWDF instead of the recirculated water.

3.2. IR Spectra Analysis of Profile Control Agents.
The structures of PCAWDF and HPAM were confirmed by IR spectra as illustrated in Figure 3. The PCAWDF prepared using HPAM, formaldehyde, and resorcinol was confirmed by strong absorption at 3438.5 cm^{-1} (-NH stretching vibration), 2938.7 cm^{-1} (-CH$_2$ stretching vibration), 1672.2 cm^{-1} (C=O stretching vibration), and 1562.4 cm^{-1} (C-N stretching vibration and -NH bending vibration) in the IR spectroscopy of the profile control agent. The characteristic absorption at 1125.6 cm^{-1} and 848.9 cm^{-1} was assigned to in-plane C-H bending vibration and out-plane C-H bending vibration of resorcinol, respectively. The characteristic absorption at 615.8 cm^{-1} belonged to out-plane N-H bending vibration of thiourea. As expected, the IR spectra demonstrated that PCAWDF was successfully synthesized.

3.3. Reaction Process and Profile Control Mechanism of PCAWDF.
PCAWDF was mainly composed of HPAM, formaldehyde, resorcinol, and thiourea. Hydroxymethyl resorcinol was produced by condensation reaction between formaldehyde and resorcinol. The methylol of hydroxymethyl resorcinol could react with the amino groups on HPAM. So PCAWDF had the complex spatial network structure.

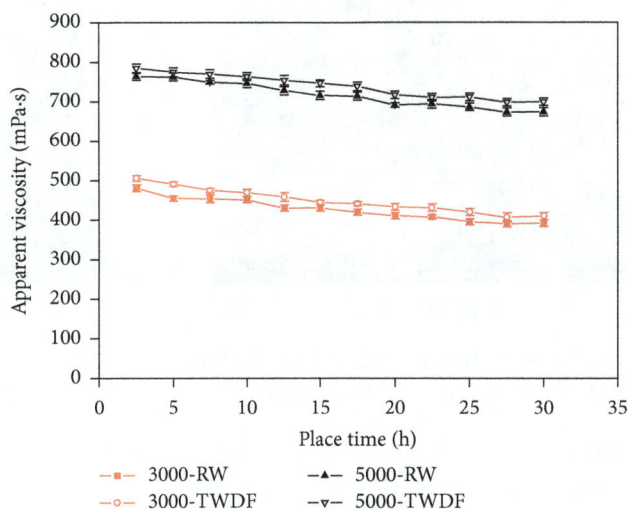

FIGURE 2: Effect of water quality on the viscosity of polymer solution with different concentration.

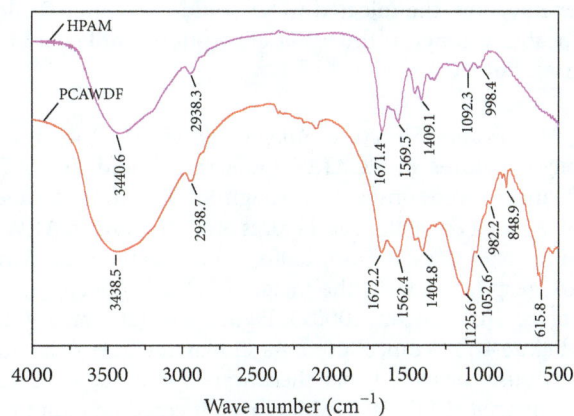

FIGURE 3: IR spectra of PCAWDF and HPAM.

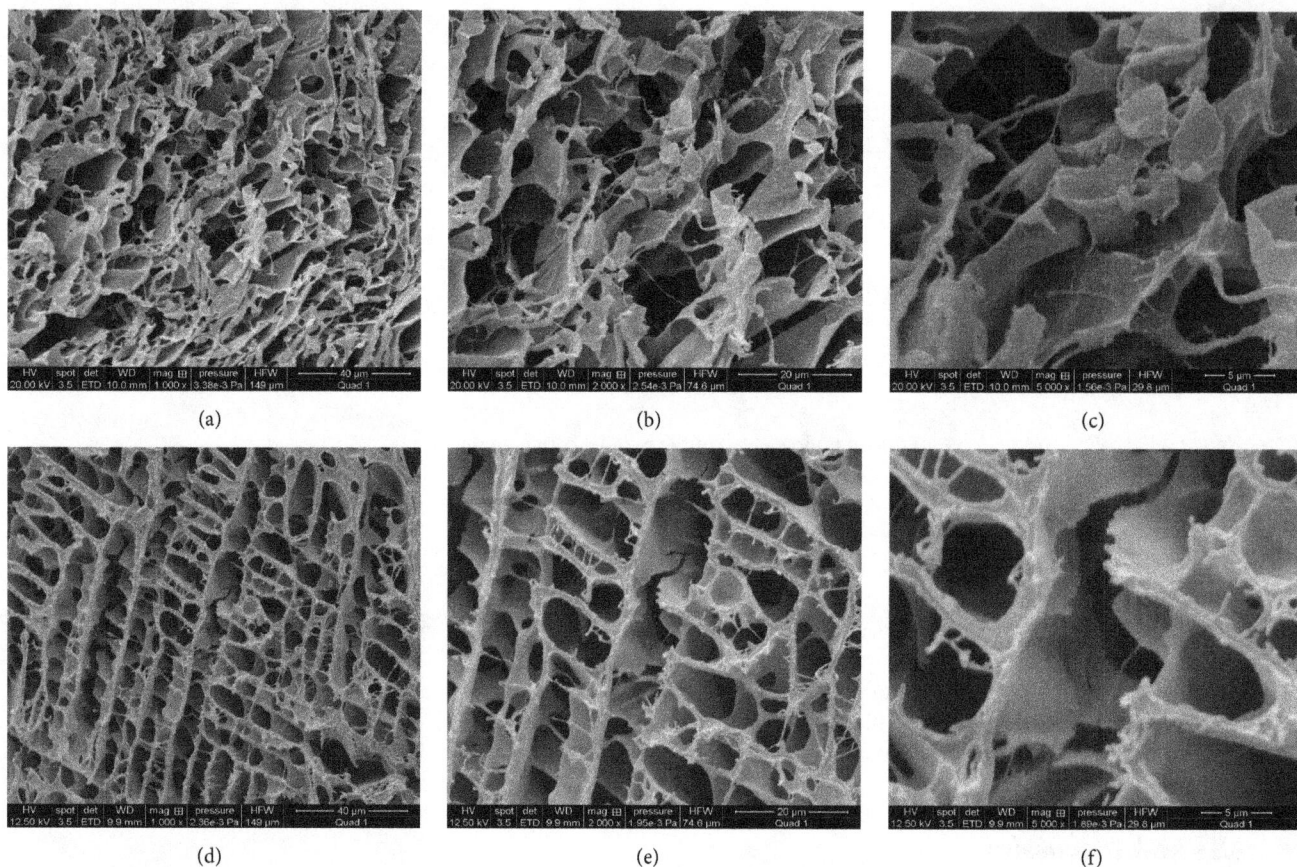

FIGURE 4: SEM images of PCARW and PCAWDF: (a) PCARW at 40 μm, 1000x; (b) PCARW at 20 μm, 2000x; (c) PCARW at 5 μm, 5000x; (d) PCAWDF at 40 μm, 1000x; (e) PCAWDF at 20 μm, 2000x; (f) PCAWDF at 5 μm, 5000x.

In the injection course, PCAWDF preferentially entered the high permeability zone, and it could form effective plugging for the high permeability zone. In the process of water injection, the injected water could only enter the low permeability zone, so the water absorption profile could be improved by PCAWDF.

3.4. Microscopic Structure Analysis by SEM. The microscopic structures of PCARW (3000 mg/L) and PCAWDF (3000 mg/L) were observed through SEM at room temperature. Among these images, Figures 4(a)–4(c) are PCARW at different scan sizes (40 μm, 1000x; 20 μm, 2000x; and 5 μm, 5000x, resp.). Similarly, the images of PCAWDF are shown in Figure 4(d) (40 μm, 1000x), Figure 4(e) (20 μm, 2000x), and Figure 4(f) (5 μm, 5000x). As shown in Figure 4, it could be obviously observed that there were space net structures in the images of PCAWDF. Moreover, it could be found that the microscopic reticular structures of PCAWDF were much more compact than that of PCARW in the same scan size. The much denser spatial network structure of PCAWDF might be due to the effect of flocculant on water quality, because good water quality was conducive to the stretch of polymer molecular chain.

3.5. Strength of Profile Control Agents. The apparent viscosity versus place time curves of 3000-PCAWDF, 5000-PCAWDF, 3000-PCARW, and 5000-PCARW are shown in Figure 5. Glue time of both 3000-PCAWDF and 5000-PCAWDF was 15 h. Strength of 3000-PCARW and 5000-PCARW was 35500 mPa·s and 26100 mPa·s, respectively. Compared with 3000-PCARW and 5000-PCARW, 3000-PCAWDF and 5000-PCAWDF displayed almost the same gel time and gel strength. This phenomenon proved once again the feasibility of replacing the recirculated water with TWDF.

3.6. Stability. The apparent viscosity versus temperature curves of profile control agents are shown in Figure 6. At 65°C, 3000-PCAWDF and 3000-PCARW had the same good stability. After 180 days, the apparent viscosity of 3000-PCAWDF was decreased by only 10% at 65°C. The experimental results showed that the flocculant and pH modifier could reduce the influence of additives in the waste drilling fluid on the polymer chain. Two valence cations (such as calcium ion, magnesium ion) concentration decreased significantly in the role of flocculant, and polymers were more conducive to maintaining the integrity of molecular chains under neutral conditions. The polymer molecular chain could

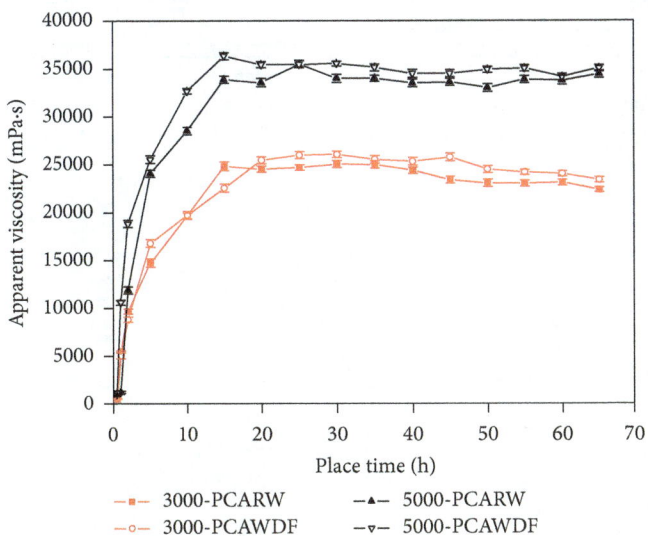

FIGURE 5: The apparent viscosity versus place time curves.

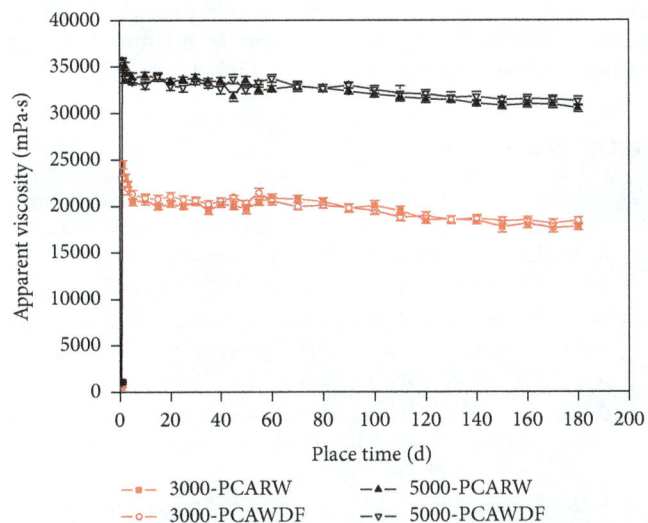

FIGURE 6: Apparent viscosity versus temperature curves of profile control agents.

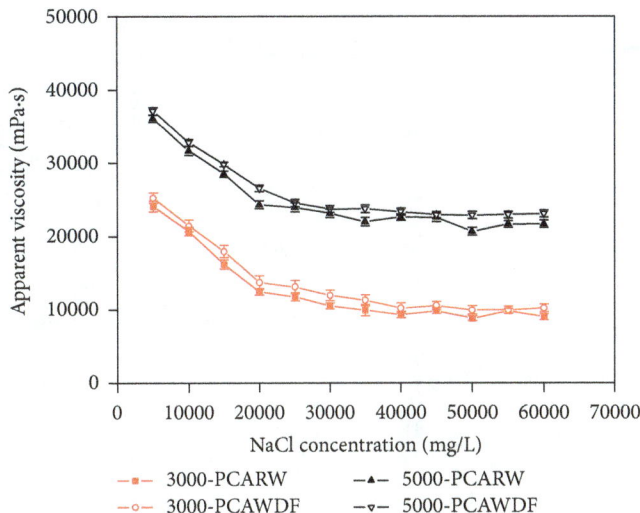

FIGURE 7: Apparent viscosity versus NaCl concentration curves at 65°C.

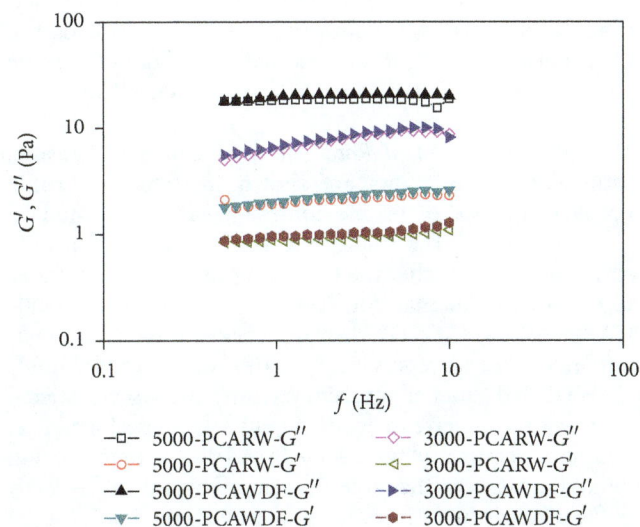

FIGURE 8: The viscosity modulus and elastic modulus versus scanning frequency curves of the profile control agents.

be stable for a long time at 65°C, so 3000-PCAWDF and 5000-PCAWDF showed satisfactory stability. This performance might ensure the validity of profile control to a certain extent.

3.7. Salt Resistance. The influences of NaCl on apparent viscosity of these profile control agents were carried out at 65°C. As shown in Figure 7, with the increase of NaCl concentration, the apparent viscosity of 3000-PCAWDF and 5000-PCAWDF decreased rapidly, and then it kept at a low value. Similarly, the measurement results of 3000-PCARW and 5000-PCARW displayed similar phenomena. However, compared with 3000-PCARW, 3000-PCAWDF displayed better antisalt under the same conditions. This phenomenon could be well explained by a more compact space mesh structure, which could reduce the effect of salt on the polymer molecules. The stronger the network structure, the stronger

the interaction between the molecular chains. Under the same concentration of salt, the molecular chain curled degree was smaller. So the apparent viscosity of 3000-PCAWDF was higher than that of 3000-PCARW in the same NaCl concentration.

3.8. Viscoelasticity. The viscosity modulus (G'') and elastic modulus (G') versus scanning frequency curves of these profile control agents are shown in Figure 8. The experimental results showed that the elastic modulus of 3000-PCAWDF was larger than that of the viscosity modulus in the whole scanning frequency range. In the linear scan range, the viscosity modulus and elastic modulus of 3000-PCAWDF were slightly higher than those of 3000-PCARW. And mass concentration of 5000 mg/L profile control agents showed the same regularity. This phenomenon might be explained by the

TABLE 3: The parameters and results of the core experiments.

Profile control agents	Sand filling model	Porosity (%)	Permeability before profile control (D)	Percentage of water absorption before profile control (%)	Percentage of water absorption after profile control (%)	Profile improvement rate (%)
3000-PCAWDF	1#	35.9	5.83	88	53	84.6
	2#	30.5	1.05	12	47	
3000-PCARW	3#	35.3	5.62	86	51	83.1
	4#	30.7	1.15	14	49	
5000-PCAWDF	5#	35.8	5.74	89	40	91.8
	6#	30.4	1.07	11	60	
5000-PCARW	7#	35.1	5.58	84	34	90.2
	8#	30.1	1.02	16	66	

space mesh structures of 3000-PCAWDF more closely than that of 3000-PCARW. The stronger the interaction between the molecular chains, the lower the degree of deformation of the molecular chain in the external force. So, in the same conditions, 3000-PCAWDF had a stronger flexibility.

3.9. Profile Improvement Rate. The parameters and results of the core experiments are shown in Table 3. Profile improvement rate of profile control agent was related to its concentration. The higher the concentration of profile control agent, the higher the profile improvement rate under the same experimental conditions. Compared with 3000-PCARW, 3000-PCAWDF displayed better profile improvement rate (84.6% versus 83.1%). This was because 3000-PCAWDF had a higher apparent viscosity and viscoelasticity, which were conducive to enhancing its sealing performance. In addition, compared with 3000-PCARW, due to the strong interaction between the molecular chains of 3000-PCAWDF, the shear failure in the injection process was small, and the gel strength of 3000-PCAWDF was higher after being injected in the core.

4. Conclusions

A novel profile control agent base on waste drilling fluid was successfully prepared using polymer, formaldehyde, resorcinol, and thiourea as raw materials under mild conditions. Compared with the profile control agent prepared by the recirculated water, PCAWDF exhibited comparable or better stability, salt resistance, viscoelasticity, and profile improvement rate due to the introduction of the flocculant and pH modifier. In addition, the crosslinking agent system had good compatibility with the polymer. This not only provided a new method for safe and environmentally friendly disposal of waste drilling fluid, but also helped to improve the effect of profile control for the water injection wells.

Conflicts of Interest

The authors declare no possible conflicts of interest.

Acknowledgments

This work was supported by the major project of Jidong Oilfield (ZC2015B01-02) and the projects of China National Petroleum Corporation (2014B-1113, KT2015-16-06).

References

[1] T. L. Chen, S. Lin, G. V. Chilingar, and Z.-S. Lin, "The reutilization of drilling fluid wastes as material for the manufacture of bricks," *Energy Sources, Part A: Recovery, Utilization and Environmental Effects*, vol. 32, no. 15, pp. 1399–1407, 2010.

[2] M. E. Hossain and M. Wajheeuddin, "The use of grass as an environmentally friendly additive in water-based drilling fluids," *Petroleum Science*, vol. 13, no. 2, pp. 292–303, 2016.

[3] S. M. Almudhhi, "Environmental impact of disposal of oil-based mud waste in Kuwait," *Petroleum Science and Technology*, vol. 34, no. 1, pp. 91–96, 2016.

[4] J. Zou, H. Zhu, F. Wang, H. Sui, and J. Fan, "Preparation of a new inorganic–organic composite flocculant used in solid–liquid separation for waste drilling fluid," *Chemical Engineering Journal*, vol. 171, no. 1, pp. 350–356, 2011.

[5] M. A. Sayyadnejad, H. R. Ghaffarian, and M. Saeidi, "Removal of hydrogen sulfide by zinc oxide nanoparticles in drilling fluid," *International Journal of Environmental Science and Technology*, vol. 5, no. 4, pp. 565–569, 2008.

[6] M. O. Benka-Coker and A. Olumagin, "Effects of waste drilling fluid on bacterial isolates from a mangrove swamp oilfield location in the Niger Delta of Nigeria," *Bioresource Technology*, vol. 55, no. 3, pp. 175–179, 1996.

[7] L. Ren, G. Zhao, T. Qiang, X. Wang, and X. Zhang, "Study on the application of cationic collagen copolymer in waste drilling fluid," *Speciality Petrochemicals*, vol. 30, no. 3, pp. 27–31, 2013.

[8] T. A. Bauder, K. A. Barbarick, J. F. Shanahan, P. D. Ayers, and P. L. Chapman, "Drilling fluid effects on crop growth and iron and zinc availability," *Journal of Environmental Quality*, vol. 28, no. 3, pp. 744–749, 1999.

[9] R. B. Brobst and P. M. Buszka, "The effect of three drilling fluids on ground water sample chemistry," *Groundwater Monitoring & Remediation*, vol. 6, no. 1, pp. 62–70, 1986.

[10] X. Wang, S. Zhang, L. Zhou, and L. Ren, "The hydrolysis of gelatin and its treatment for waste drilling fluid," *Advanced Materials Research*, vol. 281, pp. 141–146, 2011.

[11] D. Denney, "Holistic drilling-fluid and waste management in the fayetteville shale," *Journal of Petroleum Technology*, vol. 63, no. 11, pp. 68–70, 2011.

[12] B. Hou, C. Liang, H. Deng, S. Xie, M. Chen, and R. Wang, "Oil removing technology of residues from waste oil-based drilling fluid treated by solid-liquid separation," *Journal of Residuals Science and Technology*, vol. 9, no. 4, pp. 143–150, 2012.

[13] S. X. Xie, G. C. Jiang, M. Chen et al., "Treatment technology for waste drilling fluids in environmental sensitivity areas," *Energy Sources, Part A: Recovery, Utilization and Environmental Effects*, vol. 37, no. 8, pp. 817–824, 2015.

[14] C. Zou, M. Liang, X. Chen, and X. Yan, "β-Cyclodextrin modified cationic acrylamide polymers for flocculating waste drilling fluids," *Journal of Applied Polymer Science*, vol. 131, no. 9, pp. 93–98, 2014.

[15] S. X. Xie, G. C. Jiang, M. Chen et al., "Harmless treatment technology of waste oil-based drilling fluids," *Petroleum Science and Technology*, vol. 32, no. 9, pp. 1043–1049, 2014.

[16] J. S. Adeyinka, U. R. Iselema, and M. K. Oghenojoboh, "Effect of drilling fluid waste disposal on Owaza region of the Niger delta: an assessment of nitrate and sulphate ions on base metal leaching," *Journal of Scientific and Industrial Research*, vol. 63, no. 2, pp. 134–141, 2004.

[17] B. Hou, M. Chen, M. Liu, and Q. Xiong, "Safe disposal technology of waste oil-based drilling fluids," *Journal of the Japan Petroleum Institute*, vol. 56, no. 4, pp. 221–229, 2013.

[18] G. Jiang, S. Xie, M. Chen et al., "Oil dispersant preparation and mechanisms for waste oil-based drilling fluids," *Environmental Progress and Sustainable Energy*, vol. 31, no. 4, pp. 507–514, 2012.

17

Fatty Amides from Crude Rice Bran Oil as Green Corrosion Inhibitors

E. Reyes-Dorantes,[1,2] J. Zuñiga-Díaz,[1,2] A. Quinto-Hernandez,[1] J. Porcayo-Calderon,[2,3] J. G. Gonzalez-Rodriguez,[3] and L. Martinez-Gomez[2,4]

[1]Tecnológico Nacional de México, Instituto Tecnológico de Zacatepec, Calzada Instituto Tecnológico 27, 62780 Zacatepec, MOR, Mexico
[2]Instituto de Ciencias Físicas, Universidad Nacional Autónoma de México, Avenida Universidad, s/n, 62210 Cuernavaca, MOR, Mexico
[3]CIICAp, Universidad Autónoma del Estado de Morelos, Avenida Universidad 1001, 62209 Cuernavaca, MOR, Mexico
[4]Corrosion y Protección (CyP), Buffon 46, 11590 Mexico City, Mexico

Correspondence should be addressed to J. Porcayo-Calderon; jporcayoc@gmail.com

Academic Editor: Hassan Arida

Due to its high oil content, this research proposes the use of an agroindustrial byproduct (rice bran) as a sustainable option for the synthesis of corrosion inhibitors. From the crude rice bran oil, the synthesis of fatty amide-type corrosion inhibitors was carried out. The corrosion inhibitory capacity of the fatty amides was evaluated on an API X-70 steel using electrochemical techniques such as real-time corrosion monitoring and potentiodynamic polarization curves. As a corrosive medium, a CO_2-saturated solution (3.5% NaCl) was used at three temperatures (30, 50, and 70°C) and different concentrations of inhibitor (0, 5, 10, 25, 50, and 100 ppm). The results demonstrate that the sustainable use of agroindustrial byproducts is a good alternative to the synthesis of environmentally friendly inhibitors with high corrosion inhibition efficiencies.

1. Introduction

Corrosion is the degradation process of a material due to its exposure to the environment, and is one of the major problems affecting its performance, safety, and integrity. Despite the continuous advances in the field of materials with corrosion resistance, the use of corrosion inhibitors is one of the most practical and economical ways to control it [1, 2]. An inhibitor is defined as any chemical which, when added in small concentrations to an aggressive environment, significantly reduces the corrosion rate of the material in contact with the corrosive environment. A wide variety of organic and inorganic compounds have been used to control corrosion. It has been demonstrated that many organic molecules can act as corrosion inhibitors because of their affinity with the metal surfaces, replacing the adsorbed water molecules and thereby forming a molecular film that prevents the metal dissolution. In the particular case of environments where the presence of hydrocarbons predominates, the inhibitors based on amide- or imidazolines-type compounds are widely used [3–9]. The excellent performance of this type of corrosion inhibitors is associated with the fact that its molecules are constituted by two essential parts: an electron-rich polar part capable of adhering onto a metal surface through coordination bonds and a hydrophobic part which can effectively prevent the diffusion of contaminants present in the aggressive environment [2, 5, 9–12]. In addition to the above, it has also been reported that the presence of unsaturation or other functional groups of the alkyl chains also influences their inhibition efficacy [7, 9, 13]. These characteristics are provided by the nature of the vegetable oil used for the synthesis of this type of inhibitors.

From the point of view of sustainable development, it is imperative to search for new alternative nonconventional sources of vegetable oils. These new sources of vegetable oils can be used for applications as diverse as biofuels synthesis, biopolymers, corrosion inhibitors, and so forth [6, 9, 14, 15]. In this sense, rice bran is a byproduct of the rice grain milling process. In general, the overall yield of the rice milling process is about 60% white rice, 10% broken grains, 10% bran, and 20% husks [16]. The rice bran is the main oil source of the rice grain, in addition to a good content of proteins, tocopherols, and bioactive compounds. Its oil content varies from 10 to 23%, and its fatty acids are 47% monounsaturated, 33% polyunsaturated, and 20% saturated, in addition to a significant fraction of unsaponifiable components (4.3%) [17, 18]. However, only 10% of the world's rice bran production is used for oil extraction, and the remainder is used as a low-cost feed for cattle and poultry. The high oil content of the rice bran combined with the presence of the lipase enzyme causes that this byproduct has a short shelf life due to the degradation of its oil [16–18].

Therefore, the aim of this work is to explore the use of a sustainable source for the synthesis of corrosion inhibitors, where the source of vegetable oil is environmentally friendly by not affecting the food chain, does not replace crops, and does not use new crop fields for their production. The amide-type inhibitors synthesized were evaluated by two types of experimental measurements: real-time corrosion monitoring and potentiodynamic polarization curves. The corrosive medium used was a CO_2-saturated sodium chloride solution at different temperatures.

2. Experimental Procedure

2.1. Synthesis of Fatty Amides from Crude Rice Bran Oil.
The rice bran was obtained from a local mill (Puente de Ixtla, Morelos, Mexico). Rice bran was collected a few hours after its production. Crude rice bran oil (CRBO) content was determined by the Soxhlet method, where the solvent used was hexane. The extraction of the CRBO used for the synthesis process was carried out in a stirred batch system at room temperature, using a rice bran to solvent ratio of 1 : 10. Chemical and physicochemical characterization of the CRBO was made.

The method used for the synthesis of the fatty amides is a modification to the procedure suggested by Kumar et al. [19]. In general, the procedure consists of introducing one mole of crude oil with 3 moles of aminoethylethanolamine (AEEA) into the reactor (1 : 3 CRBO-AEEA ratio). The mixture is heated to 140°C with stirring at atmospheric pressure. Once the temperature is reached, the course of the reaction is monitored by thin-film chromatography (TLC), until the triglycerides disappear completely.

2.2. Electrochemical Evaluation of Fatty Amides as Corrosion Inhibitors.
The evaluation of the electrochemical performance of the synthesized compounds was performed on an API X-70 steel, commonly used in the manufacture of pipelines for the transportation of hydrocarbons. The performance of the inhibitors was evaluated by two types of electrochemical techniques, namely, real-time corrosion measurements and potentiodynamic polarization curves.

Real-time corrosion measurements were performed using an identical three-electrode probe. Steel samples with dimensions of $0.3 * 1.0 * 1.0$ cm were cut, and a conductive wire was spot-welded on one of its narrow sides. For the electrochemical tests, three steel samples were encapsulated in acrylic resin with a separation between them of 1 mm. Each encapsulation was abraded with silicon carbide abrasive paper from 120 to 600 grids; then the encapsulations were washed with alcohol and distilled water, dried, and immediately employed in the electrochemical tests. A solution of NaCl (3.5% by weight) saturated with CO_2 was used as the corrosive medium. The solution was saturated with CO_2 two hours prior to the tests and the CO_2 bubbling was maintained during the corrosion tests. Corrosion tests were for 24 hours, at a test temperature of 30, 50, and 70°C, with light stirring. Encapsulations were kept inside the electrolyte for one hour before any inhibitor was added. The concentrations of inhibitor evaluated were 0, 5, 10, 25, 50, and 100 ppm. For each electrochemical test, a volume of 400 mL of electrolyte was used. The multitechnique electrochemical monitoring equipment (SmartCET) is based on a combination of electrochemical techniques such as electrochemical noise (EN), linear polarization resistance (LPR), and harmonic distortion analysis (HDA), and as a result of this interrelationship, both the corrosion rate and the pitting corrosion behavior of the material under study are obtained. The measurement cycle is carried out for a period of 430 seconds as follows: measurements of EN in current and potential, second to second for 300 seconds, LPR/HDA measurements for 100 seconds, and electrolyte resistance measurement for 30 seconds. The details of this real-time monitoring technique are described in the scientific literature [20, 21].

For potentiodynamic polarization curves, a typical three-electrode arrangement was used where the reference electrode was a Pt wire and as counterelectrode a high-purity graphite rod was used. From this type of tests, it is possible to determine the electrochemical parameters such as corrosion potential and corrosion rate, in addition to the anodic and cathodic slopes from the extrapolation of the Tafel slopes of the obtained curves. In this case, steel samples with a reaction area of 1 cm^2 were encapsulated in acrylic resin. Encapsulations were abraded with silicon carbide abrasive paper from 120 to 600 grids; after that, they were washed with alcohol and distilled water, dried, and immediately employed in the electrochemical test. Corrosive conditions were the same as those of real-time corrosion measurements. In order to ensure that the inhibitor achieved the maximum surface coverage of the working electrode, before starting the assay, samples were allowed to stabilize for 24 hours into electrolyte prior to performing the assay. Potentiodynamic polarization tests were performed from −400 mV to 600 mV with respect to open circuit potential (E_{corr}) at a sweep rate of 1 mV/s. Electrochemical parameters were calculated using the extrapolation Tafel method considering an extrapolation potential of ±250 mV around the value of the corrosion potential. Potentiodynamic polarization curves were carried out using an ACM Instruments zero-resistance ammeter (ZRA) coupled to a personal computer.

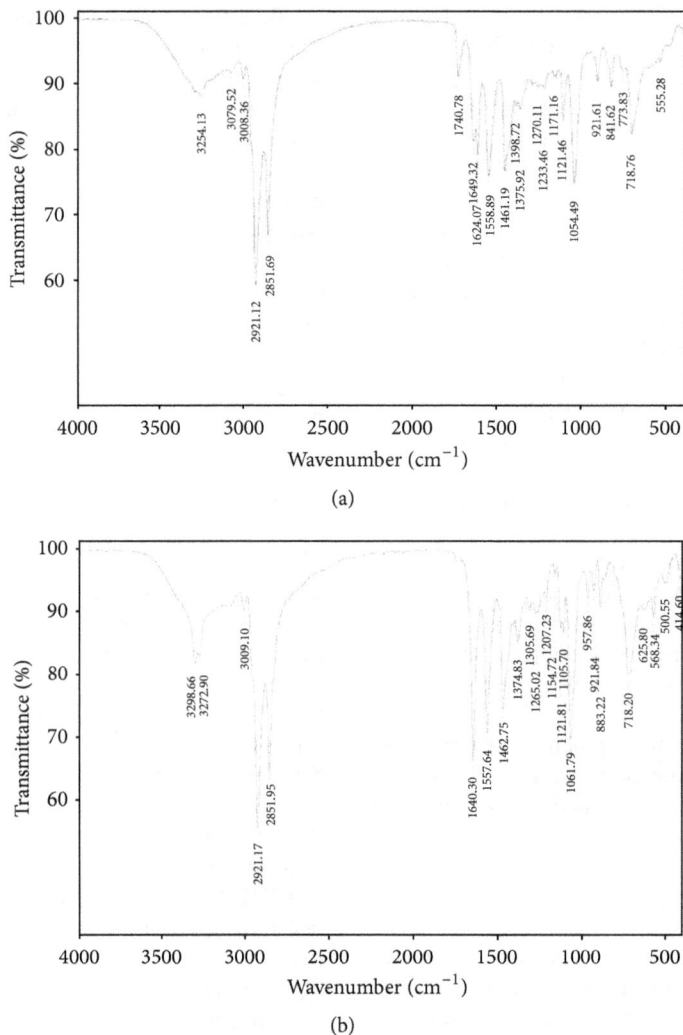

FIGURE 1: FT-IR spectra of the synthesis of the fatty amides from CRBO and AEEA: (a) initial mixture ($t = 0$ minutes reaction) and (b) final mixture (80 minutes reaction).

3. Results and Discussion

3.1. Synthesis of Fatty Amides. Rice bran analysis indicated a crude oil content of 21% (dry basis), which is in agreement with data reported in the literature [16–18]. Its molecular weight calculated from its saponification index was 923.36, and its chemical characterization showed an oleic acid content of 48.48%, 35.26% of linoleic acid, and 14.54% of palmitic acid, similar values to those reported in the literature for this type of oil [16, 17]. Knowledge of the type of fatty acids present in the oil is important, given that both the length and unsaturation of the hydrocarbon chain contribute to the mode of adsorption and inhibition efficiency of the inhibitor molecule [7].

Figure 1 shows the FT-IR spectra of the fatty amides synthesized from CRBO and AEEA. From Figure 1, it can be seen that the signal at $1740 \, cm^{-1}$ corresponding to the C=O stretching of the triglycerides disappears, and a new signal is observed at $1640 \, cm^{-1}$. It is known that the presence of

this signal is characteristic of the peaks corresponding to the amidation product (fatty amides) [10–19]. Evolution of the FT-IR spectra at different times (spectra not shown) indicated that, after 80 minutes of reaction, the complete disappearance of the triglycerides was achieved, suggesting formation of fatty amides.

3.2. Real-Time Corrosion Measurements of Synthesized Fatty Amides. Figure 2 shows the effect of the inhibitor addition (fatty amide) on the corrosion rate (based on RPL measurements) of API X-70 steel immersed in CO_2-saturated saline solution at different test temperatures. In all cases, the inhibitor was added 60 minutes after the measurements were started. RPL measurement involves measuring the polarization resistance (Rp) using a sinusoidal polarization of small amplitude ($\pm 20 \, mV$) of the working electrode (steel under study). In this case, the slope of the potential-current sweep is Rp, which is inversely proportional to the corrosion current density, and it is subsequently transformed to corrosion rate.

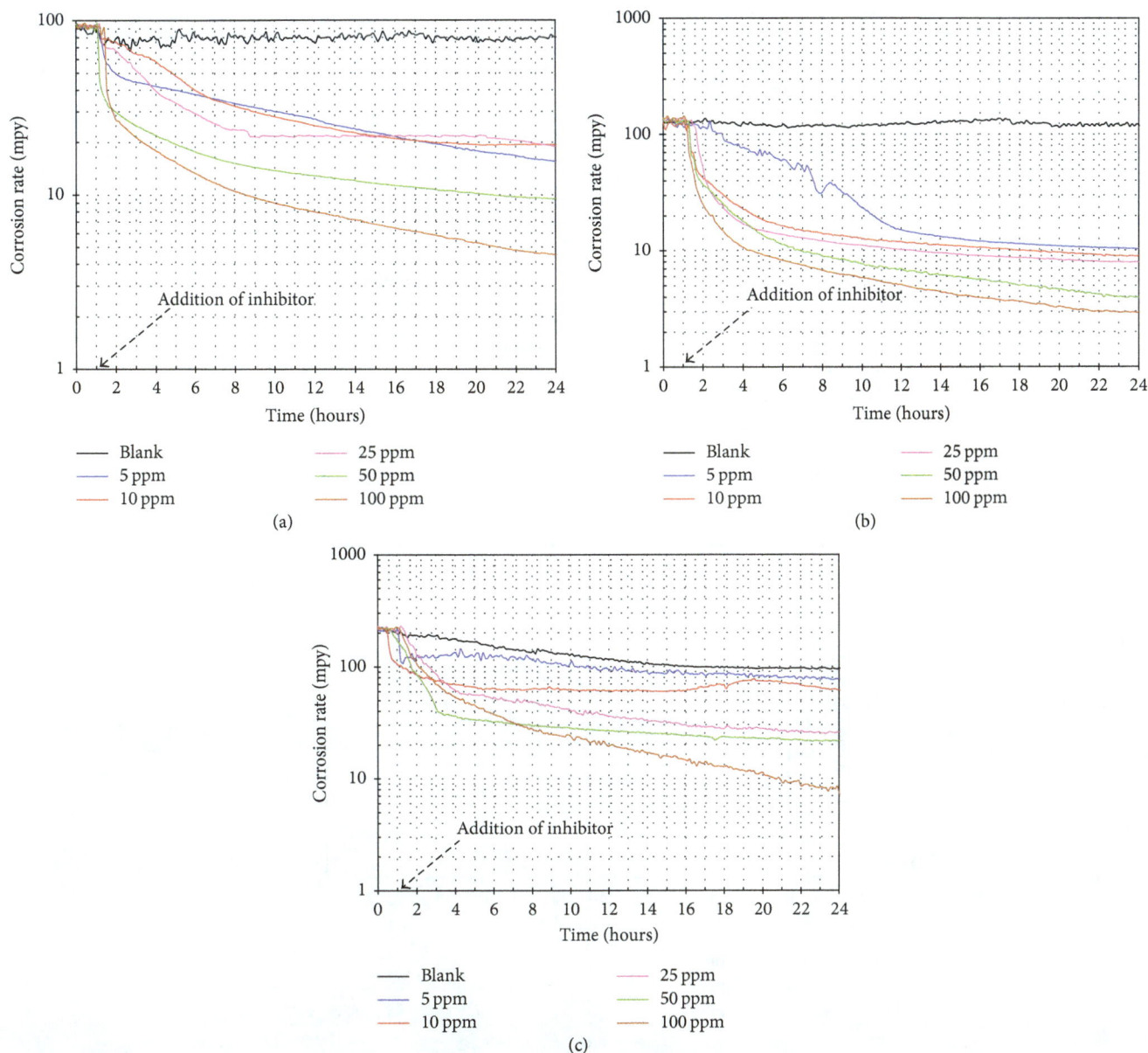

FIGURE 2: Variation of corrosion rate determined by RPL (real-time monitoring) versus time for API X-70 steel exposed to 3.5% NaCl solution saturated with CO_2 at different concentrations of fatty amide: (a) 30°C, (b) 50°C, and (c) 70°C.

It is observed that in the absence of inhibitor the corrosion rate of API X-70 steel is higher regardless of temperature. Only at 70°C, it is observed that this tends to decrease markedly as a function of time. This may be due to precipitation of $FeCO_3$ crystals onto steel surface. In general, this behavior is due to the electrochemical nature of the corrosion process of carbon steel, where both the anodic iron dissolution and the cathodic evolution of hydrogen take place [22]. The protection process (iron carbonate precipitation) is due to the CO_2 hydration which causes the formation of carbonic acid:

$$CO_2 + H_2O \longrightarrow H_2CO_3 \tag{1}$$

Since the electrolyte is deaerated, the possible dominant cathodic reactions can be H^+ ions reduction, water reduction,

and the carbonic acid dissociation:

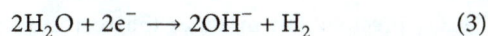

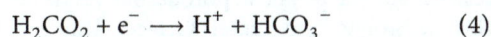

$$2H^+ + 2e^- \longrightarrow H_2 \tag{2}$$

$$2H_2O + 2e^- \longrightarrow 2OH^- + H_2 \tag{3}$$

$$H_2CO_2 + e^- \longrightarrow H^+ + HCO_3^- \tag{4}$$

$$HCO_3^- + e^- \longrightarrow H^+ + CO_3^{-2} \tag{5}$$

And, the primary anodic reaction is the dissolution of iron:

$$Fe \longrightarrow Fe^{+2} + 2e^- \tag{6}$$

(a)

(b)

(c)

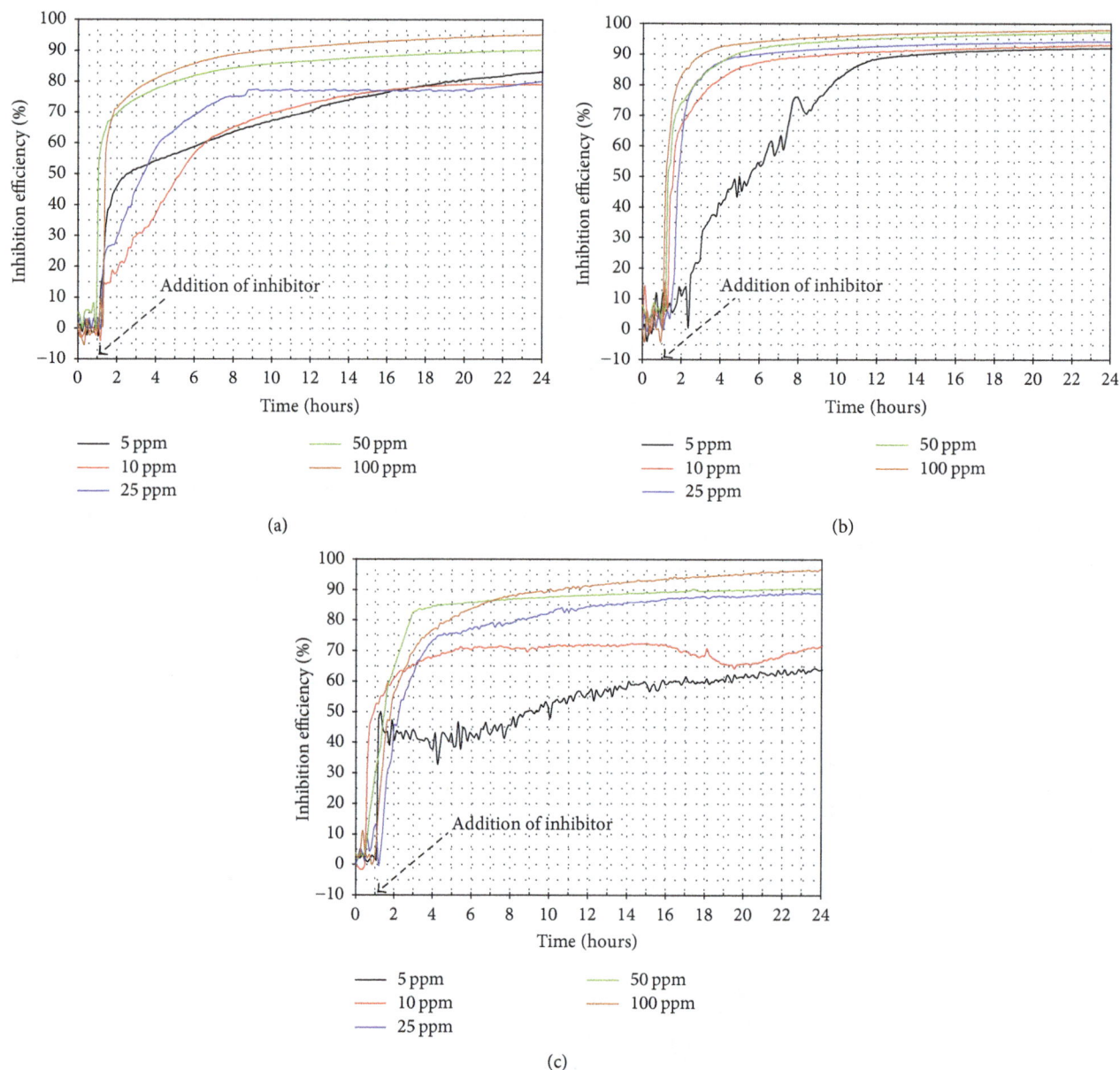

FIGURE 3: Variation of the inhibition efficiency determined from RPL measurements (real-time monitoring): (a) 30°C, (b) 50°C, and (c) 70°C.

During the steel corrosion process, the iron carbonate precipitation onto steel surface is possible, according to

$$Fe^{+2} + CO_3^{-2} \longrightarrow FeCO_3 \qquad (7)$$

$FeCO_3$ precipitation modifies the corrosion process kinetics, because a physical porous barrier is formed between the electrolyte and metal surface. This barrier influences the transport of the corrosive species of the electrolyte. However, the protection of this porous layer depends on several environmental conditions, such as iron concentration, pH solution, temperature, CO_2 partial pressure, mechanical forces due to flow conditions, and steel microstructure [23]. Due to the imperfect nature of this protective scale (cracks and porosity), the electrolyte can permeate to the metal surface and corrode the steel causing the detachment of the protective scale [24]. Therefore, the protection provided by the precipitation of iron carbonate is limited.

On the other hand, in the inhibitor's presence, the corrosion rate decreases as its concentration increases, regardless of the test temperature. At the maximum concentration of inhibitor, a reduction of one order of magnitude in the corrosion rate is observed. In all cases, a drastic drop in the corrosion rate values is observed in the first 3-4 hours after the inhibitor has been added; subsequently, the corrosion rate tends to decrease slowly without reaching a defined steady state. This reduction in the corrosion rate is due to the adsorption of an inhibitor film onto steel surface which acts as a more effective barrier to permeation of the corrosive electrolyte towards the metal surface. The effectiveness of an

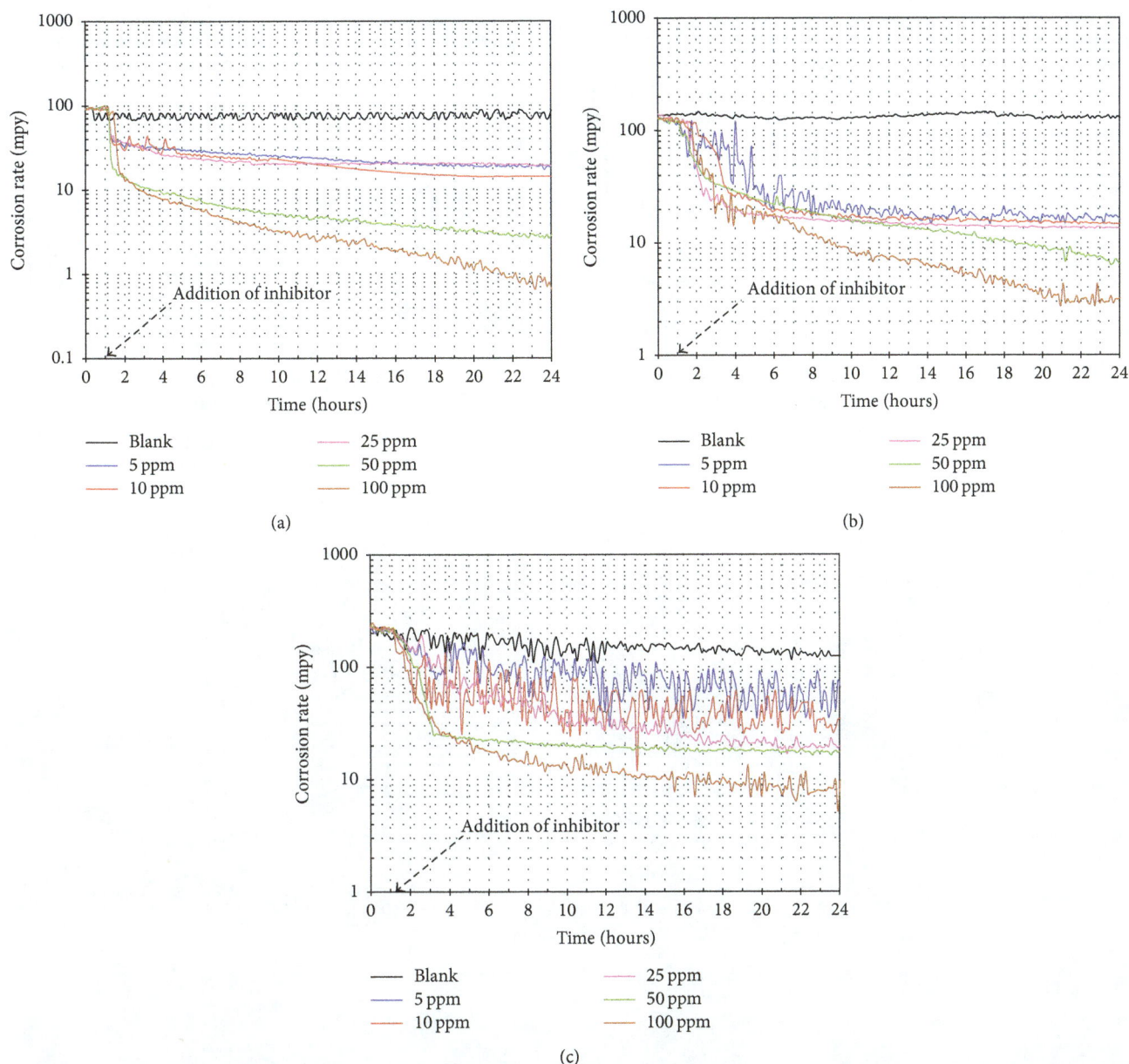

FIGURE 4: Variation of corrosion rate as determined by HDA (real-time monitoring) versus time for API X-70 steel exposed to CO_2-saturated saline solution at different inhibitor concentration: (a) 30°C, (b) 50°C, and (c) 70°C.

inhibitor of organic nature is the result of the physicochemical properties of the molecule, the nature of its functional groups, possible steric effects, and the electron density of the donor atoms [9]. Inhibitor adsorption depends on the interaction of the π orbitals of the inhibitor molecule with the orbitals of the atoms of the metal surface [9].

Figure 3 shows the variation of the inhibition efficiency of fatty amides as a function of time (from the RPL measurements of Figure 2). The inhibition efficiency was determined according to the following relation:

$$E\,(\%) = \frac{CR_i - CR_b}{CR_i} * 100, \qquad (8)$$

where CR_b is the corrosion rate in inhibitor absence and CR_i when the inhibitor is present. In all cases, the inhibition efficiency increased with the added inhibitor concentration, regardless of test temperature. The inhibition efficiency was greater than 95% at the maximum inhibitor concentration evaluated, and, in all cases, the steady state was not reached. This implies that, in longer periods of exposure, the inhibition efficiency would tend to increase.

Figure 4 shows the variation in the corrosion rate, determined from the HDA measurements, as a function of time for the API X-70 steel exposed in CO_2-saturated saline solution at 30, 50, and 70°C at different concentrations of inhibitor. The analysis of harmonics and the nonlinear response are based on the disturbance of an AC signal and the

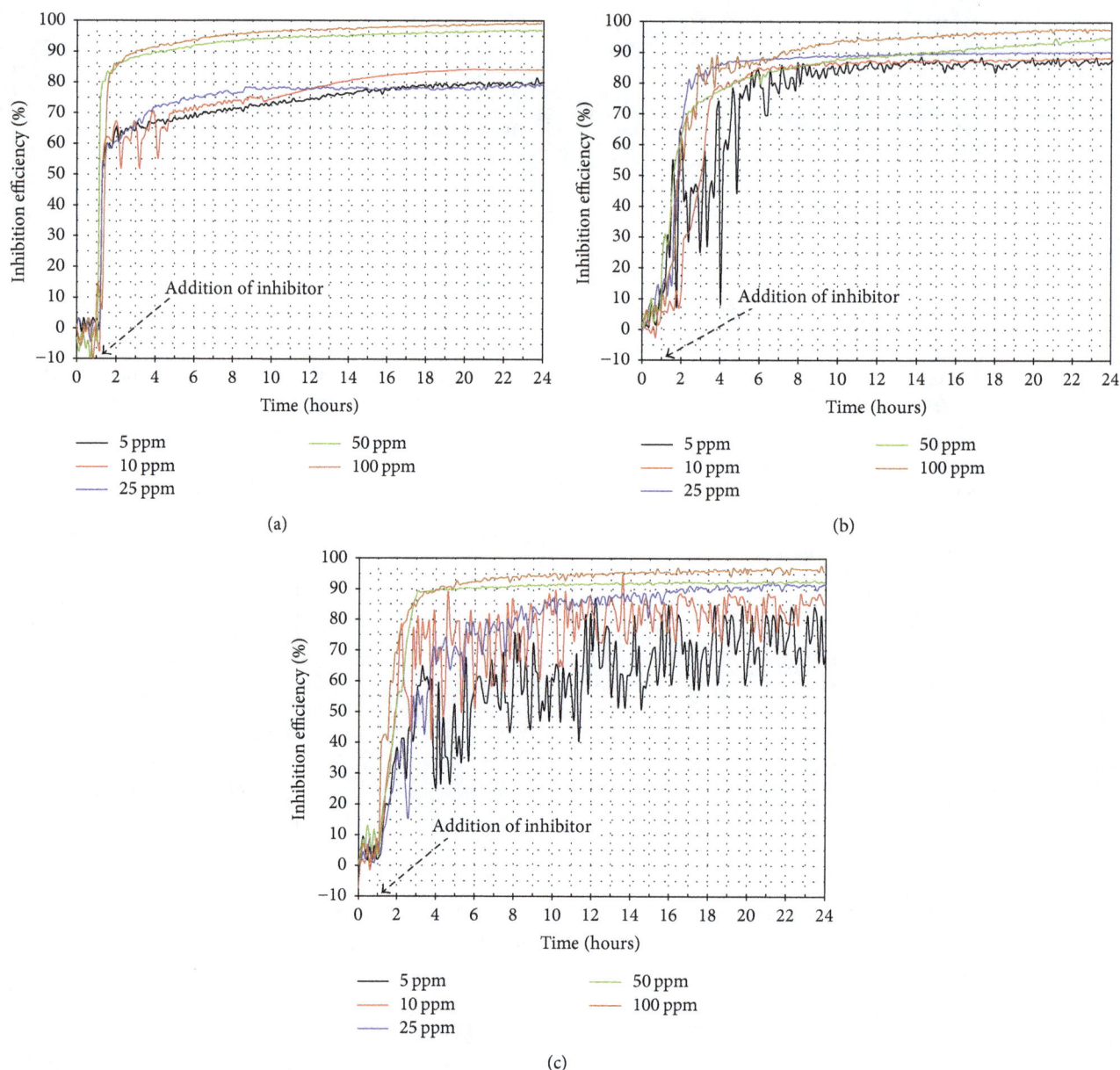

FIGURE 5: Variation of inhibition efficiency determined from HDA-based measurements (real-time monitoring): (a) 30°C, (b) 50°C, and (c) 70°C.

frequency domain, correspondingly. The latter is a measure of the nonlinear current distortion, originated during the RPL measurement, using a 10 mHz sine wave (50 mV peak-to-peak), and analyzes current signals at 10, 20, and 30 mHz. This allows obtaining the kinetic parameters of the corrosion process. The harmonics analysis in the current response offers the possibility of obtaining the corrosion rate and the anodic and cathodic parameters of Tafel.

In general, the corrosion rates obtained from the HDA-based measurements suggest a similar trend to those obtained from the RPL measurements. The possible differences observed are due to the fact that, for the calculation of the corrosion rate, the HDA technique uses instantaneous values of the anodic and cathodic Tafel slopes and, in RPL-based

measurements (Figure 2), the technique uses fixed values (120 mV). Observation of disturbances in the measurements is notorious, and their magnitude increases with the temperature and even more when the inhibitor is present. Since such perturbations can be interpreted as adsorption-desorption processes of the inhibitor molecules and these processes promote the presence of active sites onto metal surface, both corrosion potential values (E_{corr}) and Tafel kinetic parameters can be affected [7].

Thus, the calculation of the corrosion rate to use instantaneous Tafel values is advisable. Employing constant values will show a smoothed trend which suppresses the transient processes. In this sense, it can be observed that the amplitude of the transients increases with the temperature and, for

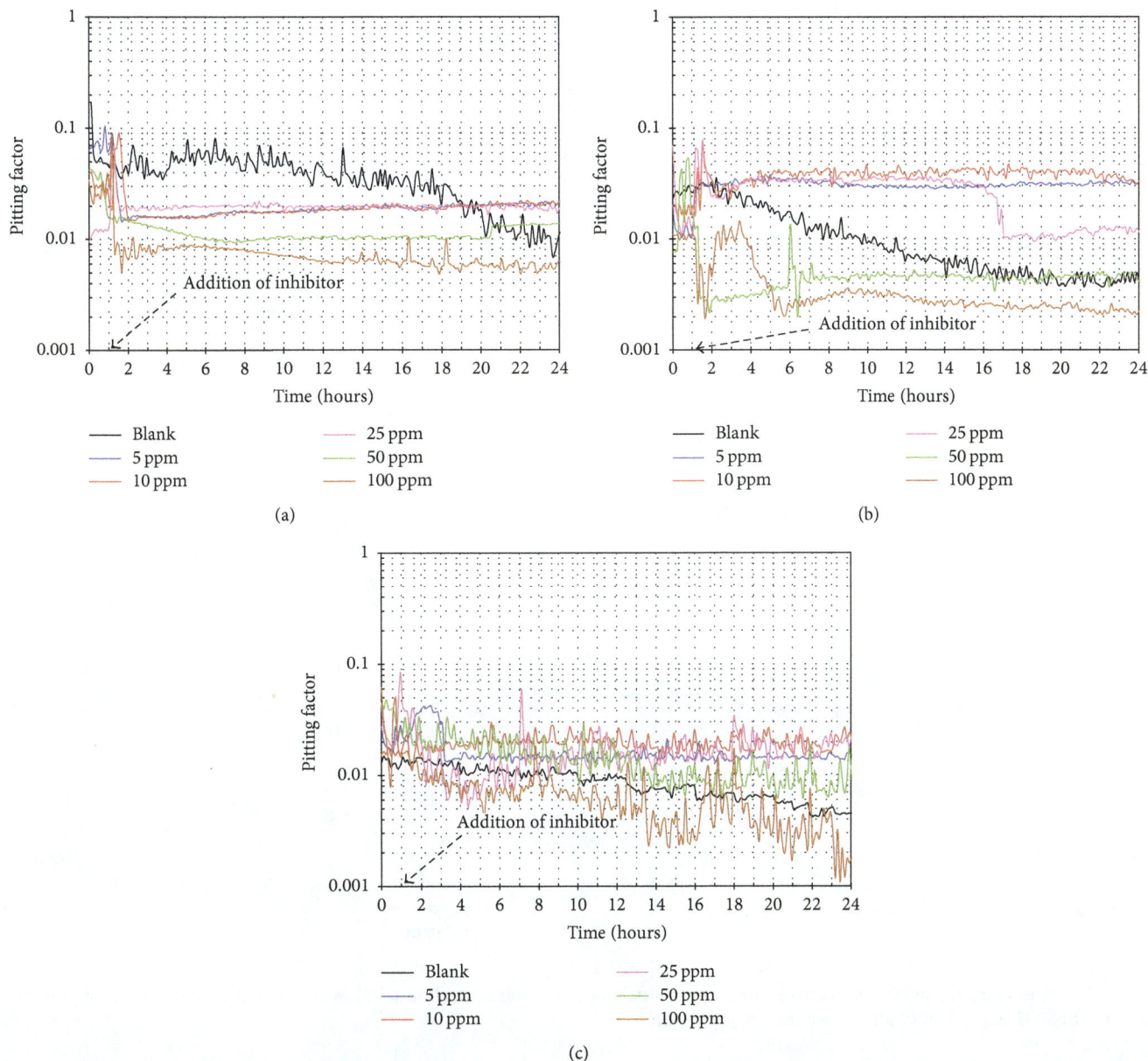

FIGURE 6: Variation in PF values as a function of time for API X-70 steel exposed to CO$_2$-saturated saline solution at different inhibitor concentrations: (a) 30°C, (b) 50°C, and (c) 70°C.

a given temperature, decreases with the inhibitor concentration. This suggests two facts: (1) the inhibitor molecules adsorption can be compromised by increasing temperature and (2) when the inhibitor concentration is increased, there are a greater number of molecules occupying the active sites left by the desorption of the inhibitor molecules.

Similarly, the evolution of the inhibition efficiency from the HDA-based corrosion rate measurements (Figure 4) is shown in Figure 5. As indicated, the observed transients can be interpreted as a measure of the adsorption stability of the inhibitor molecules onto metal surface. Mainly, increasing the inhibitor concentration decreases both the amplitude and frequency of the instabilities.

Figure 6 shows the variation of pitting factor (PF) as a function of time for API X-70 steel exposed to CO$_2$-saturated saline solution at different test temperatures, with and without inhibitor addition. This parameter is obtained from EN and HDA measurements, where EN refers to current and potential fluctuations that occur onto steel surface without disturbance of its potential. EN measurements provide information on the activity level of the corrosion process and the dominant corrosion mechanism. The pitting factor is defined as follows [25]:

$$\text{PF} = \frac{\sigma I_{\text{EN}}}{A_{\text{WE}} I_{\text{HA}}}, \tag{9}$$

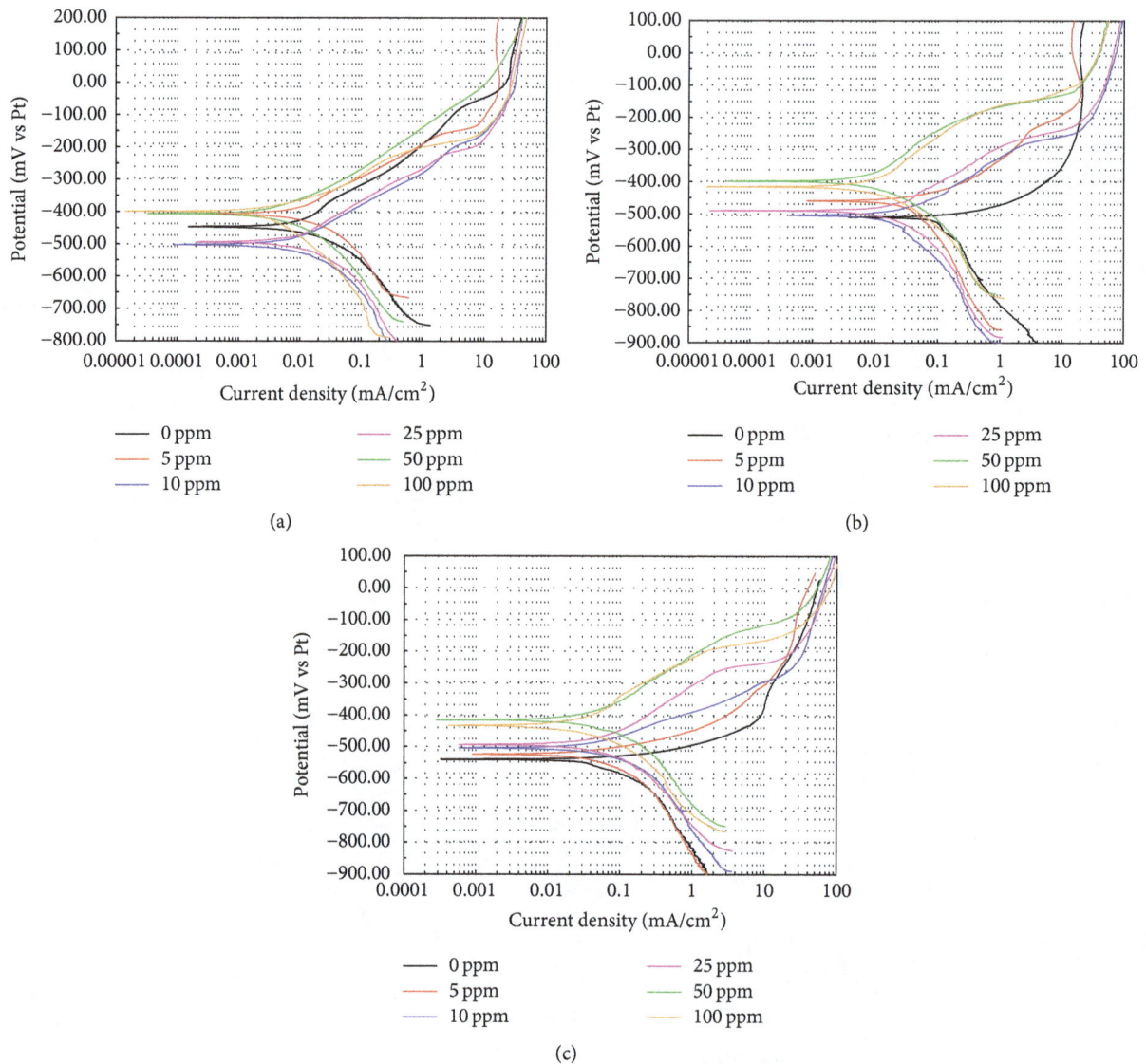

FIGURE 7: Potentiodynamic polarization curves for API X-70 steel in CO_2-saturated saline solution at different concentrations of inhibitor: (a) 30°C, (b) 50°C, and (c) 70°C, after 24 hours of stabilization.

where σI_{EN} is the standard deviation of the current (determined by electrochemical noise), A_{WE} is the steel reaction area, cm^2, and I_{HA} is the corrosion current density obtained from harmonic analysis, in A/cm^2. The PF is an index which varies within 0-1 and is an estimate of the localized attack. PF values are interpreted as follows. PF values < 0.01 suggest generalized corrosion, in the range of 0.01–0.10, generalized corrosion dominates in an intermediate zone, and values > 0.1 imply localized corrosion.

According to Figure 6, in the absence of inhibitors, the PF values decrease as the temperature increases, whereas, at a given temperature, the PF values tendency decreases as the immersion time increases. In all cases, PF values move from the region of generalized corrosion with pitting attack to the region of uniform corrosion. This is expected because the $FeCO_3$ precipitation onto steel surface gives it some protection, and this precipitation is favored by increasing

the temperature. However, in the inhibitor's presence, we observed that the PF values are lower as the inhibitor concentration increases in every case (lower values than those observed without presence of inhibitor), reaching values of uniform corrosion. In addition, it was observed that the fluctuations in PF values were higher if the temperature increased. This is possible due to the adsorption-desorption processes of the inhibitor molecules.

3.3. *Polarization Curves.* Figure 7 shows the polarization curves of API X-70 steel in CO_2-saturated saline solution and different inhibitor concentrations at 30°C, 50°C, and 70°C, respectively. These tests show whether the material under study shows active, passive, or transpassive behavior [26]. In the absence of inhibitor, the tendency of the anodic branch shows that the steel undergoes a continuous dissolution process and that the steel dissolution increases with the test

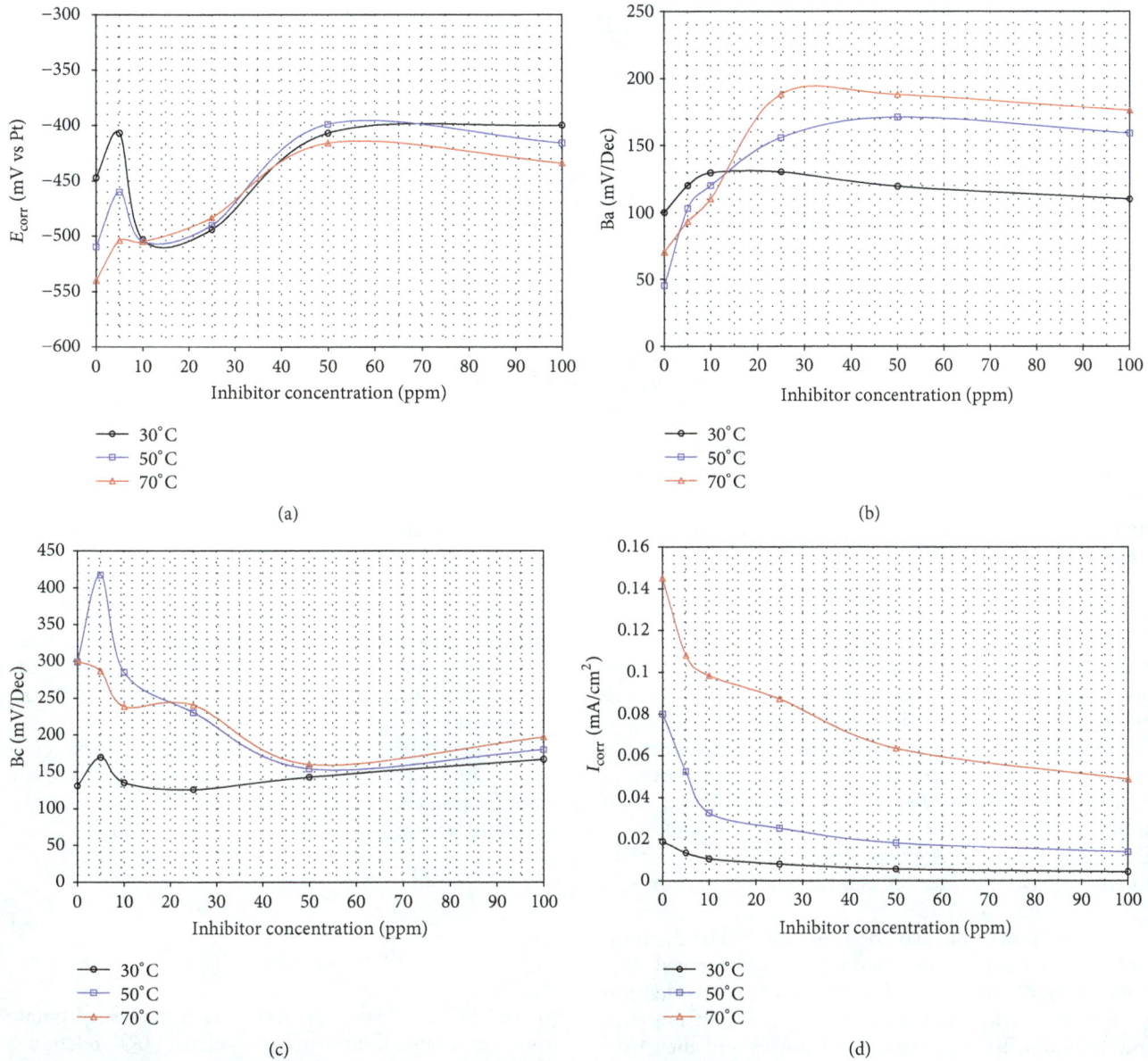

FIGURE 8: Electrochemical parameters for API X-70 steel in CO_2-saturated saline solution as a function of inhibitor concentration and temperature: (a) corrosion potential, (b) anodic slope, (c) cathodic slope, and (d) corrosion current density.

temperature. It has been suggested that the Fe dissolution can occur according to the following reactions [27, 28]:

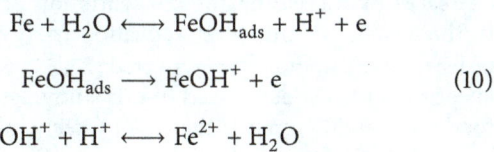

$$Fe + H_2O \longleftrightarrow FeOH_{ads} + H^+ + e$$

$$FeOH_{ads} \longrightarrow FeOH^+ + e \qquad (10)$$

$$FeOH^+ + H^+ \longleftrightarrow Fe^{2+} + H_2O$$

where the global reaction of anodic dissolution is that represented by (6) [29, 30].

The observed behavior shows that the steel does not have the ability to form a passive layer that protects it from the aggressive environment; that is, its corrosion products are not protective. At temperatures higher than 50°C and longest exposure times, the precipitation of an iron carbonate layer

onto steel surface is possible, according to (7). However, the $FeCO_3$ layer offers limited protection due to the presence of imperfections (porosity and cracks), whereby the electrolyte can penetrate and corrode the steel surface causing its detachment [24].

However, when the inhibitor is added, there is an increase in the slope of the anodic branch indicating a decrease in the active behavior, and E_{corr} values move towards nobler values, except at 30°C where, with 10 ppm and 25 ppm of inhibitor, a shift towards more active values was observed. This may be due to the low test temperature which did not facilitate the optimal dispersion of the inhibitor. The dispersion of the inhibitor is favored by increasing the temperature, facilitating its transport towards the metal surface. Similarly, in the presence of the inhibitor, the polarization curves moved towards

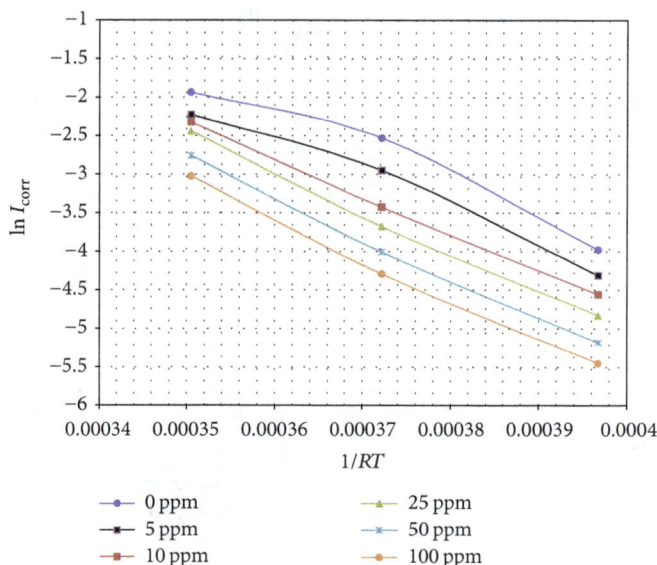

FIGURE 9: $\ln(I_{corr})$ - $1/RT$ relationship for the API X-70 steel evaluated in CO_2-saturated saline solution as a function of the inhibitor concentration.

the lower current density region, indicating a decrease in the corrosion rate of the steel. On the other hand, it is observed that the cathodic branch shows the same shape and trend at all temperatures. This indicates that the reduction reaction is the same, and it is activated thermally, since, by increasing the temperature, the cathodic branch moves to higher current densities. Because the solution is deaerated (continuous bubbling of CO_2, $CO_2 + H_2O = H_2CO_3$), the main possible cathodic reactions are the H^+ ions reduction, the carbonic acid dissociation, and the water reduction [22, 29, 30], according to (2)–(5).

Figure 8 shows the evolution of the electrochemical parameters obtained from Figure 7. It is observed that the tendency of the E_{corr} values (Figure 8(a)) is identical regardless of the test temperature. In general, in presence of the inhibitor, the E_{corr} values are nobler and they tend to increase as a function of the inhibitor concentration. At the same time, E_{corr} values are more active by increasing the temperature; this is because the corrosion phenomenon is a thermally activated process. The tendency in the E_{corr} values allows assuming that the inhibitor evaluated is of the anodic type and that its main action is to suppress the anodic reaction (Fe dissolution). This is supported by the evolution of the anodic slope values (Figure 8(b)) where it is observed that these values are higher than those observed in the absence of inhibitor. Figure 8(c) shows a similar trend in the cathodic slope values; that is, the values increase by increasing temperature. This is congruent with the dissolved CO_2 content. The content of dissolved CO_2 is a function of temperature, solutes concentration, and pressure; that is, increasing the temperature decreases the amount of dissolved CO_2 and therefore increases the cathodic slope. This is a feature of processes controlled by mass transfer [31, 32]. The same occurs with the dissolved O_2 in the solution. In general, at $30°C$, the values tend to increase with the inhibitor concentration, and, at higher temperatures, the opposite occurs.

Figure 8(d) shows that the steel corrosion rate decreases as the inhibitor concentration increases. This is congruent with the observed from the real-time measurements discussed previously.

The dependence of the metal dissolution rate on temperature can be explained using the Arrhenius equation, where the activation energy of the metal dissolution process can be calculated according to [27]

$$I_{corr} = k \exp\left(-\frac{\Delta E}{RT}\right),$$

$$\ln(I_{corr}) = \ln k - \frac{\Delta E}{RT}, \tag{11}$$

where ΔE ($J\,mol^{-1}$) is the activation energy, which is obtained from the slope of the $\ln(I_{corr})$ versus $1/RT$ relationship, k is a constant, and R is the gas constant ($8.314472\,J/K$-mol). Figure 9 shows the $\ln(I_{corr})$ versus $1/RT$ relationship as a function of inhibitor concentration. The calculated activation energies were 44.5, 45.35, 48.31, 51.55, 52.30, and $52.31\,kJ\,mol^{-1}$ for 0, 5, 10, 25, 50, and 100 ppm, respectively. This indicates that the anodic reaction is affected by the presence of the inhibitor and that the Fe dissolution rate decreases by increasing the concentration of the inhibitor. In the absence of inhibitor, activation energy values of the order of $14–30\,kJ\,mol^{-1}$ are reported [27, 33]. Such values are lower than those determined in this study; nevertheless, the fundamental difference is the type of steel evaluated and the NaCl content of the electrolyte.

In order to verify the decrease in Fe dissolution rate, additional experiments based on electrochemical impedance spectroscopy (EIS) measurements were performed. Figure 10 shows the evolution of the EIS spectra as a function of inhibitor concentration after 24 hours of immersion in the electrolyte. From the figure, it is clear that, in the presence of the inhibitor, the magnitude of the capacitive semicircle

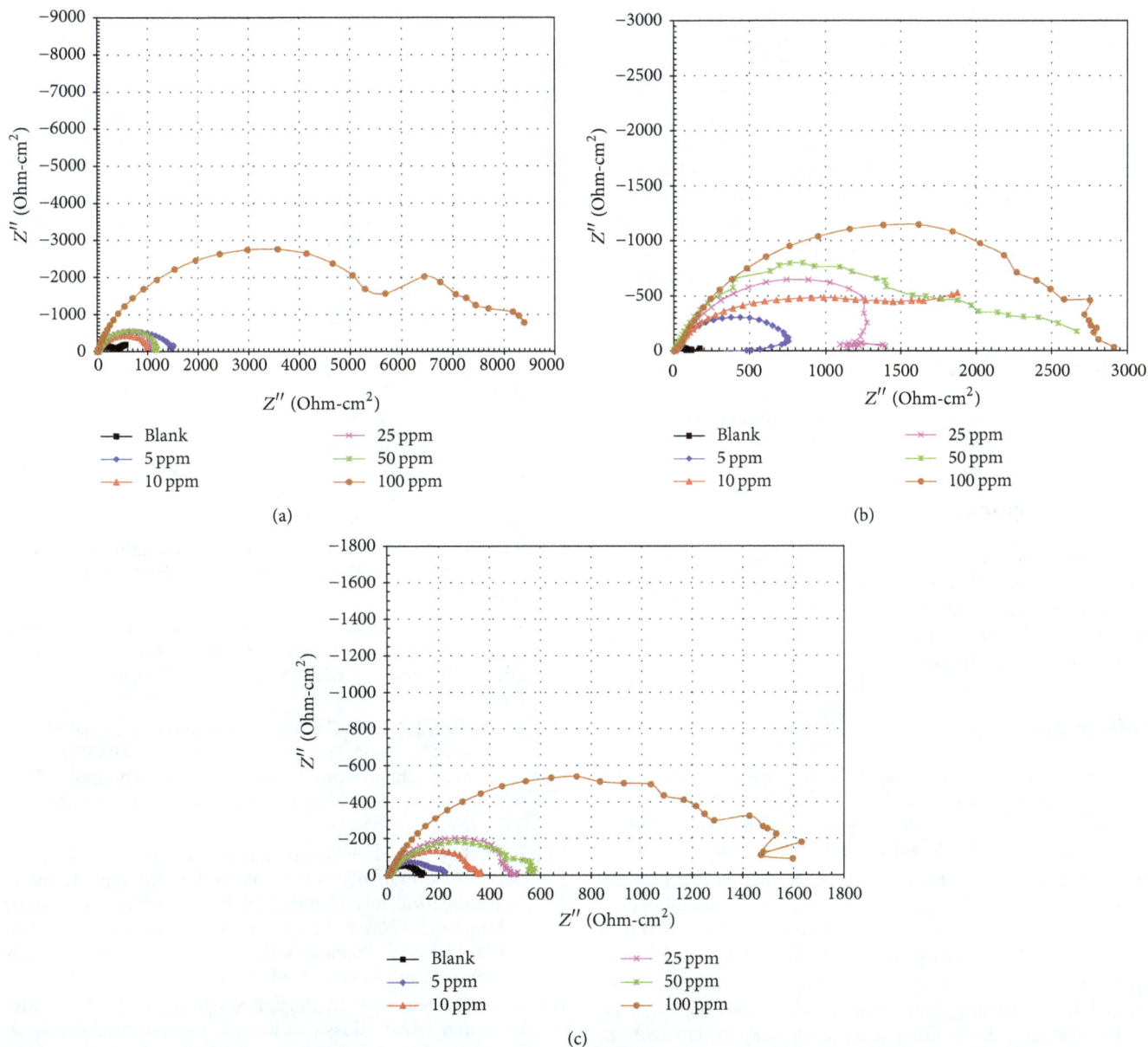

FIGURE 10: EIS spectra for API X-70 steel in CO_2-saturated saline solution at different concentrations of inhibitor: (a) 30°C, (b) 50°C, and (c) 70°C, after 24 hours of immersion.

increases, and likewise its diameter also increases by increasing the inhibitor concentration. This shows that the charge transfer resistance of the steel increases and consequently decreases its dissolution rate [6, 7].

The fatty amides synthesized from the crude rice bran oil act as corrosion inhibitors, due to the high unsaturation content of their alkyl chains. In short, the inhibitory efficacy was excellent. This allows the adsorption of the inhibitor molecule onto metal surface, thereby decreasing its corrosion rate independently of the test temperature. The above results demonstrate that the use of agroindustrial byproducts (rice bran) is a sustainable source for the synthesis of corrosion inhibitors. Notwithstanding the foregoing in the scientific literature, there are few studies that take into account the

byproducts of the rice milling process as an important source for the synthesis of green corrosion inhibitors; for example, the use of the rice husk as raw material for obtaining nanosilicate [34], tannins [35], and momilactone A [36] and the use of rice bran for the extraction of phytic acid have been reported [37–39], and only one study reported the use of rice bran oil for synthesis of fatty acid triazoles [40].

4. Conclusions

Crude rice bran oil is a promising source for the synthesis of bioproducts such as fatty acid amides. The unsaturated fatty acid content of crude rice bran oil makes it an ideal candidate for the synthesis of corrosion inhibitors. The electrochemical evaluations performed showed that the fatty amides acted as corrosion inhibitors at all test temperatures. In all cases, the inhibition efficiency was at least 95% for the optimum inhibitor concentration. From the real-time measurements, it was observed that, in the presence of the inhibitor, the pitting factor values were lower than those observed in the absence of inhibitor. This is attributed to the adsorption of the inhibitor molecule onto metal surface which prevented the diffusion of the aggressive ions of the electrolyte. Calculation of the activation energy of the corrosion process showed that the inhibitor suppresses the anodic reaction and that this effect increases with the concentration of added inhibitor.

Conflicts of Interest

The authors declare that there are no conflicts of interest regarding the publication of this paper.

Acknowledgments

Financial support from Consejo Nacional de Ciencia y Tecnología (CONACYT, Mexico) (Project 159898) is gratefully acknowledged. E. Reyes-Dorantes and J. Zuñiga-Díaz thank CONACYT, Mexico, for the financial support given to realize their postgraduate studies.

References

[1] M. Heydari and M. Javidi, "Corrosion inhibition and adsorption behaviour of an amido-imidazoline derivative on API 5L X52 steel in CO_2-saturated solution and synergistic effect of iodide ions," *Corrosion Science*, vol. 61, pp. 148–155, 2012.

[2] M. P. Desimone, G. Gordillo, and S. N. Simison, "The effect of temperature and concentration on the corrosion inhibition mechanism of an amphiphilic amido-amine in CO_2 saturated solution," *Corrosion Science*, vol. 53, no. 12, pp. 4033–4043, 2011.

[3] H. M. Abd El-Lateef, V. M. Abbasov, L. I. Aliyeva, E. E. Qasimov, and I. T. Ismayilov, "Inhibition of carbon steel corrosion in CO_2-saturated brine using some newly surfactants based on palm oil: experimental and theoretical investigations," *Materials Chemistry and Physics*, vol. 142, no. 2-3, pp. 502–512, 2013.

[4] P. B. Raja and M. G. Sethuraman, "Natural products as corrosion inhibitor for metals in corrosive media—a review," *Materials Letters*, vol. 62, no. 1, pp. 113–116, 2008.

[5] V. Jovancicevic, S. Ramachandran, and P. Prince, "Inhibition of carbon dioxide corrosion of mild steel by imidazolines and their precursors," *Corrosion*, vol. 55, pp. 449–455, 1999.

[6] J. Porcayo-Calderon, L. M. Martínez De La Escalera, J. Canto, and M. Casales-Diaz, "Imidazoline derivatives based on coffee oil as CO_2 corrosion inhibitor," *International Journal of Electrochemical Science*, vol. 10, pp. 3160–3176, 2015.

[7] J. Porcayo-Calderon, I. Regla, E. Vazquez-Velez et al., "Effect of the unsaturation of the hydrocarbon chain of fatty-amides on the CO_2-corrosion of carbon steel using EIS and real-time corrosion measurement," *Journal of Spectroscopy*, vol. 2015, Article ID 184140, 10 pages, 2015.

[8] M. Lopez, J. Porcayo-Calderon, M. Casales-Diaz et al., "Internal corrosion solution for gathering production gas pipelines involving palm oil amide based corrosion inhibitors," *International Journal of Electrochemical Science*, vol. 10, pp. 7166–7179, 2015.

[9] S. Godavarthi, J. Porcayo-Calderon, M. Casales-Diaz, E. Vazquez-Velez, A. Neri, and L. Martinez-Gomez, "Electrochemical analysis and quantum chemistry of castor oil-based corrosion inhibitors," *Current Analytical Chemistry*, vol. 12, pp. 476–488, 2016.

[10] S. H. Yoo, Y. W. Kim, K. Chung, S. Y. Baik, and J. S. Kim, "Synthesis and corrosion inhibition behavior of imidazoline derivatives based on vegetable oil," *Corrosion Science*, vol. 59, pp. 42–54, 2012.

[11] E. E. Ebenso, U. J. Ekpe, B. I. Ita, O. E. Offiong, and U. J. Ibok, "Effect of molecular structure on the efficiency of amides and thiosemicarbazones used for corrosion inhibition of mild steel in hydrochloric acid," *Materials Chemistry and Physics*, vol. 60, no. 1, pp. 79–90, 1999.

[12] S. Ramachandran, B.-L. Tsai, M. Blanco, H. Chen, Y. Tang, and W. A. Goddard, "Self-assembled monolayer mechanism for corrosion inhibition of iron by imidazolines," *Langmuir*, vol. 12, no. 26, pp. 6419–6428, 1996.

[13] S. Vyas and S. Soni, "Castor oil as corrosion inhibitor for iron in hydrochloric acid," *Oriental Journal of Chemistry*, vol. 27, pp. 1743–1746, 2011.

[14] M. Alam, D. Akram, E. Sharmin, F. Zafar, and S. Ahmad, "Vegetable oil based eco-friendly coating materials: a review article," *Arabian Journal of Chemistry*, vol. 7, no. 4, pp. 469–479, 2014.

[15] L. Canoira, J. García Galeán, R. Alcántara, M. Lapuerta, and R. García-Contreras, "Fatty acid methyl esters (FAMEs) from castor oil: production process assessment and synergistic effects in its properties," *Renewable Energy*, vol. 35, no. 1, pp. 208–217, 2010.

[16] L. A. Rigo, A. R. Pohlmann, S. S. Guterres, and R. C. R. Beck, "Rice bran oil: benefits to health and applications in pharmaceutical formulations," in *Wheat and Rice in Disease Prevention and Health: Benefits, Risks and Mechanisms of Whole Grains in Health Promotion*, R. R. Watson, V. Preedy, and S. Zibadi, Eds., pp. 311–322, Academic Press, 2014.

[17] K. Gul, B. Yousuf, A. K. Singh, P. Singh, and A. A. Wani, "Rice bran: nutritional values and its emerging potential for development of functional food—a review," *Bioactive Carbohydrates and Dietary Fibre*, vol. 6, no. 1, pp. 24–30, 2015.

[18] S. Bhosale and D. Vijayakshmi, "Processing and nutritional composition of rice bran," *Current Research in Nutrition and Food Science*, vol. 3, pp. 74–80, 2015.

[19] D. Kumar, S. M. Kim, and A. Ali, "One step synthesis of fatty acid diethanolamides and methyl esters from triglycerides using sodium doped calcium hydroxide as a nanocrystalline heterogeneous catalyst," *New Journal of Chemistry*, vol. 39, pp. 7097–7104, 2015.

[20] R. D. Kane and S. Campbell, "Real-time corrosion monitoring of steel influenced by microbial activity (SRB) in simulated seawater injection environments," Corrosion/2004, TX: NACE International, Houston, Texas.

[21] D. C. Eden, D. A. Eden, I. G. Winning, and D. Fell, "On-line, real-time optimization of corrosion inhibitors in the field," Corrosion/2006, TX: NACE International, San Diego, California, 12–16 March 2006.

[22] X. Liu, Y. G. Zheng, and P. C. Okafor, "Carbon dioxide corrosion inhibition of N80 carbon steel in single liquid phase and

liquid/particle two-phase flow by hydroxyethyl imidazoline derivatives," *Materials and Corrosion*, vol. 60, no. 7, pp. 507–513, 2009.

[23] F. Farelas, M. Galicia, B. Brown, S. Nesic, and H. Castaneda, "Evolution of dissolution processes at the interface of carbon steel corroding in a CO_2 environment studied by EIS," *Corrosion Science*, vol. 52, no. 2, pp. 509–517, 2010.

[24] S. L. Wu, Z. D. Cui, F. He, Z. Q. Bai, S. L. Zhu, and X. J. Yang, "Characterization of the surface film formed from carbon dioxide corrosion on N80 steel," *Materials Letters*, vol. 58, no. 6, pp. 1076–1081, 2004.

[25] J. Tinnea, B. S. Covino, S. J. Bullard et al., "Electrochemical techniques: investigation of corrosion in a major metropolitan wastewater treatment facility," Corrosion/2004, NACE International, Houston, Texas, March 2004.

[26] B. G. Ateya, F. M. Al Kharafi, and R. M. Abdalla, "Electrochemical behavior of low carbon steel in slightly acidic brines," *Materials Chemistry and Physics*, vol. 78, no. 2, pp. 534–541, 2002.

[27] G. Zhang, C. Chen, M. Lu, C. Chai, and Y. Wu, "Evaluation of inhibition efficiency of an imidazoline derivative in CO_2-containing aqueous solution," *Materials Chemistry and Physics*, vol. 105, no. 2-3, pp. 331–340, 2007.

[28] J. O. Bockris and D. Drazic, "The kinetics of deposition and dissolution of iron: effect of alloying impurities," *Electrochimica Acta*, vol. 7, no. 3, pp. 293–313, 1962.

[29] D. A. López, T. Pérez, and S. N. Simison, "The influence of microstructure and chemical composition of carbon and low alloy steels in CO_2 corrosion. A state-of-the-art appraisal," *Materials and Design*, vol. 24, no. 8, pp. 561–575, 2003.

[30] H. M. Ezuber, "Influence of temperature and thiosulfate on the corrosion behavior of steel in chloride solutions saturated in CO_2," *Materials and Design*, vol. 30, no. 9, pp. 3420–3427, 2009.

[31] X. Liu, P. C. Okafor, and Y. G. Zheng, "The inhibition of CO_2 corrosion of N80 mild steel in single liquid phase and liquid/particle two-phase flow by aminoethyl imidazoline derivatives," *Corrosion Science*, vol. 51, no. 4, pp. 744–751, 2009.

[32] L. M. Rivera-Grau, M. Casales, I. Regla et al., "Effect of organic corrosion inhibitors on the corrosion performance of 1018 carbon steel in 3% NaCl solution," *International Journal of Electrochemical Science*, vol. 8, pp. 2491–2503, 2013.

[33] J. Porcayo-Calderon, L. M. Martínez de la Escalera, J. Canto, M. Casales-Diaz, and V. M. Salinas-Bravo, "Effect of the Temperature on the CO_2-Corrosion of Ni_3Al," in *Proceedings of International Journal of Electrochemical Science*, vol. 10, pp. 3136–3151, 2015.

[34] D. A. Awizar, N. K. Othman, A. Jalar, A. R. Daud, I. A. Rahman, and N. H. Al-Hardan, "Nanosilicate extraction from rice husk ash as green corrosion inhibitor," *International Journal of Electrochemical Science*, vol. 8, pp. 1759–1769, 2013.

[35] K. K. Alaneme, Y. S. Daramola, S. J. Olusegun, and A. S. Afolabi, "Corrosion inhibition and adsorption characteristics of rice husk extracts on mild steel immersed in 1M H_2SO_4 and HCl solutions," *International Journal of Electrochemical Science*, vol. 10, no. 4, pp. 3553–3567, 2015.

[36] M. Prabakaran, S. H. Kim, Y. T. Oh, V. Raj, and I. M. Chung, "Anticorrosion properties of momilactone a isolated from rice hulls," *The Journal of Industrial and Engineering Chemistry*, vol. 45, pp. 380–386, 2017.

[37] H. Sheng, K. Xiaoyang, and F. Chaoyang, "Rice bran extraction used as pickling inhibitor in hydrogen chloride acid," *Journal of the Chinese Society of Corrosion and Protection*, vol. 29, pp. 149–153, 2009.

[38] W. Zhang, H. Gu, L. Xi, Y. Zhang, Y. Hu, and T. Zhang, "Preparation of phytic acid and its characteristics as copper inhibitor," *Energy Procedia*, vol. 17, pp. 1641–1647, 2012.

[39] L. Dong, L. Yuanhua, D. Yigang, and Z. Dezhi, "Corrosion inhibition of carbon steel in hydrochloric acid solution by rice bran extracts," *Anti-Corrosion Methods and Materials*, vol. 58, no. 4, pp. 205–210, 2011.

[40] S. D. Toliwal and K. Jadav, "Fatty acid triazolcs derived from neem, rice bran and karanja oils as corrosion inhibitors for mild steel," *Indian Journal of Chemical Technology*, vol. 16, pp. 32–37, 2009.

18

Characterization of Recharge Mechanisms and Sources of Groundwater Salinization in Ras Jbel Coastal Aquifer (Northeast Tunisia) Using Hydrogeochemical Tools, Environmental Isotopes, GIS, and Statistics

Jamila Hammami Abidi,[1] **Boutheina Farhat,**[1]
Abdallah Ben Mammou,[1] **and Naceur Oueslati**[2]

[1]*Faculty of Sciences of Tunis, Laboratory of Mineral Resources and Environment, University of Tunis El Manar, 2092 Tunis, Tunisia*
[2]*Regional Commission for Agricultural Development, av. Hassen Nouri, 7000 Bizerte, Tunisia*

Correspondence should be addressed to Jamila Hammami Abidi; hammami.jamila@yahoo.fr

Academic Editor: Franco Frau

Groundwater is among the most available water resources in Tunisia; it is a vital natural resource in arid and semiarid regions. Located in north-eastern Tunisia, the Metline-Ras Jbel-Raf Raf aquifer is a mio-plio-quaternary shallow coastal aquifer, where groundwater is the most important source of water supply. The major ion hydrochemistry and environmental isotope composition (δ^{18}O, δ^{2}H) were investigated to identify the recharge sources and processes that affect the groundwater salinization. The combination of hydrogeochemical, isotopic, statistical, and GIS approaches demonstrates that the salinity and the groundwater composition are largely controlled by the water-rock interaction particularly the dissolution of evaporate minerals and the ion exchange process, the return flow of the irrigation water, agricultural fertilizers, and finally saltwater intrusion which started before 1980 and which is partially mitigated by the artificial recharge since 1993. As for the stable isotope signatures, results showed that groundwater samples lay on and around the local meteoric water line LMWL; hence, this arrangement signifies that the recharge of the Ras Jbel aquifer is ensured by recent recharge from Mediterranean air masses.

1. Introduction

The hydrogeology of coastal aquifers has been studied intensively during the past decades, stimulated by both scientific interest and societal relevance [1]. Coastal areas throughout the Mediterranean face salinization problems of groundwater which is the major source of water supply especially for drinking and agricultural sector. The imbalance between abstraction and natural recharge rates causes an overexploitation of groundwater resources resulting in declining groundwater table, water quality degradation, and crop damage.

A number of aquifers in coastal zones are being increasingly exploited and affected ([2–6]). For instance, groundwater contamination and decline of water levels have been reported in Tunisia [7–13] and in many countries around the world. It has been reported in India [14], Jordan [15], Australia [16], USA [17], China ([18, 19]), Netherland [1], and among many others.

In semiarid coastal regions of north-eastern Tunisia, such as Ras Jbel plain, the groundwater is usually the main resource used for irrigation and drinking purposes. Nevertheless, salinization is becoming a common problem affecting groundwater resources. Groundwater exploitation of the Ras Jbel aquifer began in 1949 and has increased each year since 1980. Under pressures of population, climate change, and pollution, the aquifer faces substantial challenges in the management of scarce freshwater resources. During the last four decades, the shallow aquifer groundwater has been overexploited through excessive, uncontrolled pumping mainly for domestic and agricultural purposes [20]. Salinization

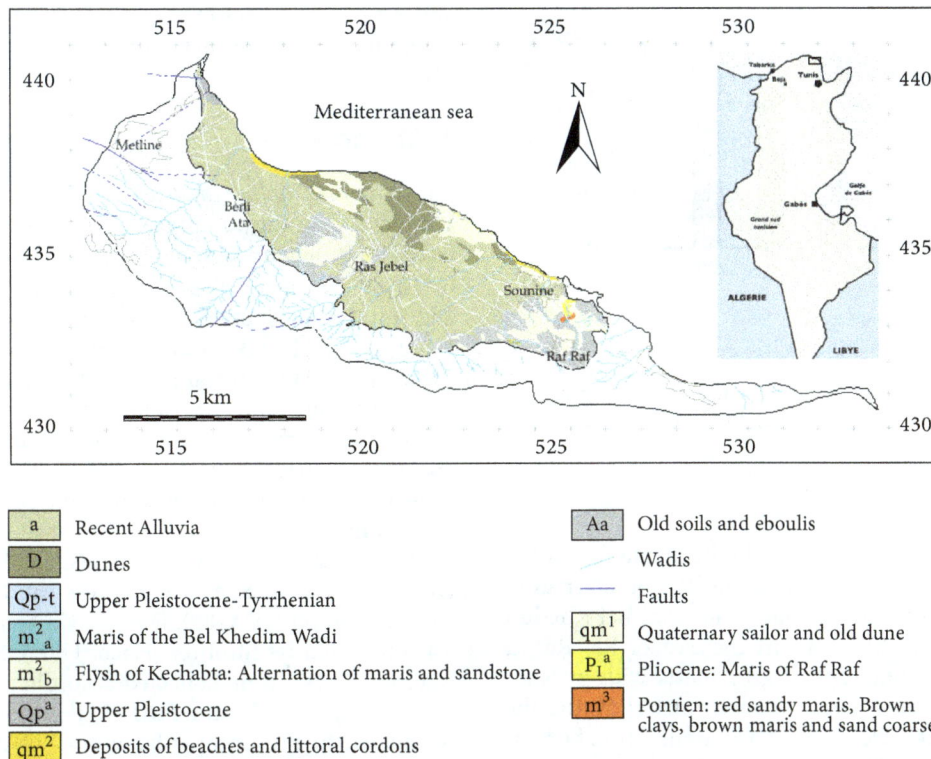

FIGURE 1: Geological and location map of the study area [28–30].

due to seawater intrusion and decreasing groundwater levels has recently been identified [21]. Unplanned and substantial withdrawals of groundwater from the shallow aquifer of Ras Jbel have resulted in severe water level decline of up to 7 m in some areas and high total dissolved solids (TDS) contents reaching 8000 mg/l. Reports of increasing salinity of groundwater supplies in the area suggest a need to define the sources of salt water. It is also useful to study the recharge mechanisms as well as the mixing fresh water/saline water.

Stable isotope and geochemical techniques have been used in groundwater studies of coastal aquifers worldwide [22–25] for determining the origins of groundwater salinization in aquifers and processes that affect water chemistry, such as rock weathering, evaporation, atmospheric precipitation, and cation exchange. Consequently, studying stable isotope and geochemical techniques can significantly improve our understanding of groundwater hydrodynamical processes and chemical evolution [26]. In the present study, environmental isotopes ($\delta^{18}O$, δ^2H) in conjunction with hydrochemistry (major ions) were employed (1) to define the potential sources and different mechanisms of groundwater salinization in the study area (2) to discuss the chemical evolution of groundwater and (3) to explain groundwater recharge and discharge in the coastal plain of Metline- Ras Jbel- Raf Raf.

2. Study Area

The Metline-Ras Jbel-Raf Raf plain, which covers a total area of about 50 km^2, represents one of the most important

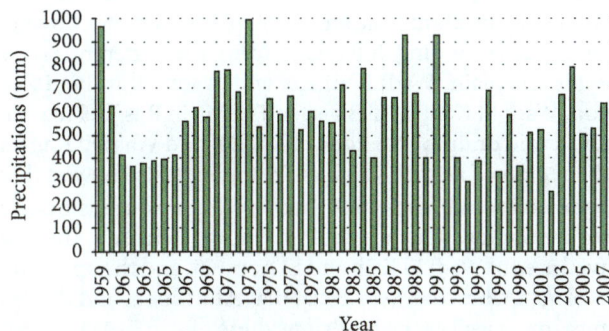

FIGURE 2: Yearly precipitation in Metline-Ras Jbel-Raf Raf plain [64, 65].

agricultural regions in north-eastern Tunisia (Figure 1). It is characterized by a semiarid, "Mediterranean" climate with mild, wet winters and warm, dry summers [27, 28]. The average annual precipitation ranges from approximately 258 to 993 mm (Figure 2). Geologically, it is limited to the south by Jbel Djaouf, En Nadhour, and Ed Demina, to the southwest by Jbel Sidi Saleh, Hakima, and El Faouar, to northwest by Jbel Bab Banzart, Sidi Bou Choucha, and Touchela, to the north and northeast by the Mediterranean sea. The plain of Metline-Ras Jbel-Raf Raf is a wide basin of collapse, formed by a subsidence followed by an alluvial and recent sedimentation. It is affected by folding and faults in NW-SE and SSW-NNE directions. Sedimentary series extend from Miocene to Quaternary. The lithological description of these sediments [28–30] reveals that the Miocene is represented by the Kechabta

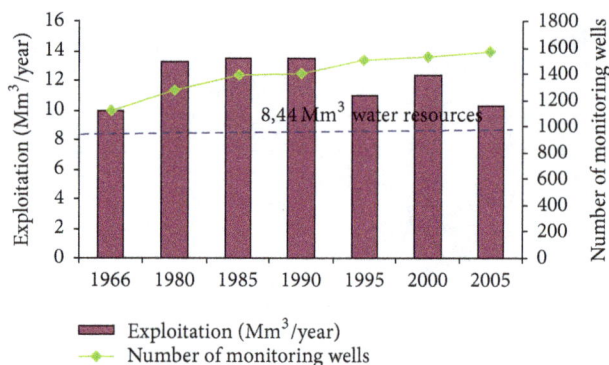

FIGURE 3: Evolution of the exploitation rate and the monitoring well number of the shallow aquifer of Ras Jbel during 1966 and 2005 [30, 31].

and Wadi Bel Khedim formations. The Kechabta formation has a thickness of about 1000 m and consists of alternation of marls and fine sandstone benches. The Wadi Bel Khedim formation is of 250 m thickness. At the east of Raf Raf, it is composed of large outcrops of gray marls with gypsum benches. The Quaternary series unconformably overlie the Miocene series and is divided into seven units (from bottom to the top qm^1, Aa, qm^2, Qp^a, Qp-t, D and a).

Hydrologically, the Quaternary series and the current formations, characterized by their extension and their good permeability, host the shallow aquifer of Ras Jbel. This aquifer is recharged by local infiltration on the plain, by water flowing from the surrounding hills, and from the different rivers crossing the plain: Wadi Beni Ata and Wadi Ali in Beni Ata region, Wadi El Kantra, El Blaat, and El Ma in Ras Jbel region, Wadi Sandid draining the zone of Raf Raf and Wadi El Kantra in the region of Dar El Khaddar. Pumping tests showed that transmissivity can reach 15.10^{-4} m^2/s. The most transmissive areas are the low alluvial plain of Bahirat Beni Ata, the low alluvial area of Wadi El Krib and El Aouinet, and the upstream area of Ain Cherchar, Ain Ezzaouia, Ain Kassa, and Ain El Hammam, as well as the sandstone dune of Ain El Mestir and Ain Mahloul.

The shallow aquifer of Ras Jbel is affected by natural and anthropogenic factors like evaporation, irrigation, pumping, and so forth. The aquifer is tapped by several private and state owned wells. In the period between 1985 and 1990, the exploitation rate was estimated at 13,5 Mm^3/year, which exceeded the renewable resources evaluated at 8,44 Mm^3/year [31]. The total number of dug wells has been estimated to be 1387 in 1985, 1396 in 1990, and 1563 in 2005. In 2005, the exploitation rate was estimated at 10,27 Mm^3/year (Figure 3). The massive exploitation of aquifer resources in response to the heavy pumping caused a drop in the water level (Figure 4) and the deterioration of water quality. In 1949, the salinity of the groundwater varied between 648 and 1692 mg/l [32]. The investigations carried out by the DGRE between 1985 and 1993 through several monitoring wells revealed the existence of a very saline groundwater. The degradation of the water quality has been detected mainly in coastal areas where salt concentrations exceeded 15 g/l in 1985

and 8 g/l in 1993 and in the depression of Bahirat Beni Ata where the ante-quaternary substratum is below the sea level [21].

3. Methods

3.1. *Water Sampling and Chemical Analysis.* Ninety-four samples (including pumping wells and piezometers) were collected for geochemical analysis (major elements) and isotopes (δ^2H, δ^{18}O) during the wet and the dry season of March and July 2007 (Figure 5). Sampling locations were recorded using a potable GPS device. Prior to sampling, all wells were pumped for several minutes to eliminate the influence from stagnant water. Samples were collected in cleaned polyethylene bottles, tightly capped and stored at 4°C until analysis. Electrical conductivity (EC), salinity, and pH were measured in the field using a portable conductivity, salinity, and pH meter.

Chemical analyses were done using a Varian 730-ES ICP Optical Emission Spectrometer for cations and using ion chromatography (DX-120, Dionex, USA) for anions. HCO$_3^-$ was measured by titration (Hach, USA). Samples for stable isotope analysis were collected according to the procedures described by Clark and Fritz [33]. Isotopic analyses were conducted in the isotopic laboratory of the department of Hydrology and Geo-environmental Sciences of the Faculty of Earth and Live Sciences of the Free University of Amsterdam. Isotopic ratios are expressed in per mil (δ) and oxygen and hydrogen isotope analyses were reported to d notation relative to Vienna Standard Mean Oceanic Water (V-SMOW), where $d = [(\text{Rs/RVSMOW}) - 1] \times 1000$, Rs represents either the ^{18}O/^{16}O or the ^{2}H/^{1}H ratio of the sample, and RSMOW is ^{18}O/^{16}O or the ^{2}H/^{1}H ratio of the SMOW. Typical precisions are ±0.1 and ±1% for oxygen-18 and deuterium, respectively.

3.2. *Statistical Analysis.* The physicochemical parameters and chemical composition of the groundwater samples are presented in Table 1. All the acquired data were integrated into hydrogeochemical database in order to study the groundwater quality and to identify the groundwater salinization processes that contributed to the acquisition of the actual chemical composition.

Piper plots [34], considered as the common method for a multiple analyses on the same graph, are used to represent the different water samples and to distinguish graphically between different water types defined by the Stuyfzand classification (1993). Sample points with similar hydrochemistry tend to cluster together in the diagram [35].

Principal component analysis method (PCA) and correlations are a popular method to assess groundwater quality. One of the principle advantages of multivariate techniques such as principal component analysis (PCA) is that they are able to rapidly reveal relationships between a large number of variables. In this study, PCA and correlations are used to identify the possible sources of major ions in groundwater, hydrogeological reactions that may occur in the study area, and dominant factors that control groundwater quality.

Gibbs diagrams which are a simple plot of the TDS versus the weight ratio of Na$^+$/(Na$^+$ + Ca^{2+}) or Cl$^-$/(Cl$^-$ + HCO$_3^-$)

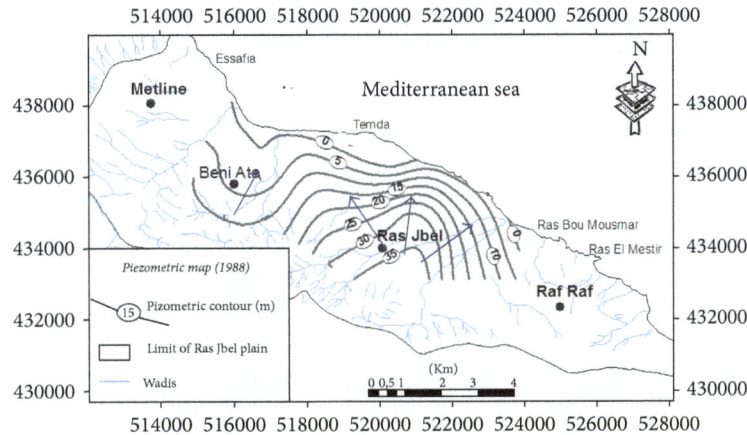

FIGURE 4: Piezometric contour maps of the study area [31].

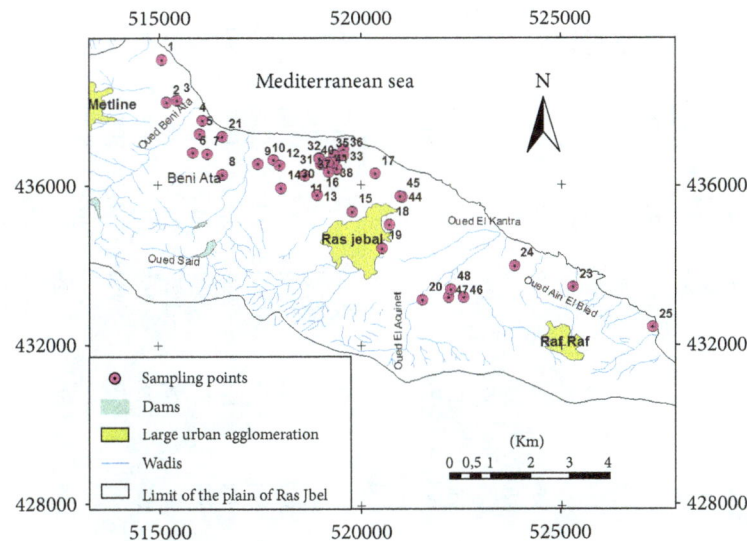

FIGURE 5: Location of sampled wells and piezometers in the study area.

are widely used to establish the relationships between the water composition and the lithological characteristics of the aquifer [36]. Three distinct fields, including precipitation dominance, evaporation dominance, and rock weathering dominance, constitute the segments in the Gibbs diagram.

Hydrogeochemical Modeling. The equilibrium state of the water with respect to a mineral phase can be determined by calculating saturation index (SI) using analytical data. In this study, saturation indices (SI) were calculated in terms of the following equation [37]:

$$SI = \log\left(\frac{IAP}{k_s(T)}\right), \tag{1}$$

where IAP is the relevant ion activity product, which can be calculated by multiplying the ion activity coefficient γi and the composition concentration m_i, and $k_s(T)$ is the equilibrium constant of the reaction considered at the sample temperature [35]. The geochemical modeling program PHREEQC has been used to evaluate the water chemistry.

SI > 0 indicates oversaturation and minerals may be subject to precipitation, SI < 0 means undersaturation and minerals will dissolve, and SI = 0 suggests saturation and minerals are in equilibrium status with respect to the solution [38].

4. Results and Discussion

4.1. Hydrogeochemical Characterization. Groundwater quality depends on various chemical constituents and their concentrations, which are mostly derived from the geological stratum of the particular region [6]. The pH was one of the primary indicators of the water chemistry evolution. The aquifer groundwater was neutral to slightly alkaline water, with a mean pH value of 7,23 and 7,15 in the wet and dry season, respectively. Electrical conductivity (EC) of the water samples was medium to high, suggestive of very highly mineralized waters. EC values ranged between 1240 and 6300 $\mu S/cm$ in the wet season and between 1131 and 5430 $\mu S/$cm in the dry season. A problem to the water supply development in the area is the increasing electrical conductivity

TABLE 1: Chemical data of sampled wells and piezometers in the Ras Jbel aquifer.

Well	Wet season EC	pH	TDS	Ca^{2+}	Mg^{2+}	Na^+	K^+	Cl^-	SO_4^{2-}	HCO_3^-	Dry season EC	pH	TDS	Ca^{2+}	Mg^{2+}	Na^+	K^+	Cl^-	SO_4^{2-}	HCO_3^-
1	3670	7,41	2369	225	89	443	10	999	298	305	5430	7,2	3630	299	140	757	12	1792	343	287
2	3990	7,05	2519	447	57	413	4	1022	307	270	3760	7,21	2446	447	58	376	1	989	326	250
3	3840	6,8	2649	442	54	425	11	997	384	338	3480	7,05	2451	421	48	392	9	918	356	308
4	4120	7,1	3166	368	72	647	5	1021	742	311	3900	7,2	2612	327	63	506	2	789	626	299
5	6300	6,9	4110	520	89	738	7	1390	1005	361	5530	7,06	4323	501	94	731	4	1599	1040	354
6	5080	6,95	3345	403	81	666	8	1160	661	366	3850	7,15	2832	363	66	499	5	953	598	348
7	5790	6,9	3630	560	115	550	4	1457	635	309	4740	7	3231	523	103	471	3	1181	633	317
8	3730	6,9	2306	286	115	284	20	895	377	329	3560	7,11	2407	353	135	294	12	876	421	317
9	5660	7,03	3800	368	129	881	11	1487	563	361	4660	7,16	3126	339	96	609	7	1248	498	329
10	3530	7,3	2384	249	59	388	25	788	589	287	2840	7,63	1989	258	46	347	20	610	489	220
11	4480	7,12	3260	379	72	646	18	989	820	337	3650	7,19	2674	334	59	500	13	774	683	311
12	3250	7	2142	230	55	346	37	705	477	293	3160	7,35	2360	285	55	434	37	720	543	287
13	2770	7	1658	216	45	264	31	474	352	277	2540	7,35	1710	243	41	262	29	487	380	268
14	2480	7,65	1972	278	46	337	24	659	382	246	2860	7,49	2040	263	50	343	23	662	462	238
15	6250	7,96	4089	466	86	857	96	1463	811	310	4230	7,35	3049	353	71	605	26	866	684	445
16	4190	7,6	2480	258	78	444	22	830	619	231	3520	7,38	2392	276	70	421	19	766	598	244
17	3990	7,5	2715	274	49	550	107	775	460	500	3290	7,7	2497	280	50	426	80	652	413	598
18	2920	7,2	2046	221	56	330	79	472	444	445	2240	7,29	1841	199	45	289	75	444	363	427
19	2730	7,75	1817	236	54	285	25	451	382	381	2250	7,3	1680	215	46	256	31	427	395	311
20	4020	7,23	2747	208	131	525	6	1011	481	386	3360	7,33	2389	206	124	410	3	872	384	390
21	1240	7,78	819	48	36	165	7	289	102	173	2120	7,55	1317	104	57	270	5	639	108	134
22	3280	7,05	2358	258	55	377	35	844	515	275	2910	7,65	1835	230	42	338	32	589	366	238
23	1573	7,68	924	84	19	200	6	276	89	250	1131	7,82	783	73	12	182	3	217	88	207
24	5170	7,15	2891	505	75	510	8	1185	407	201	4060	7,05	2697	490	67	433	5	1066	436	201
25	4610	7,3	2910	189	113	576	7	916	652	458	3880	7,33	2889	211	123	590	4	878	650	433
26	5050	7,3									2330	7,33	1666	216	55	270	12	464	264	387
27	3400	7,55	1938	255	65	287	27	594	415	296	2720	7,33	1928	257	62	312	35	568	371	323
28	3160	7,6	1662	178	48	289	87	478	356	226	2520	7,39	1910	279	57	280	16	593	393	293
29											2950	7,89	2176	187	66	419	6	713	587	198
30	3580	7,2	2353	199	80	437	11	819	613	195	3240	7,5	2217	205	72	427	11	732	582	189
31	3090	7,5	2187	225	71	394	10	710	560	217	3240	7,54	2330	228	68	424	7	766	637	201
32	3490	7,6	2317	231	69	479	10	749	560	221	3830	7,35	2804	269	80	572	7	866	723	287
33	4290	7,1	2886	358	70	404	20	954	729	351	5080	7,75	3608	454	90	654	23	1107	884	397
34	3640	7,25	2648	292	66	520	15	851	591	315	3410	7,35	2429	305	67	401	12	764	575	305

TABLE 1: Continued.

Well				Wet season										Dry season						
	EC	pH	TDS	Ca^{2+}	Mg^{2+}	Na^+	K^+	Cl^-	SO_4^{2-}	HCO_3^-	EC	pH	TDS	Ca^{2+}	Mg^{2+}	Na^+	K^+	Cl^-	SO_4^{2-}	HCO_3^-
35	3860	7,25	2836	326	81	550	27	874	634	344	3420	7,51	2487	304	71	455	14	756	571	317
36	4820	7,95	3253	398	81	472	36	1084	789	393	4110	7,43	3015	348	74	539	38	922	729	366
37	3740	7,45	2390	230	78	412	17	829	607	218	3320	7,32	2441	265	70	474	16	757	609	250
38	2720	7,7	1881	197	55	332	9	591	465	232	3650	7,5	2628	292	71	491	6	821	654	293
39	2080	7,7	1378	169	32	251	11	470	336	110	1817	7,49	1269	158	30	213	8	391	319	153
40	3540	7,4	2454	231	75	419	8	823	620	278	3770	7,19	2727	273	81	489	7	865	696	317
41	5030	7,1	3010	402	81	456	14	1166	689	201	3130	7,67	2233	232	68	431	7	734	578	183
42											2820	7,93	2061	179	65	371	6	683	580	177
44	2980	7,12	1817	236	43	324	78	531	337	270	2090	7,52	1694	198	42	234	64	502	355	299
45	2670	7,1	1694	215	43	277	69	493	316	281	2130	7,4	1553	203	40	258	67	437	294	256
46	2840	7,4	1741	200	55	267	9	555	375	281	3050	7,3	2277	285	69	390	3	770	516	244
47	3310	7,05	1749	300	59	217	3	680	233	256	3170	7,3	2257	389	57	249	2	807	522	232
48	4990	7,06	2856	555	82	400	5	1210	355	250	4120	7,1	2601	560	73	327	2	1084	372	183

EC (μs/cm); TDS, Ca^{2+}, Mg^{2+}, Na^+, K^+, Cl^-, SO_4^{2-}, HCO_3^- (mg/l).

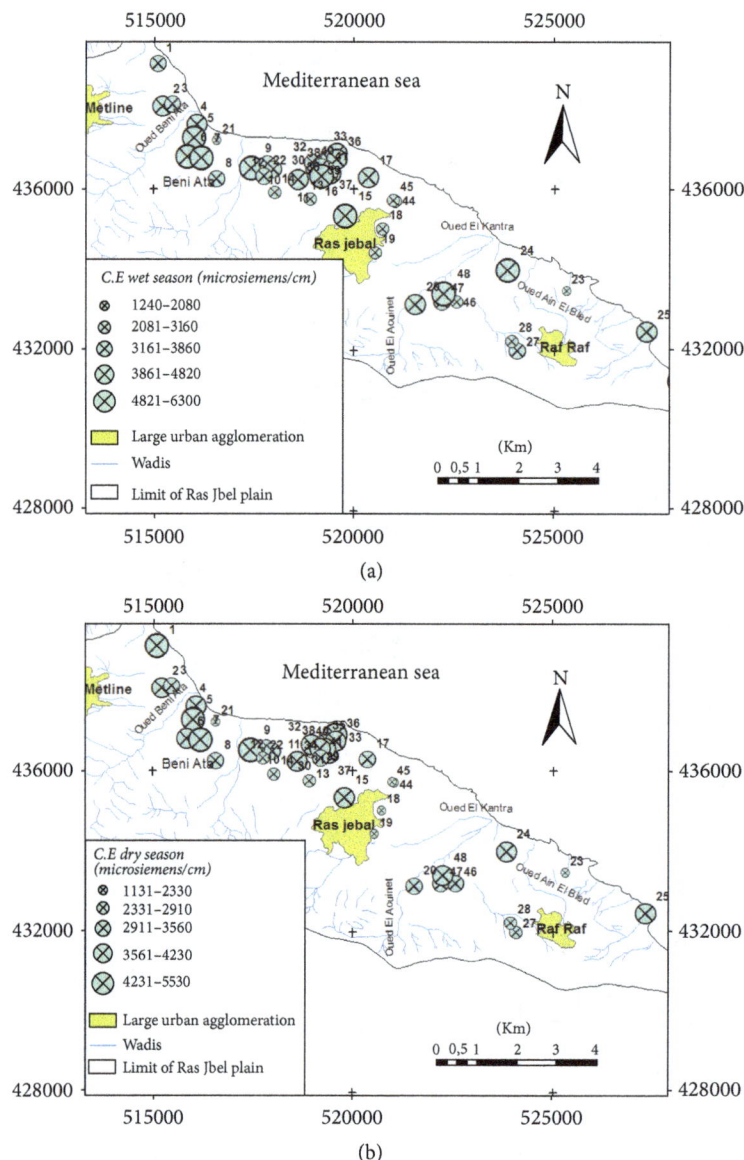

FIGURE 6: Spatial distribution of electrical conductivity (a) wet season and (b) dry season in Ras Jbel aquifer.

(EC) values of near-coastal shallow groundwater. The high EC of groundwater from wells located along the coast and in Bahirat Beni Ata can be explained by the seawater intrusion effect caused by the groundwater level drawdown due to overexploitation (Figures 6(a) and 6(b)).

Based on the conductivity values, the groundwater system could be classified into four groups: fresh water (<500 $\mu S/$ cm), marginal water ($500–1,500$ $\mu S/cm$), brackish water ($1,500–5,000$ $\mu S/cm$), and saline water ($>5,000$ $\mu S/cm$). Based on our conductivity values it is evident that groundwater in Ras Jbel aquifer is in marginal and brackish waters. Few samples are of saline water type.

The study area is characterized by a wide range of salinities. Salinity values ranged between 500 and 3600 mg/l in the wet season and between 500 and 3500 mg/l in the dry season. The higher values of EC and salinity are indicators of higher ionic concentrations, probably due to the high anthropogenic

activities in the region and geological weathering conditions but also due to the intrusion of sea water into the groundwater system.

The relative content of a cation or an anion is defined as the percentage of the relative amount of that ion to the total cations or anions, respectively [39]. In the study area, the strong acid anions (Cl^- and SO_4^{2-}) exceed weak acid anions (HCO_3^- and CO_3^{2-}). On the other hand, sodium and calcium concentrations exceed magnesium and potassium contents. The triangular diagram (Figure 7) shows that the groundwater chemistry was mainly characterized by two groups. The first group was $Cl^-–Na^+$ type, in which Na^+ accounted for more than 52–65% of the total cations; the second group was $Cl^-–Na^+/Ca^{2+}$ type, in which Na^+ and Ca^{2+} accounted for 37–48% and 36–53% of the total cations, respectively. Chemical facies in the phreatic aquifer of Ras Jbel seem to be directly related to the configuration of the ante-quaternary

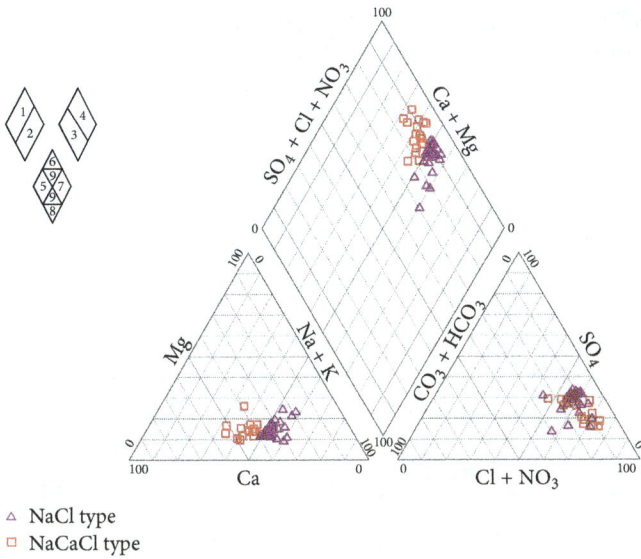

FIGURE 7: Piper diagram of groundwater: zone 1: alkaline earths (Ca + Mg) exceed alkalis (Na + K); zone 2: alkalis exceed alkaline earths; zone 3: weak acids (CO_3 + HCO_3) exceed strong acids (SO_4 + Cl); zone 4: strong acids exceed weak acids; zone 5: carbonate hardness > 50% (alkaline earths and weak acid dominate); zone 6: noncarbonate hardness > 50%; zone 7: noncarbonate alkali > 50%; zone 8: carbonate alkali > 50% (groundwater is inordinately soft in proportion to the content of TDS); zone 9: no cation–anion pair >50% [34].

TABLE 2: Principal component matrix.

	$F1$	$F2$	$F3$
CE	**0,97**	−0,03	0,05
TDS	**0,98**	−0,04	0,08
pH	−0,41	−0,42	**0,75**
Ca^{2+}	**0,82**	0,17	−0,10
Mg^{2+}	**0,78**	0,13	0,02
Na^+	**0,89**	−0,20	0,17
K^+	−0,11	**−0,86**	−0,16
Cl^-	**0,96**	0,14	0,07
SO_4^{2-}	**0,77**	−0,20	0,28
HCO_3^-	0,43	**−0,57**	−0,49
Eigenvalue	5,86	1,40	0,95
% Explication	58,64	14	9,53
% Cumulative	**58,64**	**72,64**	**82,17**

Bold values: loadings ≥ 0.5.

substratum and the proximity of several sampled wells from the sea.

The Na–Cl-type indicated the presence of high chloride concentrations in the aquifer which may originate from the dissolution of halite, influx of sewage or waste water, and mainly intrusion of sea water [40]. In the downstream part of the plain near the sea and in Bahirat Beni Ata, where a marine intrusion was detected since 1980, the groundwater is mainly of Cl^-–Na^+ type (Figure 8). The distribution maps of Cl^- in both wet and dry seasons (Figures 9(a) and 9(b)) are well correlated with those of electrical conductivity and facies.

4.2. Processes Controlling Groundwater Salinization.

Understanding the water salinization mechanism is the basis for regional salt management. Groundwater salinization is largely a function of the mineral composition of the aquifer through which it flows and hydrogeochemical processes such as mineral dissolution, precipitation, evaporation and transpiration, ion exchange, and the residence time along the flow path. It is also linked to various anthropogenic activities such as agriculture, overexploitation of groundwater resources, and sewage disposal.

4.2.1. Water-Rock Interaction and Origin of Groundwater Mineralization.

Reactions between groundwater and aquifer minerals have a significant role on water quality. The $p \times p$ correlation matrix, revealing the existence of bivariate linear correlations between variables, allows a better understanding of the dominant water-rock interactions or source of the ions over the study area. Additionally, the use of multivariate

statistics in hydr(geo)logical studies is a very common practice and numerous applications can be found in the literature ([10, 19, 41]) though most hydrologists consider that values larger than 0.5 indicate significant correlation. In the present study, PCA was carried out for 10 parameters (Ca, Mg, Na, K, HCO_3, SO_4, Cl, pH, TDS, and EC) and more than 90 observations (Tables 2 and 3). The first two factors $F1$ and $F2$ were always retained, explaining about 73% of the total variance (Table 2). Factors of a higher order generally explained the variance of a single parameter or established poorer and less significant correlations with two parameters [42].

The correlations established between the TDS and concentrations of major elements (Table 3) show that the TDS is well correlated with the concentrations of chloride (r^2 = 0,96), sodium (r^2 = 0,87), calcium (r^2 = 0,83), magnesium (r^2 = 0,73), and sulphates (r^2 = 0,76). The high correlation of TDS with chloride, sodium, magnesium, sulphate, and calcium indicated that these elements are mostly contributed by mineralization. These ions have been dissolved into groundwater continuously and resulted in the rise of TDS. The contribution of carbonates and potassium is negligible (r^2 = 0,37 and r^2 = −0,05, resp.). The low correlation between TDS and pH suggests that the dissolution of the salts is not related to acidic conditions of groundwater but it is related to their degrees of solubility. HCO_3^- and pH apparently have little association with the other variables.

Bicarbonates are not correlated to calcium $r(HCO_3 Ca)$ = 0,20 indicating another source other than the calcite dissolution. However, considerable correlation coefficients between sodium and chlorides $r(NaCl)$ = 0,84 and between calcium and sulphates $r(CaSO_4)$ = 0,50 suggest halite and gypsum dissolution, respectively.

The first factor $F1$ accounts for 58% of the total variance, and it is contributed by the following variables: EC, TDS, Mg, Ca, Na, Cl, and SO_4. This factor is associated with the salinity component (NaCl salt source with Ca and SO_4 enrichment) and the cation exchange. The second factor $F2$ accounts for 14% of the total variance, and it is negatively determined by K

TABLE 3: Correlation matrix of dissolved species and the TDS for the study area in wet season.

	CE	TDS	pH	Ca^{2+}	Mg^{2+}	Na^+	K^+	Cl^-	SO_4^{2-}	HCO_3^-
CE	1									
TDS	**0,98**	1								
pH	−0,34	−0,32	1							
Ca^{2+}	**0,84**	**0,83**	−0,44	1						
Mg^{2+}	**0,73**	**0,73**	−0,30	**0,43**	1					
Na^+	**0,84**	**0,87**	−0,18	**0,60**	**0,65**	1				
K^+	−0,04	−0,05	0,23	−0,11	−0,28	0,01	1			
Cl^-	**0,95**	**0,96**	−0,37	**0,84**	**0,76**	**0,84**	−0,19	1		
SO_4^{2-}	**0,73**	**0,76**	−0,12	**0,50**	**0,52**	**0,75**	−0,02	**0,65**	1	
HCO_3^-	0,36	0,37	−0,20	0,20	0,37	0,42	0,30	0,27	0,32	1

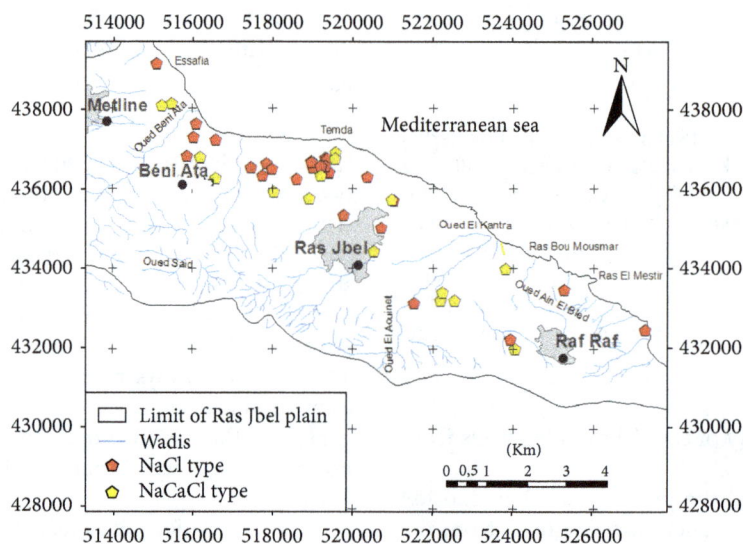

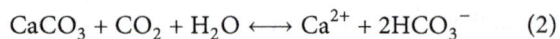

FIGURE 8: Distribution map of the groundwater facies in the shallow aquifer of Ras Jbel.

and HCO_3 (Figure 10). It suggests carbonates weathering and pollution by fertilizer application.

In order to understand the origin of groundwater mineralization in Ras Jbel plain, the saturation index (SI) was calculated. The mineral facies are chosen based on the analysis result of groundwater quality, the main components of groundwater, and the occurrences conditions ([6, 43, 44]). In the study area, the main cations are Na^+, Ca^{2+}, and Mg^{2+} and the main anions are HCO_3^-, SO_4^{2-}, and Cl^-; thus gypsum, anhydrite, calcite, dolomite, aragonite, and halite are chosen to be the mineral facies.

The positive values of the calculated SI with respect to calcite and dolomite for all groundwater samples (Figure 11) suggest their oversaturation in respect to these minerals ($0,05 < SI_{calcite} < 1,32$ and $0,08 < SI_{calcite} < 1,18$ in the wet and dry seasons, resp., and $-0,10 < SI_{dolomite} < 2,29$ and $0,06 < SI_{dolomite} < 1,98$ in the wet and dry seasons, resp.). As described by Appelo and Postma [45], the dissolution of calcite and dolomite is as follows:

Calcite:

$$CaCO_3 + CO_2 + H_2O \longleftrightarrow Ca^{2+} + 2HCO_3^- \quad (2)$$

Dolomite:

$$CaMg(CO_3)_2 + 2CO_2 + 2H_2O$$
$$\longleftrightarrow Ca^{2+} + Mg^{2+} + 4HCO_3^- \quad (3)$$

However, focusing on the scatter plots of bicarbonate versus calcium and calcium + magnesium versus bicarbonate we notice that groundwater samples are not plotted on the 1:1 straight lines of calcite and dolomite dissolution (Figures 12(a) and 12(b)). Groundwater samples show an excess of Ca^{2+} that can be explained by the gypsum dissolution.

The plot of SIGypsum and SIAnhydrite versus TDS exhibits a proportional and parabolic shape evolution with negative values of the saturation indexes (Figure 11) ($-1,86 < SI_{gypsum} < -0,34$ and $-1,72 < SI_{gypsum} < -0,35$ in the wet and dry seasons, resp., and $-2,08 < SI_{anhydrite} < -0,56$ and $-1,94 < SI_{anhydrite} < -0,56$ in the wet and dry seasons, resp.). Thus, both calcium and sulphate are derived from the same origin, which is the dissolution of gypsum and anhydrite.

Gypsum: $CaSO_4 \cdot 2H_2O \longleftrightarrow Ca^{2+} + SO_4^{2-} + 2H_2O \quad (4)$

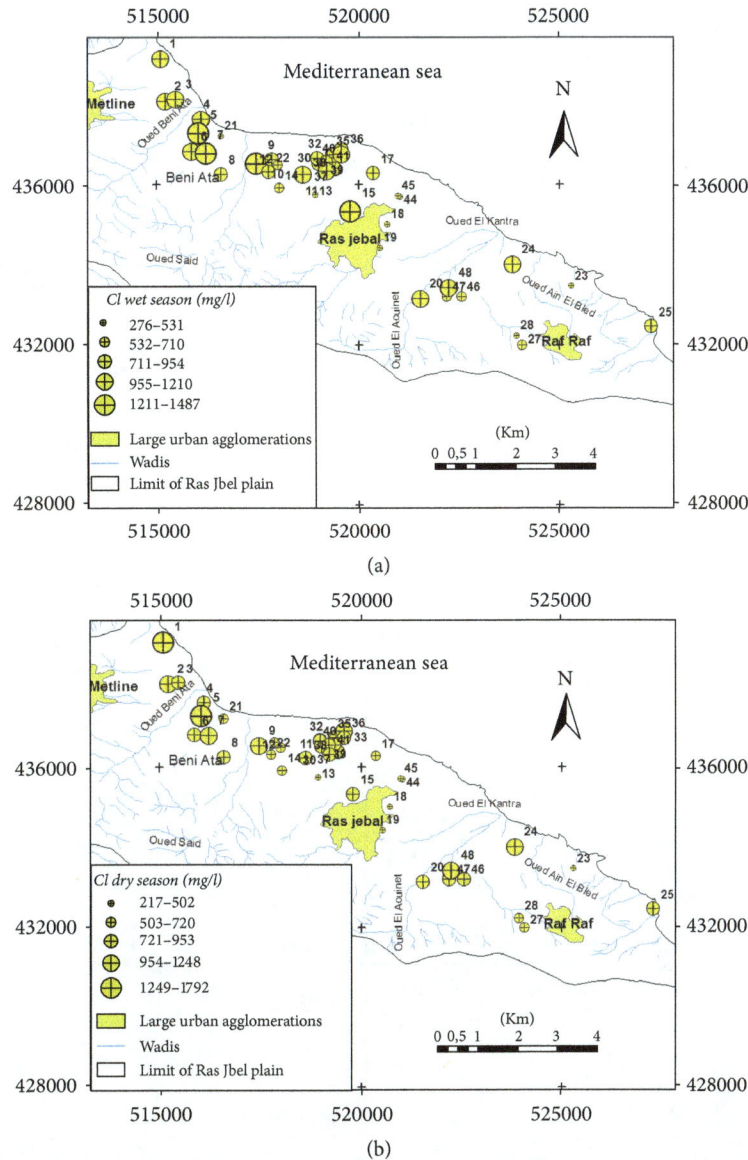

FIGURE 9: Spatial distribution of chlorides (a) wet season and (b) dry season in Ras Jbel aquifer.

However, the plot of sulphate versus calcium (Figure 12(c)) shows an excess of Ca^{2+} ions for the majority of the groundwater samples. For these samples, the $(Ca/(Ca + SO_4))$ ionic ratio greater than 0,5 ratio (from 0,53 and 0,79) confirms the ionic exchange process [46]. The $(Ca/(Ca + SO_4))$ ionic ratio close to 0,5 confirms that the main source of Ca^{2+} is the gypsum dissolution [46].

A bivariate diagram of sodium versus chloride (Figure 12(d)) reveals two main groups: for the first group halite dissolution was maintained for a slope equal to unity where majority of the samples are situated on the 1:1 straight of halite dissolution given by the following reactions [45]:

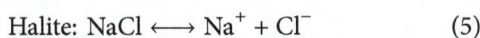

$$\text{Halite: } NaCl \longleftrightarrow Na^+ + Cl^- \qquad (5)$$

The second group includes the high-salinity samples (Na–Cl type), which do not follow the halite dissolution line and show enrichment in chloride compared to sodium. Thus, another phenomenon other than geological effect is controlling their salinization, and this may be the salt water intrusion.

Water samples were plotted in the Gibbs diagrams, which takes into account the major role of natural mechanisms (rock weathering, evaporation, and precipitation). Figure 13 clearly shows that the mechanism controlling water chemistry seems to be a combination of the weathering of carbonates minerals as well as the evaporation-precipitation processes. However, low rates of the groundwater samples were obtained in areas that were dominated by rock-water interactions.

Samples with $Na^+/(Na^+ + Ca^{2+})$ or $Cl^-/(Cl^- + HCO_3^-)$ ratios greater than 0,5 and TDS levels between 783 and 4323 mg/l showed that the groundwater chemistry was controlled mainly by the saline water mixing or evaporation. Evaporation results in increased TDS in relation to high ratios

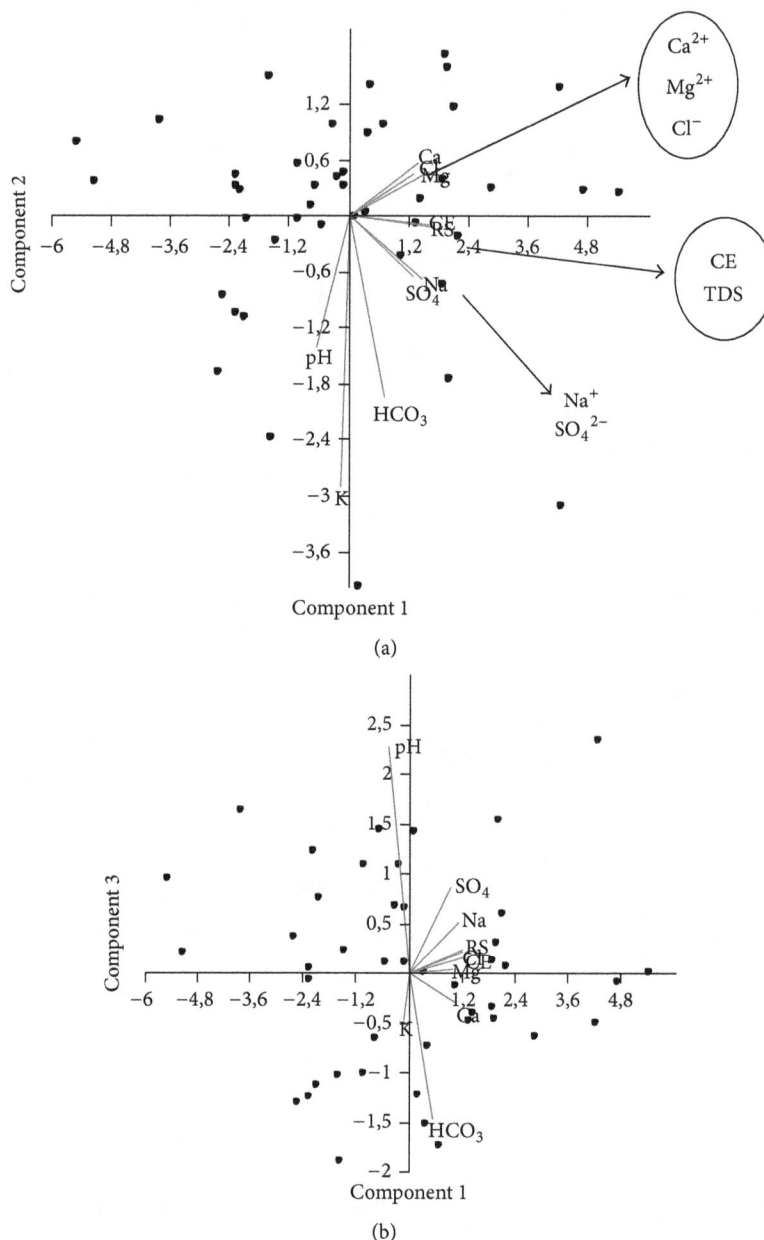

FIGURE 10: Representation of variables into two principal factors: $F1$-$F2$ (a) and $F1$-$F3$ (b).

of dominant cations and anions and $CaCO_3$ precipitates by losing Ca^{2+} and HCO_3^-.

4.2.2. Ionic Exchange Processes and Freshwater-Saline Water Mixing. Ion exchange is one of the important natural processes responsible for the concentration of ions in groundwater and has significant impact on the evolution of groundwater chemistry [18]. The dominance of salty groundwater dominated by sodium and chloride ions in Ras Jbel shallow aquifer provides evidence of mixing with an external salinity source, which could be the seawater from the coastal part of the aquifer. Cation exchange, responsible for the salinity signature, is described by two mixing mechanisms (freshening and saline water intrusion). Equations (6) and (7) show the gain

or loss related to Na^+ and $(Ca^{2+} + Mg^{2+})$ within the exchanger X.

The freshening process or direct ion exchange: where Ca^{2+} from freshwater displaced the marine cations Na^+ and Mg^{2+} from the exchanger complex. The resulting loss of Ca^{2+} from solution decreases the saturation state for calcite and possibly causes calcite dissolution.

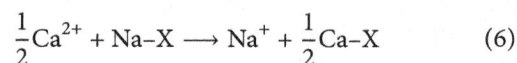

$$\frac{1}{2}Ca^{2+} + Na-X \longrightarrow Na^+ + \frac{1}{2}Ca-X \qquad (6)$$

The intrusion of seawater or reverse ion exchange also triggered cation exchange reactions where Ca^{2+} was expelled from the exchanger by seawater Na^+ and Mg^{2+}. The released

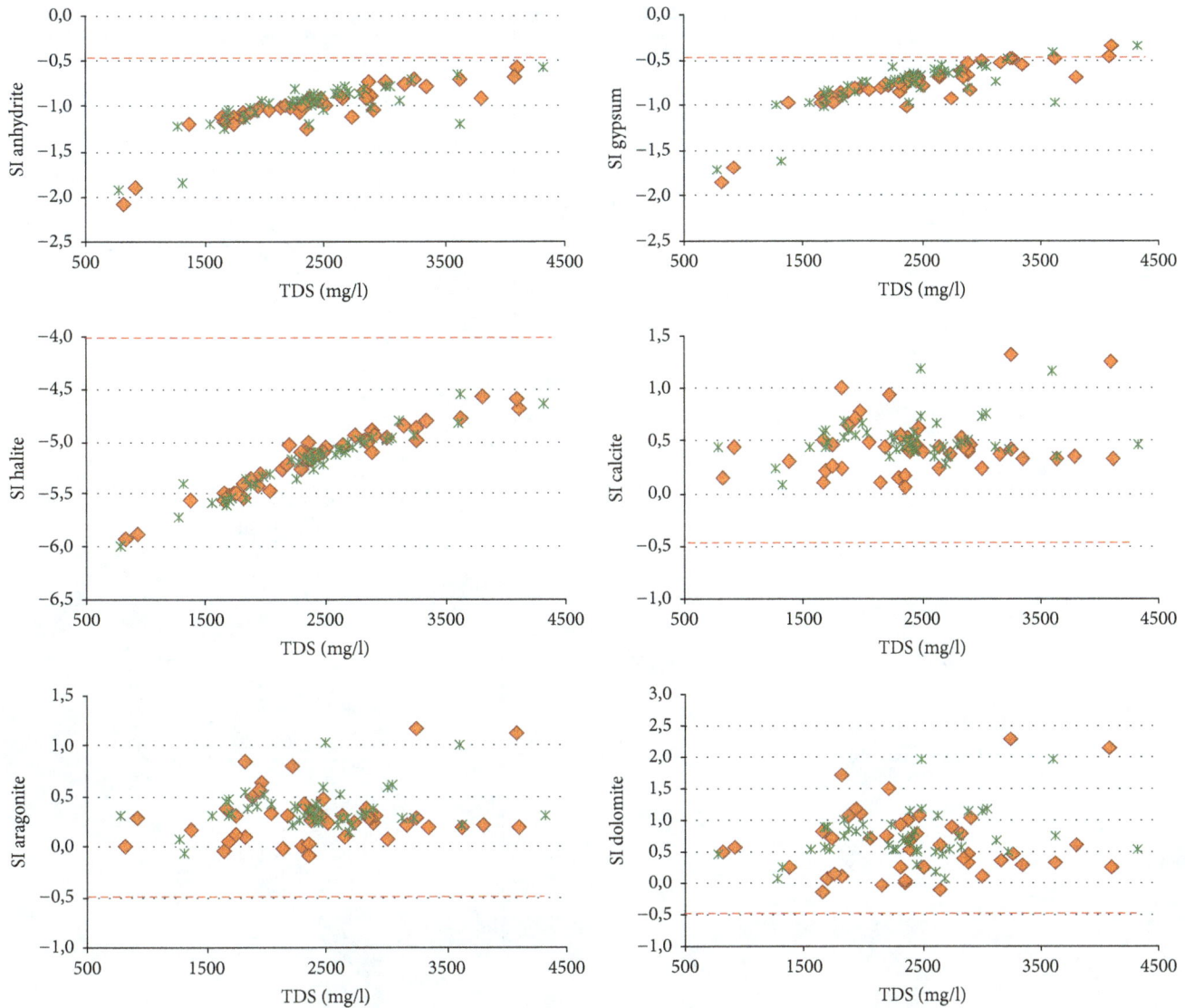

FIGURE 11: Variation of saturation indices of selected minerals in wet and dry seasons.

Ca^{2+} is being flushed from the aquifer by groundwater flow [45, 47].

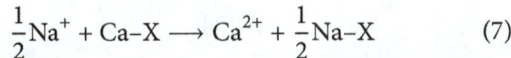

$$\frac{1}{2}Na^+ + Ca\text{-}X \longrightarrow Ca^{2+} + \frac{1}{2}Na\text{-}X \qquad (7)$$

The plot of $[(Ca^{2+} + Mg^{2+})-(HCO_3^- + SO_4^{2-})]$ versus (Na^+-Cl^-) determines the exchange of Na^+ against $(Ca^{2+}$ or $Mg^{2+})$ through the clay matrix. The relationship $[(Ca^{2+} + Mg^{2+})-(HCO_3^- + SO_4^{2-})]$ is the gain or loss of $(Ca^{2+} + Mg^{2+})$ due to the carbonates and gypsum dissolution. The relationship of (Na^+-Cl^-) determines the gain or loss of Na^+ relative to the halite dissolution. If there is no ion exchange, all water samples will be placed in the origin of diagram [46].

Figure 14 shows that reverse ion exchange is a dominant process. To confirm the effect of reverse ion exchange, chloroalkaline index CAI-1 was calculated in milliequivalents

per liter according to the relationship proposed by Schoeller [48]:

$$\text{CAI-1} = \frac{[Cl^- - (Na^+ + K^+)]}{Cl^-} \qquad (8)$$

If reverse ion exchange occurs in groundwater, CAI-1 values are positive. The calculated CAI-1 values are positive for more than 70% of the water samples which confirmed that reverse ion exchange is a dominant process. This shows that the interaction between the seawater and groundwater in the study area is playing a major role in the contamination of the aquifer by seawater intrusion. These results indicate that seawater/freshwater interface is in a continuous evolution despite the artificial recharge operations since 1993, and this probably because of the permanent heavy pumping.

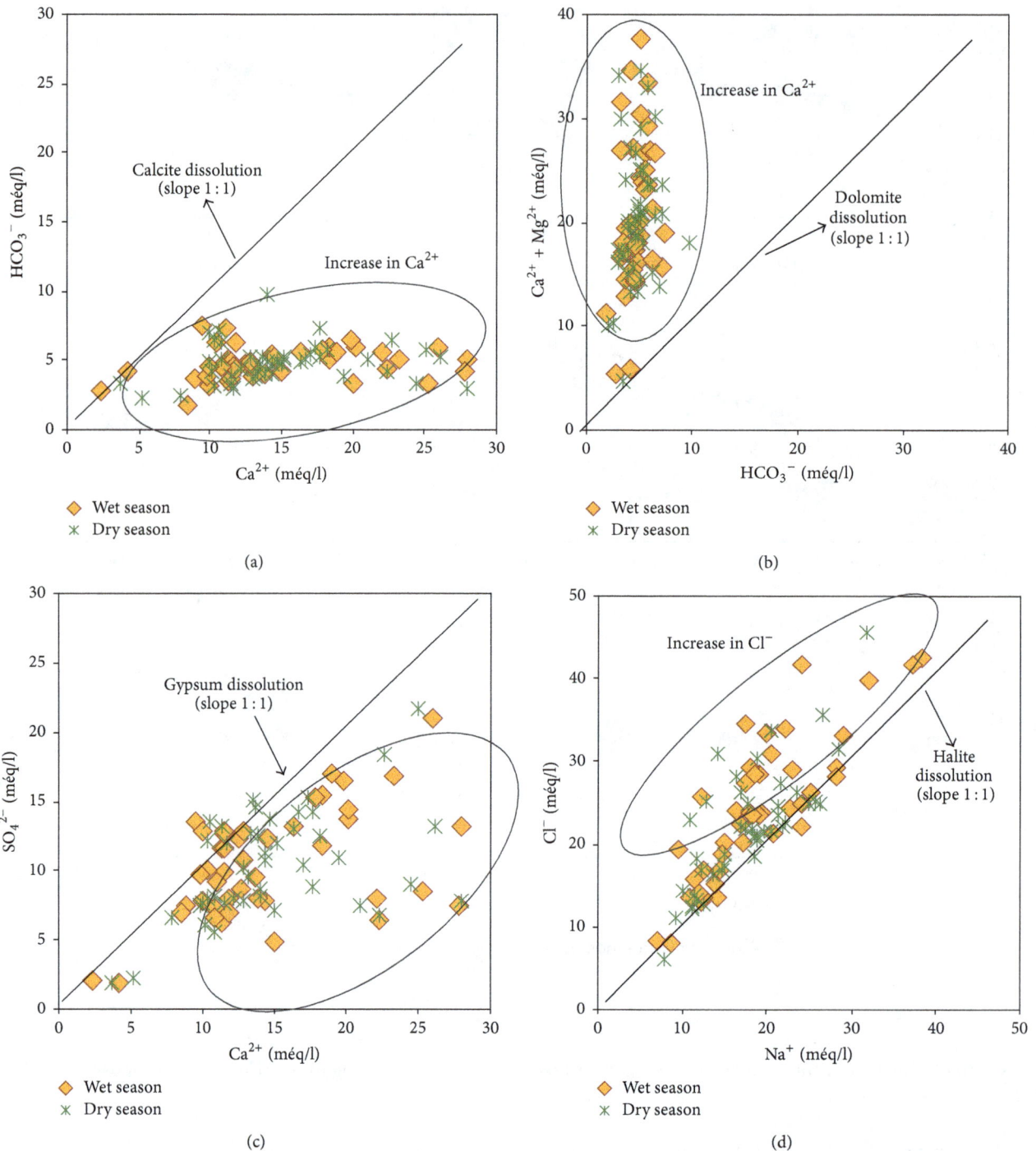

FIGURE 12: Water-rock interaction: relationships between different solutes of Na^+, Ca^{2+}, Mg^{2+}, Cl^-, SO_4^{2-}, and HCO_3^- (a, b, c, d).

Water samples concentrated in the origin of the plot $[(Ca^{2+} + Mg^{2+})–(HCO_3^- + SO_4^{2-})]$ versus $(Na^+–Cl^-)$ indicated the absence of the ion exchange process which can be attributed to the evaporation process followed by carbonate precipitation [49]. This result confirmed the results obtained using saturation states of minerals and Gibbs diagrams.

The seawater fraction in the groundwater is often estimated using chloride concentration [50]. Chloride ion has

been considered as a conservative tracer not affected by ion exchange [51]. For conservative mass balance of the mixture, the equation used is as follows [45]:

$$f = \frac{(Cl_{mix} - Cl_{freshwater})}{(Cl_{seawater} - Cl_{freshwater})} \times 100, \quad (9)$$

where Cl_{mix} is the Cl^- concentration of the sample, $Cl_{seawater}$ is the Cl^- concentration of the Mediterranean Sea, and

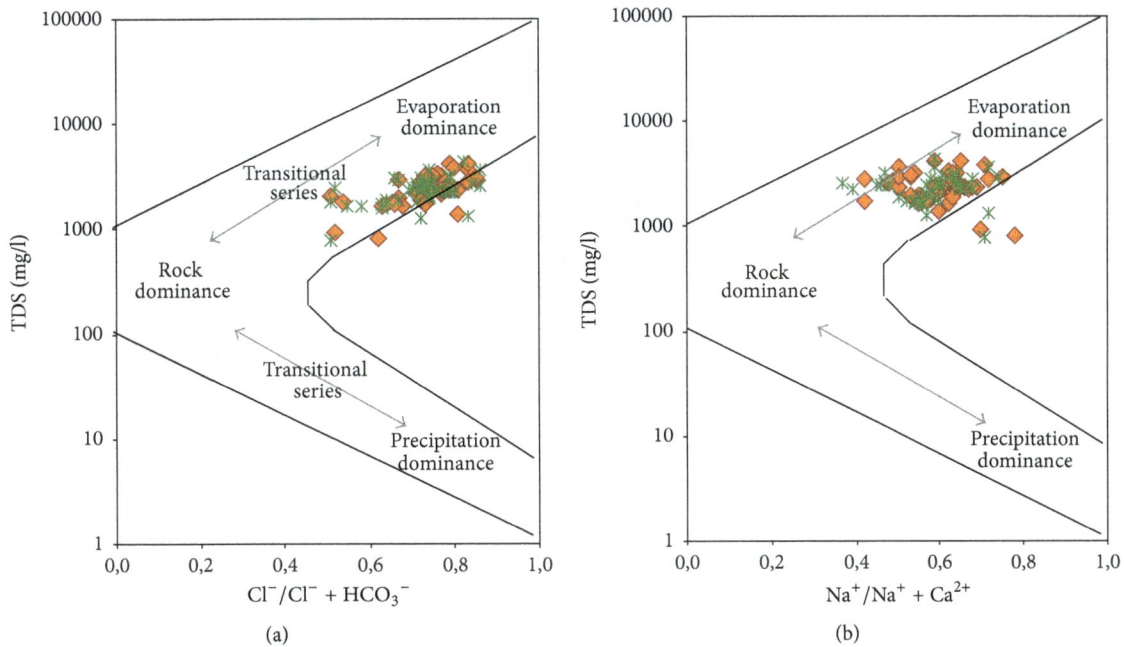

FIGURE 13: Gibbs diagram showing TDS versus (a) Na/(Na + Ca) and (b) Cl/(Cl + HCO_3).

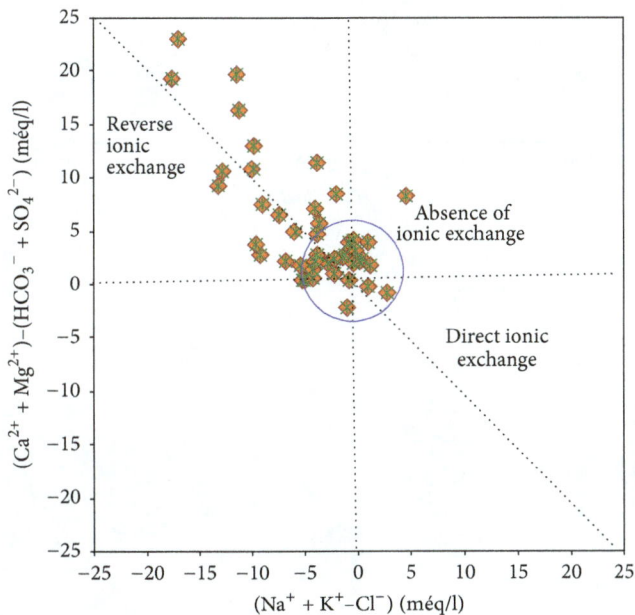

FIGURE 14: Plot of $[(Ca^{2+} + Mg^{2+})-(HCO_3^- + SO_4^{2-})]$ versus $(Na^+ + K^+-Cl^-)$ for ion exchange.

$Cl_{freshwater}$ represents the Cl^- concentration of the fresh water. The fresh water sample will be chosen considering the lowest measured value of the electrical conductivity.

The rate of mixture varies from 0,32% (well n°23) in the north of the Raf Raf region and where the configuration of the ante-quaternary substratum prevents the marine intrusion to 13% (well n°1) near the shoreline (close to the coast). The highest value of the mixing fraction corresponds to the highest measured values of Cl^- and EC (1792 mg/l and 5430 μS/cm, resp.).

4.3. Isotopes and Groundwater Origin.

The stable isotope ratios of oxygen and hydrogen in the groundwater are useful tools to differentiate between salinity origins [52, 53] and to help us understand various sources of recharge processes to groundwater because they are sensitive to physical processes such as atmospheric circulation, groundwater mixing, and evaporation ([33, 54]).

In arid and semiarid regions evaporation could be an important process influencing groundwater chemistry [19]. To understand the relationship between isotopic composition of groundwater of the shallow aquifer of Ras Jbel and those of precipitation measured at the station of Tunis Carthage situated at 50 km from the plain of Metline-Ras Jbel- Raf Raf, a bivariate diagram δ^2H versus $\delta^{18}O$ is plotted in Figure 15(a).

A local meteoric water line (LMWL) for Tunis Carthage was used to interpret the data in this study. The local meteoric water line (LMWL) is controlled by local hydrometeorological factors, including the origin of the vapor mass, reevaporation during rainfall, and the seasonality of precipitation [33]. The isotope composition of the precipitation was plotted along the LMWL using the following equation: δ^2H (‰) = $8 * \delta^{18}O$ (‰) + 12,4 (which had a correlation coefficient $R^2 = 0,99$) [55, 56].

Figure 15(a) shows that the isotopic composition of most of the groundwater samples collected in the wet season (except for sampling site number 35) lies within a narrow range, confirming that these groundwater samples had the same recharge source. Furthermore, all groundwater samples are scattered around the LMWL indicating that the recharge of the Ras Jbel shallow aquifer originates from infiltration of recent precipitation from Mediterranean vapor masses. Based on their isotopic composition, two groups of groundwater samples were identified (Table 4). The first group is relatively depleted in isotopic values and includes samples with $\delta^{18}O$

(a)

(b)

(c)

(d)

(e)

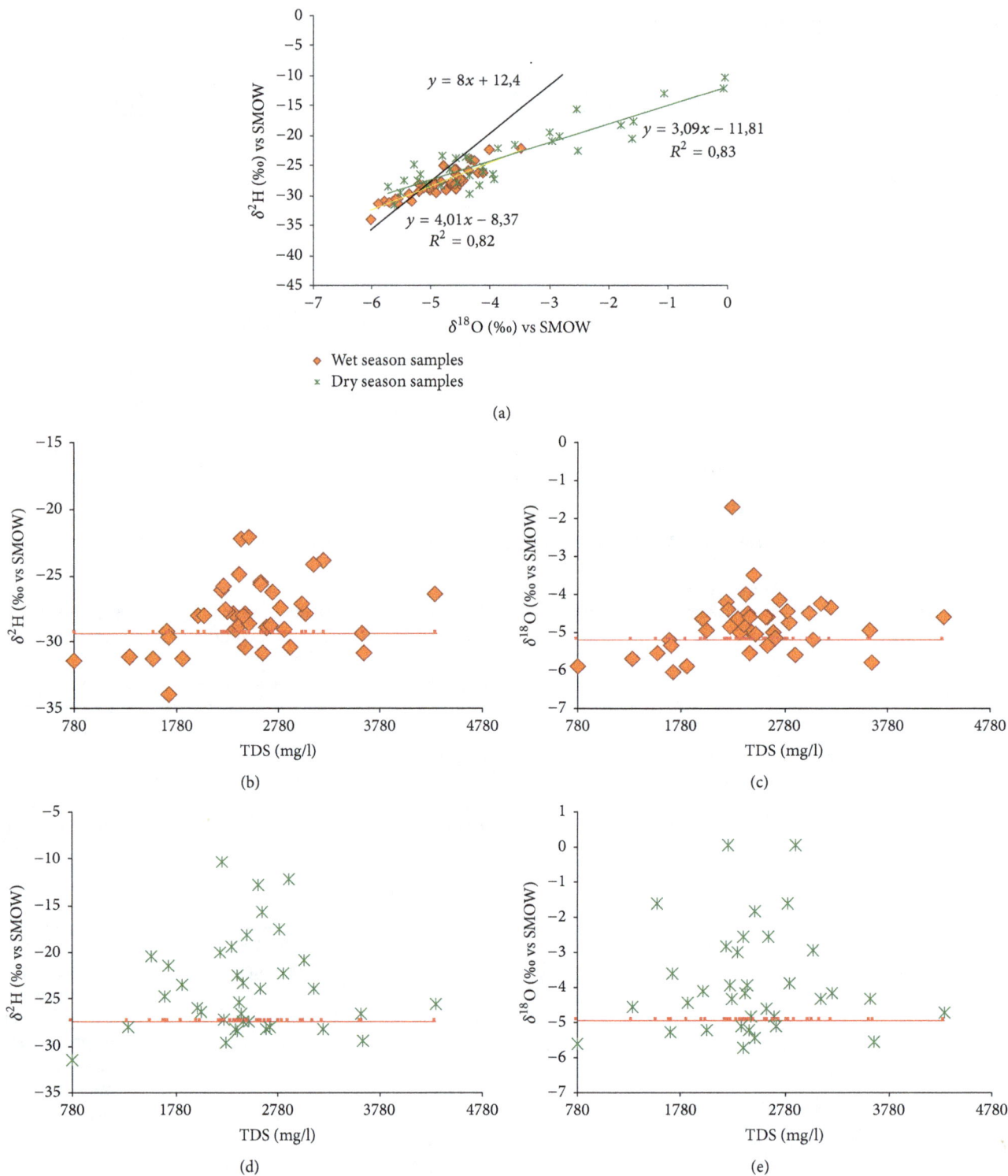

FIGURE 15: Isotopic relationships of groundwater in the study area.

and δ^2H values ranging from −6,03 to −5,37‰ versus V-SMOW and from −33,94 to −29,66‰ versus V-SMOW, respectively. This may be explained by the fact that nonevaporated water is rapidly infiltrated to the saturated zone.

As per the second group, values vary from −5,35 to −3,50‰ versus V-SMOW and from −29,37 to −22,15‰ versus

V-SMOW, for $\delta^{18}O$ and δ^2H, respectively. The relatively enriched isotopic values of the group 2 samples demonstrate that this groundwater is affected by evaporated open water or soil water.

The groundwater samples collected in the dry season were enriched compared to those collected in the wet season. The

TABLE 4: Isotopic data of sampled wells and piezometers in Ras Jbel aquifer.

Parameters	$\delta^{18}O$ (‰ versus SMOW)	δ^2H (‰ versus SMOW)	$\delta^{18}O$ (‰ versus SMOW)	δ^2H (‰ versus SMOW)
	Wet season		Dry season	
Minimum	−6,03	−33,94	−5,74	−31,46
Maximum	−3,5	−22,15	−0,06	−10,31
Average	−4,91	−28,06	−3,78	−23,33

results show isotopic content ranging from −5,74 to −0,06‰ versus V-SMOW for $\delta^{18}O$ and from −31,46 to −10,31‰ versus V-SMOW for δ^2H. This may be explained by heavy isotope enrichment in the groundwater caused by strong evaporation given that there was effectively no precipitation in the study area between the two sampling periods; meanwhile, the groundwater may have been mainly recharged by lateral inflow from outside the study area, resulting in seasonal fluctuations in the isotopic values. The enriched groundwater samples (n°15, 20, 25, 30, 32, 35, 38, 41, 45, and 48) might be an indicator for evaporation of the recharge water before its infiltration to the aquifer.

The isotopic data for the groundwater collected in the wet season were linearly fit using the regression equation δ^2H (‰) = 4,01 * $\delta^{18}O$ (‰) − 8,37 (R^2 = 0,82). This regression line can be interpreted as the groundwater evaporation line (GEL). The gel has a $\delta^2H/\delta^{18}O$ slope <8, which reflects evaporation during or after rainfall and/or mixing with an external water source (e.g., return of irrigation water) with high $\delta^{18}O$ and δ^2H values. Furthermore, the GEL of the wet season intersected the LMWL at values of $\delta^{18}O$ = −5,22‰ versus V-SMOW and δ^2H = −29,37‰ versus V-SMOW. These values are estimated as baseline for δ^2H and $\delta^{18}O$ in recharging rainfall (Figures 15(b) and 15(c)). If samples are plotted above the lines, significant groundwater evaporation process can be confirmed.

Additionally, the isotopic data from the dry season were linearly fit using the regression equation δ^2H (‰) = 3,09 * $\delta^{18}O$ (‰) − 11,81 (with a correlation coefficient R^2 = 0,83). The GEL has a smaller slope than the LMWL, because evaporation tends to enrich heavy isotopes in water ([19, 57]). The increase in groundwater salinity due to evaporation can thus result in simultaneous increase in heavy isotopes ([19, 35]). The GEL of the dry season intersects the LMWL at values of $\delta^{18}O$ = −4,93‰ versus V-SMOW and δ^2H = −27,32‰ versus V-SMOW, which are chosen as baselines (Figures 15(d) and 15(e)). It is observed that 80% of the groundwater samples were plotted above the baselines and this demonstrates that evaporation has a significant contribution to groundwater salinity in the study area.

Furthermore, the deuterium excess calculated as d-excess = δ^2H − 8$\delta^{18}O$ [54] has been widely used in hydrological studies. The d-excess is used to identify secondary processes that influence the atmospheric vapor content in the evaporation–condensation cycle in nature ([54, 58]). The d-excess plotted against $\delta^{18}O$ shows a negative correlation for the whole set of samples (Figure 16). The decrease in d-excess

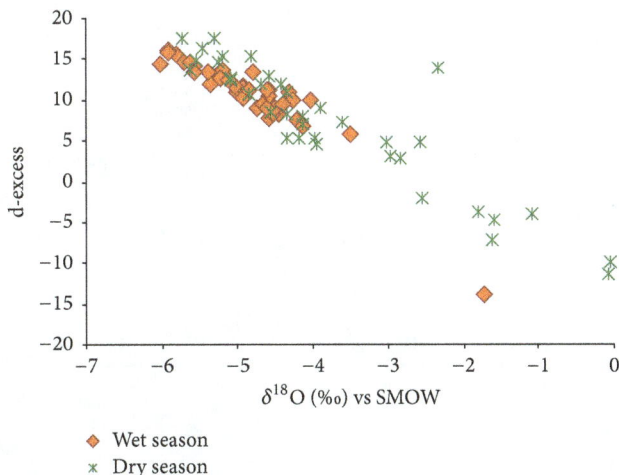

FIGURE 16: Plot of d-excess versus $\delta^{18}O$.

is an indication that evaporation has occurred during the recharge process which again confirms the previous results.

4.4. *Irrigation Return Flow.* Irrigation return flow is defined as the excess of irrigation water that is not evapotranspirated or evacuated by direct surface drainage and which returns to an aquifer or surface water [59, 60]. Irrigation return flows may induce salt and nitrate pollution of receiving water bodies [61]. Indeed, NO_3^- is the most common water contaminant, and NO_3^- pollution is increasing because the number of anthropogenic sources is increasing [26].

71% of groundwater samples are contaminated by nitrates where the concentration exceeds the permissible value of 50 mg/l set by WHO [62]. The spatial distribution of nitrates (Figure 17) shows that high nitrate contents are observed especially in the upstream of Ras Jbel aquifer. Groundwater contamination by nitrate is due to the intensive use of nitrogen fertilizers ($Ca(NO_3)_2$, KNO_3, and $MgSO_4$). In recent years, the agricultural land area in Ras Jbel plain has increased and copious amounts of nitrogenous fertilizer have been used, which have increased the groundwater NO_3^- concentrations.

Furthermore, return flow from irrigation water also seems to contribute notably to the recharge process. Most of groundwater samples shows a correlation between NO_3 and $\delta^{18}O$, reflecting the significant role of evaporated and contaminated irrigation water to the groundwater salinization (Figure 18(a)). Huge quantities of irrigation return flow

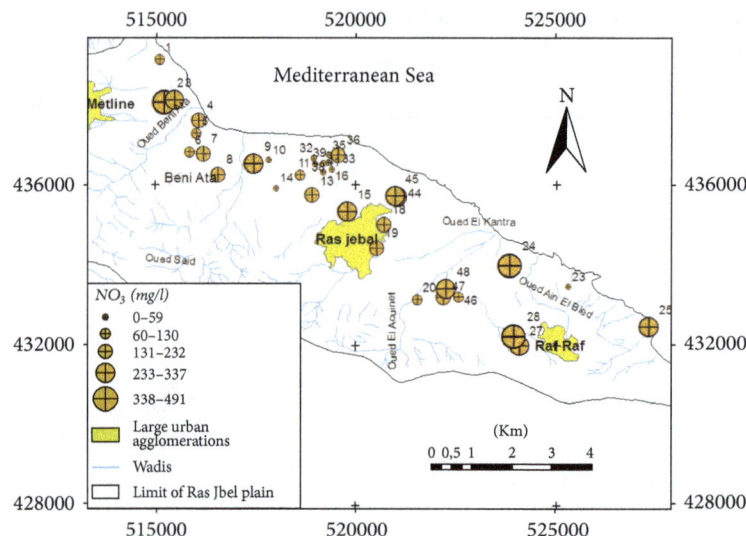

FIGURE 17: Spatial distribution of nitrate concentrations in the shallow aquifer of Ras Jbel during the dry season.

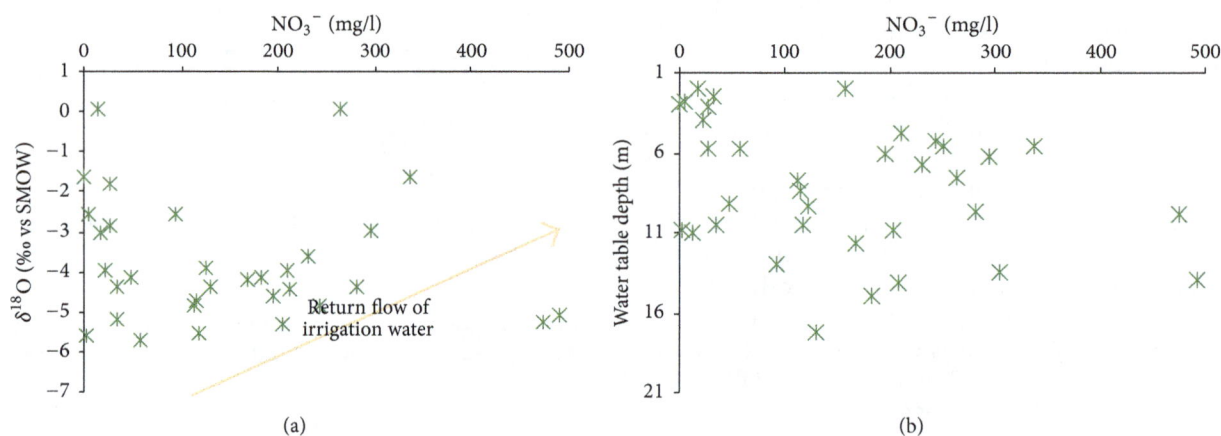

FIGURE 18: Plots of NO_3^- versus water table depth and $\delta^{18}O$ versus NO_3^-.

elevated groundwater level, hence increasing evaporation and inducing salinization [63]. As highlighted in Figure 18(b), the contamination by return of irrigation water is observed in the shallower horizons (depth ≤ 13 m).

5. Conclusions

This paper aimed to discuss the origin, processes, and mechanisms of groundwater salinization, as well as the chemical evolution of groundwater in the Ras Jbel coastal aquifer using isotopic tools and hydrochemical tracers.

Most of the groundwater is considered to be of brackish to saline water and contains high ion concentrations. The groundwater in the study area is influenced by both natural and anthropogenic factors. The major geochemical processes controlling hydrochemical evolution are the inverse cationic exchange due to the phenomena of seawater intrusion, dissolution of evaporates minerals (halite, gypsum, and/or anhydrite), irrigation return flow, water-rock interactions, and

evapo(transpi)ration. (Figure 19). The mixing rate among freshwater and saline water ranges between 1 and 13%.

In addition, groundwater in the shallow aquifer of Ras Jbel is also contaminated by agricultural fertilizers containing high amounts of nitrates. Nitrates are transported to the aquifer by natural recharge process and by return flow from irrigation water.

Hydrogen and oxygen-18 stable isotopes signatures of groundwater have identified recent groundwater recharge by infiltration of local precipitations. The enrichment in stable isotope of groundwater confirms that return flow of irrigation waters is an important factor influencing groundwater quality.

The results of this study can be used to improve our understanding of hydrogeochemical processes and enable the protection and sustainable use of water resources. It, therefore, calls for more comprehensive research for better water resources management.

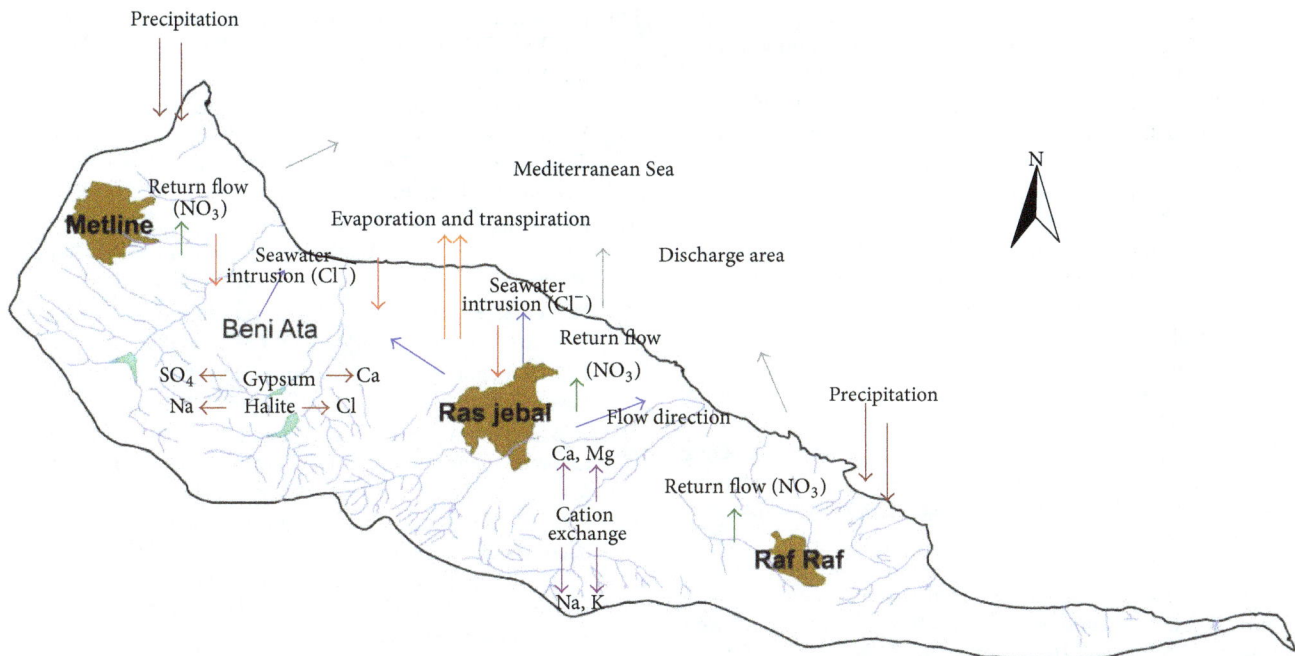

FIGURE 19: Schematic conceptual model summarizing salinization sources of groundwater in the plain of Ras Jbel.

Conflicts of Interest

The authors declare no conflicts of interest.

Acknowledgments

The authors would like to gratefully acknowledge all members of the Regional Commission for Agricultural Development of Bizerte for their guidance and support in field campaigns. Special thanks are due to Dr. Maarten Waterloo, Senior hydrologist, Acacia Water (Formerly Professor in Hydrology, Free University Amsterdam), Netherlands.

References

[1] V. E. A. Post, *Groundwater salinization processes in the coastal area of the Netherlands due to transgressions during the Holocene [Ph.D. thesis]*, University of Amsterdam, Amsterdam, the Netherlands, 2004.

[2] G. de Marsily, "Importance of the maintenance of temporary ponds in arid climates for the recharge of groundwater," *Comptes Rendus—Geoscience*, vol. 335, no. 13, pp. 933–934, 2003.

[3] D. Han, X. Song, M. J. Currell, G. Cao, Y. Zhang, and Y. Kang, "A survey of groundwater levels and hydrogeochemistry in irrigated fields in the Karamay Agricultural Development Area, northwest China: implications for soil and groundwater salinity resulting from surface water transfer for irrigation," *Journal of Hydrology*, vol. 405, no. 3-4, pp. 217–234, 2011.

[4] M. U. Igboekwe and A. Ruth, "Groundwater recharge through infiltration process: A Case Study of Umudike, Southeastern Nigeria," *Journal of Water Resource and Protection*, vol. 3, no. 5, pp. 295–299, 2011.

[5] P. Banerjee and V. S. Singh, "Statistical approach for comprehensive planning of watershed development through artificial recharge," *Water Resources Management*, vol. 26, no. 10, pp. 2817–2831, 2012.

[6] J. Jing, Q. Hui, C. Yu-Fei, and X. Wen-Juan, "Assessment of groundwater quality based on matter element extension model," *Journal of Chemistry*, vol. 2013, Article ID 715647, 7 pages, 2013.

[7] L. Kouzana, R. Benassi, A. Ben mammou, and M. Sfar felfoul, "Geophysical and hydrochemical study of the seawater intrusion in Mediterranean semi arid zones. Case of the Korba coastal aquifer (Cap-Bon, Tunisia)," *Journal of African Earth Sciences*, vol. 58, no. 2, pp. 242–254, 2010.

[8] R. Trabelsi, K. Abid, K. Zouari, and H. Yahyaoui, "Groundwater salinization processes in shallow coastal aquifer of Djeffara plain of Medenine, Southeastern Tunisia," *Environmental Earth Sciences*, vol. 66, no. 2, pp. 641–653, 2012.

[9] A. Chekirbane, M. Tsujimura, A. Kawachi, H. Isoda, J. Tarhouni, and A. Benalaya, "Hydrogeochemistry and groundwater salinization in an ephemeral coastal flood plain: Cap Bon, Tunisia," *Hydrological Sciences Journal*, vol. 58, no. 5, pp. 1097–1110, 2013.

[10] H. Bouzourra, R. Bouhlila, L. Elango, F. Slama, and N. Ouslati, "Characterization of mechanisms and processes of groundwater salinization in irrigated coastal area using statistics, GIS, and hydrogeochemical investigations," *Environmental Science and Pollution Research*, vol. 22, no. 4, pp. 2643–2660, 2015.

[11] A. Kharroubi, S. Farhat, B. Agoubi, and Z. Lakhbir, "Assessment of water qualities and evidence of seawater intrusion in a deep confined aquifer: case of the coastal Djeffara aquifer (Southern Tunisia)," *Journal of Water Supply*, vol. 63, no. 1, pp. 76–84, 2014.

[12] I. Triki, N. Trabelsi, M. Zairi, and H. Ben Dhia, "Multivariate statistical and geostatistical techniques for assessing groundwater salinization in Sfax, a coastal region of eastern Tunisia," *Desalination and Water Treatment*, vol. 52, no. 10–12, pp. 1980–1989, 2014.

[13] M. F. B. Hamouda, A. J. Kondash, N. Lauer, L. Mejri, J. Tarhouni, and A. Vengosh, "Assessment of groundwater salinity mechanisms in the coastal aquifer of el haouaria, Northern Tunisia," *Procedia Earth and Planetary Science*, vol. 13, pp. 194–198, 2015.

[14] S. K. Ambast, N. K. Tyagi, and S. K. Raul, "Management of declining groundwater in the Trans Indo-Gangetic Plain (India): some options," *Agricultural Water Management*, vol. 82, no. 3, pp. 279–296, 2006.

[15] A. El-Naqa and A. Al-Shayeb, "Groundwater protection and management strategy in Jordan," *Water Resources Management*, vol. 23, no. 12, pp. 2379–2394, 2009.

[16] J. L. McCallum, R. S. Crosbie, G. R. Walker, and W. R. Dawes, "Impacts of climate change on groundwater in Australia: a sensitivity analysis of recharge," *Hydrogeology Journal*, vol. 18, no. 7, pp. 1625–1638, 2010.

[17] J. D. Ayotte, M. Belaval, S. A. Olson et al., "Factors affecting temporal variability of arsenic in groundwater used for drinking water supply in the United States," *Science of the Total Environment*, vol. 505, pp. 1370–1379, 2015.

[18] J. Wu and Z. Sun, "Evaluation of shallow groundwater contamination and associated human health risk in an alluvial plain impacted by agricultural and industrial activities, Mid-west China," *Exposure and Health*, vol. 8, no. 3, pp. 311–329, 2016.

[19] P. Li, J. Wu, and H. Qian, "Hydrochemical appraisal of groundwater quality for drinking and irrigation purposes and the major influencing factors: a case study in and around Hua County, China," *Arabian Journal of Geosciences*, vol. 9, no. 1, article 15, 2016.

[20] DGRE, *Annuaire de Surveillance Piézométrique*, Rapp. Int. Ministère de l'Agriculture, Tunis, Tunisia, 2007.

[21] A. Choura, *Impact de la surexploitation et de la recharge artificielle de la nappe de Ras Jbel par les systèmes d'information géographiques [M.S. thesis]*, Faculty of Sciences of Tunis, 1993.

[22] L. Bouchaou, J. L. Michelot, A. Vengosh et al., "Application of multiple isotopic and geochemical tracers for investigation of recharge, salinization, and residence time of water in the Souss-Massa aquifer, southwest of Morocco," *Journal of Hydrology*, vol. 352, no. 3-4, pp. 267–287, 2008.

[23] F. El Yaouti, A. El Mandour, D. Khattach, J. Benavente, and O. Kaufmann, "Salinization processes in the unconfined aquifer of Bou-Areg (NE Morocco): a geostatistical, geochemical, and tomographic study," *Applied Geochemistry*, vol. 24, no. 1, pp. 16–31, 2009.

[24] F. Sdao, S. Parisi, D. Kalisperi et al., "Geochemistry and quality of the groundwater from the karstic and coastal aquifer of Geropotamos River Basin at north-central Crete, Greece," *Environmental Earth Sciences*, vol. 67, no. 4, pp. 1145–1153, 2012.

[25] G. Mongelli, S. Monni, G. Oggiano, M. Paternoster, and R. Sinisi, "Tracing groundwater salinization processes in coastal aquifers: a hydrogeochemical and isotopic approach in the Na-Cl brackish waters of northwestern Sardinia, Italy," *Hydrology and Earth System Sciences*, vol. 17, no. 7, pp. 2917–2928, 2013.

[26] Y. Lu, C. Tang, J. Chen, and J. Chen, "Groundwater recharge and hydrogeochemical evolution in leizhou peninsula, China," *Journal of Chemistry*, vol. 2015, Article ID 427579, 12 pages, 2015.

[27] M. Ennabli, "Hydrogéologie de la plaine de Ras Jbel-Raf Raf," *Annales des Mines et de la Géologie*, vol. 26, pp. 537–561, 1973.

[28] M. H. Hamza, *Evaluation de la vulnérabilité a la pollution des nappes phréatiques de Ras Jbel et de Guenniche par les méthodes paramétriques DRASTIC, SINTACS, et SI appliquées par les systèmes d'informations géographiques [Ph.D. thesis]*, Faculty of Sciences of Tunis, Tunis, Tunisia, 2007.

[29] P. F. Burollet, "Etude géologique des bassins Mio-Pliocènes du Nord Est de la Tunisie," *Annales des Mines et de la Géologie*, no. 7, p. 82, 1951.

[30] M. Ennabli, "Etat des travaux réalisés dans la plaine de Metline—Ras Jbel—Raf Raf en vue de l'étude hydrogéologique de la plaine côtière," Rapp. Interne, Bureau de l'Inventaire et des Ressources Hydrauliques, 1969.

[31] DGRE, "Annuaires de surveillance piézométrique," DGRE. Rapports Techniques des Piézomètres et des Forages, Rapp. Int. Ministère de l'Agriculture, Tunis, Tunisia, 1986–2006.

[32] Pimienta, Etude hydrogéologiquede Ras Jebel. Fasc.1 et 2, Service Géologique, (BIRH 3-10 et 3-11), 1949.

[33] I. D. Clark and P. Fritz, *Environmental Isotopes in Hydrogeology*, Lewis Publishers, New York, NY, USA, 1997.

[34] A. M. Piper, "A graphic procedure in the geochemical interpretation of water-analyses," *Eos, Transactions American Geophysical Union*, vol. 25, no. 6, pp. 914–928, 1944.

[35] F. Liu, X. Song, L. Yang et al., "Identifying the origin and geochemical evolution of groundwater using hydrochemistry and stable isotopes in the Subei Lake basin, Ordos energy base, Northwestern China," *Hydrology and Earth System Sciences*, vol. 19, no. 1, pp. 551–565, 2015.

[36] R. J. Gibbs, "Mechanisms controlling world water chemistry," *Science*, vol. 170, no. 3962, pp. 1088–1090, 1970.

[37] J. W. Lloyd and J. Heathcote, *Natural Inorganic Hydrochemistry in Relation to Groundwater*, Oxford University Press, New York, NY, USA, 1985.

[38] D. L. Parkhurst and C. Appelo, "PHREEQC2 user's manual and program," Water-Resources Investigations Report, US Geological Survey, Denver, Colo, USA, 2004.

[39] K. Pazand, A. Hezarkhani, Y. Ghanbari, and N. Aghavali, "Groundwater geochemistry in the Meshkinshahr basin of Ardabil province in Iran," *Environmental Earth Sciences*, vol. 65, no. 3, pp. 871–879, 2012.

[40] D. Sujatha and B. R. Reddy, "Quality characterization of groundwater in the south-eastern part of the Ranga Reddy district, Andhra Pradesh, India," *Environmental Geology*, vol. 44, no. 5, pp. 579–586, 2003.

[41] A. Kharroubi, F. Tlahigue, B. Agoubi, C. Azri, and S. Bouri, "Hydrochemical and statistical studies of the groundwater salinization in Mediterranean arid zones: case of the Jerba coastal aquifer in southeast Tunisia," *Environmental Earth Sciences*, vol. 67, no. 7, pp. 2089–2100, 2012.

[42] J. C. Rozemeijer, *Dynamics in groundwater and surface water quality: from field-scale processes to catchment-scale monitoring [Ph.D. thesis]*, Utrecht University, Utrecht, The Netherlands, 2010.

[43] Q. B. Luo, W. D. Kang, Y. L. Xie, and B. F. Zhao, "Groundwater hydro-geochemistry simulation in the Jingbian area of the Luohe of Cretaceous," *Ground Water*, vol. 30, no. 6, pp. 22–24, 2008.

[44] N. Ettayfi, L. Bouchaou, J. L. Michelot et al., "Geochemical and isotopic (oxygen, hydrogen, carbon, strontium) constraints for the origin, salinity, and residence time of groundwater from a carbonate aquifer in the Western Anti-Atlas Mountains, Morocco," *Journal of Hydrology*, vol. 438-439, pp. 97–111, 2012.

[45] C. A. J. Appelo and D. Postma, *Geochemistry, Groundwater and Pollution*, Balkema, Rotterdam, The Netherlands, 2nd edition, 1993.

[46] A. W. Hounslow, *Water Quality Data. Analysis and Interpretation*, Lewis Publishers, Boca Raton, Fla, USA, 1995.

[47] M. S. Andersen, V. Nyvang, R. Jakobsen, and D. Postma, "Geochemical processes and solute transport at the seawater/freshwater interface of a sandy aquifer," *Geochimica et Cosmochimica Acta*, vol. 69, no. 16, pp. 3979–3994, 2005.

[48] H. Schoeller, "Qualitative evaluation of groundwater resources," in *Methods and Techniques of Groundwater Investigations and Development*, pp. 53–83, UNESCO, Paris, France, 1965.

[49] B. C. Richter and C. W. Kreitler, "Identification of sources of ground-water salinization using geochemical techniques," p. 273, 1993.

[50] E. Custodio and K. A. Bruggeman, *Groundwater Problems in Coastal Areas. Studies and Reports in Hydrology*, UNESCO, Paris, France, 1987.

[51] E. Custodio, *Groundwater Problems in Coastal Areas*, Studies and Reports in Hydrology, UNESCO, Paris, France, 1987.

[52] W. M. Edmunds, A. H. Guendouz, A. Mamou, A. Moulla, P. Shand, and K. Zouari, "Groundwater evolution in the Continental Intercalaire aquifer of southern Algeria and Tunisia: trace element and isotopic indicators," *Applied Geochemistry*, vol. 18, no. 6, pp. 805–822, 2003.

[53] T. W. Butler II, "Application of multiple geochemical indicators, including the stable isotopes of water, to differentiate water quality evolution in a region influenced by various agricultural practices and domestic wastewater treatment and disposal," *Science of the Total Environment*, vol. 388, no. 1–3, pp. 149–167, 2007.

[54] W. Dansgaard, "Stable isotopes in precipitation," *Tellus*, vol. 16, no. 4, pp. 436–468, 1964.

[55] M. Ahmed Maliki, M. Krimissa, J. Michelot, and K. Zouari, "Relation entre nappes superficielles et aquifère profond dans le bassin de Sfax (Tunisie)," *Comptes Rendus de l'Académie des Sciences. Series IIA*, vol. 331, no. 1, pp. 1–6, 2000.

[56] A. Ben Moussa, S. B. H. Salem, K. Zouari, and F. Jlassi, "Hydrochemical and isotopic investigation of the groundwater composition of an alluvial aquifer, Cap Bon Peninsula, Tunisia," *Carbonates and Evaporites*, vol. 25, no. 3, pp. 161–176, 2010.

[57] H. Qian, P. Li, J. Wu, and Y. Zhou, "Isotopic characteristics of precipitation, surface and ground waters in the Yinchuan plain, Northwest China," *Environmental Earth Sciences*, vol. 70, no. 1, pp. 57–70, 2013.

[58] H. Craig, "Isotopic variations in meteoric waters," *Science*, vol. 133, no. 3465, pp. 1702–1703, 1961.

[59] B. Dewandel, J.-M. Gandolfi, D. de Condappa, and S. Ahmed, "An efficient methodology for estimating irrigation return flow coefficients of irrigated crops at watershed and seasonal scale," *Hydrological Processes*, vol. 22, no. 11, pp. 1700–1712, 2008.

[60] Z. Kattan, "Estimation of evaporation and irrigation return flow in arid zones using stable isotope ratios and chloride mass-balance analysis: case of the Euphrates River, Syria," *Journal of Arid Environments*, vol. 72, no. 5, pp. 730–747, 2008.

[61] J. Causapé, D. Quílez, and R. Aragüés, "Assessment of irrigation and environmental quality at the hydrological basin level: II. Salt and nitrate loads in irrigation return flows," *Agricultural Water Management*, vol. 70, no. 3, pp. 211–228, 2004.

[62] WHO, *Guidelines for Drinking Water Quality*, World Health Organization, Geneva, Switzerland, 3rd edition, 2004.

[63] H. Wu, J. Chen, H. Qian, and X. Zhang, "Chemical characteristics and quality assessment of groundwater of exploited aquifers in Beijiao water source of Yinchuan, China: a case study for drinking, irrigation, and industrial purposes," *Journal of Chemistry*, vol. 2015, Article ID 726340, 14 pages, 2015.

[64] M. A. Haddad, *Evolution de l'État de la Nappe de Ras Jebel de 1949 à 2005 et éValuation des Impacts de la Recharge Artificielle (Période 1992–2005)*, P.F.E du Cycle d'Ingénieur en Géosciences, Faculté des Sciences de Tunis, 2006.

[65] DGRE, "Annuaires d'exploitation des nappes phréatiques en Tunisie," DGRE, Tunis, Tunisia, 1993–2008.

19

NO$_x$ Removal from Simulated Marine Exhaust Gas by Wet Scrubbing Using NaClO Solution

Zhitao Han, Bojun Liu, Shaolong Yang, Xinxiang Pan, and Zhijun Yan

Marine Engineering College, Dalian Maritime University, Dalian 116026, China

Correspondence should be addressed to Zhitao Han; hanzt@dlmu.edu.cn

Academic Editor: Andrea Gambaro

The experiments were performed in a lab-scale countercurrent spraying reactor to study the NO$_x$ removal from simulated gas stream by cyclic scrubbing using NaClO solution. The effects of NaClO concentration, initial solution pH, coexisting gases (5% CO$_2$ and 13% O$_2$), NO$_x$ concentration, SO$_2$ concentration, and absorbent temperature on NO$_x$ removal efficiency were investigated in regard to marine exhaust gas. When NaClO concentration was higher than 0.05 M and initial solution pH was below 8, NO$_x$ removal efficiency was relatively stable and it was higher than 60%. The coexisting CO$_2$ (5%) had little effect on NO$_x$ removal efficiency, but the outlet CO$_2$ concentration decreased slowly with the initial pH increasing from 6 to 8. A complete removal of SO$_2$ and NO could be achieved simultaneously at 293 K, initial pH of 6, and NaClO concentration of 0.05 M, while the outlet NO$_2$ concentration increased slightly with the increase of inlet SO$_2$ concentration. NO$_x$ removal efficiency increased slightly with the increase of absorbent temperature. The relevant reaction mechanisms for the oxidation and absorption of NO with NaClO were also discussed. The results indicated that it was of great potential for NO$_x$ removal from marine exhaust gas by wet scrubbing using NaClO solution.

1. Introduction

The exhaust gas emitted from marine diesel engines contains a large number of atmospheric pollutants, such as sulphur oxides (SO$_x$), nitrogen oxides (NO$_x$), and particulate matters (PMs), which has caused serious damage to the ecological environment [1]. A direct way to reduce SO$_x$ emission is to adopt low sulphur fuel oil (LSFO), but this will increase the transportation cost greatly. An alternative way is to install an exhaust gas cleaning system (EGCS) on board to achieve the abatement of SO$_x$ emission equivalently. But it is very difficult for SO$_x$ scrubbers to remove NO$_x$ effectively at the same time. Generally, the methods for reducing marine NO$_x$ emission can be divided into combustion control and postcombustion control techniques. The combustion control techniques include exhaust gas recirculation (EGR) [2], fuel-water emulsion (FWE) [3], and direct water injection (DWI) [4]. They aim at reducing the formation of NO$_x$ during the combustion process, but it will result in the decrease of total heat efficiency. At present, the typical postcombustion control

approach for ocean-going ships is selective catalytic reduction (SCR), which can remove NO$_x$ with an efficiency of 80–95%. As a relatively mature denitrification technology, SCR has been extensively applied in power plants and mobile vehicles. But there are still some problems that limit the application of SCR in marine industry. It requires large additional space and high investment/operation cost. The ash content or sulfate salts may result in the inactivation of SCR catalyst. A complex control system is also required to reduce the ammonia slip [5–7]. Therefore, ship owners around the world are still seeking a better way to reduce NO$_x$ emission. It is of great importance to develop an efficient denitrification technology that can cater for the needs of ocean-going ships.

During the past decades, wet scrubbing technique becomes more and more attractive for the simultaneous removal of SO$_x$, NO$_x$, and other pollutants. As to NO$_x$ removal, a preoxidation process is required to oxidize NO into NO$_2$ or other nitrogen oxides of higher values. That is due to the low solubility (1.93×10^{-3} mol·L^{-1}·atm^{-1} at 25°C) of NO in water. The nonthermal plasma (NTP) can be used to oxidize NO effectively, and then an absorption process is followed for NO_2 absorption. However, this method requires a high energy consumption [8, 9]. Similarly, ozone oxidation method encounters the same limitation [10–12]. Another feasible approach is to add the oxidants into the absorbent. For this purpose, various oxidants such as hydrogen peroxide (H_2O_2) [13–15], potassium permanganate ($KMnO_4$) [16], sodium chlorite ($NaClO_2$) [17, 18], calcium hypochlorite ($Ca(ClO)_2$) [19], and sodium hypochlorite (NaClO) [20–24] have been investigated to enhance the NO removal efficiency of the scrubbing solution. Compared with other oxidants, NaClO has some distinct advantages, such as low cost, easy availability, strong oxidative ability, easy to storage, good stability, and low toxicity. So it is attractive for researchers to investigate the simultaneous removal of NO and SO_2 by wet scrubbing using NaClO solution [25–28]. A previous study implied that NO could be effectively removed by wet scrubbing using NaClO solution in a cyclic mode, and the utilization of NaClO oxidant in solution was extremely high [29]. But the effect of the operating parameters (such as NaClO concentration, solution pH, absorbent temperature, NO, and SO_2 concentrations) on NO removal efficiency by cyclic scrubbing using NaClO solution had not been investigated. In this paper, marine exhaust gas was chosen as the treatment objective. Though the components of marine exhaust gas might vary with the engine load and fuel type, the typical compositions of marine exhaust gas contained ~13% O_2 and ~5% CO_2. The effect of coexisting gases on NO_x removal efficiency had also been studied preliminarily, and the relevant reaction mechanism was discussed.

2. Experimental Section

2.1. Materials.
As mentioned in [11], exhaust gas scrubbers are designed in accordance with the maximum power and exhaust amount of target engine for the practical application in marine industry. The exhaust gas compositions are also measured at maximum load of engine. In this study, the concern is mainly focused on the NO_x removal efficiency, and the concentrations of gas components of a typical marine slow-speed 2-stroke diesel engine are considered as the reference. The fuel type is heavy fuel oil with 2.4% sulphur content. Thus, the concentrations of O_2, CO_2, SO_2, and NO_x are ~13%, ~5%, ~600 ppm, and ~1000 ppm, respectively. Here the simulated exhaust gas was prepared by blending various kinds of synthetic gases. Five kinds of gases, N_2 (99.999%), O_2 (99.995%), CO_2 (99.999%), NO (10.04% NO with N_2 as the balance gas), and SO_2 (10.1% SO_2 with N_2 as the balance gas) (Dalian Date Gas Co., Ltd), were used to make the simulated flue gas. As NO accounted for more than 95% of NO_x in marine exhaust gas, only NO span gas was used to prepare the NO_x components in simulated gas stream.

The NaClO solutions were prepared using the commercial NaClO solution (5% available chlorine, Shanghai Aladdin Bio-Chem Technology Co., Ltd) and the deionized water. The volume of NaClO solution was 1 L for each test. The pH value was adjusted by adding 0.5 M H_2SO_4 solution and determined using an acidimeter (Mettler-Toledo International Trading Co., Ltd).

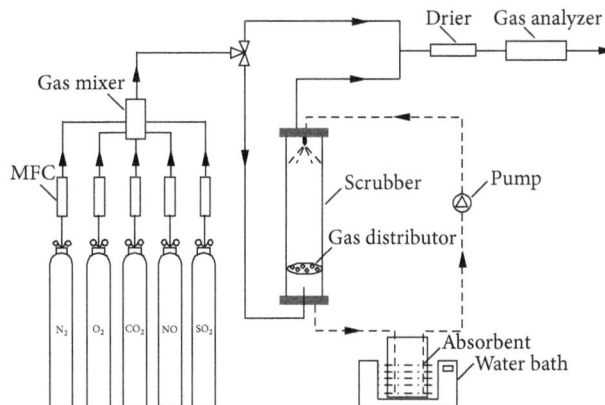

FIGURE 1: Schematic diagram of the experimental setup.

2.2. Experimental Apparatus.
A schematic diagram of the experimental apparatus is shown in Figure 1. It consists of a gas distributing system, a gas-liquid countercurrent scrubbing reactor, and a gas analyzer.

N_2, O_2, CO_2, NO, and SO_2 were provided from separate air bottles and metered through mass flow controllers (MFC, Beijing Sevenstar Electronics Co., Ltd). The simulated flue gas was obtained from the feed gases by blending with an on-line mixer, and then it was introduced into the spraying column. The height and inner diameter of the column were 25 cm and 5 cm, respectively. A spraying nozzle (B1/4TT-SS+TG-SS0.4, Spraying System Co., Ltd) was located at the top of the column. The size of liquid droplet sprayed from the nozzle was in the range of 80–100 μm. The flow rate of the simulated flue gas was fixed at 1.25 L/min. The calculated residence time of flue gas in the column was ~23 s.

When the initial gas concentrations were adjusted to the required level, the simulated flue gas was introduced into the scrubber from the bottom of the column. The NaClO absorbent was sprayed from top to bottom. A peristaltic pump was used to pump the scrubbing solution cyclically. The flow rate of the scrubbing solution was ~0.27 L/min. Each run of the test was 20 min. The outlet concentrations of flue gas were measured at an interval of 10 s. The solution temperatures were adjusted by the constant water bath (F34-MA, Julabo Labortechnik GmbH) and measured with a mercury thermometer. A MRU MGA-5 gas analyzer was used to determinate the gas concentrations of O_2, CO_2, NO, and NO_2 in flue gas.

2.3. Data Process.
The gas concentrations measured by the bypass are taken as the inlet concentrations. The average concentrations within 20 min measured by the gas outlet are considered as the outlet concentrations. The removal efficiencies of NO_x and SO_2 are calculated by the following equation:

$$\eta = \frac{C_{in} - C_{out}}{C_{in}} \times 100\% \tag{1}$$

in which η is the removal efficiency and C_{in} and C_{out} are the inlet and outlet concentrations, respectively. Here NO_x refers to the mixture of NO and NO_2 in flue gas.

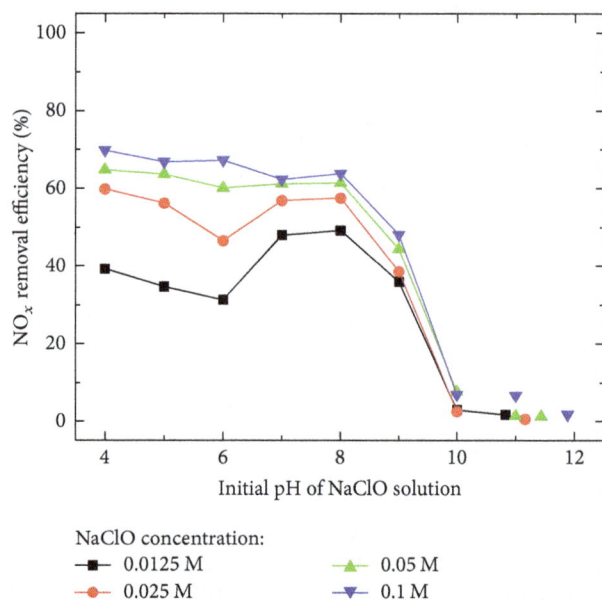

FIGURE 2: Effect of initial solution pH on NO_x removal efficiency (gas flow is 1.25 L/min, inlet NO_x concentration is 1000 ppm, NaClO concentration is 0.0125–0.1 M, and solution temperature is 293 K).

3. Results and Discussion

3.1. Effect of Initial pH and NaClO Concentration. The active components in NaClO solution were available chlorine, which mainly includes HClO, ClO^-, and Cl_2. The compositions of NaClO solution depended greatly on the solution pH. Firstly, it was necessary to investigate the effect of initial solution pH on the NO_x removal efficiency. Figure 2 showed that NO_x removal efficiency was very low when the solution pH was higher than 10. That was because little HClO existed in the solution when solution pH was higher than 10, but HClO was considered to be the main component in NaClO solution to oxidize NO [22]. It demonstrated that NaClO solution without optimizing the initial pH was not suitable for removing NO_x. With the decrease of initial pH from 10 to 8, NO_x removal efficiency increased quickly, which was due to the increase of fractional composition of HClO in solution. HClO oxidized NO into NO_2, N_2O_3, N_2O_4, and NO_3^- in chain reactions of (2)–(9) [30, 31]. The possible reaction pathways were summarized as shown in Figure 3.

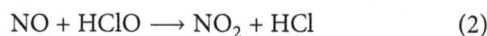

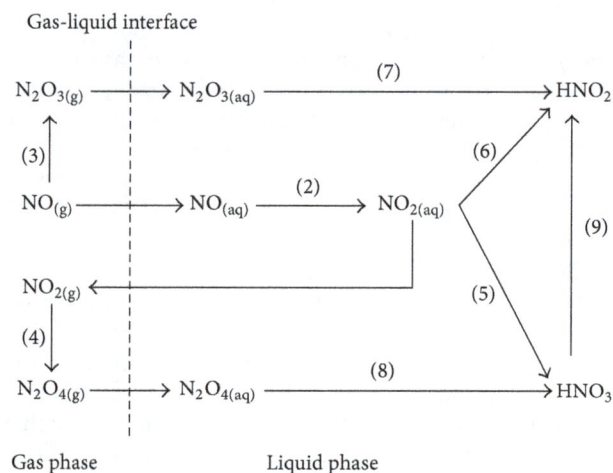

$$NO + HClO \longrightarrow NO_2 + HCl \qquad (2)$$

$$NO + NO_2 \longrightarrow N_2O_3 \qquad (3)$$

$$2NO_2 \longleftrightarrow N_2O_4 \qquad (4)$$

$$3NO_2 + H_2O \longrightarrow 2HNO_3 + NO \qquad (5)$$

$$2NO_2 + H_2O \longrightarrow HNO_3 + HNO_2 \qquad (6)$$

$$N_2O_3 + H_2O \longrightarrow 2HNO_2 \qquad (7)$$

$$N_2O_4 + H_2O \longrightarrow HNO_3 + HNO_2 \qquad (8)$$

$$HNO_2 + HClO \longrightarrow HNO_3 + HCl \qquad (9)$$

FIGURE 3: Reaction pathways for NO_x removal by NaClO solution.

As shown in Figure 2, with the initial pH decreasing from 8 to 7, a slight drop of NO_x removal efficiency appeared. On further reducing the initial pH, the changes of NO_x removal efficiency depended on NaClO concentration. When initial NaClO concentration was below 0.05 M, NO_x removal efficiency continued to decrease with initial pH decreasing from 7 to 6. This was because the oxidation power of ClO^- was stronger than that of HClO in neutral or weak acidic medium. As shown in Figure 4, HClO concentration increased while ClO^- concentration decreased with solution pH decreasing from 7 to 6. However, when NaClO concentration was higher than 0.05 M, the oxidation power of active chlorine was high enough to oxidize NO effectively at pH 6 and 7 [32]. At the moment, the effect of the change of active chlorine concentration was not obvious enough. The result indicated that a high NaClO concentration might be appropriate for practical application for it was easy to obtain a high and stable NO_x removal efficiency in a relatively wide range of solution pH.

Figure 5 showed the effect of initial solution pH on NO_x concentration in flue gas. When NaClO concentration was higher than 0.05 M and solution pH was in the range of 4–8, outlet NO concentration was lower than 124 ppm. It suggested that the majority of NO had been oxidized into NO_2 by NaClO. When NaClO concentration was 0.05 M, outlet NO concentration decreased sharply to 35 ppm with initial pH decreasing from 10 to 8. At the same time, NO_2 concentration in outlet gas increased from 14 ppm to 353 ppm. On further decreasing the initial pH down to 6, a complete removal of NO could be achieved with 0.05 M NaClO while outlet NO_2 concentration reached about 400 ppm. With the initial pH decreasing from 6 to 4, NO in outlet gas increased to 124 ppm while NO_2 in outlet gas decreased to 270 ppm. NO_x removal efficiency changed little in the range of pH 4–8. The results implied that, to a certain extent, NO_x removal efficiency for wet scrubbing using NaClO solution was mainly limited to the absorption of NO_2 rather than the oxidation of NO.

3.2. Effect of Coexisting CO_2. During wet scrubbing process, CO_2 in flue gas would react with absorbent thus influencing

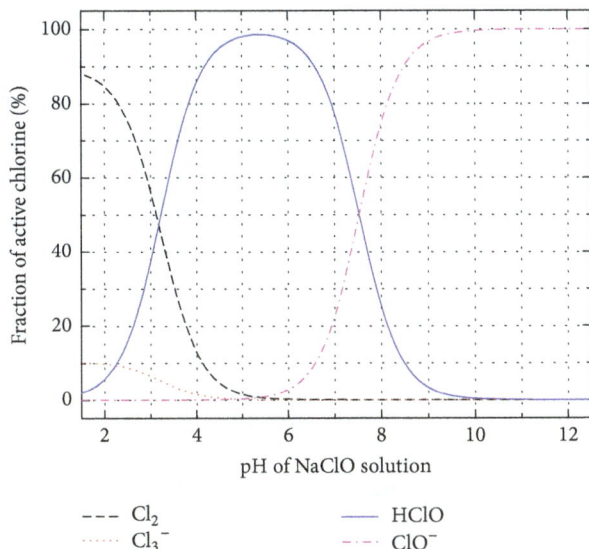

FIGURE 4: Equilibrium concentrations of active chlorine species in NaClO solution as a function of solution pH.

the oxidation and absorption of targeted pollutants. As an acidic oxide, CO_2 was sensitive to the solution pH. The effect of coexisting CO_2 on NO_x removal efficiency was investigated, and results were shown in Figure 6. It can be seen that, with initial solution pH increasing from 4 to 8, NO_x removal efficiency was relatively stable. A complete removal of NO had been achieved and outlet NO_2 concentration was ~350 ppm. The coexistence of CO_2 had not affected the NO_x removal efficiency so much. But the average CO_2 concentration in outlet gas changed obviously with initial solution pH. When initial solution pH was below 6, the average CO_2 concentration kept stable at 5%. However, it began to decrease quickly with initial pH increasing from 6 to 8.

The change of outlet CO_2 concentration during the cyclic scrubbing process was shown in Figure 7. When initial pH was in the range of 4–6, CO_2 concentration decreased to 4.4–4.5% at the start of the scrubbing process. Then it recovered to the initial level due to the hydrolysis equilibrium between CO_2 and absorbent solution. The hydrolysis reactions of CO_2 were described in (10) and (11). Furthermore, a certain amount of H^+ might be produced during the hydrolysis process, which would affect the chlorine hydrolysis equilibrium reactions as shown in (12) and (13). Since HCO_3^- and CO_3^{2-} had buffering ability to some extent, the hydrolysis of CO_2 would not influence the solution pH obviously when initial pH was in the range of 4–6. But with initial pH increasing from 6 to 8, the absorption of CO_2 might reduce the solution alkalinity, resulting in the decrease of the solution pH. Thus it was necessary to keep the solution pH at 6 in order to reduce the consumption of solution alkalinity when NaClO solution was adopted to remove NO_x from flue gas.

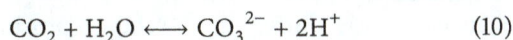

$$CO_2 + H_2O \longleftrightarrow CO_3^{2-} + 2H^+ \tag{10}$$

$$CO_2 + H_2O \longleftrightarrow HCO_3^- + H^+ \tag{11}$$

$$Cl_{2(aq)} + H_2O \longleftrightarrow HClO + H^+ + Cl^- \tag{12}$$

$$HClO \longleftrightarrow ClO^- + H^+ \tag{13}$$

When initial pH was in the range of 7-8, CO_2 concentration decreased largely at the very beginning of the cyclic scrubbing process. It suggested that much more CO_2 had been absorbed by the scrubbing solution due to the weak alkaline medium. With the proceeding of the scrubbing process, CO_2 concentration began to increase slowly. Although no evidence showed that CO_2 would react with NaClO directly, the hydrolyzation and absorption of CO_2 would increase the consumption of solution alkalinity obviously. It meant that extra alkaline solution was required to maintain the solution pH during cyclic scrubbing process, which would largely increase the operation complexity and cost at the same time. Thus initial pH of 6 might be appropriate for practical application.

Figure 8 showed the change of NaClO solution pH after scrubbing for 20 min. With the proceeding of cyclic scrubbing process, the solution pH would decrease slowly due to the absorption of NO_x and CO_2. It was worth noting that the solution pH increased from 4 to 4.63 for NaClO solution with initial pH 4. That was because active chlorine species in NaClO solution were mainly HClO and Cl_2 at pH 4, as shown in Figure 4. During the scrubbing process, Cl_2 would be purged out easily from the solution and reacted with NO effectively in gas phase as (14) and (15) [25]. The decrease of Cl_2 in NaClO solution would lead to a left shift of the hydrolysis equilibrium of active chlorine as shown in (16), resulting in the increase of solution pH for NaClO solution with initial pH 4.

$$Cl_{2(aq)} \longleftrightarrow Cl_{2(g)} \tag{14}$$

$$Cl_{2(g)} + H_2O + NO_{(g)} \longleftrightarrow NO_{2(g)} + 2HCl \tag{15}$$

$$Cl_{2(g)} + H_2O \longleftrightarrow HClO_{(aq)} + H^+ + Cl^- \tag{16}$$

It was possible that excessive Cl_2 escaped from NaClO solution might result in secondary pollution. In addition, acidic mist formed in the flue gas might result in the corrosion of operation system. In view of this, NaClO solution with pH 6 was appropriate for NO_x removal in cyclic scrubbing mode.

3.3. Effect of Coexisting O_2. For marine diesel engines, there was typical ~13% O_2 in exhaust gas. O_2 could partially oxidize NO under certain conditions, so it was necessary to investigate the effect of coexisting O_2 on NO_x removal efficiency. In the experiments, only NO standard gas was used to prepare NO_x in the initial simulated flue gas. The introduction of O_2 has oxidized a little of NO into NO_2 in the gas mixer. As shown in Figure 9, the initial NO_2 concentration in inlet gas increased almost linearly with the increase of NO concentration. When inlet NO_x concentration was 1000 ppm, the initial NO_2 concentration reached 112 ppm.

$$O_2 + 2NO \longrightarrow 2NO_2 \tag{17}$$

With the NO_x increasing from 250 to 700 ppm, the outlet NO concentration decreased gradually to 0. When inlet

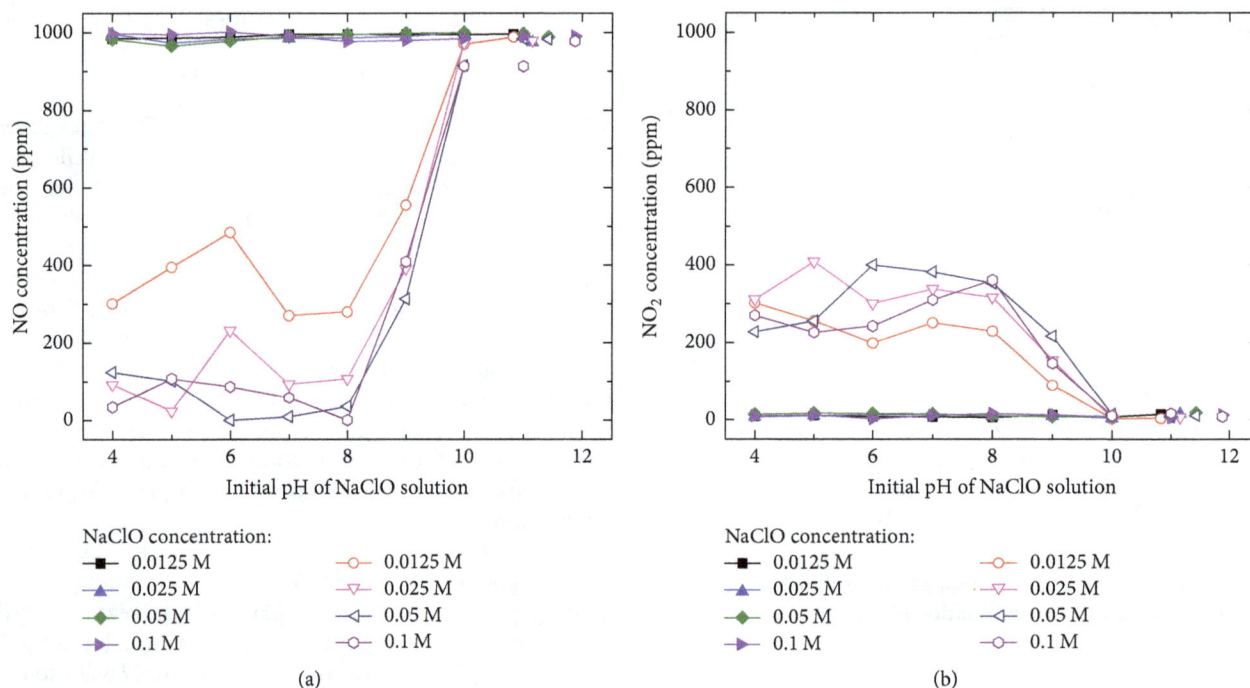

FIGURE 5: The change of (a) NO concentrations and (b) NO_2 concentrations in flue gas with initial pH of NaClO solution (gas flow is 1.25 L/min, inlet NO_x concentration is 1000 ppm, NaClO concentration is in the range of 0.0125–0.1 M, and solution temperature is 293 K; solid points refer to inlet NO_x concentrations and open points refer to outlet NO_x concentrations).

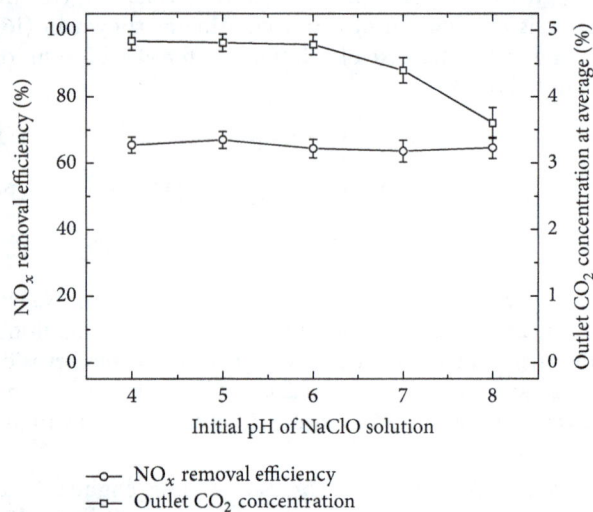

FIGURE 6: The change of NO_x removal efficiency and averaged CO_2 concentration in flue gas with initial pH (gas flow is 1.25 L/min, inlet NO_x concentration is 1000 ppm, initial CO_2 concentration is 5%, NaClO concentration is 0.05 M, and solution temperature is 293 K).

FIGURE 7: The change of CO_2 concentration during the cyclic scrubbing duration (gas flow is 1.25 L/min, inlet NO concentration is 1000 ppm, initial CO_2 concentration is 5%, NaClO concentration is 0.05 M, and solution temperature is 293 K).

NO_x concentration was higher than 700 ppm, NO could be removed completely. As expected, the outlet NO_2 concentration increased almost linearly with the increase of inlet NO_x concentration. It indicated that NO_x removal efficiency depended to a great extent on the absorption of NO_2 during the scrubbing process.

Figure 10 presented the change of NO_x removal efficiency and outlet O_2 concentration with the initial NO_x

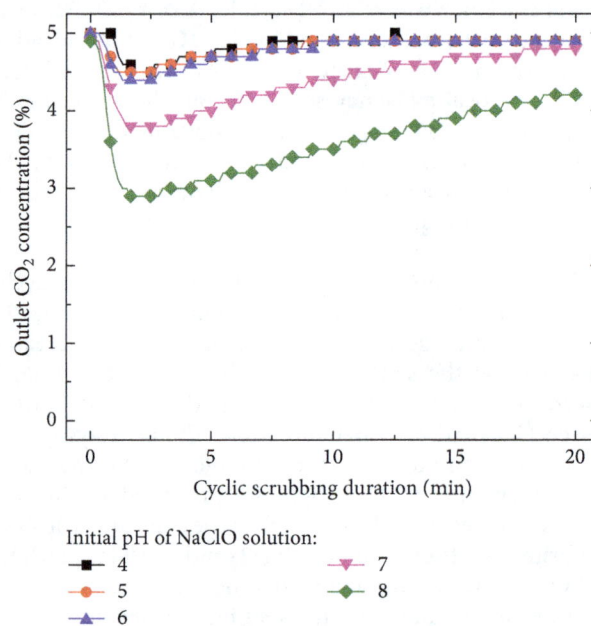

concentration. Since O_2 concentration was relatively high, it changed little during the scrubbing process. With the increase of NO_x concentration from 250 ppm to 700 ppm, NO_x removal efficiency increased from 47% to 74%. It could

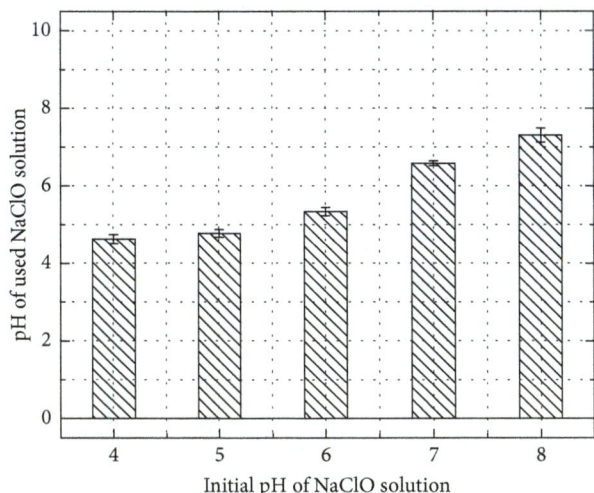

FIGURE 8: NaClO solution pH after scrubbing for 20 min (gas flow is 1.25 L/min, inlet NO_x concentration is 1000 ppm, initial CO_2 concentration is 5%, NaClO concentration is 0.05 M, and solution temperature is 293 K).

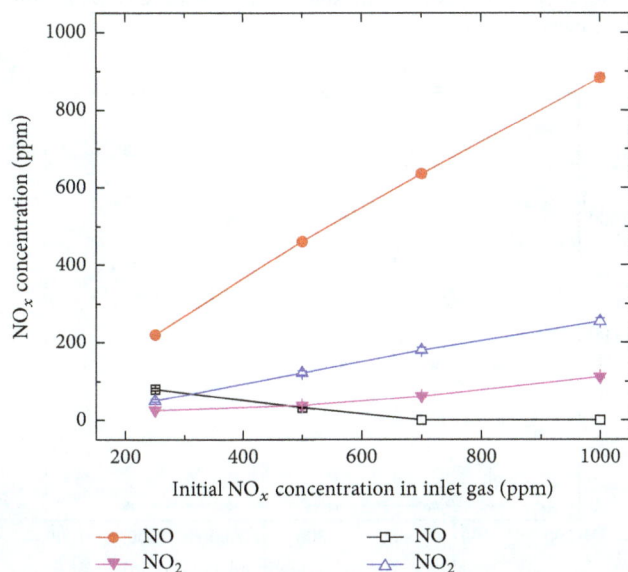

FIGURE 10: The change of NO_x removal efficiency and outlet O_2 concentration with the initial NO_x concentration (conditions: gas flow: 1.25 L/min; inlet O_2 concentration: 13%; inlet NO_x concentration: 250–1000 ppm; NaClO concentration: 0.05 M; initial solution pH: 6; solution temperature: 293 K).

FIGURE 9: The change of NO_x concentration with the initial NO_x concentration (gas flow is 1.25 L/min, inlet O_2 concentration is 13%, inlet NO_x concentration is in the range of 250–1000 ppm, NaClO concentration is 0.05 M, initial solution pH is 6, and solution temperature is 293 K; solid points refer to inlet NO_x concentrations and open points refer to outlet NO_x concentrations).

be ascribed to the improvement of the mass transfer at the gas-liquid interface. When O_2 was present in flue gas, NO_x removal efficiency was a little higher than that without O_2 in flue gas. The existence of O_2 improved the NO_x removal efficiency in way of partly oxidizing NO in initial flue gas.

3.4. Effect of NO_x Concentration. The effects of NO_x concentration on NO_x removal are investigated, and the results are shown in Figure 11. It can be seen that, with initial NO

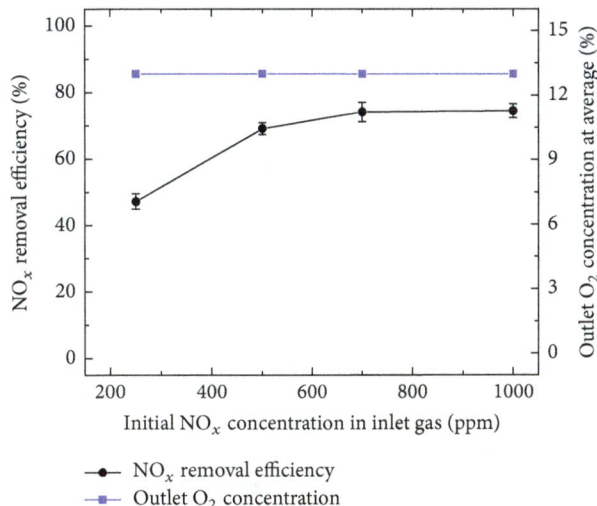

concentration increasing from 250 ppm to 500 ppm, NO_x removal efficiency increased from 43% to ~63%. NO could be removed completely when inlet NO concentration was higher than 500 ppm. Outlet NO_2 concentration increased quickly with the increase of inlet NO concentration, resulting in a relatively stable NO_x removal efficiency.

3.5. Simultaneous Removal of NO_x and SO_2. Marine diesel engines usually burn heavy fuel oil (HFO) in order to save the operating cost. At present, the mass concentration of sulphur (S) content in HFO was about 2.5% at average. The combustion of HFO fuel would produce a large amount of SO_2 in exhaust gas. Experiments were conducted to investigate the effect of coexisting SO_2 on NO_x removal efficiency, and the results are shown in Figures 12 and 13.

Figure 12 depicted the change of NO_x removal efficiency and NO_x concentration with inlet NO_x concentration. A complete removal of SO_2 and NO had been achieved simultaneously. Due to its high solubility, SO_2 could be absorbed effectively by scrubbing solution. Then SO_2 was removed through the hydrolysis reaction as (18) and (19). The hydrolysis products of SO_3^{2-} would be oxidized by active chlorine species into SO_4^{2-} quickly. Thus the removal of SO_2 would consume the solution alkalinity and oxidants at the same time. The result demonstrated that NaClO solution could be used to remove NO and SO_2 simultaneously from marine exhaust gas. However, outlet NO_2 concentration increased gradually with the increase of inlet NO_x concentration.

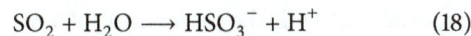

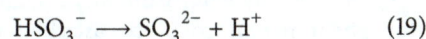

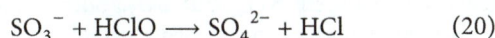

$$SO_2 + H_2O \longrightarrow HSO_3^- + H^+ \qquad (18)$$

$$HSO_3^- \longrightarrow SO_3^{2-} + H^+ \qquad (19)$$

$$SO_3^- + HClO \longrightarrow SO_4^{2-} + HCl \qquad (20)$$

(a)

(b)

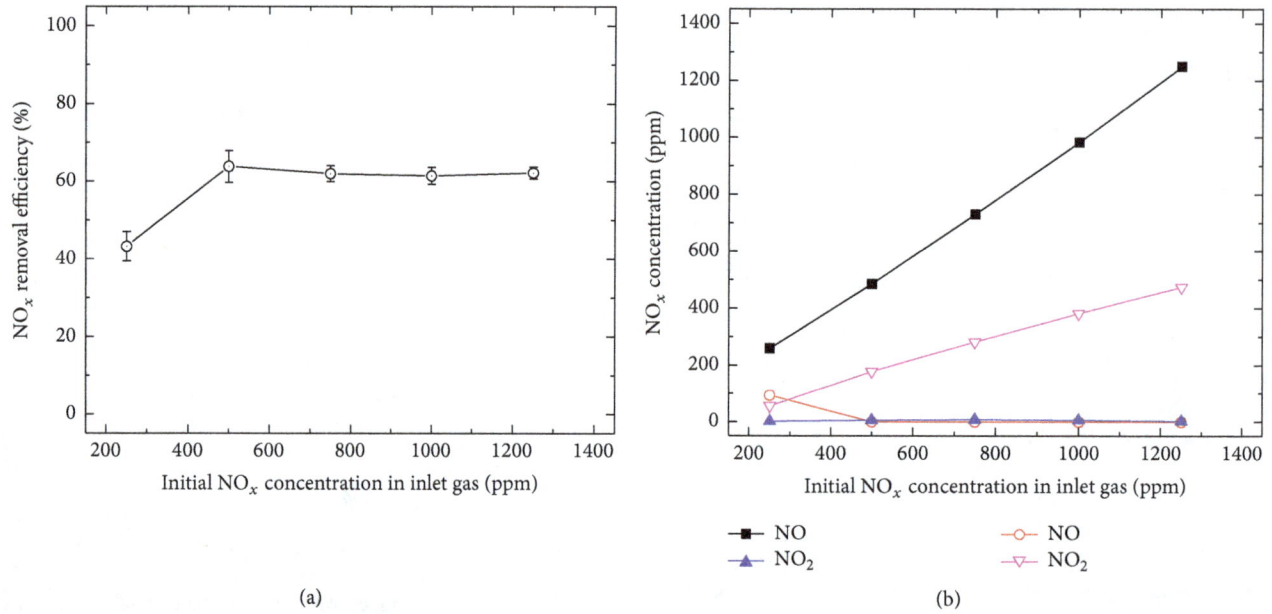

FIGURE 11: The change of (a) NO_x removal efficiency and (b) NO_x concentrations with the initial NO_x concentration (gas flow is 1.25 L/min, inlet NO_x concentration is 250–1250 ppm, NaClO concentration is 0.05 M, initial solution pH is 6, and solution temperature is 293 K; solid points refer to inlet NO_x concentrations and open points refer to outlet NO_x concentrations).

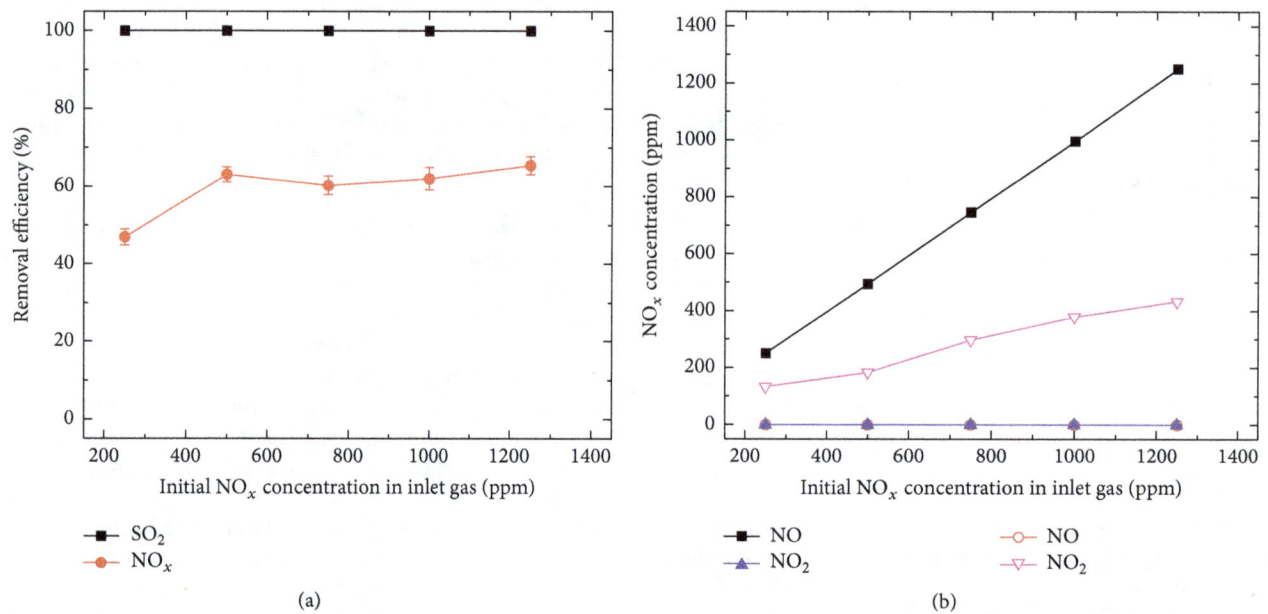

(a)

(b)

FIGURE 12: The change of (a) NO_x removal efficiency and (b) NO_x concentration with the inlet NO_x concentration (gas flow is 1.25 L/min, inlet SO_2 concentration is 600 ppm, inlet NO_x concentration is 250–1250 ppm, NaClO concentration is 0.05 M, initial solution pH is 6, and solution temperature is 293 K; solid points refer to inlet NO_x concentrations and open points refer to outlet NO_x concentrations).

Figure 13 exhibited the effect of SO_2 concentration on NO_x removal efficiency and NO_x concentration in flue gas. The result showed that SO_2 together with NO was completely removed. But with the increase of inlet SO_2 concentration, NO_x removal efficiency exhibited a slightly downside trend. This phenomenon could be ascribed to the competition reactions between NO_x and SO_2. Some oxidants in the absorbent would be consumed through the hydrolyzation

and absorption of SO_2. With inlet SO_2 concentration increasing from 200 ppm to 600 ppm, the pH of the scrubbed NaClO solution decreased from 5.25 to 4.52. The decrease of solution pH was negative for the absorption of NO_2 through (5) and (6).

3.6. *Effect of Absorbent Temperature.* The reaction temperature could greatly influence the diffusion, dissolution, and

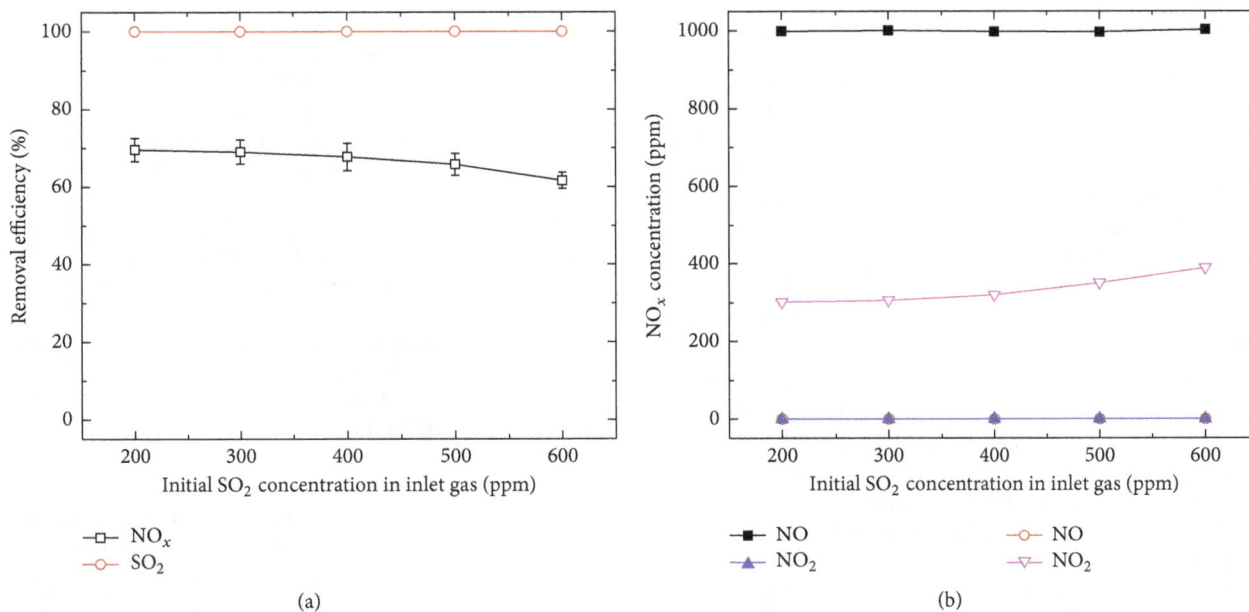

(a)   (b)

FIGURE 13: The change of (a) NO_x removal efficiency and (b) NO_x concentration with the initial SO_2 concentration (gas flow is 1.25 L/min, inlet SO_2 concentration is 200–600 ppm, inlet NO_x concentration is 1000 ppm, NaClO concentration is 0.05 M, initial solution pH is 6, and solution temperature is 293 K; solid points refer to inlet NO_x concentrations and open points refer to outlet NO_x concentrations).

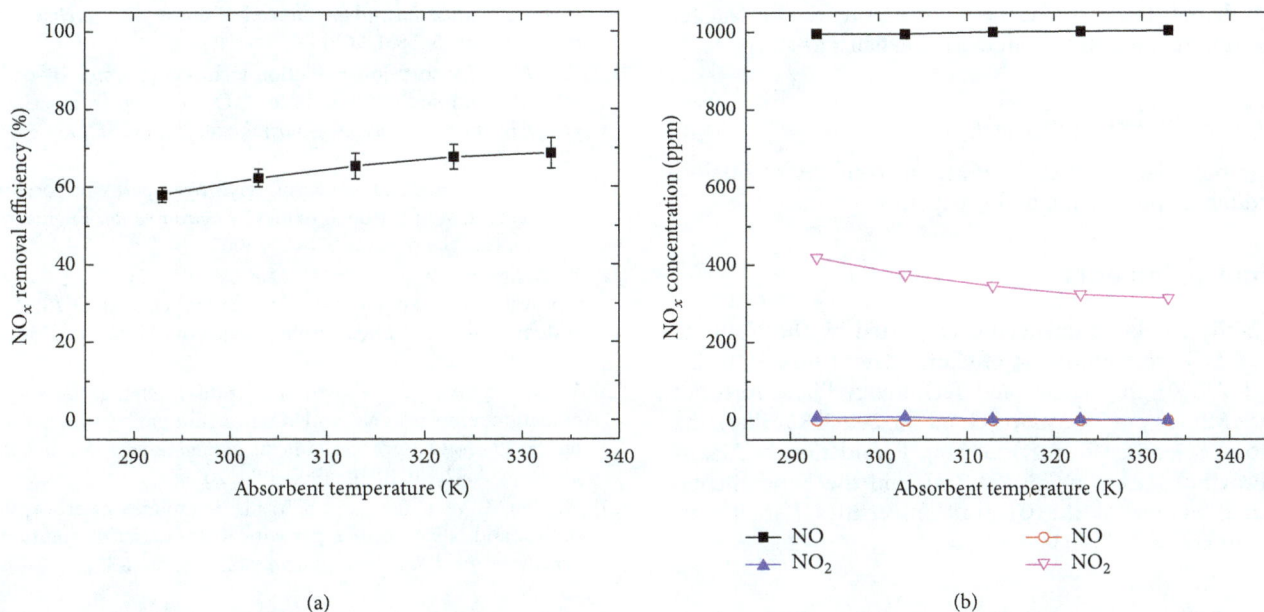

(a)   (b)

FIGURE 14: The change of (a) NO_x removal efficiency and (b) NO_x concentration with the absorbent temperature (gas flow is 1.25 L/min, inlet NO_x concentration is 1000 ppm, NaClO concentration is 0.05 M, initial solution pH is 6, and solution temperature is 293–333 K; solid points refer to inlet NO_x concentrations and open points refer to outlet NO_x concentrations).

reaction characteristics of molecules or ions in liquid phase [19]. The effect of absorbent temperature on NO_x removal was investigated, and the results were shown in Figure 14. Generally, the absorbent temperature was kept below 333 K in industrial application in order to reduce the supply of make-up water. So the absorbent temperature was chosen to be in the range of 293–333 K in the experiments. Figure 14 showed that NO_x removal efficiency increased gradually with the increase of absorbent temperature. The highest NO_x removal efficiency of 69% was achieved at 333 K. As NO could be removed by 100% easily, the increase of NO_x removal efficiency resulted from the improvement of NO_2 absorption. This could be explained by the Arrhenius equation of the reaction rate constant. The increase of temperature could enhance the mass transfer of NO_2 to the absorbent, thus improving the absorption of NO_2 accordingly.

4. Conclusion

NO_x removal by wet scrubbing using NaClO solution was studied based on a spraying reactor in a cyclic mode. The results showed that when NaClO concentration was higher than 0.05 M and initial solution pH was below 8, NO_x removal efficiency was relatively stable, which was higher than 60%. The coexisting CO_2 (5%) had little effect on NO_x removal efficiency, but the solution pH began to decrease with the proceeding of cyclic scrubbing process when initial pH was higher than 6. The coexisting O_2 (13%) could oxidize NO partially in the gas mixer, resulting in a little improvement of NO_x removal efficiency. When initial NO_x concentration was higher than 500 ppm, NO could be removed completely while outlet NO_2 concentration increased almost linearly with the increase of initial NO_x concentration, resulting in a relative stable NO_x removal efficiency. A complete removal of SO_2 and NO could be achieved simultaneously at 293 K, initial NaClO solution pH 6, and 0.05 M NaClO concentration. With the increase of inlet SO_2 concentration, the outlet NO_2 concentration increased slightly due to the decrease of solution pH. NO_x removal efficiency increased with the increase of absorbent temperature. The relevant reaction mechanism for the removal of NO_x and SO_2 by wet scrubbing using NaClO solution was also discussed. The results demonstrated that it was of great potential for NaClO solution to remove NO_x from marine exhaust gas.

Conflicts of Interest

The authors declare that there are no conflicts of interest regarding the publication of this paper.

Acknowledgments

This study has been financially supported by the National Natural Science Foundation of China (Grant nos. 51402033 and 51479020), the Science and Technology Plan Project of China's Ministry of Transport (Grant no. 2015328225150), the Doctoral Scientific Research Staring Foundation of Liaoning Province (Grant no. 201601073), and the Fundamental Research Funds for the Central Universities (Grant nos. 3132017003 and 3132016326).

References

[1] M. Guo, Z. Fu, D. Ma, N. Ji, C. Song, and Q. Liu, "A short review of treatment methods of marine diesel engine exhaust gases," *Procedia Engineering*, vol. 121, pp. 938–943, 2015.

[2] M.-W. Bae, "A study on the effects of recirculated exhaust gas upon NO_x and soot emissions in diesel engines with scrubber EGR system," *SAE Technical Papers*, 1999.

[3] A. M. A. Attia and A. R. Kulchitskiy, "Influence of the structure of water-in-fuel emulsion on diesel engine performance," *Fuel*, vol. 116, pp. 703–708, 2014.

[4] X. Tauzia, A. Maiboom, and S. R. Shah, "Experimental study of inlet manifold water injection on combustion and emissions of an automotive direct injection diesel engine," *Energy*, vol. 35, no. 9, pp. 3628–3639, 2010.

[5] Å. M. Hallquist, E. Fridell, J. Westerlund, and M. Hallquist, "Onboard measurements of nanoparticles from a SCR-equipped marine diesel engine," *Environmental Science and Technology*, vol. 47, no. 2, pp. 773–780, 2013.

[6] J. Herdzik, "Emissions from marine engines versus IMO certification and requirements of tier 3," *Journal of KONES*, vol. 18, pp. 161–167, 2011.

[7] M. Magnusson, E. Fridell, and H. H. Ingelsten, "The influence of sulfur dioxide and water on the performance of a marine SCR catalyst," *Applied Catalysis B: Environmental*, vol. 111-112, pp. 20–26, 2012.

[8] T. Kuwahara, K. Yoshida, Y. Kannaka, T. Kuroki, and M. Okubo, "Improvement of NOx reduction efficiency in diesel emission control using nonthermal plasma combined exhaust gas recirculation process," *IEEE Transactions on Industry Applications*, vol. 47, no. 6, pp. 2359–2366, 2011.

[9] T. Kuwahara, K. Yoshida, T. Kuroki, K. Hanamoto, K. Sato, and M. Okubo, "Pilot-scale aftertreatment using nonthermal plasma reduction of adsorbed NO_x in marine diesel-engine exhaust gas," *Plasma Chemistry and Plasma Processing*, vol. 34, no. 1, pp. 65–81, 2014.

[10] D. Xie, Y. Sun, T. Zhu, and L. Ding, "Removal of NO in mist by the combination of plasma oxidation and chemical absorption," *Energy and Fuels*, vol. 30, no. 6, pp. 5071–5076, 2016.

[11] S. Zhou, J. Zhou, Y. Feng, and Y. Zhu, "Marine Emission Pollution Abatement Using Ozone Oxidation by a Wet Scrubbing Method," *Industrial and Engineering Chemistry Research*, vol. 55, no. 20, pp. 5825–5831, 2016.

[12] Y. S. Mok, "Absorption-reduction technique assisted by ozone injection and sodium sulfide for NO_x removal from exhaust gas," *Chemical Engineering Journal*, vol. 118, no. 1-2, pp. 63–67, 2006.

[13] E. B. Myers and T. J. Overcamp, "Hydrogen peroxide scrubber for the control of nitrogen oxides," *Environmental Engineering Science*, vol. 19, no. 5, pp. 321–327, 2002.

[14] Z. Wang, Z. Wang, Y. Ye, N. Chen, and H. Li, "Study on the removal of nitric oxide (NO) by dual oxidant ($H_2O_2/S_2O_{82}^-$) system," *Chemical Engineering Science*, vol. 145, pp. 133–140, 2016.

[15] Y. Liu, Q. Wang, Y. Yin, J. Pan, and J. Zhang, "Advanced oxidation removal of NO and SO_2 from flue gas by using ultraviolet/H_2O_2/NaOH process," *Chemical Engineering Research and Design*, vol. 92, no. 10, pp. 1907–1914, 2014.

[16] H. Chu, T. W. Chien, and S. Y. Li, "Simultaneous absorption of SO_2 and NO from flue gas with $KMnO_4$/NaOH solutions," *Science of the Total Environment*, vol. 275, no. 1-3, pp. 127–135, 2001.

[17] R. Hao, Y. Zhang, Z. Wang et al., "An advanced wet method for simultaneous removal of SO_2 and NO from coal-fired flue gas by utilizing a complex absorbent," *Chemical Engineering Journal*, vol. 307, pp. 562–571, 2017.

[18] Z. Han, S. Yang, D. Zheng, X. Pan, and Z. Yan, "An investigation on NO removal by wet scrubbing using $NaClO_2$ seawater solution," *SpringerPlus*, vol. 5, no. 1, article 751, 2016.

[19] Y. Zhou, C. Li, C. Fan et al., "Wet removal of sulfur dioxide and nitrogen oxides from simulated flue gas by $Ca(ClO)_2$ solution," *Environmental Progress and Sustainable Energy*, vol. 34, no. 6, pp. 1586–1595, 2015.

[20] S.-L. Yang, Z.-T. Han, J.-M. Dong, Z.-S. Zheng, and X.-X. Pan, "UV-enhanced naclo oxidation of nitric oxide from simulated flue gas," *Journal of Chemistry*, vol. 2016, Article ID 6065019, 8 pages, 2016.

[21] L. Chen, C.-H. Hsu, and C.-L. Yang, "Oxidation and absorption of nitric oxide in a packed tower with sodium hypochlorite aqueous solutions," *Environmental Progress*, vol. 24, no. 3, pp. 279–288, 2005.

[22] E. Ghibaudi, J. R. Barker, and S. W. Benson, "Reaction of NO with hypochlorous acid," *International Journal of Chemical Kinetics*, vol. 11, no. 8, pp. 843–851, 1979.

[23] R.-T. Guo, W.-G. Pan, X.-B. Zhang et al., "The absorption kinetics of NO into weakly acidic NaClO solution," *Separation Science and Technology*, vol. 48, no. 18, pp. 2871–2875, 2013.

[24] S. Yang, Z. Han, X. Pan, Z. Yan, and J. Yu, "Nitrogen oxide removal using seawater electrolysis in an undivided cell for ocean-going vessels," *RSC Advances*, vol. 6, no. 115, pp. 114623–114631, 2016.

[25] M. K. Mondal and V. R. Chelluboyana, "New experimental results of combined SO_2 and NO removal from simulated gas stream by NaClO as low-cost absorbent," *Chemical Engineering Journal*, vol. 217, pp. 48–53, 2013.

[26] C. V. Raghunath and M. K. Mondal, "Experimental scale multi component absorption of SO_2 and NO by NH_3/NaClO scrubbing," *Chemical Engineering Journal*, vol. 314, pp. 537–547, 2017.

[27] J. Wang and W. Zhong, "Simultaneous desulfurization and denitrification of sintering flue gas via composite absorbent," *Chinese Journal of Chemical Engineering*, vol. 24, no. 8, pp. 1104–1111, 2016.

[28] S. An and O. Nishida, "New application of seawater and electrolyzed seawater in air pollution control of marine diesel engine," *JSME International Journal, Series B: Fluids and Thermal Engineering*, vol. 46, no. 1, pp. 206–213, 2003.

[29] Z. Han, S. Yang, X. Pan et al., "New experimental results of NO removal from simulated flue gas by wet scrubbing using NaClO solution," *Energy & Fuels*, vol. 31, no. 3, pp. 3047–3054, 2017.

[30] N. Lahoutifard, P. Lagrange, and J. Lagrange, "Kinetics and mechanism of nitrite oxidation by hypochlorous acid in the aqueous phase," *Chemosphere*, vol. 50, no. 10, pp. 1349–1357, 2003.

[31] Y. G. Adewuyi, X. He, H. Shaw, and W. Lolertpihop, "Simultaneous absorption and oxidation of NO and SO_2 by aqueous solutions of sodium chlorite," *Chemical Engineering Communications*, vol. 174, no. 1, pp. 21–51, 1999.

[32] A. Mukimin, K. Wijaya, and A. Kuncaka, "Electro-degradation of reactive blue dyes using cylinder modified electrode: Ti/β-PbO_2 as dimensionally stable anode," *Indonesian Journal of Chemistry*, vol. 10, no. 3, pp. 285–289, 2010.

20

Ceramic-Based 3D Printed Supports for Photocatalytic Treatment of Wastewater

Lorena Hernández-Afonso,[1] Ricardo Fernández-González,[1] Pedro Esparza,[1] M. Emma Borges,[2] Selene Díaz González,[1] Jesús Canales-Vázquez,[3] and Juan Carlos Ruiz-Morales[1]

[1]*Chemistry Department, University of La Laguna, San Cristóbal de La Laguna, Tenerife, 38200 Canary Islands, Spain*
[2]*Chemical Engineering Department, University of La Laguna, San Cristóbal de La Laguna, Tenerife, 38200 Canary Islands, Spain*
[3]*Instituto de Energías Renovables, Print3D Solutions, University of Castilla-La Mancha, 02006 Albacete, Spain*

Correspondence should be addressed to Juan Carlos Ruiz-Morales; jcruiz@ull.edu.es

Academic Editor: Roberto Comparelli

3D printing technology has become a powerful tool to produce 3D structures in any type of materials. In this work, 3D printing technology is used to produce 3D porous structures in $CaSO_4$ which can be later activated with an appropriate photocatalyst. TiO_2 was selected as an ideal photocatalyst producing activated 3D structures which can be used to study their effectiveness in the degradation of pollutants in wastewater. Methylene blue was used as a model molecule in these studies. The photocatalytic studies showed that TiO_2-activated 3D structures using nanoparticles of SiO_2 in the process produce more than 50% of conversion of methylene blue in just 1 h of irradiation and almost 90% in 5 h.

1. Introduction

Global warming, energy crisis, and pollution are serious concerns affecting both the human health and environment. The environmental pollution includes a wide range of hazardous chemicals which are harmful even at extremely low concentrations.

The common treatment of these pollutants involves the use of pyrolytic methods which consume large amounts of fossil fuels producing elevated levels of CO_2 and thus contributing again to climate change and energy crisis.

Another severe issue [1–3] is water pollution from emerging contaminants (ECs) such as endocrine disrupting chemicals, pharmaceuticals, and personal care products even at trace levels. Some adverse potential effects caused by ECs are water toxicity, resistance development in pathogenic bacteria, genotoxicity, and endocrine disruption [4–6].

Wastewater treatment plants are not designed to remove low concentrations of synthetic pollutants such as pharmaceuticals and hence alternatives such us the Advanced Oxidation Technology (AOT) have been used to solve this environmental problem [7, 8].

Heterogeneous photocatalytic processes constitute one of the most important AOTs and are based on the oxidation of polluting compounds which can be found in air or water by means of a reaction occurring on a semiconductor catalytic surface activated by light with a specific wavelength.

TiO_2 is the most investigated semiconductor catalyst particularly because of its great potential in the treatment of environmental pollution [9] and it is chemically stable, nontoxic, and inexpensive [10–12]. However, TiO_2 has an important disadvantage as it is usually commercially available as powder and, therefore, a posttreatment separation stage is needed [13, 14] and hence it would be highly desirable to have the possibility of supporting the TiO_2 active phase on other structured inorganic materials.

In this sense, a new exciting approach for the production of structured materials on demand may be the use of a 3D printing (3DP) system. Until the past few years, this type of technology has been restricted to medium- and big-sized

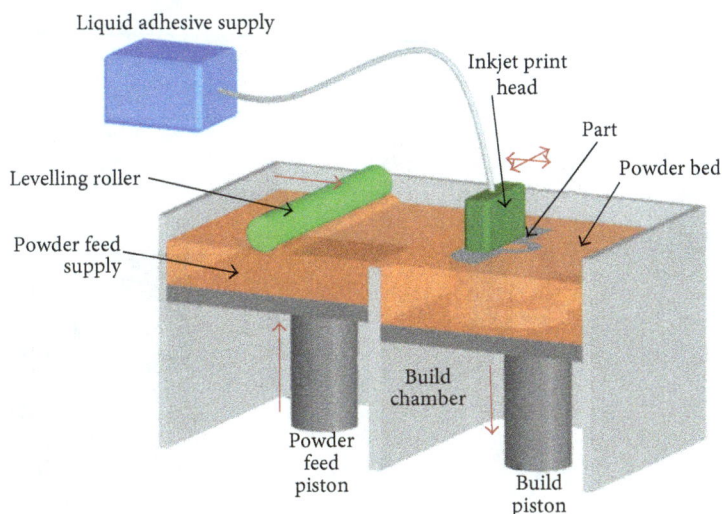

FIGURE 1: Indirect Inject Print (IIP) is a layer-by-layer process of depositing liquid binder onto thin layers of powder to create a 3D object [17].

companies devoted to the fabrication of prototypes. Nowadays, they are increasingly available for small businesses (research [15]) and even for direct manufacturing. 3D micro fabrication remained a challenge until the introduction of free-forming fabrication (FFF) technology. Free-forming fabrication is any fabrication technology that fabricates 3D complex structures by assembling small elements together and usually starts with a computer aided design (CAD) model. FFF includes, but is not limited to, rapid prototyping, 3D printing, and direct writing for macro scale fabrication [16].

A digital model of the object is created in a computer. Using adequate software, the user may control several relevant parameters such as the number of layers which are piled up to generate the full item depending on the resolution required, thickness of the layers, and porosity. In the second step, each digital layer is printed in an appropriate substrate. Different 3D printing techniques can be selected depending on the material required and the way to fuse layers together. One of the 3D printing options is the Indirect Inject Print, where a powder is spread from a well (Figure 1), levelling it to produce a thin layer. Then the printer heads will dispense a thin layer of a binder in the required pattern of the cross section. When the layer is finished, the "build tray" will be lowered by a fraction of a millimetre, typically between 10 and 100 μ and then the process is repeated again for the next cross section. Finally, after printing the whole 3D structure, the loose powder is blown away with compressed air, revealing the full structure [16].

This type of technology has the potential to allow the fabrication of monolithic porous structures that can be covered with a layer of catalytically active material such as TiO_2 [18]. Calcium sulphate hemihydrate ($CaSO_4 \cdot 1/2H_2O$) was one of the first materials to be used for IIP. It can be wetted using a commercially formulated binder (98% content water) and then forms a gypsum paste ($CaSO_4 \cdot 2H_2O$) by activating self-hydration [19] which can be used to produce 3D porous supporting structures.

The aim of this study is to verify that the 3D printing technology can be used to produce porous monolithic structures of calcium sulphate which can be then activated with commercial TiO_2 photocatalyst (Degussa P25) through an impregnation process. The whole 3D activated system will be tested in a fixed-bed photoreactor to verify the potential use of these heterostructures for removing wastewater contaminants using methylene blue (MB) as a model molecule.

2. Materials and Methods

The activity of the TiO_2 photocatalyst supported over a 3D printed ceramic structure was studied evaluating the photodegradation of methylene blue (Sigma-Aldrich) as a model molecule for wastewater treatments. The photocatalytically active material, TiO_2 (Degussa P25, Evonik Industries), was examined by X-ray diffraction (XRD) in order to obtain the percentage of anatase and rutile phase. Nitrogen adsorption-desorption porosimetry and mercury porosimetry techniques were used to study its specific surface area and textural properties [20]. TiO_2 (Degussa P25) was mainly composed by the photocatalytically active anatase phase, 81% (wt). The textural parameters of TiO_2 powder were 51.1 m^2/g of BET specific area, a total pore area of 63.7 m^2/g with 92.5% of porosity, and a density of 2.58 g/ml.

2.1. 3D Printing Structures. The materials used in the 3D printing process were calcium sulphate hemihydrate (VisiJet PXL Core, from S.A.T.ÉLITE) and a water-based binder (VisiJet Clear, from S.A.T.ÉLITE). These materials were used in a 3D printer (ProJet® 360, from S.A.T.ÉLITE), with a 300 × 450 dpi resolution and layer thickness of 0.1 mm.

First, the structure is digitally designed using free software Tinkercad [22] and the corresponding digital model is saved as a *stl* file and then sent to the 3D printer. During the three printing processes (Figures 2(a)–2(d)), a roller spread a thin powder layer from the feed area to the build area

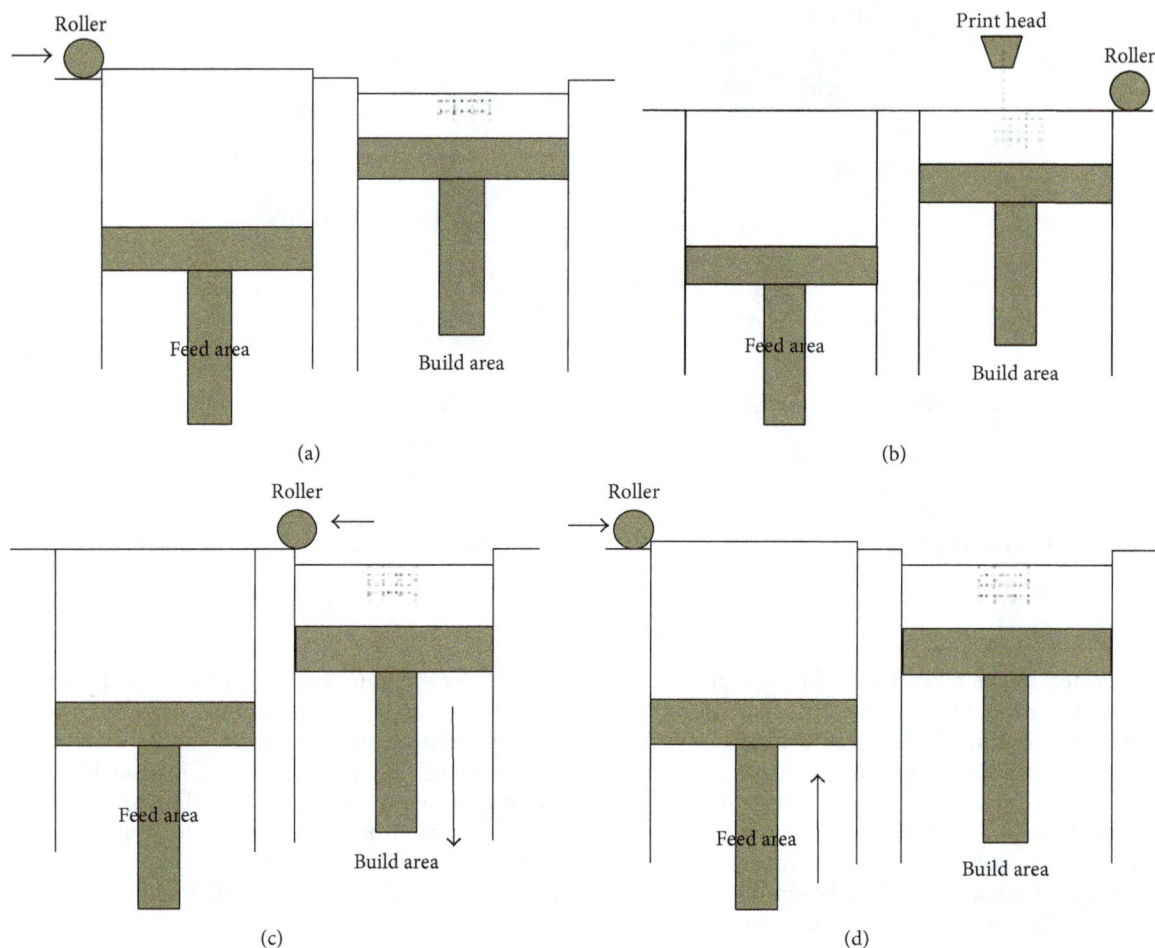

FIGURE 2: 3DP process scheme. (a) The roller spreads a thin layer of powder from the feed area to the build area. (b) The print head injects binder droplets on the powder bed. (c) After printing a layer, the roller returns to the feed area. (d) Powder in the feed area is raised, while that in the build area is lowered. The roller then spreads another layer of powder [21].

and then the print head deposits binder droplets selectively within the build area. When the first layer is printed, the roller returns to the feed area, spreading another powder layer to the build area. This procedure is repeated until the fabrication of the whole 3D structure is completed [21].

Following this procedure, several microtubes of ceramic supporting structures were 3D printed (Figures 3(a) and 3(b)), and several small rectangular and square pieces of 2 × 2 cm were also 3D printed as testing samples in mechanical stability versus temperature studies (Figures 3(c) and 3(d)). 150°C was the selected prefiring temperature to give enough mechanical stability for withstanding the conditions of the impregnation process.

2.2. Improving the Mechanical Stability of the 3D Printed Structures.
Two routes were followed to impregnate the 3D structures with another inorganic material enhancing the mechanical stability up to 1200°C. The 3D printed samples were previously fired at 150°C for 2 h to improve the mechanical stability in aqueous solution needed for the impregnation processes.

(i) Route 1. Impregnation was done with an aqueous dispersion of alumina (Al_2O_3, Almatis GmbH) using Dolapix CE64 (Zschimmer & Schwarz) as dispersant. Several ratios of the components and impregnation times (Table 1) were considered to optimize the process. The impregnation times were also optimized ranging from 10 s up to 30 min. The so-impregnated samples were left to dry for 45 min under 70°C and then were finally fired at 1200°C for 4 h with heating and cooling ramp rates of 5°C/min (Figure 4(a)).

(ii) Route 2. Impregnation was done with an aqueous solution of Ludox® AS-30 colloidal silica (Sigma-Aldrich). As in the previous method, the printed 3D structures were initially fired at 150°C for 2 h. After that, the specimens were impregnated in Ludox (1.1 g) and then fired at 600°C for 2 h, with heating and cooling ramp rates of 5°C/min (Figure 4(b)); this step is performed twice.

2.3. Photocatalytic Studies.
The methylene blue (MB) photodegradation process with the activated 3D printed structures (A3DS) was carried out in a fixed-bed photoreactor system

TABLE 1: Ratio of components and impregnation times used in the optimization of the impregnation process of 3D structures with Al_2O_3.

Procedure	Al_2O_3 (g)	Water (g)	Dolapix (g)	Impregnation times (s)
1	30,0	8,39	0,161	
2	30,0	16,2	0,423	10, 20, 30, 60, 300, 600, and 1800
3	30,0	33,8	0,644	

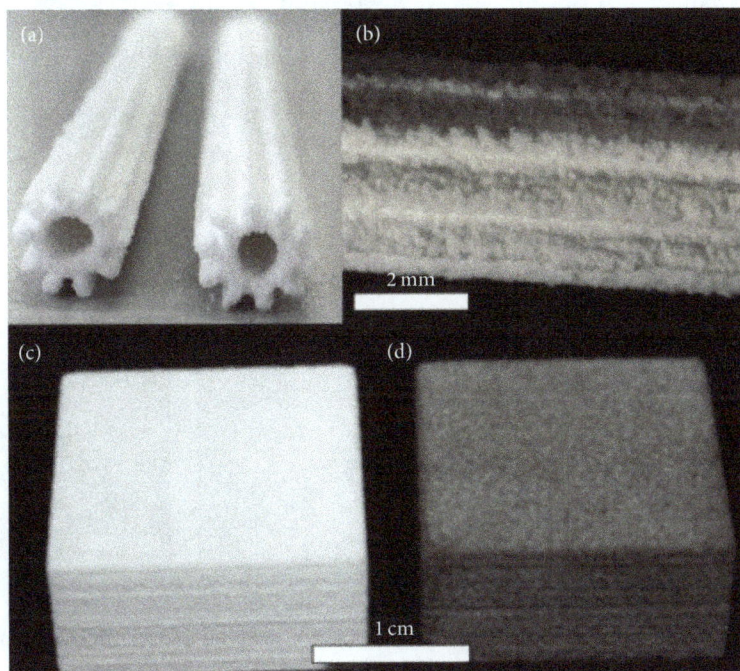

FIGURE 3: ((a) and (b)) 3D printed supporting structures in $CaSO_4$. (c) 3D printed square testing pieces in the green state and (d) after firing at 250°C for 3 h.

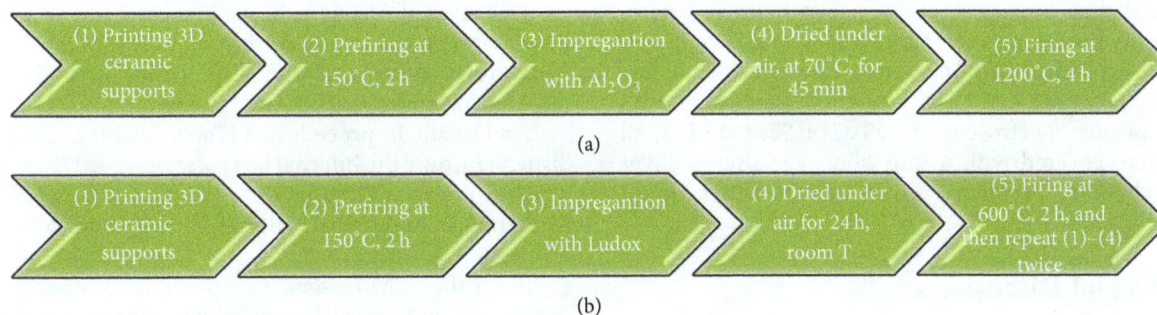

FIGURE 4: Schemes for improving the mechanical stability of 3D printed supporting structures using two alternative impregnation routes: (a) route 1 with Al_2O_3 and (b) route 2 with Ludox dispersions.

(Figure 5(a)). A3DS were placed in the photocatalytic reactor (18 cm length and 0.6 cm internal diameter) (Figure 5(b)), and a solar radiation sodium vapor lamp (Philips, model 400-WG/92/2), placed at a distance of 50 cm from the fixed-bed reactor, was used as light source (total radiation flux measured on the fixed-bed surface was 160 mW/cm²). MB solution in water (20 ppm) was introduced in a 250 ml wastewater photoreactor tank (Figure 5(c)), keeping the temperature constant at 25°C. The MB solution was recirculated along the system using a peristaltic pump (Watson-Marlow, model 302S). Several samples were taken during several hours of irradiation time and they were analyzed by UV-Vis spectrophotometry in order to follow the evolution of MB concentration into the reactor. Moreover, MB photolysis (with light source and without photocatalyst) and MB adsorption onto A3DS (with photocatalyst and without light source) experiments were developed in order to evaluate its contribution to the global wastewater decontamination process.

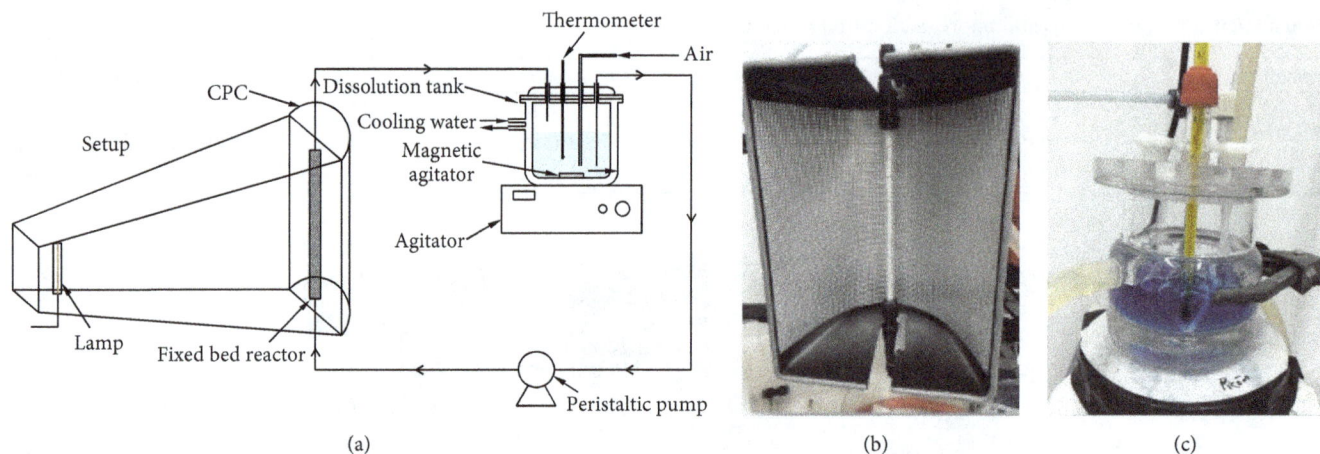

FIGURE 5: (a) Fixed-bed photoreactor scheme. (b) Fixed-bed reactor. (c) Wastewater photoreactor tank.

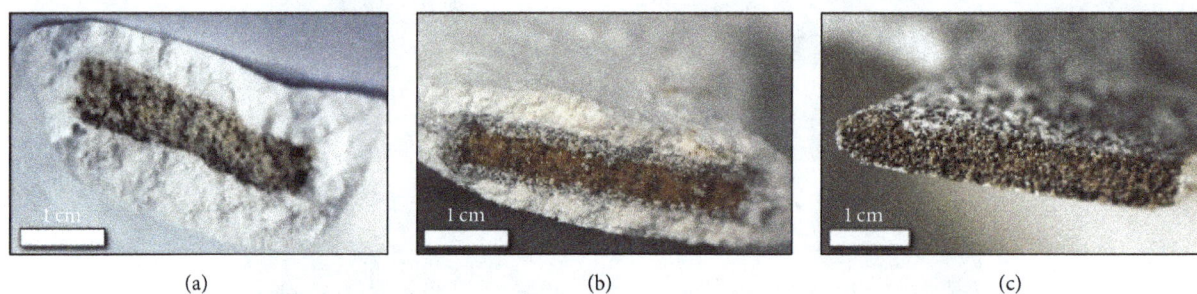

FIGURE 6: Optical images of impregnated 3D structures with dispersions of Al_2O_3, using an Al_2O_3 concentration of (a) 0.77 g/ml, (b) 0.65 g/ml, and (c) 0.47 g/ml.

2.4. Microstructural Characterization. The morphologies of the 3D structured samples in the green and sintered states were examined using a stereomicroscope Leica Zoom 2000 (Leica Microsystems, Inc.) and a scanning electron microscope (SEM) (model Jeol LTD, JSM-6300) combined with energy dispersive spectroscopy (EDS). For SEM studies, all samples were covered with a thin film of sputtered silver to avoid charging problems and to obtain better image definition.

3. Results and Discussion

As commented in the text, before the activation of the 3D supporting structures, the mechanical stability of the structures must be improved and we can use two routes.

3.1. Route 1: Impregnation with Alumina. Several impregnation studies (Table 1) with alumina dispersions were tested. The impregnation time was also optimized and it was concluded that no significant differences were observed between samples impregnated for 30 s or 5 min and, hence, 30 s was the time used in all the procedures tested. Procedure 1 (Table 1) produces a 0.8 mm thick external covering layer of Al_2O_3 over the 3D structure; however, the porous 3D structure seems to be alumina-free (Figure 6(a)), and hence the inner structure will collapse at high temperature.

Procedure 2 (Table 1) produces two types of covering: an external layer of Al_2O_3, as in the first case, and an additional inner impregnated layer covering about 70% of the 3D porous structure (Figure 6(b)). The combination of both layers is ideal for improved mechanical stability.

And finally in procedure 3 (Table 1) with a more diluted dispersion, only the internal layer was observed (Figure 6(c)), which is not enough to provide mechanical stability at high temperatures and hence procedure 2 was used to test all the 3D printed structures.

From the SEM images, it is possible to observe a good distribution of 1 μm Al_2O_3 particles in the surface of the covered 3D structure (Figure 7(a)) and also a good distribution inside of the porous structure (Figures 7(b) and 7(c)). The presence of Al_2O_3 is negligible in the inner part of the 3D porous structure (Figure 7(d)), with a value of 0,6% (wt) of Al shown in the EDX studies compared to 25,1% (wt) of Al obtained in the structures impregnated shown in Figures 7(b) and 7(c).

Once the 3D printed structures (Figures 3(a) and 3(b)) have been impregnated with Al_2O_3 and fired at 1200°C, an activation process with the photocatalytic material must be performed. A cement-based material was used to fix the TiO_2 powder to the 3D structure producing an A3DS and then 3 photocatalytic experiments were carried to analyze the photocatalytic behavior of the 3D-TiO_2 structure under UV light.

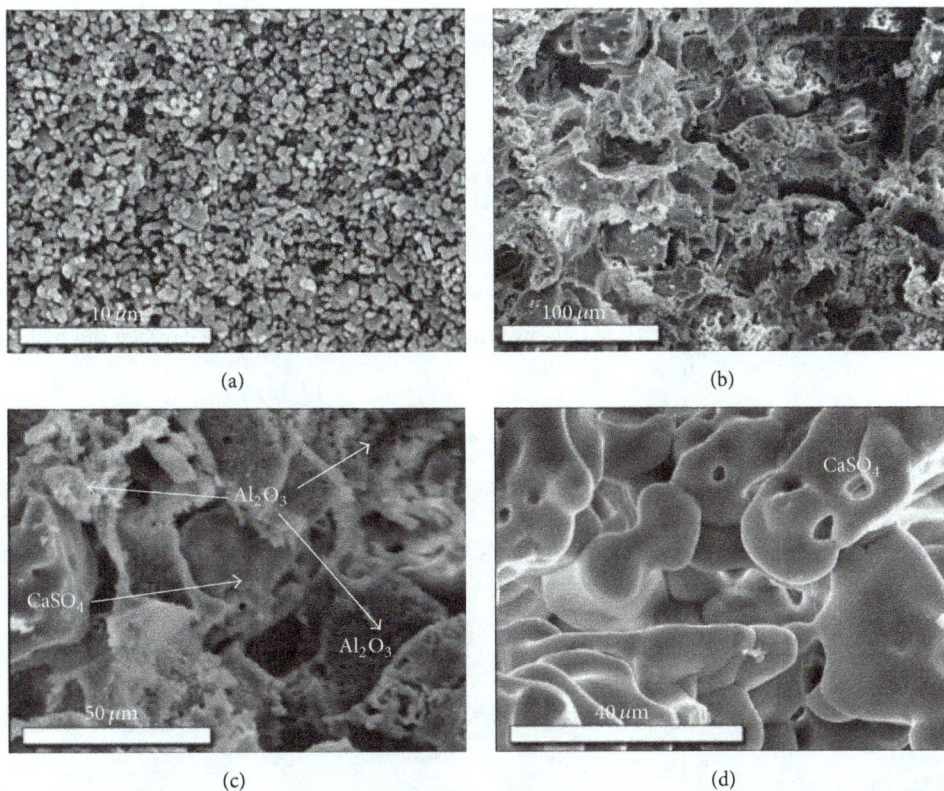

FIGURE 7: SEM images and EDX data of Al_2O_3-impregnated samples following procedure 2 (Table 1) with 30 s of impregnation time. (a) SEM image of surface of the covered 3D structure. ((b) and (c)) SEM image showing a good distribution of Al_2O_3 inside the porous structure. (d) SEM image of the inner part of the supporting $CaSO_4$ porous structure.

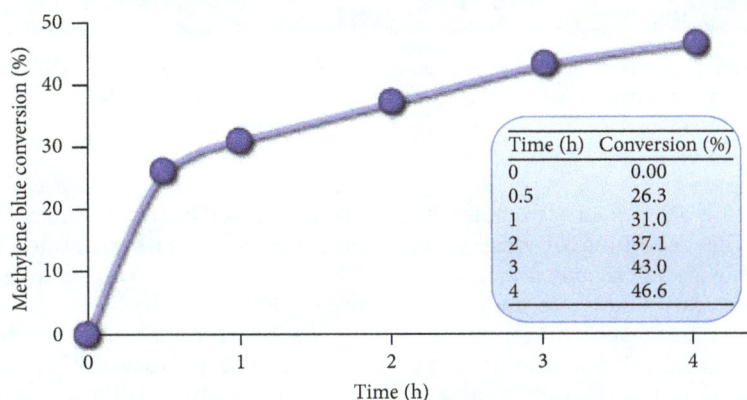

FIGURE 8: Temporal evolution of MB removal during photocatalytic experiments under UV light irradiation using 3DAS fabricated with the first route (Section 2.2).

Figure 8 shows the result of the photocatalytic performance under UV light radiation. The temporal evolution of the concentration of the MB as contaminant model showed that MB achieves 46.6% of conversion with the irradiation time after 4 h. This value proves that the 3D supporting structure can be effectively used as a support for photocatalytic materials; however, the degree of conversion can be considered relatively low and hence a more optimized structure, possibly with geometries maximizing illumination of the active area and/or improvement of the active area, should be fabricated.

Indeed, just modifying the geometrical surface from plain surface to twisted one produces an enhancement of more than 6% in the conversion of MB (Figure 9), that is, from 46.6 to 52.6%, after 4 h.

3.2. Route 2: Impregnation with Silica Nanoparticles (Ludox).
Another way to improve the efficiency is to minimize the

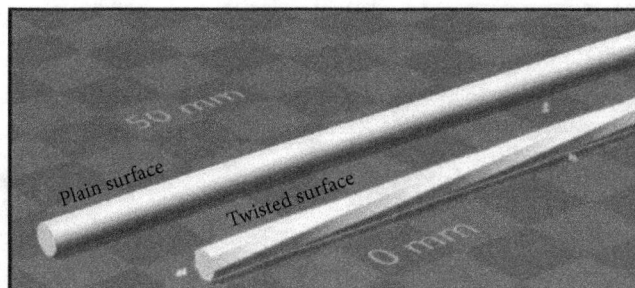

FIGURE 9: Simple 3D models of two types of geometrical surfaces printed to verify how the surface design can modify the photocatalytic experiments.

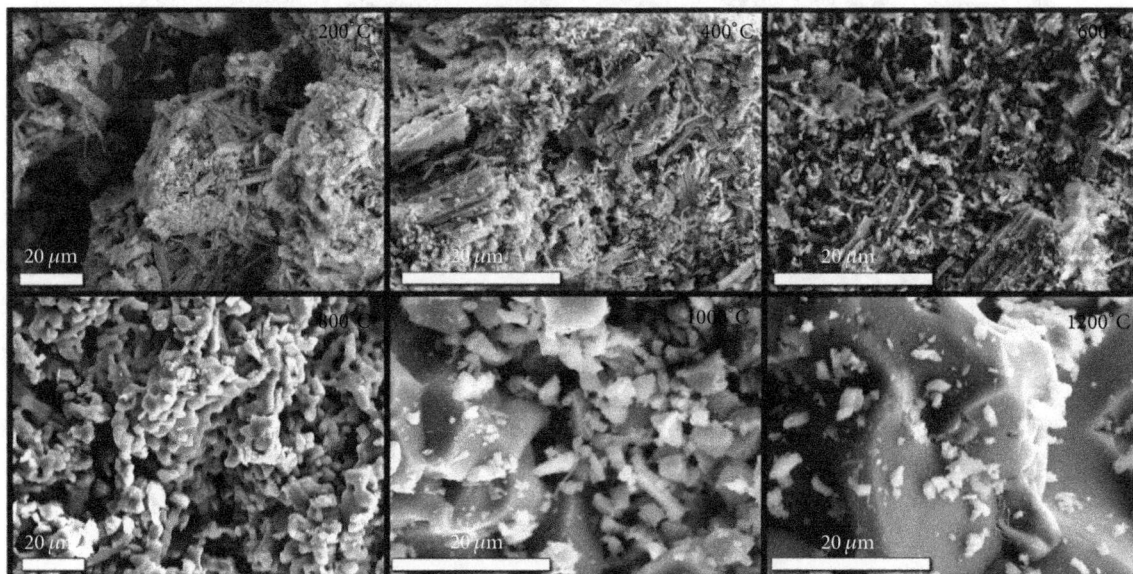

FIGURE 10: Several 3D printed square pieces of $CaSO_4$ were used to optimize the infiltration process with Ludox. All the samples were infiltrated with a solution of Ludox : water with a ratio of 1 : 5 (wt) for 30 s and then dried for 10 min at 70°C and finally they were fired at the temperatures shown in the images for 2 h.

distortion of the surface of the 3D printed structures by the impregnated particles. The deposition of microsized Al_2O_3 solid particles, previously described, may decrease the available surface area to be covered with the photocatalyst and hence an alternative way to improve the mechanical stability of the 3D printed structures with a minimal distortion of the 3D surface is required. This can be achieved by using a solution of silica nanoparticles (Ludox) which can be easily infiltrated through the whole porous structure as described in the experimental section, Section 2.2, route 2.

Several tests were performed to optimize the Ludox : water ratio, drying steps, and firing temperature. For example, Figure 10 shows 3D impregnated samples exhibiting a good distribution of micro/nanoparticles of silica when the final firing temperature was maintained below 600°C. At higher temperatures, the 3D porous $CaSO_4$ structure disappears as well as the silica agglomerates; thus the final temperature was kept at 600°C.

The 3D structures were printed following the second route described in Section 2.2 and then they were activated

producing A3DS. In this case, a dispersion of TiO_2 was prepared using a Ludox solution in a 1 : 1 (TiO_2 : SiO_2) ratio (weight). The sample was left to dry for 1 h at 70°C and finally fired at 300°C for 2 h.

The photoreactor tube was filled with A3DS (Figure 11) and then the photocatalytic performance under UV light radiation was studied. The temporal evolution of the concentration of the MB showed that it achieves 50% of conversion with the irradiation time in just 1 h, which is more than the conversion obtained with the first route after 4 h. A value close to the 90% of conversion was obtained after 5 h. This proves again that the 3D supporting structure can be used as a support of photocatalysts materials and that the optimization of the previous impregnation process can be critical to enhance the activity of the optimized A3DS.

4. Conclusion

3D printing technology has been successfully used to create 3D supporting structures of $CaSO_4$ which can be activated with TiO_2 showing photocatalytic activity.

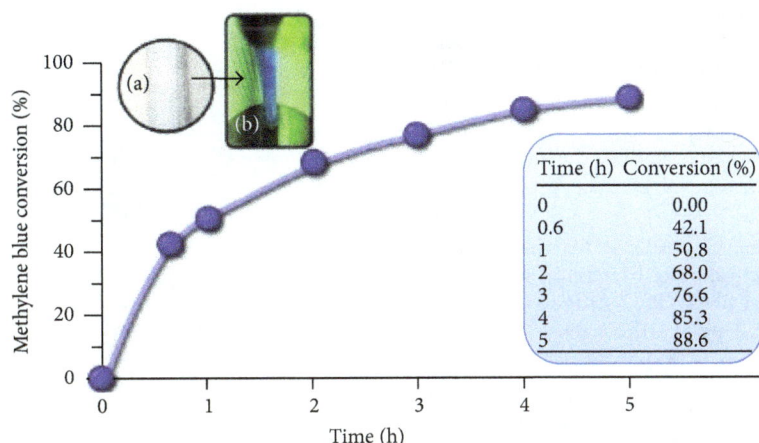

FIGURE 11: Temporal evolution of MB removal during photocatalytic experiments under UV light irradiation using 3D printed supports fabricated with the second route (Section 2.2). The inset image shows (a) the 3D printed support covered with the photocatalyst TiO_2 and (b) the photoreactor filled with the 3D structures and recirculating MB.

Two approaches have been proposed and optimized to enhance the mechanical stability of the 3D supports up to 1200°C. In one of them, Al_2O_3 is used as a binder of the whole 3D structure; in the other method, a dispersion of nanoparticles of SiO_2 (Ludox) acts as a binder.

In both cases, the A3DS show photocatalytic activity for the MB removal from wastewater, reaching more than 50% of conversion in just 1 h of irradiation when using Ludox in the activation process. This approach reaches a high photodegradation rate of about 90% after 5 h.

The A3DS offer clear advantages for the industrial treatment of wastewater given that the photocatalytic materials remain confined to the packed bed avoiding the need of separating the catalyst from the decontaminated water effluent.

Conflicts of Interest

The authors declare that there are no conflicts of interest regarding the publication of this paper.

Acknowledgments

The authors wish to acknowledge the financial support provided by "Ministerio de Economía y Competitividad" (MINECO), "Agencia Estatal de Investigación" (AEI), and "Fondo Europeo de Desarrollo Regional" (FEDER) through the Project ENE2016-74889-C4-2-R.

References

[1] M. D. Hernando, M. Mezcua, A. R. Fernández-Alba, and D. Barceló, "Environmental risk assessment of pharmaceutical residues in wastewater effluents, surface waters and sediments," *Talanta*, vol. 69, no. 2, pp. 334–342, 2006.

[2] N. Miranda-García, S. Suárez, B. Sánchez, J. M. Coronado, S. Malato, and M. I. Maldonado, "Photocatalytic degradation of emerging contaminants in municipal wastewater treatment plant effluents using immobilized TiO_2 in a solar pilot plant,"

Applied Catalysis B: Environmental, vol. 103, no. 3-4, pp. 294–301, 2011.

[3] Z. Zhang, A. Hibberd, and J. L. Zhou, "Analysis of emerging contaminants in sewage effluent and river water: Comparison between spot and passive sampling," *Analytica Chimica Acta*, vol. 607, no. 1, pp. 37–44, 2008.

[4] N. Miranda-García, M. I. Maldonado, J. M. Coronado, and S. Malato, "Degradation study of 15 emerging contaminants at low concentration by immobilized TiO_2 in a pilot plant," *Catalysis Today*, vol. 151, no. 1-2, pp. 107–113, 2010.

[5] M. Schriks, M. B. Heringa, M. M. E. van der Kooi, P. de Voogt, and A. P. van Wezel, "Toxicological relevance of emerging contaminants for drinking water quality," *Water Research*, vol. 44, no. 2, pp. 461–476, 2010.

[6] L. Yang, L. E. Yu, and M. B. Ray, "Degradation of paracetamol in aqueous solutions by TiO_2 photocatalysis," *Water Research*, vol. 42, no. 13, pp. 3480–3488, 2008.

[7] M. Pelaez, N. T. Nolan, S. C. Pillai et al., "A review on the visible light active titanium dioxide photocatalysts for environmental applications," *Applied Catalysis B: Environmental*, vol. 125, pp. 331–349, 2012.

[8] M. A. Sousa, C. Gonçalves, V. J. P. Vilar, R. A. R. Boaventura, and M. F. Alpendurada, "Suspended TiO2-assisted photocatalytic degradation of emerging contaminants in a municipal WWTP effluent using a solar pilot plant with CPCs," *Chemical Engineering Journal*, vol. 198-199, pp. 301–309, 2012.

[9] Y.-L. Wei, K.-W. Chen, and H. P. Wang, "Study of chromium modified TiO_2 nano catalyst under visible light irradiation," *Journal of Nanoscience and Nanotechnology*, vol. 10, no. 8, pp. 1–5, 2010.

[10] N. Bahadur, K. Jain, A. K. Srivastavaa, R. Govinda, D. Haranatha, and M. S. Dulat, "Effect of nominal doping of Ag and Ni on the crystalline structure and photo-catalytic properties of mesoporous titania," *Materials Chemistry and Physics*, vol. 124, no. 1, pp. 600–608, 2010.

[11] Y.-H. Peng, G.-F. Huang, and W.-Q. Huang, "Visible-light absorption and photocatalytic activity of Cr-doped TiO_2 nanocrystal films," *Advanced Powder Technology*, vol. 23, no. 1, pp. 8–12, 2012.

[12] K. Soutsas, V. Karayannis, I. Poulios et al., "Decolorization and degradation of reactive azo dyes via heterogeneous photocatalytic processes," *Desalination*, vol. 250, no. 1, pp. 345–350, 2010.

[13] Y. Li, J. Chen, J. Liu, M. Ma, W. Chen, and L. Li, "Activated carbon supported TiO_2-photocatalysis doped with Fe ions fo continuous treatment of dye wastewater in a dynamic reactor," *Journal of Environmental Sciences*, vol. 22, no. 8, pp. 1290–1296, 2010.

[14] R. van Grieken, J. Marugán, C. Sordo, P. Martínez, and C. Pablos, "Photocatalytic inactivation of bacteria in water using suspended and immobilized silver-TiO_2," *Applied Catalysis B: Environmental*, vol. 93, no. 1-2, pp. 112–118, 2009.

[15] E. M. Hernández-Rodríguez, P. Acosta-Mora, J. Méndez-Ramos et al., "Prospective use of the 3D printing technology for the microstructural engineering of solid oxide fuel cell components," *Boletin de la Sociedad Espanola de Ceramica y Vidrio*, vol. 53, no. 5, pp. 213–216, 2014.

[16] J. C. Ruiz-Morales, D. Marrero-López, J. Canales-Vázquez, and J. T. S. Irvine, "Symmetric and reversible solid oxide fuel cells," *RSC Advances*, vol. 1, no. 8, pp. 1403–1414, 2011.

[17] J. C. Ruiz-Morales, A. Taranc, J. Canales-Vázquez et al., "Three dimensional printing of components and functional devices for energy and environmental applications," *Energy & Environmental Science*, vol. 10, pp. 846–859, 2017.

[18] A. Cybulsky and J. A. Moulin, *Structured Catalysts and Reactors*, Marcell and Dekker, Inc, 1998.

[19] B. Utela, D. Storti, R. Anderson, and M. Ganter, "A review of process development steps for new material systems in three dimensional printing (3DP)," *Journal of Manufacturing Processes*, vol. 10, no. 2, pp. 96–104, 2008.

[20] M. E. Borges, D. M. García, T. Hernández, J. C. Ruiz-Morales, and P. Esparza, "Supported photocatalyst for removal of emerging contaminants from wastewater in a continuous packed-bed photoreactor configuration," *Catalysts*, vol. 5, no. 1, pp. 77–87, 2015.

[21] Z. Zhou, C. A. Mitchell, F. J. Buchanan, and N. J. Dunne, "Effects of heat treatment on the mechanical and degradation properties of 3D-printed calcium-sulphate-based scaffolds," *ISRN Biomaterials*, vol. 2013, 10 pages, 2013.

[22] https://www.tinkercad.com/.

21

Seasonal Changes and Spatial Variation in Water Quality of a Large Young Tropical Reservoir and Its Downstream River

Teck-Yee Ling,[1] Norliza Gerunsin,[2] Chen-Lin Soo,[1] Lee Nyanti,[1] Siong-Fong Sim,[1] and Jongkar Grinang[3]

[1]Faculty of Resource Science and Technology, Universiti Malaysia Sarawak, 94300 Kota Samarahan, Sarawak, Malaysia
[2]Universiti Teknologi MARA, Kota Samarahan Campus, Jalan Meranek, 94300 Kota Samarahan, Sarawak, Malaysia
[3]Institute of Biodiversity and Environmental Conservation, Universiti Malaysia Sarawak, 94300 Kota Samarahan, Sarawak, Malaysia

Correspondence should be addressed to Teck-Yee Ling; teckyee60@gmail.com

Academic Editor: Wenshan Guo

This study examined the water quality of the large young tropical Bakun hydroelectric reservoir in Sarawak, Malaysia, and the influence of the outflow on the downstream river during wet and dry seasons. Water quality was determined at five stations in the reservoir at three different depths and one downstream station. The results show that seasons impacted the water quality of the Bakun Reservoir, particularly in the deeper water column. Significantly lower turbidity, SRP, and TP were found during the wet season. At 3–6 m, the oxygen content fell below 5 mg/L and hypoxia was also recorded. Low NO_2^--N, NO_3^--N, and SRP and high BOD_5, OKN, and TP were observed in the reservoir indicating organic pollution. Active logging activities and the dam construction upstream resulted in water quality deterioration. The outflow decreased the temperature, DO, and pH and increased the turbidity and TSS downstream. Elevated organic matter and nutrients downstream are attributable to domestic discharge along the river. This study shows that the downstream river was affected by the discharge through the turbines, the spillway operations, and domestic waste. Therefore, all these factors should be taken into consideration in the downstream river management for the health of the aquatic organisms.

1. Introduction

The creation of a large-scale dam and its associated reservoir often has significant negative impacts on the hydrological, biological, and chemical processes of the reservoir, upstream, and downstream of the dam [1–9]. The Bakun hydroelectric dam, which was impounded from 2010 to 2012 on the Balui River in Malaysia, produced these effects. The dam which is one of the tallest concrete rock filled dams (205 m) in the world created a reservoir covering a surface area of 695 km^2. A few pre- and postimpoundment studies on the physicochemical parameters of the Bakun Dam reservoir have been performed [10–12]. However, the reservoir water quality is likely changing as the reservoir is receiving loads of pollutants from adjacent anthropogenic activities during its operation [13, 14]. Water quality deterioration is a common problem in reservoirs surrounded with anthropogenic activities receiving high loads of suspended solids, organic matter, and nutrients [15, 16].

The water quality of reservoirs has been observed to vary seasonally in tandem with changes in temperature and rainfall [17–19]. The low and high precipitation during dry and wet seasons in a tropical country like Malaysia can greatly change the water quality of the reservoir. The high precipitation during the wet season can either decrease the pollutant concentration by dilution or deteriorate the reservoir water quality due to increased surface runoff from anthropogenic activities. Reference [20] demonstrated that the levels of total phosphorus in Batang Ai Reservoir during the rainy season and high water levels were lower than those observed during the dry season and low water levels. Besides, high volume of inflow following heavy rainfall promotes mixing and disturbs stratification in the reservoir. The increase of bottom

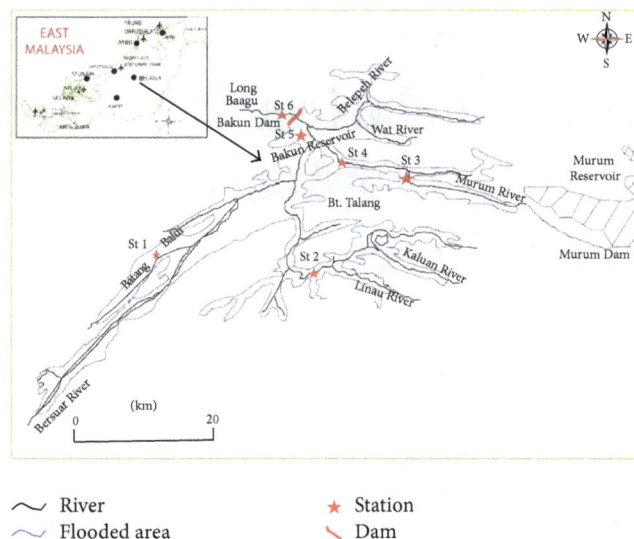

~ River   ★ Station
~ Flooded area   ↘ Dam

FIGURE 1: The study area and sampling stations in the present study.

dissolved oxygen level in the well-mixed reservoir inhibits the release of nutrients from sediments causing a rapid reduction of phytoplankton concentration in the reservoir [17].

On the other hand, the reservoir outflow has a great influence on the downstream river. Studies have shown that the downstream river is subjected to major environmental impacts which range from downstream morphology changes to loss of biodiversity of the ecosystem [1, 5, 7, 8, 21]. The reservoir outflow is often controlled by the electrical demand and operation cost, independent of ecological considerations in the downstream river. Differences in structure and operation scheme of a dam may result in differences in water quality downstream. Recently, [22] demonstrated that the physicochemical characteristics of the river downstream of the Bakun Dam changed when the spillway was opened.

As a young reservoir in a tropical country, changes continue to occur in the reservoir and it is important to monitor the water quality in order to evaluate its suitability for secondary purposes such as aquaculture and recreation. The knowledge of the seasonal variation of the reservoir's water quality is important for dam operation and management decision. The impact of the dam on the water quality of its downstream river during the wet and dry seasons remains unknown. Hence, the aim of this study was to assess the water quality of the Bakun Reservoir and the influence of its outflow on the water quality of the downstream river during wet and dry seasons.

2. Materials and Methods

2.1. Study Area and Sampling Stations.
The present study was conducted at Bakun Reservoir and its downstream river in Sarawak, Malaysia, as illustrated in Figure 1. The Bakun hydroelectric dam was built across the Batang Balui with a total of eight installed turbines and a spillway weir located at 209 m above sea level. The reservoir covers mainly the Balui River that is fed by three major tributaries, namely,

the Murum River, Linau River, and Bahau River. A total of five stations were selected at the Bakun Reservoir and one station was selected at the downstream river. Stations 1 and 2 were located at the Batang Balui and Linau River, respectively. Stations 3 and 4 were located at the Murum River where the upstream Murum hydroelectric dam was under ongoing construction during the time of sampling. Station 5 was located in the proximity of the Bakun hydroelectric dam and downstream of active logging activities while Station 6 was located at the downstream river approximately 4.3 km from the dam.

Sampling was conducted in February and September 2014 corresponding to the wet and dry seasons in Sarawak (Table 1). There was no rain recorded during the two and three weeks prior to the first and second samplings, respectively. The water level during the second sampling in the dry season was approximately 7 m lower than the water level during the wet season. The water release during hydropower generation is drawn from the top 10 m of the reservoir using selective withdrawal intake structures. Occasionally, additional water is released from the spillway with intake at a depth of approximately 15 m. At the end of the spillway, the water hits the concrete barrier before entering Balui River downstream. Sampling was conducted during electrical power generation where the downstream river received the water discharged from the reservoir after the water passed through the turbines. During the first sampling, additional water was discharged from the spillway at a rate of 501 m^3/s in addition to the turbine outflow (536 m^3/s). The spillway was closed during the second sampling; hence, Station 6 was receiving solely the turbine outflow at a rate of 730 m^3/s.

2.2. Field Collection and Laboratory Analysis.
Depth profiles of temperature and dissolved oxygen (DO) were measured using a YSI 6820 V2 multiparameter water quality sonde during the first sampling in February 2014. The pH and turbidity were measured at 0 m, 10 m, and 20 m depths in Bakun Reservoir in both samplings by using a pH meter (EcoScan, Eutech) and a turbidity meter (Martini Instruments, Mi415), respectively. Triplicate water samples were collected at 0 m, 10 m, and 20 m depths in Bakun Reservoir (Stations 1 to 5) using a Van Dorn water sampler whereas triplicate water samples were collected at 0 m depth at the downstream river of the dam (Station 6). The depth of the reservoir was measured using a portable depth sounder (Speedtech). All sampling bottles were acid-washed, cleaned, and dried before use. Water samples were acidified to pH < 2 for total phosphorus (TP) analysis. All samples were placed in an ice box and transported to the laboratory for further analysis [23].

All the analyses were conducted according to standard methods [23, 24]. Chlorophyll a (Chl a) was determined from adequate samples filtered through 0.45 μm glass fiber filter (Whatman GF/F) and extracted for 24 h using 90% (v/v) acetone. The absorbance was read using a DR 2800 spectrophotometer and concentration of Chl a was calculated according to [25]. Total suspended solid (TSS) was calculated as the difference between the initial and final weights of the

TABLE 1: The details of the sampling location and sampling regime in the present study.

Station	Coordinates	Date	Location
Bakun hydroelectric reservoir			
St. 1	N 02°43′34.4″ E 114°01′44.2″	26 Feb. 2014, 1:15 p.m. 24 Sept. 2014, 8:15 a.m.	Batang Balui Sunny during both sampling trips
St. 2	N 02°39′32.2″ E 114°03′29.5″	26 Feb. 2014, 9:45 a.m. 24 Sept. 2014, 10:45 a.m.	Linau River Sunny during both sampling trips
St. 3	N 02°42′59.8″ E 114°09′43.8″	27 Feb. 2014, 12:55 p.m. 25 Sept. 2014, 9:51 a.m.	Upper part of Murum River Sunny during both sampling trips Soil erosion was observed in the upper Murum River bank
St. 4	N 02°44′15.3″ E 114°05′16.6″	26 Feb. 2014, 3:06 p.m. 24 Sept. 2014, 1:42 p.m.	Lower part of Murum River Sunny during both sampling trips
St. 5	N 02°45′09.8″ E 114°02′32.9″	27 Feb. 2014, 3:00 p.m. 25 Sept. 2014, 2:08 p.m.	Near the intake point and the dam Cloudy during both sampling trips
Downstream river of Bakun hydroelectric dam			
St. 6	N 02°46′21.8″ E 114°01′41.6″	26 Feb. 2014, 3:00 p.m. 24 Sept. 2014, 1:35 pm	Long Baagu (4.3 km downstream of the Bakun hydroelectric dam) Sunny during both sampling trips

$0.45\,\mu m$ glass fiber filter (Whatman GF/F), after filtration of an adequate sample volume and drying at 105°C. Five-day biochemical oxygen demand (BOD_5) was determined as the difference between the initial and five-day DO content, after five-day-long incubation of the sample. The initial DO content was determined in the field and increased by vigorous aeration if the DO value was low. $NO_2^- $-N and NO_3^--N levels were determined by the diazotization method (low range) and the cadmium reduction method, respectively, after filtering through a $0.45\,\mu m$ glass fiber filter (Whatman GF/F). Organic Kjeldahl nitrogen (OKN) was determined by the Macro-Kjeldahl Method where ammonia was removed from the water sample before digestion and distillation, followed by Nessler's method. SRP was determined by the colorimetric ascorbic acid method after filtering through a $0.45\,\mu m$ glass fiber filter (Whatman GF/F). TP was determined by the ascorbic acid method after persulfate digestion of samples. The estimated detection limits of NO_2^--N, NO_3^--N, and SRP were 0.005 mg/L NO_2^--N, 0.01 mg/L NO_3^--N, and 0.02 mg/L PO_4^{3-}, respectively.

Quality control steps were taken throughout the study. Sample bottles and glassware were washed using phosphate-free detergent followed by the standard acid wash procedure. Sample preparation and storage were performed according to the standard methods [23]. Triplicate blank water that was free of the analytes of interest was used in the same procedure for each of the aforementioned analyses.

2.3. Statistical Analysis. Comparison of water quality parameters between the stations and the depths in the Bakun hydroelectric reservoir was conducted using one-way ANOVA and Tukey's pairwise comparisons with 5% significance level. Student's t-test was used to compare the water quality of the reservoir between the wet and dry seasons. Pearson's correlation analysis was performed to determine the relationship among all the parameters in the reservoir during each season. The water quality of the downstream river between the wet and dry seasons and the results between the intake point

of the dam and the downstream river were compared using Student's t-test. Cluster analysis (CA) was used to investigate the grouping of the sampling stations with different depths by using the water quality parameters collected in the reservoir and the downstream river. Z-score standardization of the variables and Ward's method using Euclidean distances as a measure of similarity were used. All the statistical analyses were carried out by using the Statistical Package for the Social Sciences (SPSS Version 22, SPSS Inc., 1995).

3. Results and Discussion

3.1. Water Quality of Bakun Reservoir. Figure 2 illustrates the vertical stratification in Bakun Reservoir, indicating poor water mixing in the reservoir. Among the five sampling stations in the Bakun Reservoir, Station 2, which is located at Linau River, is stratified into three distinct layers of different temperatures. The thermocline layer observed at 3 m to 7 m separates the epilimnion ($\approx$30.5°C) and hypolimnion ($\approx$25.5°C) at Station 2. Similarly, [10, 11] reported that the thermocline started at a depth of 4-5 m and between 6 m and 9 m in Bakun Reservoir during the filling phase and 13 months after reaching the full-supply level, respectively. Thermal stratification in reservoirs has been widely reported in tropical and subtropical reservoirs [19, 26-28]. The temperature gradient within the thermocline layer in the Bakun Reservoir is in agreement with the range of thermal stratification of 0.5°C to 5°C for a tropical reservoir [29].

Dissolved oxygen was relatively consistent in the surface water of the Bakun Reservoir, with a mean value of 7.22 mg/L. The DO level started to decrease rapidly from a depth of 2 m to less than 0.2 mg/L at a depth of 4 m at Station 1 which is located at Batang Balui. The DO level at Stations 2, 3, and 5 started to decrease rapidly from the depth of around 3 m whereas the DO level at Station 4 started to decrease from 5 m depth. In other words, the healthy level of DO content above 5 mg/L was only observed at the water column above 3-6 m in Bakun Reservoir. Similarly, [26]

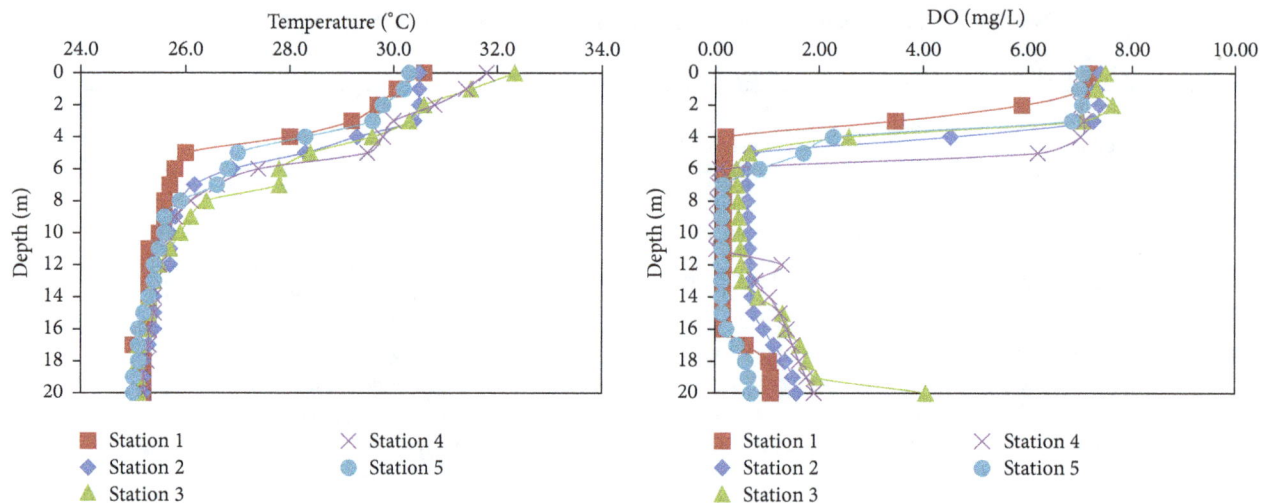

FIGURE 2: Depth profile of temperature and DO in Bakun Reservoir in February 2014.

showed that oxygen depletion is a common phenomenon in the hypolimnia of Indonesian lakes and reservoirs with different oxycline depths. The authors attributed the shallow oxycline depth and thick anoxic layer in the Cirata Reservoir to the weak wind-induced mixing and high organic loads that lead to rapid decomposition and oxygen depletion in the reservoir. On the other hand, the DO concentration never fell below 2 mg/L in Qiandaohu Lake, China, where the DO depth profiles were closely linked to the water temperature depth profiles [19]. The decrease of DO with depth is commonly observed in reservoirs as photosynthesis increases oxygen level in the surface water while respiration of bacteria decomposing dead organic matter consumes all the dissolved oxygen in the bottom water column coupled with insufficient exchange with oxygenated surface water [30]. However, a slight increase of DO content was observed at the water column of the Bakun Reservoir between 12 m and 20 m which is most likely due to the additional water discharged from the spillway where the water intake was at a depth of approximately 15 m. The rapid water movement due to the additional water withdrawal at the particular water column promotes the mixing of the low DO water with a large volume of oxygenated colder water inflow from tributaries around the reservoir [14]. This phenomenon was not observed in the study in [11] where the DO content was reported as undetectable from a depth of 7 m up to a depth of 30 m as the reservoir water was not discharged from the spillway during this study.

The pH value of the Bakun Reservoir ranged from 4.93 ± 0.06 to 8.06 ± 0.05 during the wet season with the lowest and highest pH value being observed at Station 5 and Station 2, respectively. On the other hand, the pH value of the Bakun Reservoir is relatively consistent during the dry season with a mean value of 7.30. Vertical distribution of pH values in Bakun Reservoir differed between the wet and dry seasons although this was not significantly different (p value > 0.05) (Table 2). During the dry season, the pH value of the Bakun Reservoir decreased as depth increased up to a depth of

10 m and remained at a similar value up to a depth of 20 m as illustrated in Figure 3. The vertical distribution of pH values during the dry season in the present study is in good agreement with the previous study in the Bakun Reservoir [11] and the Batang Ai Reservoir [31] where the pH value of the reservoir water decreased as depth increased. However, the pH value tends to increase with depth when the surface pH value is low as demonstrated by Stations 3 and 5 during the wet season. The results showed that the low pH value at the surface water was diluted by the reservoir water with higher pH value as depth increased. The dilution in the water column improved the pH at Station 3 from 6.3 to 6.8. However, despite the dilution in the water column, Station 5, which was the closest station to the dam, still showed pH values of less than 6.5 mg/L. On the other hand, when the pH value was high (>7), the pH value decreased as depth increased which is similar to vertical pH distribution during the dry season. The surface pH value was classified as Class I but was changed to Class II as depth increased according to the National Water Quality Standard (NWQS) for Malaysia [32] during the dry season. During the wet season, the pH values of the Bakun Reservoir were classified as Class I except for Stations 3 and 5. The surface water at Station 3 was classified as Class II while the extremely low surface pH value of 4.9 at Station 5 exceeded the NWQS. Besides, the pH values at Station 5 at depths of 10 m and 20 m were classified as Classes II and III, respectively.

Table 3 shows that no significant correlation (p value > 0.05) was found between the pH value of the Bakun Reservoir and the other parameters during the wet season suggesting that the pH value of the Bakun Reservoir, particularly Stations 3 and 5, was mainly influenced by the low pH surface runoff from the anthropogenic activities in the surrounding area. Stations 3 and 5 were located downstream of the construction site of the Murum hydroelectric dam and active logging activities. The decomposition of organic matter derived from anthropogenic activity acidified the upstream rivers that flow into the reservoir and caused the acidification at the stations.

TABLE 2: Mean difference of water quality parameters of the Bakun Reservoir during the wet season and dry season conducted in February and September 2014, respectively ($N = 3$).

Parameter	Sampling	Depth	Station 1	2	3	4	5	Mean	Difference	p value
pH	Wet season	0 m	$7.52 \pm 0.02^{d,3}$	$8.06 \pm 0.05^{e,3}$	$6.34 \pm 0.02^{b,1}$	$7.37 \pm 0.02^{c,3}$	$4.93 \pm 0.06^{a,1}$	6.67	+0.20	0.422
		10 m	$6.81 \pm 0.02^{b,2}$	$6.83 \pm 0.01^{bc,2}$	$6.78 \pm 0.02^{b,2}$	$6.90 \pm 0.02^{c,2}$	$6.22 \pm 0.06^{a,3}$			
		20 m	$6.52 \pm 0.01^{b,1}$	$6.63 \pm 0.02^{c,1}$	$6.76 \pm 0.01^{d,2}$	$6.66 \pm 0.02^{c,1}$	$5.65 \pm 0.02^{a,2}$			
	Dry season	0 m	$7.21 \pm 0.01^{b,3}$	$7.01 \pm 0.02^{a,2}$	$7.56 \pm 0.01^{c,3}$	$7.36 \pm 0.01^{c,3}$	$7.36 \pm 0.00^{c,3}$	6.47		
		10 m	$5.96 \pm 0.01^{a,1}$	$5.95 \pm 0.00^{a,1}$	$6.17 \pm 0.00^{d,1}$	$6.11 \pm 0.01^{c,1}$	$6.03 \pm 0.01^{b,2}$			
		20 m	$6.03 \pm 0.00^{b,2}$	$5.95 \pm 0.01^{a,1}$	$6.24 \pm 0.00^{d,2}$	$6.14 \pm 0.00^{c,2}$	$5.95 \pm 0.01^{a,1}$			
Turbidity, FNU	Wet season	0 m	$0.06 \pm 0.01^{a,1}$	$2.65 \pm 0.04^{b,1}$	$5.16 \pm 0.29^{d,1}$	$3.77 \pm 0.12^{c,1}$	$3.49 \pm 0.19^{c,1}$	38.89	−42.35	**0.020**
		10 m	$34.63 \pm 0.12^{b,2}$	$33.81 \pm 0.49^{b,2}$	$40.63 \pm 0.82^{c,2}$	$63.00 \pm 1.00^{d,2}$	$16.09 \pm 2.27^{a,2}$			
		20 m	$48.29 \pm 0.72^{b,3}$	$51.00 \pm 0.00^{b,3}$	$128.00 \pm 4.36^{c,3}$	$131.33 \pm 0.58^{c,3}$	$21.46 \pm 1.56^{a,3}$			
	Dry season	0 m	$2.39 \pm 0.05^{a,1}$	$3.66 \pm 0.01^{b,1}$	$4.07 \pm 0.03^{c,1}$	$4.35 \pm 0.03^{d,1}$	$3.84 \pm 0.13^{b,1}$	81.24		
		10 m	$63.33 \pm 0.58^{b,2}$	$38.33 \pm 0.07^{a,2}$	$261.33 \pm 1.53^{e,2}$	$102.00 \pm 1.00^{d,2}$	$67.00 \pm 0.00^{c,2}$			
		20 m	$78.33 \pm 0.58^{b,3}$	$45.58 \pm 0.16^{a,3}$	$263.67 \pm 1.53^{e,2}$	$199.67 \pm 0.58^{d,3}$	$81.00 \pm 0.58^{c,3}$			
Chl a, µg/L	Wet season	0 m	$1.41 \pm 0.23^{ab,2}$	$1.45 \pm 0.22^{ab,1}$	$2.30 \pm 0.81^{b,2}$	$0.62 \pm 0.21^{a,1}$	$0.64 \pm 0.15^{a,2}$	0.93	−0.18	0.655
		10 m	$0.90 \pm 0.00^{b,1}$	$1.50 \pm 0.45^{c,1}$	$0.26 \pm 0.06^{a,1}$	$0.71 \pm 0.06^{ab,1}$	$0.37 \pm 0.10^{ab,12}$			
		20 m	$0.76 \pm 0.06^{b,1}$	$1.59 \pm 0.02^{c,1}$	$0.57 \pm 0.22^{b,1}$	$0.75 \pm 0.03^{b,1}$	$0.17 \pm 0.12^{a,1}$			
	Dry season	0 m	$1.30 \pm 0.09^{ab,3}$	$1.76 \pm 0.57^{ab,2}$	$2.57 \pm 0.55^{b,3}$	$0.64 \pm 0.02^{a,3}$	$5.87 \pm 0.99^{c,2}$	1.11		
		10 m	$0.62 \pm 0.04^{d,2}$	$0.38 \pm 0.06^{bc,1}$	$0.18 \pm 0.05^{a,1}$	$0.35 \pm 0.07^{ab,2}$	$0.55 \pm 0.11^{cd,1}$			
		20 m	$0.40 \pm 0.08^{b,1}$	$0.33 \pm 0.01^{b,1}$	$1.20 \pm 0.11^{c,2}$	$0.10 \pm 0.00^{a,1}$	$0.39 \pm 0.07^{b,1}$			
TSS, mg/L	Wet season	0 m	$7.8 \pm 1.0^{a,1}$	$6.7 \pm 1.7^{a,1}$	$13.3 \pm 0.0^{b,1}$	$8.0 \pm 2.0^{a,1}$	$9.1 \pm 0.8^{a,1}$	40.1	−29.6	0.147
		10 m	$43.2 \pm 1.1^{a,2}$	$49.9 \pm 1.8^{b,2}$	$55.3 \pm 1.2^{c,2}$	$64.5 \pm 0.8^{d,2}$	$40.3 \pm 0.6^{a,2}$			
		20 m	$63.3 \pm 1.2^{bc,3}$	$60.7 \pm 1.2^{b,3}$	$60.7 \pm 1.2^{b,3}$	$66.0 \pm 0.7^{c,2}$	$53.0 \pm 1.0^{a,3}$			
	Dry season	0 m	$6.0 \pm 0.0^{b,1}$	$10.2 \pm 0.3^{c,1}$	$14.2 \pm 0.5^{d,1}$	$4.1 \pm 0.2^{a,1}$	$6.1 \pm 0.5^{b,1}$	69.7		
		10 m	$45.4 \pm 0.9^{b,2}$	$30.4 \pm 1.2^{a,2}$	$170.2 \pm 1.1^{c,2}$	$82.7 \pm 2.3^{c,2}$	$48.6 \pm 1.1^{b,2}$			
		20 m	$60.2 \pm 0.9^{b,3}$	$38.1 \pm 0.3^{a,3}$	$328.9 \pm 1.9^{e,3}$	$131.8 \pm 1.4^{d,3}$	$68.5 \pm 0.7^{c,3}$			
BOD$_5$, mg/L	Wet season	0 m	$3.24 \pm 0.19^{a,2}$	$3.42 \pm 0.26^{a,1}$	$3.97 \pm 0.08^{b,1}$	$3.36 \pm 0.22^{a,1}$	$4.30 \pm 0.15^{b,1}$	4.02	−0.26	0.135
		10 m	$3.79 \pm 0.19^{a,3}$	$4.11 \pm 0.06^{ab,2}$	$5.32 \pm 0.12^{c,3}$	$4.30 \pm 0.19^{b,2}$	$4.44 \pm 0.06^{b,1}$			
		20 m	$2.23 \pm 0.01^{a,1}$	$4.63 \pm 0.24^{c,3}$	$4.55 \pm 0.03^{c,2}$	$4.11 \pm 0.07^{b,2}$	$4.50 \pm 0.08^{c,1}$			
	Dry season	0 m	$3.90 \pm 0.10^{b,2}$	$4.27 \pm 0.06^{cd,1}$	$4.13 \pm 0.06^{c,1}$	$3.47 \pm 0.06^{a,1}$	$4.43 \pm 0.06^{d,1}$	4.28		
		10 m	$5.13 \pm 0.06^{d,3}$	$4.70 \pm 0.10^{c,2}$	$4.27 \pm 0.06^{b,1}$	$3.87 \pm 0.06^{a,2}$	$4.63 \pm 0.06^{c,2}$			
		20 m	$3.60 \pm 0.10^{a,1}$	$4.50 \pm 0.10^{c,2}$	$5.07 \pm 0.06^{d,2}$	$3.87 \pm 0.10^{b,2}$	$4.40 \pm 0.10^{c,1}$			

TABLE 2: Continued.

Parameter	Sampling	Depth	Station 1	2	3	4	5	Mean	Difference	p value
NO$_2^-$-N, mg/L	Wet season	0 m	0.002 ± 0.001[a,1]	0.003 ± 0.001[a,1]	0.004 ± 0.001[a,1]	0.003 ± 0.001[a,1]	0.003 ± 0.001[a,1]	0.004	+0.001	0.282
		10 m	0.001 ± 0.001[a,1]	0.004 ± 0.001[b,1]	0.007 ± 0.001[c,2]	0.003 ± 0.001[b,1]	0.004 ± 0.000[b,1]			
		20 m	0.002 ± 0.001[a,1]	0.003 ± 0.001[ab,1]	0.005 ± 0.001[bc,12]	0.003 ± 0.001[ab,1]	0.007 ± 0.001[c,2]			
	Dry season	0 m	0.002 ± 0.001[a,1]	0.001 ± 0.000[a,1]	0.007 ± 0.001[b,2]	0.008 ± 0.001[b,2]	0.002 ± 0.001[a,1]	0.003		
		10 m	0.001 ± 0.001[a,1]	0.001 ± 0.000[a,1]	0.001 ± 0.001[a,1]	0.001 ± 0.001[a,1]	0.001 ± 0.000[a,1]			
		20 m	0.001 ± 0.001[a,1]	0.001 ± 0.001[a,1]	0.002 ± 0.001[a,1]	0.009 ± 0.001[b,2]	0.001 ± 0.000[a,1]			
NO$_3^-$-N, mg/L	Wet season	0 m	0.018 ± 0.001[a,1]	0.037 ± 0.001[b,2]	0.026 ± 0.009[ab,1]	0.014 ± 0.006[a,1]	0.020 ± 0.006[a,1]	0.021	+0.006	0.233
		10 m	0.012 ± 0.005[a,1]	0.009 ± 0.007[a,1]	0.023 ± 0.009[a,1]	0.020 ± 0.005[a,1]	0.019 ± 0.006[a,1]			
		20 m	0.012 ± 0.006[a,1]	0.050 ± 0.005[b,3]	0.018 ± 0.007[a,1]	0.027 ± 0.011[a,1]	0.016 ± 0.005[a,1]			
	Dry season	0 m	0.008 ± 0.001[ab,1]	0.009 ± 0.000[ab,1]	0.006 ± 0.006[a,1]	0.022 ± 0.011[b,1]	0.008 ± 0.001[ab,1]	0.015		
		10 m	0.009 ± 0.001[a,1]	0.016 ± 0.006[a,1]	0.009 ± 0.001[a,1]	0.059 ± 0.011[b,2]	0.009 ± 0.000[a,1]			
		20 m	0.009 ± 0.001[a,1]	0.009 ± 0.001[a,1]	0.008 ± 0.001[a,1]	0.041 ± 0.010[b,12]	0.009 ± 0.000[a,1]			
OKN, mg/L	Wet season	0 m	0.39 ± 0.01[c,2]	0.35 ± 0.01[b,2]	0.32 ± 0.01[a,1]	0.34 ± 0.01[ab,1]	0.42 ± 0.01[d,3]	0.34	+0.01	0.783
		10 m	0.22 ± 0.01[a,1]	0.32 ± 0.01[b,1]	0.36 ± 0.01[c,2]	0.35 ± 0.00[c,1]	0.31 ± 0.01[b,1]			
		20 m	0.22 ± 0.01[a,1]	0.43 ± 0.01[d,3]	0.38 ± 0.01[c,3]	0.34 ± 0.01[b,1]	0.33 ± 0.01[b,2]			
	Dry season	0 m	0.35 ± 0.00[c,3]	0.29 ± 0.00[a,1]	0.33 ± 0.01[b,1]	0.32 ± 0.00[b,1]	0.34 ± 0.01[bc,1]	0.33		
		10 m	0.30 ± 0.01[a,2]	0.33 ± 0.01[b,2]	0.34 ± 0.01[bc,1]	0.36 ± 0.01[c,2]	0.34 ± 0.01[b,1]			
		20 m	0.26 ± 0.00[a,1]	0.35 ± 0.00[b,3]	0.41 ± 0.01[d,2]	0.37 ± 0.01[c,3]	0.34 ± 0.01[b,1]			
SRP, µg/L	Wet season	0 m	8.0 ± 1.8[a,12]	6.9 ± 0.0[a,1]	10.1 ± 3.1[a,1]	5.9 ± 1.8[a,1]	10.1 ± 3.1[a,1]	8.6	−29.4	**0.000**
		10 m	5.9 ± 1.8[a,1]	4.8 ± 1.8[a,1]	12.2 ± 1.8[b,1]	6.9 ± 3.1[ab,1]	6.9 ± 0.0[ab,1]			
		20 m	13.2 ± 3.1[a,2]	10.1 ± 3.1[a,1]	10.1 ± 3.1[a,1]	9.0 ± 1.8[a,1]	9.0 ± 1.8[a,1]			
	Dry season	0 m	44.7 ± 3.1[b,3]	54.1 ± 3.1[c,3]	54.1 ± 3.1[c,3]	17.4 ± 1.8[a,1]	107.5 ± 3.1[d,3]	37.9		
		10 m	14.3 ± 1.8[a,1]	32.1 ± 3.1[b,2]	36.3 ± 1.8[b,2]	34.2 ± 1.8[b,2]	44.7 ± 3.1[c,2]			
		20 m	22.6 ± 3.1[a,2]	22.6 ± 3.1[a,1]	17.4 ± 1.8[a,1]	32.1 ± 3.1[b,2]	35.2 ± 3.1[b,1]			
TP, µg/L	Wet season	0 m	78.2 ± 3.6[a,1]	107.5 ± 12.6[b,1]	136.9 ± 9.6[c,1]	109.6 ± 9.6[b,1]	109.6 ± 9.6[b,2]	112.4	−122.6	**0.000**
		10 m	78.2 ± 13.1[a,1]	95.0 ± 6.3[a,1]	149.5 ± 3.6[b,1]	90.8 ± 15.8[a,1]	88.7 ± 6.3[a,1]			
		20 m	107.5 ± 6.3[a,2]	109.6 ± 13.1[a,1]	191.4 ± 9.6[b,2]	113.8 ± 6.3[a,1]	120.1 ± 6.3[a,2]			
	Dry season	0 m	212.8 ± 19.3[bc,1]	193.5 ± 19.3[b,1]	135.6 ± 19.3[a,1]	232.1 ± 19.3[bc,1]	244.9 ± 11.1[c,12]	235.1		
		10 m	270.6 ± 19.3[ab,2]	289.9 ± 19.3[bc,2]	244.9 ± 11.1[a,2]	322.1 ± 11.1[c,2]	270.6 ± 19.3[ab,2]			
		20 m	232.1 ± 19.3[a,12]	212.8 ± 19.3[a,1]	232.1 ± 19.3[a,2]	225.6 ± 29.5[a,1]	206.4 ± 22.3[a,1]			

Means in the same row with the same letters or column with the same numbers are not significantly different at 5% level. The positive value of mean difference indicates that the parameter studied is higher during the wet season whereas the negative value indicates that the parameter studied is higher during the dry season. The significant difference at p value ≤ 0.05 is indicated in bold.

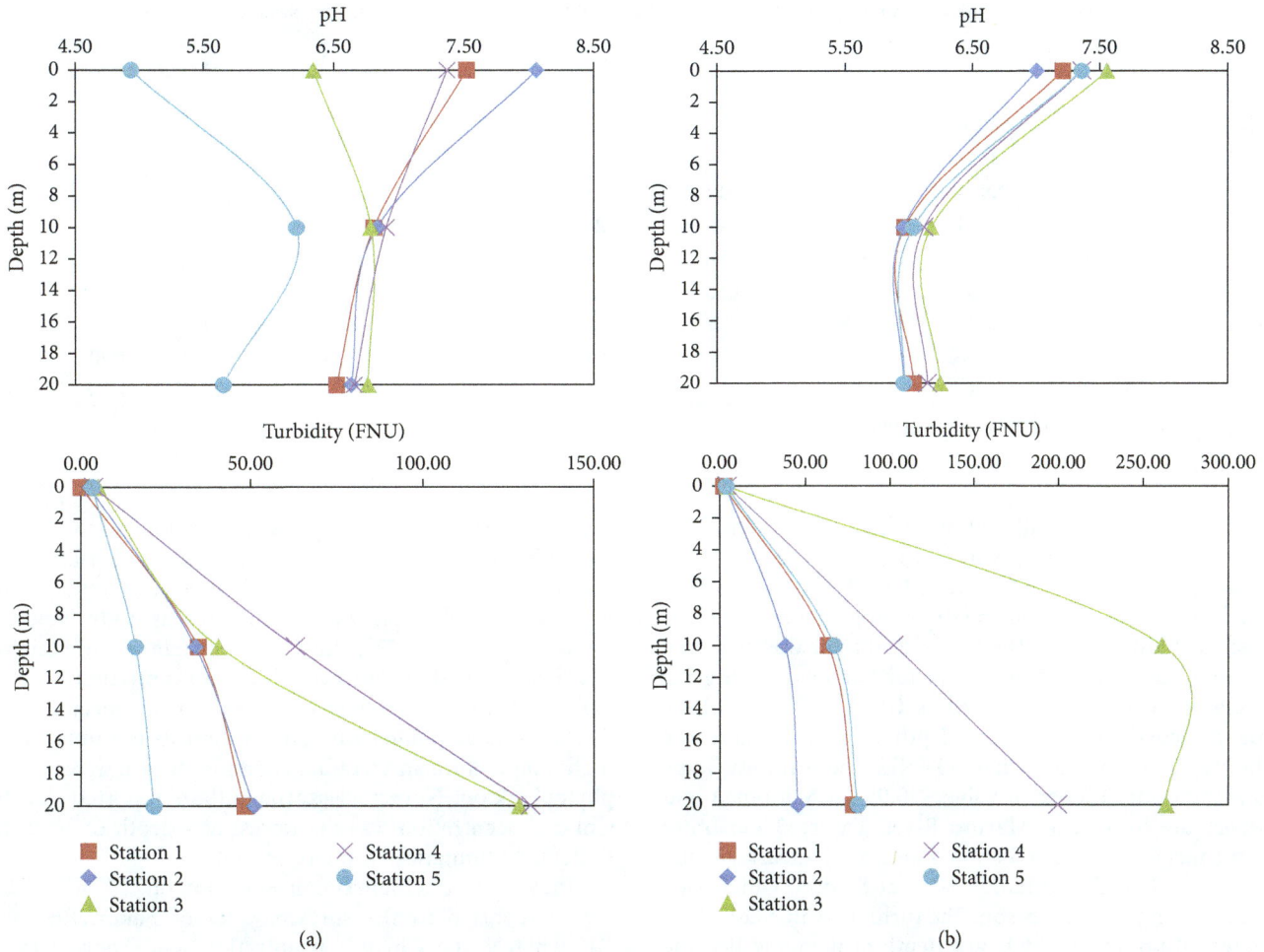

FIGURE 3: The distribution of pH and turbidity at three different depths (0, 10, and 20 m) of Bakun Reservoir in February (a) and September (b) 2014.

TABLE 3: Correlation of water quality parameters of the Bakun Reservoir during the wet season ($N = 15$).

Parameter	pH	Turbidity	Chl a	TSS	BOD$_5$	NO$_2^-$-N	NO$_3^-$-N	OKN	SRP	TP
pH	1.000									
Turbidity	0.005	1.000								
Chl a	0.314	−0.254	1.000							
TSS	−0.162	**0.755***	−0.368	1.000						
BOD$_5$	−0.345	0.242	−0.263	0.310	1.000					
NO$_2^-$-N	−0.288	0.093	−0.383	0.247	**0.674***	1.000				
NO$_3^-$-N	0.161	0.067	0.389	−0.011	0.272	−0.002	1.000			
OKN	−0.081	0.023	0.091	−0.208	0.499	0.265	**0.552***	1.000		
SRP	−0.365	0.237	−0.130	0.281	0.044	0.288	0.223	0.093	1.000	
TP	−0.128	0.474	−0.144	0.217	0.400	**0.620***	0.121	0.271	**0.539***	1.000

*The significant correlation at p value ≤ 0.05 is indicated in bold.

On the other hand, during the dry season, the pH value of the Bakun Reservoir was significantly positively correlated with Chl a and SRP but negatively correlated with turbidity and TP (p value ≤ 0.05) as shown in Table 4. The relationship revealed that the pH value of the Bakun Reservoir was regulated by the process of photosynthesis and decomposition of organic matter in the reservoir during the dry season. Photosynthesis increases pH value in the surface water but the rate decreases with depth due to light limitation in the water column [33, 34].

The surface turbidity value of the Bakun Reservoir was low (<6 FNU) and increased significantly (p value ≤ 0.05)

TABLE 4: Correlation of water quality parameters of the Bakun Reservoir during the dry season ($N = 15$).

Parameter	pH	Turbidity	Chl a	TSS	BOD$_5$	NO$_2^-$-N	NO$_3^-$-N	OKN	SRP	TP
pH	1.000									
Turbidity	**−0.516***	1.000								
Chl a	**0.676***	−0.385	1.000							
TSS	−0.407	**0.923***	−0.259	1.000						
BOD$_5$	−0.378	0.217	0.090	0.346	1.000					
NO$_2^-$-N	0.465	−0.008	0.041	−0.048	−0.479	1.000				
NO$_3^-$-N	−0.192	0.197	−0.293	0.085	−0.405	0.289	1.000			
OKN	−0.085	**0.524***	0.001	**0.622***	0.283	0.166	0.276	1.000		
SRP	**0.556***	−0.361	**0.884***	−0.359	−0.024	−0.041	−0.163	−0.009	1.000	
TP	**−0.539***	0.205	−0.255	0.128	0.176	−0.406	0.513	0.103	−0.168	1.000

*The significant correlation at p value ≤ 0.05 is indicated in bold.

as depth increased at all stations which agrees with the previous study on the reservoir [10, 11]. The turbidity value increased up to 131.33 FNU and 263.67 FNU at a depth of 20 m during the wet season and dry season, respectively. The surface turbidity value in the Bakun Reservoir was classified as Class I during both trips except for Station 3 during the wet season which was classified as Class II. The turbidity value exceeded the NWQS as depth increased where the turbidity value was more than 50 FNU. The turbidity value was significantly higher (p value ≤ 0.05) at Stations 3 and 4 which are located at Murum River. Figure 3 illustrates that the turbidity value at Station 4 increased linearly during both trips while the turbidity value at Station 3 increased linearly during the wet season. The turbidity value at Station 3 increased up to 261.3 FNU at a depth of 10 m and became stagnant up to a depth of 20 m during the dry season. The significant positive correlation between turbidity and TSS (p value ≤ 0.05) suggested that the turbidity resulted from the suspended solids in the water column. The land clearing coupled with the construction at the upstream area of the Murum River accelerated the soil erosion and sedimentation and the resulting suspended solids were transported into the reservoir during surface runoff and were deposited at the bottom of the reservoir. The high turbidity value which increased with depth was most likely due to the settling and resuspension of settled solids. The turbidity value at Station 3 was recorded up to 1000 FNU at a depth of 15 m and 30 m in the year 2013 [13]. The present study did demonstrate an improvement in the water turbidity over time although the value still exceeded the standard. Turbidity was also significantly positively correlated with OKN (p value ≤ 0.05) during the dry season. Many pollutants are attaching to the particles; thus, an increase in particles in the reservoir results in an increase in OKN in the present study.

The surface Chl a concentration ranged from 0.62 ± 0.21 μg/L to 2.30 ± 0.81 μg/L and from 0.64 ± 0.02 μg/L to 5.87 ± 0.99 μg/L in the Bakun Reservoir during the wet and dry seasons, respectively. The vertical distribution of Chl a in the Bakun Reservoir shows that Chl a decreased with depth or remained at similar concentrations in the water column (Figure 4). Sufficient light availability on the surface water promoted the growth of phytoplankton leading to the highest concentration of Chl a in the surface water whereas light limitation as depth increased reduced the growth of the phytoplankton in the present study. However, in studies such as [27], the Chl a concentration exhibited a different trend where the Chl a concentration was the highest at a depth of 10 m compared to the surface water. The authors attributed this observation to the unfavorable high temperature and irradiance in the surface water for the phytoplankton. Nevertheless, the authors reported that the Chl a concentration was the lowest at a depth of 30 m due to the light limitation in the reservoir.

Previously, the highest concentration of 7.25 μg/L of Chl a was reported in the surface water of Bakun Reservoir [11] whereas the Chl a concentration was reported up to 4.58 mg/m^3 [35] and 6.02 mg/m^3 [28] in the surface water of the Batang Ai Reservoir. The high Chl a in the Batang Ai Reservoir was attributed to the cage culture activities in the reservoir [35] whereas the high Chl a concentration in the present study was most likely due to the nutrient availability from the anthropogenic activities in the adjacent area. In the present study, Chl a was significantly positively correlated with SRP during the dry season (p value ≤ 0.05). The positive correlation between Chl a and SRP reveals that the growth of phytoplankton in the Bakun Reservoir was not limited by the phosphorus.

All surface TSS values in Bakun Reservoir were classified as Class I which is less than 25 mg/L. The surface TSS concentration in the present study was lower than the previously reported surface TSS concentration (66.7–100.0 mg/L) in the year 2013 [11]. The improvement of TSS concentration demonstrated the settling of the suspended solids in the reservoir over time. The old Batang Ai Reservoir also contained low TSS values which are less than 25 mg/L even at a depth of 30 m [35]. The vertical distribution of TSS exhibited a similar trend with turbidity where the TSS value increased significantly (p value ≤ 0.05) as depth increased in the present study. Figure 4 shows that Station 3 contained the highest TSS concentration as depth increased during the dry season, followed by Station 4 where both stations were located at the Murum River. The extremely high TSS concentrations at

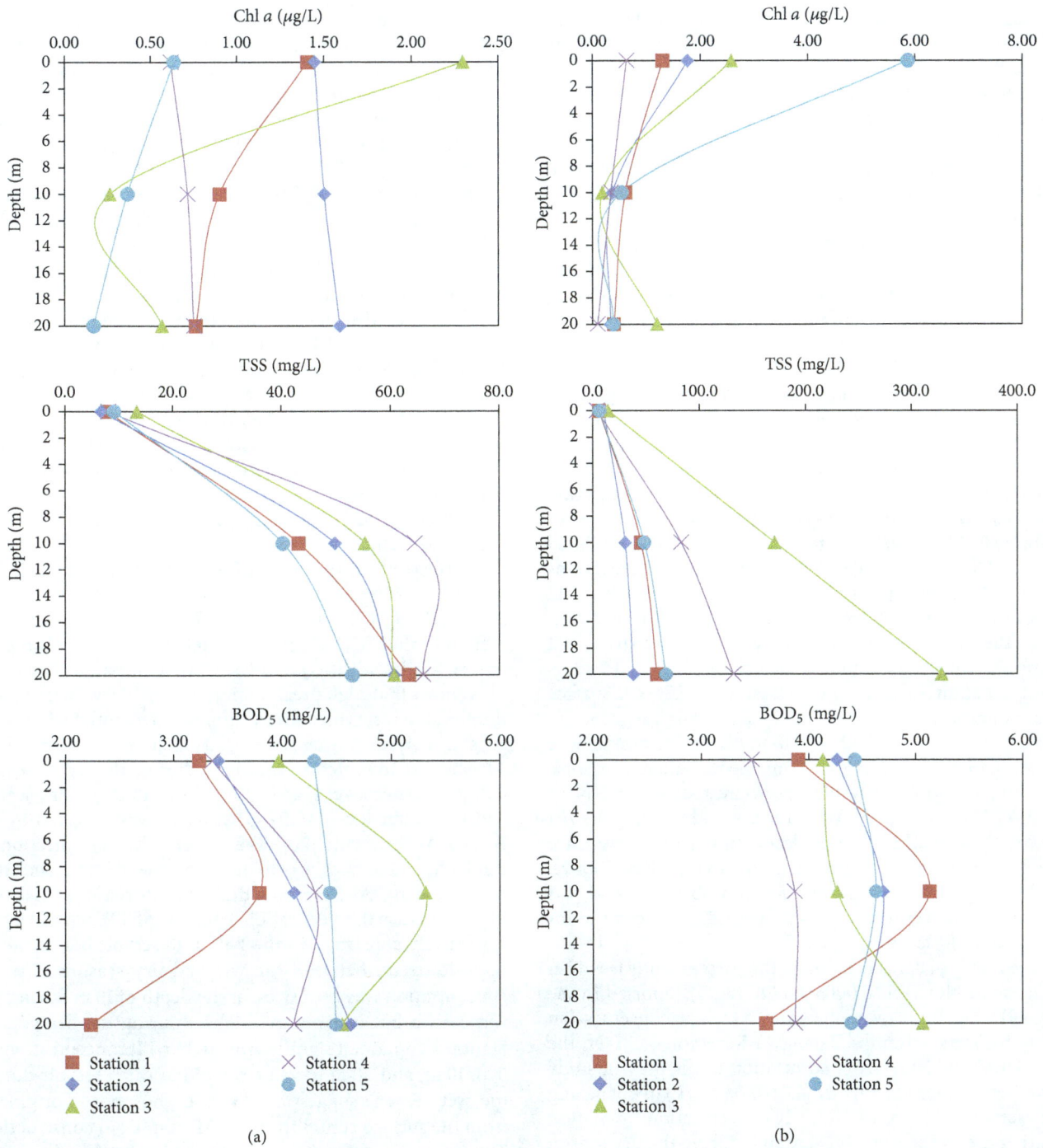

FIGURE 4: The distribution of Chl *a*, TSS, and BOD$_5$ at three different depths (0, 10, and 20 m) of Bakun Reservoir in February (a) and September (b) 2014.

Stations 3 and 4 revealed the impact of the land clearing and dam construction upstream of the river that carries eroded soil particles into the reservoir.

Surface BOD$_5$ concentration at the Bakun Reservoir ranged from 3.24 ± 0.19 mg/L to 4.30 ± 0.15 mg/L and from 3.47 ± 0.06 mg/L to 4.43 ± 0.06 mg/L during the wet and dry seasons, respectively. Table 2 shows that Stations 3 and 5 often contained significantly higher BOD$_5$ concentrations

(p value ≤ 0.05) at the three different depths of the water column. The high BOD$_5$ concentrations at the two stations downstream of the construction of the Murum hydroelectric dam and logging activities show organic matter loading from the anthropogenic activities into the reservoir. BOD$_5$ concentrations decreased to a value of 2.23 mg/L at a depth of 20 m at Station 1, classified as Class II. Other than that, BOD$_5$ concentrations in the Bakun Reservoir were classified as Class III. In contrast to the TSS concentration, the surface BOD_5 concentration in the present study was higher than the surface BOD_5 concentration (<2 mg/L) in the year 2013 [13]. The elevated BOD_5 concentration in the reservoir indicates high loading and accumulation of organic matter in the Bakun Reservoir over time. However, the present surface BOD_5 concentration was lower than the BOD_5 concentration in the Batang Ai Reservoir where up to 12 mg/L of BOD_5 concentration was reported [35].

The NO_2^--N concentration was low in the Bakun Reservoir. The surface NO_2^--N concentrations were not significantly different (p value > 0.05) among the stations during the wet season with a mean of 0.003 mg/L. During the dry season, the highest value of surface NO_2^--N was observed at Station 4 (0.008 ± 0.001 mg/L) followed by Station 3 (0.007 ± 0.001 mg/L) which were significantly higher (p value ≤ 0.05) than other stations (≈0.002 mg/L). Similarly, the surface NO_3^--N concentration was generally low in the Bakun Reservoir, ranging from 0.014 ± 0.006 mg/L to 0.037 ± 0.001 mg/L during the wet season. The mean concentration of surface NO_3^--N in the Bakun Reservoir during the dry season was 0.008 mg/L except at Station 4 which exhibited a peak of 0.022 ± 0.011 mg/L. The highest concentrations of NO_3^--N in the reservoir were observed at Station 2 at a depth of 20 m (0.050 mg/L) and Station 4 (0.059 mg/L) during the wet and dry seasons, respectively. Comparisons of the results with NWQS indicated that NO_2^--N and NO_3^--N in the Bakun Reservoir were classified as Class I. Surface OKN concentrations of the Bakun Reservoir ranged from 0.32 ± 0.01 mg/L to 0.42 ± 0.01 mg/L and from 0.29 ± 0.01 mg/L to 0.35 ± 0.01 mg/L during the wet and dry seasons, respectively. Significantly high concentrations of OKN (p value ≤ 0.05) were observed at Stations 4 and 5 during the wet and dry seasons which are predominantly from the surface runoff from the anthropogenic activities as mentioned above. Similar to the BOD_5 concentration, no obvious trend was observed in the vertical distribution of OKN concentrations in the Bakun Reservoir (Figure 5).

The NO_2^--N concentration in the present study is within the range of NO_2^--N (0.0003–0.0083 mg/L) reported in the year 2013 [11] and lower than the NO_2^--N concentration (0.001–0.053 mg/L) in the Batang Ai Reservoir [35]. On the other hand, the NO_3^--N concentration in the present study is within the range of NO_3^--N (0.01–0.06 mg/L) in the Batang Ai Reservoir [35] but the NO_3^--N concentrations in Station 2 during the wet season and Station 4 during the dry season were higher than the range of NO_3^--N (0.003–0.027 mg/L) reported in the Bakun Reservoir in the year 2013 [11]. NO_2^--N was significantly positively correlated with TP and BOD_5 (p value ≤ 0.05), and OKN was significantly positively correlated with NO_3^--N (p value ≤ 0.05) during the wet season. The relationship indicated the active decomposition and nitrification process in the reservoir. The relatively higher NO_3^--N concentration in the reservoir compared to the year 2013 indicated that nitrogen in the reservoir is being converted to NO_3^--N which is less toxic to aquatic organisms in the reservoir.

The concentration of SRP was low and relatively consistent in the Bakun Reservoir during the wet season with a mean value of 8.6 μg/L. The highest concentration of SRP (13.2 μg/L) was observed at Station 1 at a depth of 20 m. The SRP concentration was significantly higher during the dry season in the Bakun Reservoir (p value ≤ 0.05) with a mean value of 37.9 μg/L. The lowest and the highest concentrations of surface SRP were observed at Station 4 (17.4 ± 1.8 μg/L) and Station 5 (107.5 ± 3.1 μg/L), respectively, and significantly differed (p value ≤ 0.05) from other stations in the reservoir. Figure 6 illustrates that most stations in the Bakun Reservoir exhibited similar vertical distributions of SRP concentration during the dry season except for Station 4. The SRP concentration significantly decreased (p value ≤ 0.05) as depth increased in the Bakun Reservoir except at Station 4 where the surface SRP concentration was significantly lower (p value ≤ 0.05) than the SRP concentration at depths of 10 m and 20 m. Reference [26] reported that phosphate was lower in hypolimnion (>2.5 μM) compared to the surface water (0.05–0.23 μM) in the Cirata Reservoir which could be caused by enhanced loading from the sediment in the anoxic condition. The SRP concentration in the present study was lower than the SRP concentration in the year 2013 where the highest SRP concentration was 652.2 μg/L [11].

Surface TP of the Bakun Reservoir ranged from 78.2 ± 3.6 μg/L to 136.9 ± 9.6 μg/L and from 135.6 ± 19.3 μg/L to 244.9 ± 11.1 μg/L during the wet and dry seasons, respectively. Similar to the SRP, TP concentration was significantly higher during the dry season (p value ≤ 0.05) in the Bakun Reservoir. This shows that high precipitation during the wet season and the elevated reservoir water volume diluted both SRP and TP substantially in the reservoir. A similar observation where TP showed lower concentrations during the rainy season and high water level (24.90–38.59 μg/L) than the dry season and low water level (45.94–67.28 μg/L) was reported in the Batang Ai Reservoir [20]. The present TP concentration in the Bakun Reservoir was higher than the TP concentration in the Batang Ai Reservoir during both seasons. Figure 6 illustrates that the vertical distribution of TP concentration is relatively consistent in the Bakun Reservoir but shows an opposite trend between the wet and dry seasons. The TP concentration was the lowest at the depth of 10 m during the wet season but became the highest during the dry season. Station 3 contained significantly higher TP concentrations at 0 m, 10 m, and 20 m (p value ≤ 0.05) in the reservoir during the wet season, suggesting that the phosphorus originates from the surface runoff from the Murum Dam construction. The intensity of the impact increased substantially during the wet season as more phosphorus is washed down into the reservoir. In the present study, SRP concentrations complied with the 200 μg/L standard in accordance with the NWQS [32] during both trips. The TP concentration complied with the NWQS during the wet season but was noncompliant with the standard when the TP concentration increased substantially during the dry season.

3.2. Water Quality of the Downstream River of the Bakun Hydroelectric Dam. Table 5 summarizes the in situ and ex situ water quality of the downstream river of the Bakun Dam during the wet and dry seasons. The result demonstrated that

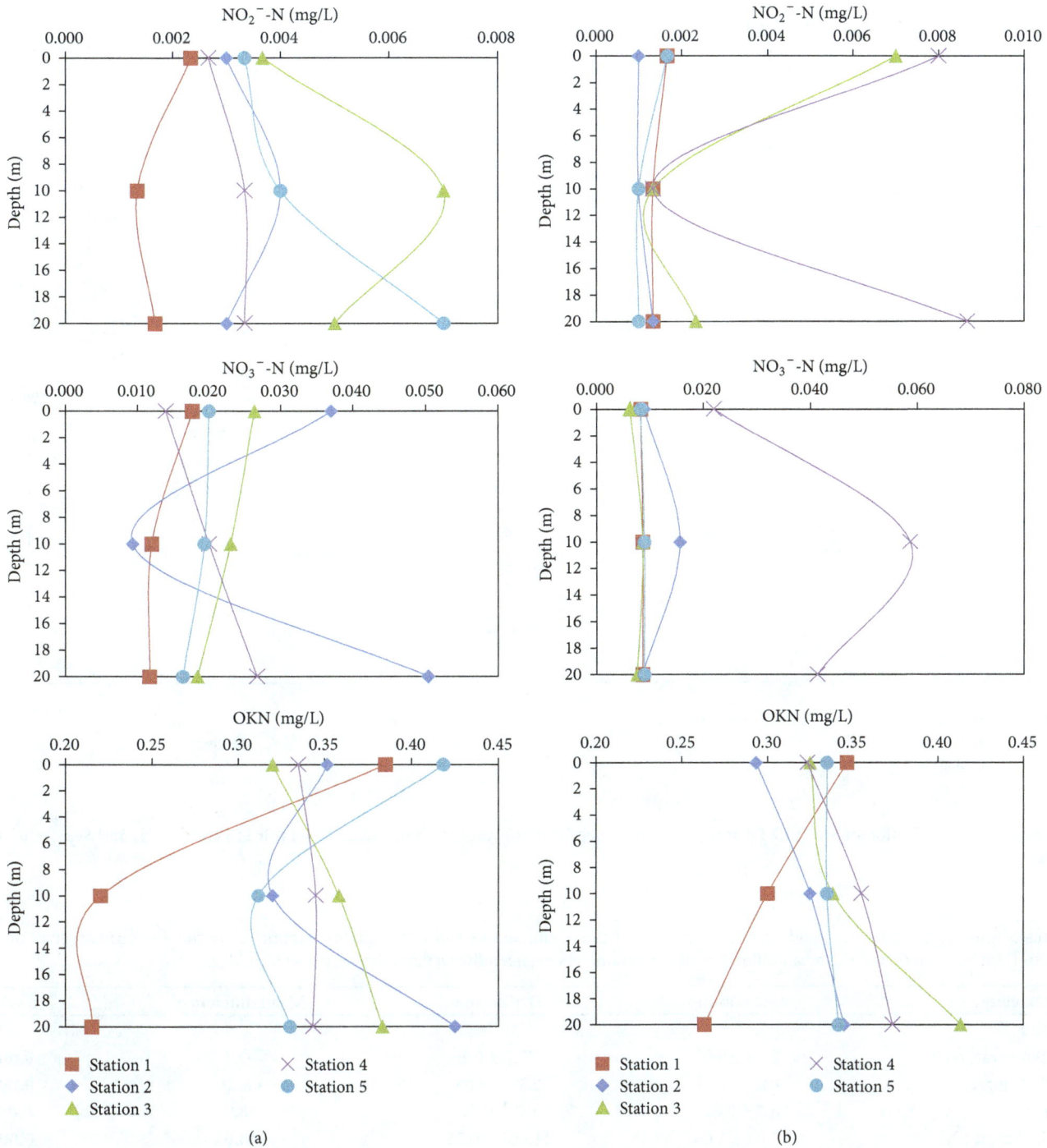

FIGURE 5: The distribution of NO_2^--N, NO_3^--N, and OKN at three different depths (0, 10, and 20 m) of Bakun Reservoir in February (a) and September (b) 2014.

the reservoir water has altered the surface water temperature of the downstream river. When the cooler reservoir water at a depth of 10 m is released into the downstream river, it decreased approximately 4°C to 5°C of the surface water temperature of its downstream river. The mean value of DO in the downstream river was 9.40 mg/L and 2.59 mg/L during the wet and dry seasons and was classified as Class I and Class IV, respectively. Both DO values in the downstream river were

higher than the DO value of the reservoir water at a depth of 10 m (<1 mg/L), particularly the DO content during the wet season because the spillway of the dam was open and additional water was discharged from the spillway during the sampling. The strong water current from the spillway coupled with the water flow from the turbines promotes aeration and increases the DO content substantially. On the other hand, the DO content was low when the spillway was

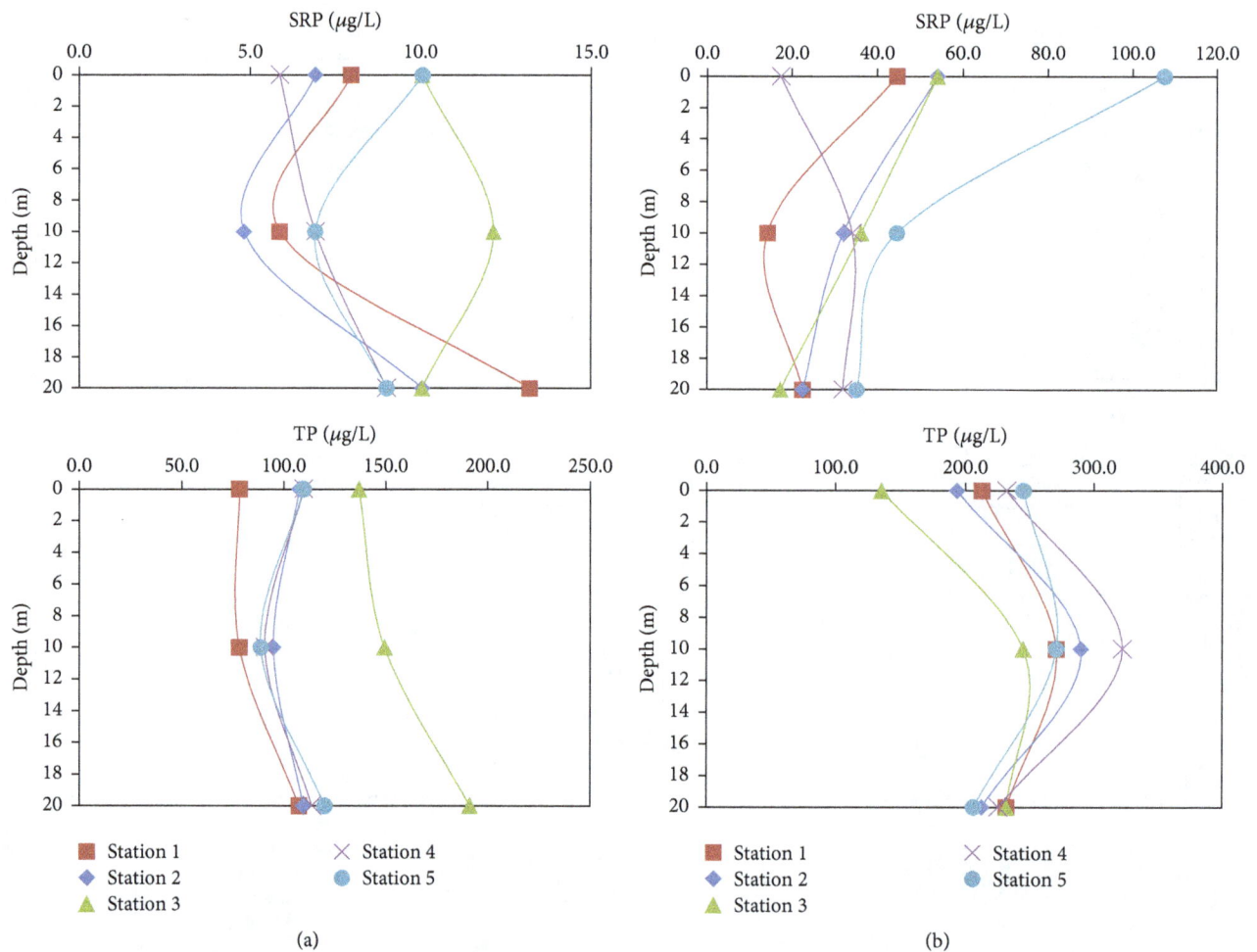

FIGURE 6: The distribution of SRP and TP at three different depths (0, 10, and 20 m) of Bakun Reservoir in February (a) and September (b) 2014.

TABLE 5: Summary of the mean and standard deviation of the in situ and ex situ water quality parameters in the downstream river of the Bakun Dam (Station 6) and the mean difference of the parameters between the wet and dry seasons ($N = 3$).

Parameter	Wet season	Dry season	Mean difference	p value
In situ				
Temperature, °C	25.0 ± 0.0	27.1 ± 0.1	−2.1	**0.000**
DO, mg/L	9.40 ± 0.10	2.59 ± 0.05	**+6.81**	**0.000**
pH	6.2 ± 0.01	6.0 ± 0.01	**+0.2**	**0.001**
Turbidity, FNU	77.00 ± 1.00	113.67 ± 0.58	**−36.67**	**0.000**
Ex situ				
Chl a, μg/L	0.64 ± 0.06	0.52 ± 0.06	+0.12	0.179
TSS, mg/L	45.0 ± 1.7	61.7 ± 1.4	**−16.7**	**0.011**
BOD$_5$, mg/L	5.70 ± 0.03	3.10 ± 0.10	**+2.60**	**0.000**
NO$_2^-$-N, mg/L	0.007 ± 0.001	0.001 ± 0.001	**+0.006**	**0.009**
NO$_3^-$-N, mg/L	0.023 ± 0.010	0.009 ± 0.001	+0.014	0.115
OKN, mg/L	0.40 ± 0.01	0.38 ± 0.00	+0.01	0.057
SRP, μg/L	10.1 ± 3.1	34.2 ± 1.8	**−24.1**	**0.008**
TP, μg/L	99.2 ± 9.6	386.4 ± 19.3	**−287.2**	**0.001**

The positive value of mean difference indicates that the parameter studied is higher during the wet season whereas the negative value indicates that the parameter studied is higher during the dry season.

TABLE 6: Mean difference of in situ and ex situ water quality parameters between the intake point of the dam at 10 m (Station 5) and its downstream river (Station 6) during wet and dry seasons ($N = 3$).

Parameter	Wet season		Dry season	
	Mean difference	p value	Mean difference	p value
In situ				
pH	−0.03	0.469	+0.01	0.288
Turbidity, FNU	**+60.9**	**0.000**	**+46.7**	**0.000**
Ex situ				
Chl a, μg/L	**+0.28**	**0.013**	−0.02	0.755
TSS, mg/L	**+4.67**	**0.011**	**+13.03**	**0.000**
BOD_5, mg/L	**+1.26**	**0.000**	**−1.53**	**0.000**
NO_2^--N, mg/L	**+0.003**	**0.001**	+0.000	0.374
NO_3^--N, mg/L	+0.003	0.631	−0.000	0.374
OKN, mg/L	**+0.08**	**0.000**	**+0.05**	**0.000**
SRP, μg/L	+3.2	0.157	**−10.5**	**0.007**
TP, μg/L	+10.5	0.189	**+115.7**	**0.002**

The positive value of mean difference indicates that the parameter studied is higher in the downstream river whereas the negative value indicates that the parameter studied is lower in the downstream river.

not open during the sampling in the dry season. When the oxygen-deprived reservoir water was released into the downstream river without additional aeration, it decreased the oxygen level of the downstream river below the minimum requirement of 5 mg/L for sensitive aquatic organisms.

Low pH values ($\approx$6.1) were observed at the downstream river and classified as Class II according to NWQS [32]. Table 6 shows that there was no significant difference in pH value between Station 6 and the dam intake point at 10 m for both seasons (p value > 0.05) revealing that the low pH in the downstream river is due to the low pH of the reservoir water that was released into the downstream river after passing through the turbines. The pH value of the downstream river was relatively lower than the pH value of tributaries that flow into the Bakun Reservoir (6.8–7.8) [14]. The turbidity value in the downstream river of the dam was high and exceeded the standard guideline of 50 FNU in Malaysia [32]. The turbidity values were also significantly higher (p value > 0.05) than the reservoir water during both seasons. When water is discharged from the spillway in addition to turbine outflow, resuspension of deposited sediments under the high flow rate increases the suspended solids downstream. The turbidity value (77.00 ± 1.00 FNU) during the wet season was significantly lower (p value > 0.05) than the dry season (113.67 ± 0.58 FNU) which is most probably due to more dilution from the tributaries along the downstream river during the wet season. Similar to the turbidity value, the TSS concentration during the wet season was significantly lower than the dry season (p value ≤ 0.05) and was classified as Class II and Class III, respectively. Both values were also significantly higher than the TSS concentration (p value ≤ 0.05) at the intake point.

There was no significant difference in Chl a between the wet and dry seasons (p value > 0.05) in the downstream river with a mean value of 0.58 μg/L. The Chl a concentration was significantly higher (p value ≤ 0.05) than the intake

point during the wet season but it was similar to the Chl a concentration at surface reservoir water (0.64 μg/L). There was no significant difference in Chl a between downstream river and the intake point (p value > 0.05) during the dry season. The mean value of BOD_5 in the downstream river was 5.70 mg/L and 3.10 mg/L during the wet and dry seasons and was classified as Class III. The BOD_5 concentration during the wet season was significantly higher than the dry season (p value ≤ 0.05). Besides, the downstream BOD_5 concentration was significantly higher than the BOD_5 concentration at the intake point (p value ≤ 0.05) during the wet season whereas the BOD_5 concentration was significantly lower than the BOD_5 concentration at the intake point (p value ≤ 0.05) during the dry season. The higher downstream BOD_5 concentration during the wet season indicates that the high BOD_5 concentration is most likely attributed to other domestic discharge and runoff in addition to the reservoir water. Several longhouses and villages located along the downstream river may have contributed substantial organic matter to the downstream river.

NO_2^--N and NO_3^--N concentrations were also low in the downstream river, similar to the reservoir water, and were classified as Class I according to NWQS [32]. Significantly higher NO_2^--N concentration (p value ≤ 0.05) was found during the wet season whereas no significant difference of NO_3^--N concentration was found between the wet and dry seasons (p value > 0.05). The downstream NO_2^--N concentration was also significantly higher than the reservoir NO_2^--N concentration at the intake point. There was no significant difference in OKN between the wet and dry seasons (p value > 0.05) in the downstream river with a mean value of 0.39 mg/L. OKN was significantly higher (p value ≤ 0.05) at the downstream river than the OKN concentration at intake point. The higher downstream NO_2^--N and OKN concentrations besides BOD_5 demonstrated

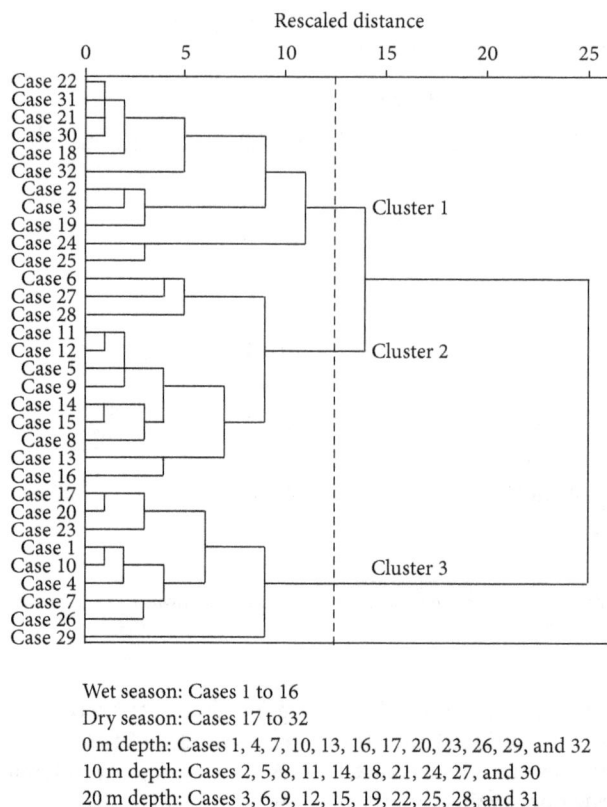

FIGURE 7: Clusters of the five sampling stations located in Bakun Reservoir at three different depths (0 m, 10 m, and 20 m) and one station located at its downstream river collected at 0 m during dry and wet seasons in Sarawak, Malaysia.

the organic pollutant contribution from adjacent domestic discharge and runoff in the downstream river.

SRP and TP concentrations in the downstream river exhibited a similar trend where the concentration during the wet season was significantly lower than during the dry season (p value ≤ 0.05). The downstream SRP concentration complied with the 200 μg/L standard in accordance with the NWQS [32] in both trips. On the other hand, the downstream TP concentration also complied with the NWQS during the wet season but changed to noncompliance with the standard when the TP concentration increased substantially during the dry season. The TP concentration was found to be significantly higher (p value ≤ 0.05) at the downstream river than the TP concentration at the intake point. The high TP and low SRP concentration indicate that phosphorus concentration mainly consisted of organic phosphorus in the present study. The higher downstream TP concentration further confirms the organic pollutant contribution from the adjacent domestic discharge.

3.3. Cluster Analysis. Figure 7 demonstrates that the water quality in the reservoir can be grouped into three clusters according to the season and the water depth of the reservoir. Cluster 1 and Cluster 2 are mostly made up of sampling stations at depths of 10 m and 20 m conducted during the dry season and wet season, respectively, indicating that the dry and wet seasons have an influence on the deeper water

column of the reservoir. The surface water quality of the downstream river during the dry (Case 32) and wet (Case 16) seasons was also grouped to Clusters 1 and 2, respectively, as it was influenced by the deeper reservoir water discharged into the river. On the other hand, surface water quality of the reservoir except at Station 5 (Case 13) during the wet season was not influenced by the season where the surface water quality of the Bakun Reservoir during both seasons was categorized as Cluster 3. This phenomenon is the most apparent based on the turbidity and TSS in the reservoir. The surface turbidity and TSS values during the wet and dry seasons were relatively similar with a mean value of 3.02 FNU and 3.66 FNU and 9.0 mg/L and 8.1 mg/L, respectively, but the turbidity (76.02 FNU versus 133.65 FNU) and TSS (60.7 mg/L versus 125.5 mg/L) values at a depth of 20 m during the wet season were around two times lower than during the dry season.

4. Conclusions

The Bakun hydroelectric reservoir is a thermally stratified reservoir with a temperature gradient of approximately 5°C within the thermocline layer. The thickness of the well oxygenated water was around 3–6 m of the surface water, whereas the oxygen content of most of the water body was below 5 mg/L or even in hypoxia. The Bakun Reservoir showed signs of organic pollution with high BOD_5, OKN, and TP concentrations observed in the reservoir. Acidification was observed in parts of the reservoir, particularly downstream of active logging activities and the Murum hydroelectric dam construction during the wet season. The water quality of the reservoir was influenced by the wet and dry seasons particularly in the deeper water column. SRP and TP concentrations were discovered to be higher during the dry season in the reservoir. This result suggests the necessity of management and conservation of the reservoir to prevent further deterioration in the reservoir's water quality where different water quality parameters should be targeted during different seasons. The present study also demonstrated that the water discharged from the Bakun Reservoir has a great impact on the water quality at the downstream river. The water released from the reservoir decreased the temperature, DO, and pH of the downstream river whereas turbidity and TSS concentration increased in the downstream river. Nevertheless, the water quality of the downstream river, particularly BOD_5, OKN, and TP concentrations, was also influenced by adjacent anthropogenic activities such as household wastewater. This result suggests that the downstream river of the Bakun Reservoir was not solely impacted by the reservoir's outflow. Therefore, all factors should be taken into account in decision-making of the management of the downstream river for the health of sensitive aquatic organisms.

Conflicts of Interest

The authors declare that there are no conflicts of interest regarding the publication of this paper.

Acknowledgments

The authors appreciate the financial support provided by the Malaysian Ministry of Higher Education through Grant no. FRGS/STWN01(04)/991/2013(32) and the facilities provided by Universiti Malaysia Sarawak.

References

[1] M. W. Beck, A. H. Claassen, and P. J. Hundt, "Environmental and livelihood impacts of dams: common lessons across development gradients that challenge sustainability," *International Journal of River Basin Management*, vol. 10, no. 1, pp. 73–92, 2012.

[2] X. Li, S. Dong, Q. Zhao, and S. Liu, "Impacts of Manwan Dam construction on aquatic habitat and community in Middle Reach of Lancang River," *Procedia Environmental Sciences*, vol. 2, no. 5, pp. 706–712, 2010.

[3] J. Li, S. Dong, S. Liu, Z. Yang, M. Peng, and C. Zhao, "Effects of cascading hydropower dams on the composition, biomass and biological integrity of phytoplankton assemblages in the middle Lancang-Mekong River," *Ecological Engineering*, vol. 60, pp. 316–324, 2013.

[4] D. D. A. Cunha and L. V. Ferreira, "Impacts of the Belo Monte hydroelectric dam construction on pioneer vegetation formations along the Xingu River, Pará State, Brazil," *Revista Brasileira de Botanica*, vol. 35, no. 2, pp. 159–167, 2012.

[5] Q. G. Wang, Y. H. Du, Y. Su, and K. Q. Chen, "Environmental impact post-assessment of dam and reservoir projects: a review," *Procedia Environmental Sciences*, vol. 13, pp. 1439–1443, 2012.

[6] G. L. Wei, Z. F. Yang, B. S. Cui et al., "Impact of dam construction on water quality and water self-purification capacity of the Lancang River, China," *Water Resources Management*, vol. 23, no. 9, pp. 1763–1780, 2009.

[7] Y. Yi, Z. Yang, and S. Zhang, "Ecological influence of dam construction and river-lake connectivity on migration fish habitat in the Yangtze River basin, China," *Procedia Environmental Sciences*, vol. 2, no. 5, pp. 1942–1954, 2010.

[8] Q. Lin, "Influence of dams on river ecosystem and its countermeasures," *Journal of Water Resource and Protection*, vol. 03, no. 01, pp. 60–66, 2011.

[9] W. L. Graf, "Downstream hydrologic and geomorphic effects of large dams on American rivers," *Geomorphology*, vol. 79, no. 3-4, pp. 336–360, 2006.

[10] L. Nyanti, T. Y. Ling, and J. Grinang, "Physico-chemical characteristics in the filling phase of Bakun hydroelectric reservoir," vol. 2, pp. 92–101, Sarawak, Malaysia, 2012.

[11] T.-Y. Ling, L. Nyanti, T. Muan, J. Grinang, S.-F. Sim, and A. Mujahid, "Physicochemical parameters of Bakun Reservoir in Belaga, Sarawak, Malaysia, 13 months after reaching full supply level," *Sains Malaysiana*, vol. 45, no. 2, pp. 157–166, 2016.

[12] S. F. Sim, T. Y. Ling, L. Nyanti, N. Gerunsin, Y. E. Wong, and L. P. Kho, "Assessment of heavy metals in water, sediment, and fishes of a large tropical hydroelectric dam in Sarawak, Malaysia," *Journal of Chemistry*, vol. 2016, Article ID 8923183, 10 pages, 2016.

[13] L. Nyanti, T. Y. Ling, and T. Muan, "Water quality of Bakun hydroelectric dam reservoir, the construction of Murum dam," *ESTEEM Academic Journal*, vol. 11, no. 1, pp. 81–88, 2015.

[14] T. Y. Ling, L. Nyanti, and A. S. Masion, "Water quality of rivers that flow into Bakun hydroelectric dam reservoir, Sarawak, Malaysia," *ESTEEM Academic Journal*, vol. 11, no. 1, pp. 9–16, 2015.

[15] V. Rossel and A. de la Fuente, "Assessing the link between environmental flow, hydropeaking operation and water quality of reservoirs," *Ecological Engineering*, vol. 85, pp. 26–38, 2015.

[16] T. da Costa Lobato, R. A. Hauser-Davis, T. F. de Oliveira et al., "Categorization of the trophic status of a hydroelectric power plant reservoir in the Brazilian Amazon by statistical analyses and fuzzy approaches," *Science of the Total Environment*, vol. 506-507, pp. 613–620, 2015.

[17] X. Li, T. Huang, W. Ma, X. Sun, and H. Zhang, "Effects of rainfall patterns on water quality in a stratified reservoir subject to eutrophication: Implications for management," *Science of the Total Environment*, vol. 521-522, pp. 27–36, 2015.

[18] M. Varol, B. Gökot, A. Bekleyen, and B. Şen, "Spatial and temporal variations in surface water quality of the dam reservoirs in the Tigris River basin, Turkey," *Catena*, vol. 92, pp. 11–21, 2012.

[19] Y. Zhang, Z. Wu, M. Liu et al., "Dissolved oxygen stratification and response to thermal structure and long-term climate change in a large and deep subtropical reservoir (Lake Qiandaohu, China)," *Water Research*, vol. 75, pp. 249–258, 2015.

[20] T. Y. Ling, T. Z. E. Lee, and L. Nyanti, "Phosphorus in batang ai hydroelectric dam Reservoir, Sarawak, Malaysia," *World Applied Sciences Journal*, vol. 28, no. 10, pp. 1348–1354, 2013.

[21] P. McCully, "Rivers no more: the environmental effects of dams," in *Silenced Rivers: The Ecology and Politics of Large Dams*, P. McCully, Ed., pp. 29–64, Zed Books, London, UK, 1996.

[22] T.-Y. Ling, C.-L. Soo, T. L.-E. Heng, L. Nyanti, S.-F. Sim, and J. Grinang, "Physicochemical characteristics of river water downstream of a large tropical hydroelectric dam," *Journal of Chemistry*, vol. 2016, Article ID 7895234, 2016.

[23] D. Jenkins, J. J. Connors, and A. E. Greenberg, *Standard Methods for the Examination of Water and Wastewater*, American Public Health Association, Washington, Wash, D.C,, USA, 21st edition edition, 205.

[24] Hach, *Hach Water Analysis Handbook*, Hach Company, USA, 2015.

[25] D. F. Goerlitz and E. Brown, "Methods for analysis of organic substances in water," in *Techniques of Water-Resources Investigations of The United States Geological Survey*, R. L. Wershaw, M. J. Fishman, R. R. Grabbe, and L. E. Lowe, Eds., pp. 1–40, U. S. Geological Survey, United States, 1972.

[26] Y. Hayami, K. Ohmori, K. Yoshino, and Y. S. Garno, "Observation of anoxic water mass in a tropical reservoir: the cirata reservoir in java, Indonesia," *Limnology*, vol. 9, no. 1, pp. 81–87, 2008.

[27] C. Ariyadej, P. Tansakul, and R. Tansakul, "Variation of phytoplankton biomass as chlorophyll a in banglang reservoir, yala province," *Songklanakarin Journal of Science and Technology*, vol. 30, no. 2, pp. 159–166, 2008.

[28] T.-Y. Ling, L. Nyanti, C.-K. Leong, and Y.-M. Wong, "Comparison of water quality at different locations at Batang Ai Reservoir, Sarawak, Malaysia," *World Applied Sciences Journal*, vol. 26, no. 11, pp. 1473–1481, 2013.

[29] G. B. Sahoo and D. Luketina, "Modeling of bubble plume design and oxygen transfer for reservoir restoration," *Water Research*, vol. 37, no. 2, pp. 393–401, 2003.

[30] Y. Zhou, D. R. Obenour, D. Scavia, T. H. Johengen, and A. M. Michalak, "Spatial and temporal trends in Lake Erie hypoxia, 1987-2007," *Environmental Science and Technology*, vol. 47, no. 2, pp. 899–905, 2013.

[31] T. Y. Ling, D. P. Debbie, N. Lee, I. Norhadi, and J. J. E. Justin, "Water quality at Batang Ai Hydroelectric Reservoir (Sarawak, Malaysia) and implications for aquaculture," vol. 2, pp. 23–30, 2012.

[32] Department of Environment, *Malaysia Environmental Quality Report 2014*, Department of Environment, Kuala Lumpur, Malaysia, 2015.

[33] T. R. Fisher, L. W. Harding Jr., D. W. Stanley, and L. G. Ward, "Phytoplankton, nutrients, and turbidity in the Chesapeake, Delaware, and Hudson estuaries," *Estuarine, Coastal and Shelf Science*, vol. 27, no. 1, pp. 61–93, 1988.

[34] P.-P. Shen, G. Li, L.-M. Huang, J.-L. Zhang, and Y.-H. Tan, "Spatio-temporal variability of phytoplankton assemblages in the Pearl River estuary, with special reference to the influence of turbidity and temperature," *Continental Shelf Research*, vol. 31, no. 16, pp. 1672–1681, 2011.

[35] L. Nyanti, K. M. Hiii, A. Sow, I. Norhadi, and T. Y. Ling, "Impacts of aquaculture at different depths and distances from cage culture sites in batang Ai hydroelectric dam reservoir, Sarawak, Malaysia," *World Applied Sciences Journal*, vol. 19, no. 4, pp. 451–456, 2012.

Permissions

List of Contributors

Chunrong Wang, Qi Zhang, Longxin Jiang and Zhifei Hou
School of Chemical and Environmental Engineering, China University of Mining and Technology (Beijing), Beijing 100083, China

Awa Kangama, Defang Zeng, Xu Tian and Jinfu Fang
School of Resource and Environmental Engineering, Wuhan University of Technology, 122 Luoshi Road, Wuhan 430070, China

Yanling Deng
Graduate School of Science and Technology, Niigata University, Niigata 950-2181, Japan

Naoki Kano and Hiroshi Imaizumi
Department of Chemistry and Chemical Engineering, Faculty of Engineering, Niigata University, Niigata 950-2181, Japan

Jingnan Chen, Xingxing Jiang, Guolong Yang, Yanlan Bi and Wei Liu
Provincal Key Laboratory for Transformation and Utilization of Cereal Resource, College of Food Science and Technology, Henan University of Technology, Zhengzhou 450001, China

Rodrigo Heleno Alves, Suzimara Rovani and Denise Alves Fungaro
Instituto de Pesquisas Energéticas e Nucleares (IPEN-CNEN/SP), Av. Prof. Lineu Prestes, No. 2242, Cidade Universit´aria, 05508-000 São Paulo, SP, Brazil

Thais Vitória da Silva Reis
Instituto de Pesquisas Energéticas e Nucleares (IPEN-CNEN/SP), Av. Prof. Lineu Prestes, No. 2242, Cidade Universit´aria, 05508-000 São Paulo, SP, Brazil
Faculdades Oswaldo Cruz, Rua Brigadeiro Galvão, 540 Barra Funda, 01151-000 São Paulo, SP, Brazil

M. Farnane, A. Machrouhi, A. Elhalil, M. Abdennouri and N. Barka
Laboratoire des Sciences des Matériaux, des Milieux et de la Modélisation (LS3M), FPK, Univ Hassan 1, B.P. 145, 25000 Khouribga, Morocco

H. Tounsadi
Laboratoire des Sciences des Matériaux, des Milieux et de la Modélisation (LS3M), FPK, Univ Hassan 1, B.P. 145, 25000 Khouribga, Morocco

Université Sidi Mohamed Ben Abdellah, Faculté des Sciences Dhar Elmehraz, Laboratoire d'Ingénierie, d'Electrochimie, de Modélisation et d'Environnement, Fés, Morocco

S. Qourzal
Equipe de Catalyse et Environnement, Département de Chimie, Faculté des Sciences, Université Ibn Zohr, B.P. 8106 Cité Dakhla, Agadir, Morocco

Haining Li, Qicheng Song and Ning Wang
School of Municipal and Environmental Engineering, Shandong Jianzhu University, Jinan 250101, China

Hongbo Wang
School of Municipal and Environmental Engineering, Shandong Jianzhu University, Jinan 250101, China
Shandong Co-Innovation Center of Green Building, Jinan, China

Lili Gao
Shandong Urban and Rural Planning Design Institute, Jinan, Shandong 250013, China

Zhu Shui Jin
Institute of Crop Science, College of Agriculture and Biotechnology, Zhejiang University, Zijingang Campus, Hangzhou, China

M. K. Daud
Institute of Crop Science, College of Agriculture and Biotechnology, Zhejiang University, Zijingang Campus, Hangzhou, China
Department of Biotechnology and Genetic Engineering, Kohat University of Science and Technology, Kohat 26000, Pakistan

Hina Rizvi, Muhammad Farhan Akram and Muhammad Rizwan
Department of Environmental Sciences and Engineering, Government College University, Allama Iqbal Road, Faisalabad 38000, Pakistan

Shafaqat Ali
Department of Environmental Sciences and Engineering, Government College University, Allama Iqbal Road, Faisalabad 38000, Pakistan
Key Laboratory of Soil Environment and Pollution Remediation, Institute of Soil Science, Chinese Academy of Sciences, Nanjing 210008, China

Muhammad Nafees
Institute of Soil & Environmental Sciences, University of Agriculture, Faisalabad 38000, Pakistan

Martina KluIáková
Materials Research Centre, Faculty of Chemistry, Brno University of Technology, Purkyňova 118, 612 00 Brno, Czech Republic

Marcela Pavlíková
Institute of Chemistry, Faculty of Civil Engineering, Brno University of Technology, Žižkova 17, 602 00 Brno, Czech Republic

Aamir Shakoor, Zahid Mahmood Khan, Hafiz Umar Farid and Muhammad Sultan
Department of Agricultural Engineering, Bahauddin Zakariya University, Multan, Pakistan

Muhammad Arshad
Department of Irrigation and Drainage, University of Agriculture, Faisalabad, Pakistan

Muhammad Azmat
Geo-Informatics Engineering, National University of Sciences and Technology (NUST), Islamabad, Pakistan

Muhammad Adnan Shahid
Water Management Research Center, University of Agriculture, Faisalabad, Pakistan

Zafar Hussain
Department of Forestry and Range Management, Bahauddin Zakariya University, Multan, Pakistan

Victor S. Ruys, Kamel Zerari, Isabelle Seyssiecq and Nicolas Roche
Aix-Marseille University, CNRS, Centrale Marseille, M2P2 UMR 7340, 13541 Marseille Cedex 13, France

Vipin Kumar Saini, Surindra Suthar and Chaudhari Karmveer
School of Environment & Natural Resources, Doon University, Dehradun, Uttarakhand 248001, India

Kapil Kumar
Department of Environmental Engineering, National Institute of Technology, New Delhi 110040, India

Z. Majbar, K. Lahlou, M. Taleb and Z. Rais
Laboratory of Engineering, Electrochemistry and Modeling Environment (LEEME), Faculty of Sciences, Fez, Morocco

M. Ben Abbou and H. Bouka
Laboratory of Natural Resources and Environment, University Sidi Mohammed Ben Abdallah, Faculty Polydisciplinary, Taza, Morocco

E. Ammar and W. Abid
Urban and Costal Environments, University of Sfax, National Engineering School of Sfax, Sfax, Tunisia

A. Triki
Improvement and Protection of Olive Tree Genetic Resources, Institute of the Olive Tree, Sfax BP1087, Tunisia

M. Nawdali
Laboratory of Chemistry of Condensed Matter (LCCM), University Sidi Mohammed Ben Abdallah, Faculty of Sciences and Technology, Fez, Morocco

M. El Haji
Laboratory Engineering Research-OSIL Team Optimization of Industrial and Logistics Systems, University Hassan II, Superior National School of Electricity and Mechanic (ENSEM), Casablanca, Morocco

Michał Sadowski, Piotr Anielak and Wojciech M. Wolf
Institute of General and Ecological Chemistry, Faculty of Chemistry, Lodz University of Technology, 116 Zeromskiego Street, 90-924 Lodz, Poland

Teck-Yee Ling, Chen-Lin Soo, Jing-Jing Liew and Siong-Fong Sim
Department of Chemistry, Faculty of Resource Science and Technology, Universiti Malaysia Sarawak, 94300 Kota Samarahan, Sarawak, Malaysia

Lee Nyanti
Department of Aquatic Science, Faculty of Resource Science and Technology, Universiti Malaysia Sarawak, 94300 Kota Samarahan, Sarawak, Malaysia

Xiaoping Qin, Haiwei Lu, Yilin Li, Tong Peng, Lijie Xing, Haixi Xue and Jing Xu
Drilling and Production Technology Research Institute, PetroChina Jidong Oilfield Company, Tangshan 063004, China

A. Quinto-Hernandez
Tecnológico Nacional de México, Instituto Tecnológico de Zacatepec, Calzada Instituto Tecnológico 27, 62780 Zacatepec, MOR, Mexico

E. Reyes-Dorantes and J. Zuñiga-Díaz
Tecnológico Nacional de México, Instituto Tecnológico de Zacatepec, Calzada Instituto Tecnológico 27, 62780 Zacatepec, MOR, Mexico
Instituto de Ciencias Físicas, Universidad Nacional Autónoma de México, Avenida Universidad, s/n, 62210 Cuernavaca, MOR, Mexico

J. Porcayo-Calderon
Instituto de Ciencias Físicas, Universidad Nacional Autónoma de México, Avenida Universidad, s/n, 62210 Cuernavaca, MOR, Mexico
CIICAp, Universidad Autónoma del Estado deMorelos, Avenida Universidad 1001, 62209 Cuernavaca, MOR, Mexico

L. Martinez-Gomez
Instituto de Ciencias Físicas, Universidad Nacional Autónoma de México, Avenida Universidad, s/n, 62210 Cuernavaca, MOR, Mexico
Corrosion y Protección (CyP), Buffon 46, 11590 Mexico City, Mexico

J. G. Gonzalez-Rodriguez
CIICAp, Universidad Autónoma del Estado deMorelos, Avenida Universidad 1001, 62209 Cuernavaca, MOR, Mexico

Jamila Hammami Abidi, Boutheina Farhat and Abdallah Ben Mammou
Faculty of Sciences of Tunis, Laboratory of Mineral Resources and Environment, University of Tunis El Manar, 2092 Tunis, Tunisia

Naceur Oueslati
Regional Commission for Agricultural Development, av. Hassen Nouri, 7000 Bizerte, Tunisia

Zhitao Han, Bojun Liu, Shaolong Yang, Xinxiang Pan and Zhijun Yan
Marine Engineering College, Dalian Maritime University, Dalian 116026, China

Lorena Hernández-Afonso, Ricardo Fernández-González, Pedro Esparza, Selene Díaz González and Juan Carlos Ruiz-Morales
Chemistry Department, University of La Laguna, San Cristóbal de La Laguna, Tenerife, 38200 Canary Islands, Spain

M. Emma Borges
Chemical Engineering Department, University of La Laguna, San Cristóbal de La Laguna, Tenerife ,38200 Canary Islands, Spain

Jesús Canales-Vázquez
Instituto de Energías Renovables, Print3D Solutions, University of Castilla-La Mancha, 02006 Albacete, Spain

Teck-Yee Ling, Chen-Lin Soo, Lee Nyanti and Siong-Fong Sim
Faculty of Resource Science and Technology, Universiti Malaysia Sarawak, 94300 Kota Samarahan, Sarawak, Malaysia

Norliza Gerunsin
Universiti Teknologi MARA, Kota Samarahan Campus, Jalan Meranek, 94300 Kota Samarahan, Sarawak, Malaysia

Jongkar Grinang
Institute of Biodiversity and Environmental Conservation, Universiti Malaysia Sarawak, 94300 Kota Samarahan, Sarawak, Malaysia

Index